HAWQ
数据仓库与数据挖掘实战

王雪迎 著

清华大学出版社
北京

内 容 简 介

Apache HAWQ 是一个 SQL-on-Hadoop 产品,它非常适合用于 Hadoop 平台上快速构建数据仓库系统。HAWQ 具有大规模并行处理、完善的 SQL 兼容性、支持存储过程和事务、出色的性能表现等特性,还可与开源数据挖掘库 MADlib 轻松整合,从而使用 SQL 就能进行数据挖掘与机器学习。

本书内容分技术解析、实战演练与数据挖掘三个部分共 27 章。技术解析部分说明 HAWQ 的基础架构与功能特性,包括安装、连接、对象与资源管理、查询优化、备份恢复、高可用性等。实战演练部分用一个完整的示例,说明如何使用 HAWQ 取代传统数据仓库,包括 ETL 处理、自动调度系统、维度表与事实表技术、OLAP 与数据的图形化表示等。数据挖掘部分用实例说明 HAWQ 与 MADlib 整合,实现降维、协同过滤、关联规则、回归、聚类、分类等常见数据挖掘与机器学习方法。

本书适合数据库管理员、大数据技术人员、Hadoop 技术人员、数据仓库技术人员,也适合高等院校和培训机构相关专业的师生教学参考。

本书封面贴有清华大学出版社防伪标签,无标签者不得销售。
版权所有,侵权必究。侵权举报电话:010-62782989 13701121933

图书在版编目(CIP)数据

HAWQ 数据仓库与数据挖掘实战 / 王雪迎著. — 北京:清华大学出版社,2018
ISBN 978-7-302-49802-5

Ⅰ. ①H… Ⅱ. ①王… Ⅲ. ①数据库系统②数据采集 Ⅳ. ①TP311.13②TP274

中国版本图书馆 CIP 数据核字(2018)第 037177 号

责任编辑:夏毓彦
封面设计:王 翔
责任校对:闫秀华
责任印制:沈 露

出版发行:清华大学出版社
网　　址:http://www.tup.com.cn,http://www.wqbook.com
地　　址:北京清华大学学研大厦 A 座　　邮　编:100084
社 总 机:010-62770175　　邮　购:010-62786544
投稿与读者服务:010-62776969,c-service@tup.tsinghua.edu.cn
质量反馈:010-62772015,zhiliang@tup.tsinghua.edu.cn

印 装 者:三河市铭诚印务有限公司
经　　销:全国新华书店
开　　本:190mm×260mm　　印　张:37　　字　数:947 千字
版　　次:2018 年 4 月第 1 版　　印　次:2018 年 4 月第 1 次印刷
印　　数:1~3000
定　　价:98.00 元

产品编号:077883-01

推荐序

回想过去几年，从我在EMC（Greenplum）启动HAWQ项目开始，到全球多个世界500强公司使用HAWQ，后来把HAWQ开源到Apache社区，现在又基于HAWQ创立"偶数"，时光荏苒。今天非常高兴能够看到雪迎这本关于HAWQ的书出现。

数据仓库的架构发展经历了几个阶段，第一代数据仓库是基于传统交易型数据库的共享存储（Share Storage）架构，比如Oracle，这种架构的缺点是基于专有高端存储，价格昂贵，可扩展性差，扩展到十几个节点往往就会撞到存储的瓶颈。

第二代数据仓库称为MPP（Massively Parallel Processing），采用无共享架构（Share Nothing），最早商业化的MPP产品为20世纪80年代出现的Teradata。Teradata当时基于大型机和专有硬件。在2000年左右又出现了几个基于普通x86服务器的MPP数据仓库创业公司，比如Greenplum、Vertica和Netezza，这几个创业公司后来分别被巨头EMC、HP和IBM收购。MPP架构解决了专有硬件的问题，可扩展性也得到了一定的提高，一般可以扩展到100节点左右。这种架构的缺点是在执行查询时，无论查询多大，所有节点都同样执行查询中均匀划分的一小部分，在节点数特别多的时候，很难协调保证所有节点的状态和工作都是均匀一致的。就像几个人一起干活，大家分工协调起来容易，如果几千人一起干活，人与人之间的不同以及协调问题就会突显起来。这也是MPP架构很难扩展到大规模的一个重要原因。

MPP之后的新一代数据仓库（New Data Warehouse）都采取了存储与计算分离架构。正是因为存储与计算分离，计算可以访问存储在任何节点的数据，并在任意节点进行调度，从而可以实现高可扩展性。存储与计算分离的另外一个好处是管理的简单性，比如扩容不再需要像MPP一样重新分布一遍数据。

新一代数据仓库根据存储实现方式的不同也可以分为三大类：SQL on Hadoop、SQL on Object Store以及SQL on Global Store。Hive、SparkSQL和HAWQ 2.x版本属于典型的SQL on Hadoop，存储为HDFS；像Amazon的Athena和Snowflake则属于SQL on Object Store，数据存储在S3对象存储中。一般SQL on Hadoop和SQL on Object Store都有着兼容性不好、性能一般或者对Update/Delete以及混合工作负载支持不好的缺点，但HAWQ因为从开始就定位为下一代的Greenplum Database和语法解析器等源于Greenplum Database，所以在兼容性和性能等方面表现得很优秀。HAWQ 社区现在正在开发的 HAWQ 新版本将会创新性地提出 SQL on Global Store 架构，HAWQ 将会具有一个可以全球规模部署、多数据中心、多活的存储。这样 HAWQ 就可以更加高效地支持各种传统数据仓库可以实现的功能，比如 Update/Delete 等，还可以更好地支持传统数据仓库做不到的功能，比如多数据中心、多活等，从而彻底取代传统数据仓库。

雪迎的这本书很好地介绍了 HAWQ 的基本技术，并从用户角度详细给出了如何使用 HAWQ 来构建数据仓库、进行机器学习和数据挖掘的方法，非常全面，是一本很好的 HAWQ 入门书籍。人工智能的流行以及数据驱动的方法是企业能够在新的数据和 AI 时代取得成功的关键，相信这本书的读者一定会从中受益，掌握最新的技术发展趋势与潮流。

Apache HAWQ 创始人
常雷
2018 年 1 月于北京

前　言

从 Bill Inmon 在 1991 年提出数据仓库的概念，至今已有 27 的时间。在这期间人们所面对的数据，以及处理数据的方法都发生了翻天覆地的变化。随着互联网和移动终端等应用的普及，运行在单机或小型集群上的传统数据仓库不再能满足数据处理要求，以 Hadoop 及其生态圈组件为代表的新一代分布式大数据处理平台逐渐流行。

尽管大多数人都在讨论某种技术或者架构可能会胜过另一种，而我更倾向于从"Hadoop 与数据仓库密切结合"这个角度来探讨问题。一方面企业级数据仓库中已经积累了大量的数据和应用程序，它们仍然在决策支持领域发挥着至关重要的作用；另一方面，传统数据仓库从业人员的技术水平和经验也在逐步提升。如何才能使积累的大量历史数据平滑过渡到 Hadoop 上，并让熟悉传统数据仓库的技术人员能够有效地利用已有的知识，可以在大数据处理平台上一展身手，才是一个亟待解决的问题。

虽然伴随着大数据的概念也出现了以 MongoDB、Cassandra 为代表的 NoSQL 产品，但不可否认，SQL 仍然是数据库、数据仓库中常使用的开发语言，也是传统数据库工程师或 DBA 的必会语言，从它出现至今一直被广泛使用。首先，SQL 有坚实的关系代数作为理论基础，经过几十年的积累，查询优化器也已经相当成熟。再者，对于开发者，SQL 作为典型的非过程语言，其语法相对简单，但语义却相当丰富。据统计 95%的数据分析问题都能用 SQL 解决，这是一个相当惊人的结论。那么 SQL 怎样才能与 Hadoop 等大数据技术结合起来，既能复用已有的技能，又能有效处理大规模数据呢？在这样的需求背景下，近年来涌现出越来越多的 SQL-on-Hadoop 软件，比如从早期的 Hive 到 Spark SQL、Impala、Kylin 等，本书所论述的就是众多 SQL-on-Hadoop 产品中的一员——HAWQ。

我最初了解到 HAWQ 是在 BDTC 2016 大会上，Apache HAWQ 的创始人常雷博士介绍了该项目。他的演讲题目是"以 HAWQ 轻松取代传统数据仓库"，这正是我的兴趣所在。HAWQ 支持事务、性能表现优良，关键是与 SQL 的兼容性非常好，甚至支持存储过程。对于传统数据仓库的开发人员，使用 HAWQ 转向大数据平台，学习成本应该是比较低的。我个人认为 HAWQ 更适合完成 Hadoop 上的数据仓库及其数据分析与挖掘工作。

本书内容

一年来，我一直在撰写 HAWQ 相关的文章和博客，并在利用 HAWQ 开发 Hadoop 数据仓库方面做了一些基础的技术实践，本书就是对这些工作的系统归纳与总结。全书分为技术解析、实战演练、数据挖掘三个部分，共 27 章。

技术解析部分说明 HAWQ 的基础架构与功能特性，包括安装部署、客户端与服务器连接、数据库对象与资源管理、查询优化、备份恢复、高可用性等。

实战演练部分通过一个简单而完整的示例，说明使用 HAWQ 设计和实现数据仓库的方法，包括初始和定期 ETL 处理、自动调度系统、维度表与事实表技术、联机分析处理与数据的图形化表示等。这部分旨在将传统数据仓库建模、SQL 开发的简单性与大数据技术相结合，快速、高效地建立可扩展的数据仓库及其应用系统。

数据挖掘部分结合应用实例，讨论将 HAWQ 与 MADlib 整合，MADlib 是一个开源机器学习库，提供了精确的数据并行实现、统计和机器学习方法，可以对结构化和非结构化数据进行分析。它的主要目的是可以非常方便地加载到数据库中，扩展数据库的分析功能。MADlib 仅用 SQL 查询就能做简单的数据挖掘与机器学习，实现矩阵分解、降维、关联规则、回归、聚类、分类、图算法等常见数据挖掘方法。这也是 HAWQ 的一大亮点。

本书读者

本书适合数据库管理员、数据仓库技术人员、Hadoop 或其他大数据技术人员，也适合高等院校和培训学校相关专业的师生教学参考。

代码、彩图下载

本书代码与彩图文件下载地址如下（注意数字与字母大小写）：

https://pan.baidu.com/s/1bpppAj1（密码：r7er）

如果下载有问题，请联系电子邮箱 booksaga@163.com，邮件主题为本书书名。

致谢

在本书编写过程中，得到了很多人的帮助与支持。感谢清华大学出版社图格事业部的老师和编辑们，他们的辛勤工作使得本书得以尽早与读者见面。感谢 CSDN 提供的技术分享平台，给我有一个将博客文章整理成书的机会。感谢我在优贝在线的所有同事，特别是技术部的同事们，他们在工作中的鼎力相助，使我有更多的时间投入到本书的写作中。感谢 Apache HAWQ 的创始人常雷先生在百忙之中为本书写推荐序。最后，感谢家人对我一如既往地支持。

因为水平有限，错漏之处在所难免，希望读者批评指正。

著　者
2018 年 1 月

目 录

第一部分 HAWQ 技术解析

第 1 章 HAWQ 概述 3
- 1.1 SQL-on-Hadoop 3
  - 1.1.1 对 SQL-on-Hadoop 的期待 3
  - 1.1.2 SQL-on-Hadoop 的实现方式 4
- 1.2 HAWQ 简介 6
  - 1.2.1 历史与现状 7
  - 1.2.2 功能特性 7
- 1.3 HAWQ 系统架构 9
  - 1.3.1 系统架构 10
  - 1.3.2 内部架构 11
- 1.4 为什么选择 HAWQ 12
  - 1.4.1 常用 SQL-on-Hadoop 产品的不足 12
  - 1.4.2 HAWQ 的可行性 13
  - 1.4.3 适合 DBA 的解决方案 18
- 1.5 小结 18

第 2 章 HAWQ 安装部署 19
- 2.1 安装规划 19
  - 2.1.1 选择安装介质 19
  - 2.1.2 选择 HAWQ 版本 20
  - 2.1.3 确认 Ambari 与 HDP 的版本兼容性 20
- 2.2 安装前准备 21
  - 2.2.1 确认最小系统需求 21
  - 2.2.2 准备系统安装环境 22
  - 2.2.3 建立本地 Repository 24
- 2.3 安装 Ambari 25
- 2.4 安装 HDP 集群 27
- 2.5 安装 HAWQ 29
- 2.6 启动与停止 HAWQ 34

2.6.1　基本概念 34
　　2.6.2　操作环境 34
　　2.6.3　基本操作 36
2.7　小结 40

第3章　连接管理 41
3.1　配置客户端身份认证 41
3.2　管理角色与权限 45
　　3.2.1　HAWQ 中的角色与权限 45
　　3.2.2　管理角色及其成员 46
　　3.2.3　管理对象权限 48
　　3.2.4　口令加密 49
3.3　psql 连接 HAWQ 50
3.4　Kettle 连接 HAWQ 52
3.5　连接常见问题 55
3.6　小结 56

第4章　数据库对象管理 57
4.1　创建和管理数据库 57
4.2　创建和管理表空间 61
4.3　创建和管理模式 65
4.4　创建和管理表 72
　　4.4.1　创建表 72
　　4.4.2　删除表 74
　　4.4.3　查看表对应的 HDFS 文件 74
4.5　创建和管理视图 76
4.6　管理其他对象 77
4.7　小结 78

第5章　分区表 79
5.1　HAWQ 中的分区表 79
5.2　确定分区策略 80
5.3　创建分区表 81
　　5.3.1　范围分区与列表分区 81
　　5.3.2　多级分区 86
　　5.3.3　对已存在的非分区表进行分区 86
5.4　分区消除 87
5.5　分区表维护 91

5.6 小结 98

第 6 章 存储管理 99

6.1 数据存储选项 99
6.2 数据分布策略 103
  6.2.1 数据分布策略概述 103
  6.2.2 选择数据分布策略 104
  6.2.3 数据分布用法 108
6.3 从已有的表创建新表 111
6.4 小结 117

第 7 章 资源管理 118

7.1 HAWQ 资源管理概述 118
  7.1.1 全局资源管理 118
  7.1.2 HAWQ 资源队列 119
  7.1.3 资源管理器配置原则 119
7.2 配置独立资源管理器 120
7.3 整合 YARN 123
7.4 管理资源队列 129
7.5 查询资源管理器状态 134
7.6 小结 137

第 8 章 数据管理 138

8.1 基本数据操作 138
8.2 数据装载与卸载 141
  8.2.1 gpfdist 协议及其外部表 141
  8.2.2 基于 Web 的外部表 148
  8.2.3 使用外部表装载数据 151
  8.2.4 外部表错误处理 151
  8.2.5 使用 hawq load 装载数据 152
  8.2.6 使用 COPY 复制数据 155
  8.2.7 卸载数据 157
  8.2.8 hawq register 159
  8.2.9 格式化数据文件 159
8.3 数据库统计 163
  8.3.1 系统统计 163
  8.3.2 统计配置 166
8.4 PXF 168

8.4.1　安装配置 PXF 168
　　8.4.2　PXF profile 168
　　8.4.3　访问 HDFS 文件 170
　　8.4.4　访问 Hive 数据 174
　　8.4.5　访问 JSON 数据 186
　　8.4.6　向 HDFS 中写入数据 190
8.5　小结 194

第 9 章　过程语言 195
9.1　HAWQ 内建 SQL 语言 195
9.2　PL/pgSQL 函数 197
9.3　给 HAWQ 内部函数起别名 198
9.4　表函数 198
9.5　参数个数可变的函数 201
9.6　多态类型 202
9.7　UDF 管理 205
9.8　UDF 实例——递归树形遍历 207
9.9　小结 214

第 10 章　查询优化 215
10.1　HAWQ 的查询处理流程 215
10.2　GPORCA 查询优化器 217
　　10.2.1　GPORCA 的改进 218
　　10.2.2　启用 GPORCA 224
　　10.2.3　使用 GPORCA 需要考虑的问题 225
　　10.2.4　GPORCA 的限制 227
10.3　性能优化 228
10.4　查询剖析 232
10.5　小结 238

第 11 章　高可用性 239
11.1　备份与恢复 239
　　11.1.1　备份方法 239
　　11.1.2　备份与恢复示例 242
11.2　高可用性 247
　　11.2.1　HAWQ 高可用简介 247
　　11.2.2　Master 节点镜像 248
　　11.2.3　HAWQ 文件空间与 HDFS 高可用 251

11.2.4　HAWQ 容错服务 260
　11.3　小结 262

第二部分　HAWQ 实战演练

第 12 章　建立数据仓库示例模型 265
　12.1　业务场景 265
　12.2　数据仓库架构 267
　12.3　实验环境 268
　12.4　HAWQ 相关配置 269
　12.5　创建示例数据库 273
　　　12.5.1　在 hdp4 上的 MySQL 中创建源库对象并生成测试数据 273
　　　12.5.2　创建目标库对象 275
　　　12.5.3　装载日期维度数据 283
　12.6　小结 284

第 13 章　初始 ETL 285
　13.1　用 Sqoop 初始数据抽取 285
　　　13.1.1　覆盖导入 286
　　　13.1.2　增量导入 286
　　　13.1.3　建立初始抽取脚本 287
　13.2　向 HAWQ 初始装载数据 288
　　　13.2.1　数据源映射 288
　　　13.2.2　确定 SCD 处理方法 288
　　　13.2.3　实现代理键 289
　　　13.2.4　建立初始装载脚本 289
　13.3　建立初始 ETL 脚本 291
　13.4　小结 293

第 14 章　定期 ETL 294
　14.1　变化数据捕获 294
　14.2　创建维度表版本视图 296
　14.3　创建时间戳表 297
　14.4　用 Sqoop 定期数据抽取 298
　14.5　建立定期装载 HAWQ 函数 298
　14.6　建立定期 ETL 脚本 303
　14.7　测试 303
　　　14.7.1　准备测试数据 303

IX

		14.7.2	执行定期 ETL 脚本	304
		14.7.3	确认 ETL 过程正确执行	305
	14.8	动态分区滚动		307
	14.9	准实时数据抽取		309
	14.10	小结		317

第 15 章 自动调度执行 ETL 作业 318

- 15.1 Oozie 简介 318
- 15.2 建立工作流前的准备 320
- 15.3 用 Oozie 建立定期 ETL 工作流 324
- 15.4 Falcon 简介 328
- 15.5 用 Falcon process 调度 Oozie 工作流 329
- 15.6 小结 332

第 16 章 维度表技术 333

- 16.1 增加列 333
- 16.2 维度子集 342
- 16.3 角色扮演维度 348
- 16.4 层次维度 354
  - 16.4.1 固定深度的层次 355
  - 16.4.2 多路径层次 357
  - 16.4.3 参差不齐的层次 359
- 16.5 退化维度 361
- 16.6 杂项维度 366
- 16.7 维度合并 374
- 16.8 分段维度 380
- 16.9 小结 386

第 17 章 事实表技术 387

- 17.1 周期快照 388
- 17.2 累积快照 394
- 17.3 无事实的事实表 404
- 17.4 迟到的事实 409
- 17.5 累积度量 416
- 17.6 小结 422

第 18 章 联机分析处理 423

- 18.1 联机分析处理简介 423
  - 18.1.1 概念 423

		18.1.2 分类 424
		18.1.3 性能 426
18.2	联机分析处理实例 427
		18.2.1 销售订单 427
		18.2.2 行列转置 433
18.3	交互查询与图形化显示 440
		18.3.1 Zeppelin 简介 440
		18.3.2 使用 Zeppelin 执行 HAWQ 查询 441
18.4	小结 448

第三部分　HAWQ 数据挖掘

第 19 章　整合 HAWQ 与 MADlib 451

19.1　MADlib 简介 452
19.2　安装与卸载 MADlib 455
19.3　MADlib 基础 458
　　　19.3.1　向量 458
　　　19.3.2　矩阵 469
19.4　小结 484

第 20 章　奇异值分解 485

20.1　奇异值分解简介 485
20.2　MADlib 奇异值分解函数 486
20.3　奇异值分解实现推荐算法 489
20.4　小结 501

第 21 章　主成分分析 502

21.1　主成分分析简介 502
21.2　MADlib 的 PCA 相关函数 504
21.3　PCA 应用示例 509
21.4　小结 513

第 22 章　关联规则方法 514

22.1　关联规则简介 514
22.2　Apriori 算法 517
　　　22.2.1　Apriori 算法基本思想 517
　　　22.2.2　Apriori 算法步骤 518
22.3　MADlib 的 Apriori 算法函数 518
22.4　Apriori 应用示例 519

XI

22.5	小结	524

第 23 章 聚类方法 ... 525

23.1	聚类方法简介	525
23.2	k-means 方法	526
	23.2.1 基本思想	527
	23.2.2 原理与步骤	527
	23.2.3 k-means 算法	527
23.3	MADlib 的 k-means 相关函数	529
23.4	k-means 应用示例	532
23.5	小结	537

第 24 章 回归方法 ... 538

24.1	回归方法简介	538
24.2	Logistic 回归	539
24.3	MADlib 的 Logistic 回归相关函数	539
24.4	Logistic 回归示例	542
24.5	小结	546

第 25 章 分类方法 ... 547

25.1	分类方法简介	547
25.2	决策树	549
	25.2.1 决策树的基本概念	549
	25.2.2 决策树的构建步骤	549
25.3	MADlib 的决策树相关函数	551
25.4	决策树示例	555
25.5	小结	561

第 26 章 图算法 ... 562

26.1	图算法简介	562
26.2	单源最短路径	565
26.3	MADlib 的单源最短路径相关函数	566
26.4	单源最短路径示例	567
26.5	小结	569

第 27 章 模型验证 ... 570

27.1	交叉验证简介	570
27.2	MADlib 的交叉验证相关函数	573
27.3	交叉验证示例	575
27.4	小结	578

第一部分

HAWQ 技术解析

第 1 章

◀ HAWQ概述 ▶

　　HAWQ 的全称为 Hadoop With Query，即带查询的 Hadoop，是一个出色的 SQL-on-Hadoop 解决方案，尤其适合构建 Hadoop 数据仓库。它最初由 Pivotal 公司开发，后来贡献给 Apache 社区，成为孵化器项目。本章是对 HAWQ 的一个概要介绍。首先对 SQL-on-Hadoop 的功能需求有个基本认识，然后以此作为参照，说明 HAWQ 的功能特性。为了更好地使用 HAWQ，我们需要了解它的整体系统架构，以及各组件所起的作用。本章最后将阐述选择 HAWQ 的理由。

1.1　SQL-on-Hadoop

　　过去几年里，许多企业和开发者已慢慢接受 Hadoop 生态系统，将它用作大数据分析堆栈的核心组件。尽管 Hadoop 生态系统的 MapReduce 组件是一个强大的典范，但随着时间的推移，MapReduce 自身不再是连接存储在 Hadoop 生态系统中的数据的最简单途径。企业需要一种更简单的方式来访问要查询、分析甚至要执行深度挖掘的数据，以便发现存储在 Hadoop 中的所有数据的真正价值。SQL 以其扎实的理论基础、简单的语法、丰富的语义得到广泛应用，在帮助各类用户发掘数据的商业价值领域具有很长历史。

　　Hadoop 上的 SQL 支持一开始是 Apache Hive，一种类似于 SQL 的查询引擎，它将有限的 SQL 方言编译到 MapReduce 中。Hive 对 MapReduce 的完全依赖会导致严重的查询延迟，因此其主要适用场景是批处理模式。另外，尽管 Hive 对于 SQL 的支持是好的开端，但对 SQL 的有限支持意味着精通 SQL 的用户忙于企业级使用场景时将遇到严重的限制。它还暗示着庞大的基于标准 SQL 的工具生态系统无法利用 Hive。值得庆幸的是，在为 SQL-on-Hadoop 提供更好的解决方案方面已取得长足进展。除 Hive 外，当前常见的框架已经有 HAWQ、Impala、Presto、Spark SQL、Drill、Kylin 等很多种。

1.1.1　对 SQL-on-Hadoop 的期待

　　表 1-1 显示了一流的 SQL-on-Hadoop 方案需要具有的功能以及这些功能给使用者带来的好处。从传统的意义上说，这些功能中的大部分在分析型数据仓库中都能找到。

表 1-1　一流 SQL-on-Hadoop 方案应有功能及带来的业务好处

功能	业务好处
丰富且合规的 SQL 支持	功能强大的可移植 SQL 应用程序，能够利用基于 SQL 的数据分析和数据可视化工具的大型生态系统
符合 TPC-DS 规格	TPC-DS 帮助确保所有级别的 SQL 查询得到处理，从而广泛支持各种使用场景并避免企业级实施期间出现意外
灵活高效的表连接	简化应用系统开发，提高数据仓库查询性能
线性可扩展性	平衡数据仓库工作负载
一体化深度挖掘与机器学习	用 SQL 实现所需的统计学、数学和机器学习算法
外部数据处理能力	有效利用多种外部数据资产，降低数据重构成本
高可用性与容错	确保业务连续性，保证数据仓库的关键业务分析
原生 Hadoop 文件格式支持	简化 ETL 过程，减少数据迁移

1.1.2　SQL-on-Hadoop 的实现方式

1. Hive

Hive 建立在 Hadoop 的分布式文件系统（HDFS）和 MapReduce 之上。为了缩小 Hive 与传统 SQL 引擎之间的性能落差，现在已经可以通过 MapReduce、Spark 或 Tez 等多种计算框架执行查询。Hive 提供一种称为 HiveQL 的语言，允许用户进行类似于 SQL 的查询。Hive 的体系结构如图 1-1 所示。

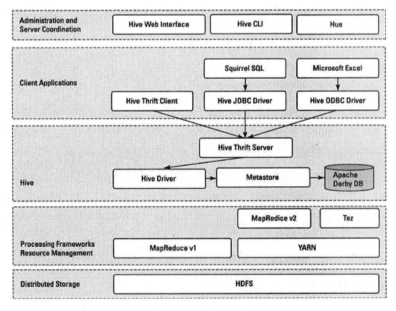

图 1-1　Hive 体系结构

2. Spark SQL

Spark SQL 是 Spark 处理结构化数据的程序模块。它将 SQL 查询与 Spark 程序无缝集成，可以将结构化数据作为 Spark 的 RDD 进行查询。RDD 的全称为 Resilient Distributed Datasets，即弹性分布式数据集，是 Spark 基本的数据结构。Spark 使用 RDD 作为分布式程序的工作集合，提供一种分布式共享内存的受限形式。RDD 是只读的，对其只能进行创建、转化和求值等操作。Spark SQL 的体系结构如图 1-2 所示。

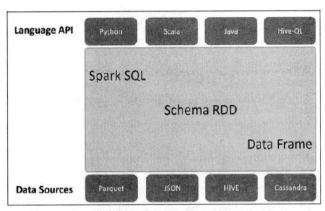

图 1-2　Spark SQL 体系结构

Dataset 是一个分布式的数据集合。Dataset API 是 Spark 1.6 中新增的编程接口，利用 Spark SQL 执行引擎优化器提供 RDD 的功能。Dataset 可以从 JVM 对象中构建，然后使用 map、flatMap、filter 等方法进行转换。DataFrame 是被组织为命名列的 Dataset，其概念与关系数据库中的表类似，但底层结构更加优化。DataFrame 可以从结构化的数据文件、Hive 表、外部数据库或者已有的 RDD 中构建。

3. Impala

Impala 是一个运行在 Hadoop 上的大规模并行处理（Massively Parallel Processing，MPP）查询引擎，提供对 Hadoop 集群数据的高性能、低延迟的 SQL 查询，使用 HDFS 作为底层存储。对查询的快速响应使交互式查询和对分析查询的调优成为可能，而这些在针对处理长时间批处理作业的 SQL-on-Hadoop 传统技术上是难以完成的。Impala 可与 Hive 共享数据库表，并且 Impala 与 HiveQL 的语法兼容。

Impala 体系结构如图 1-3 所示。Impala 服务器由不同的守护进程组成，每种守护进程运行在 Hadoop 集群中的特定主机上。其中，Impalad、Statestored、Catalogd 三个守护进程在其架构中扮演主要角色。

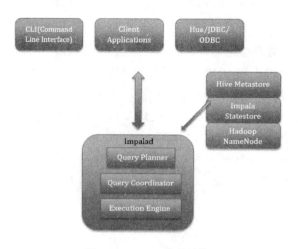

图 1-3　Impala 体系结构

4. HAWQ

HAWQ 引擎利用 Greenplum 数据仓库的代码基础和深度数据管理专业知识构建，在 HDFS 中存储底层数据。HAWQ 使用业内唯一一款专为 HDFS 量身打造的、基于成本的查询优化框架来增强其性能。与 Impala 类似，HAWQ 也采用 MPP 架构，使用户能够获益于经过锤炼的基于 MPP 的分析功能及其查询性能，同时有效利用 HDFS 的分布式存储、容错机制、机架感知等功能，兼顾了低延时与高扩展。HAWQ 可与其他传统 SQL-on-Hadoop 引擎共存于一个分析堆栈。图 1-4 取自 Pivotal 官方文档，显示了 HAWQ 与 Greenplum 的联系与区别。

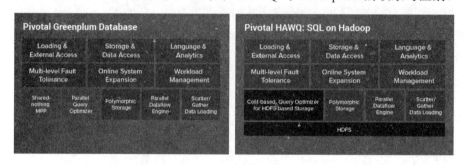

图 1-4　将基于 MPP 的分析数据仓库用于 SQL-on-Hadoop 方案

1.2　HAWQ 简介

HAWQ 是一个 Hadoop 原生大规模并行 SQL 分析引擎，针对的是分析型应用。它和其他关系型数据库类似，接受 SQL，返回结果集。

1.2.1 历史与现状

（1）想法和原型系统（2011 年）：GOH 阶段（Greenplum Database On HDFS）。

（2）HAWQ 1.0 Alpha（2012 年）：多个国外大型客户试用，当时客户性能测试是 Hive 的数百倍，促进了 HAWQ 1.0 作为正式产品发布。

（3）HAWQ 1.0 GA（2013 年初）：改变了传统 MPP 数据库架构，包括事务、容错、元数据管理等。

（4）HAWQ 1.X 版本（2014-2015 Q2）：增加了一些企业级需要的功能，比如 Parquet 存储、新的优化器、Kerberos 支持、Ambari 安装部署等。

（5）HAWQ 2.0 Alpha 发布并成为 Apache 孵化器项目：针对云环境的系统架构重新设计，新增数十个高级功能，包括弹性执行引擎、高级资源管理、YARN 集成、快速扩容等。当前最新版本是 HAWQ++ 2.2.0。

1.2.2 功能特性

虽然 HAWQ 采用 MPP 架构，但它具有很多传统大规模并行处理数据库没有的特性及功能。让我们考虑 SQL-on-Hadoop 的各个方面，并将之与 HAWQ 相比较。

1. 丰富且完全兼容的 SQL 标准

数据仓库项目的数据源往往是多种异构数据库，而且很多时候我们不能直连源库，得到的只是从源库导出的 SQL 脚本。在这种情况下，对 SQL 的兼容性要求尤为重要。HAWQ 百分之百符合 ANSI SQL 规范并且支持 SQL 92、99、2003 OLAP，以及基于 Hadoop 的 PostgreSQL。它包含关联子查询、窗口函数、分析函数、标量函数与聚合函数的功能，并且支持 SQL UDF。由于 HAWQ 系统完全符合 SQL 规范，因此使用 HAWQ 编写的分析应用程序可以轻松移植到其他符合 SQL 规范的数据引擎上，反之亦然。用户可通过 ODBC 和 JDBC 连接 HAWQ。

2. TPC-DS 合规性

TPC-DS 针对具有各种操作要求和复杂性的查询定义了 99 个模板，比如点对点、报表、迭代、OLAP、数据挖掘等。成熟的基于 Hadoop 的 SQL 系统需要支持和正确执行多数此类查询，以解决各种不同分析工作和使用场景中的问题。基准测试通过 TPC-DS 中的 99 个模板生成的 111 个查询来执行。依据符合可优化、可执行两个要求的查询个数，图 1-5 所示的条形图显示了一些基于 SQL-on-Hadoop 常见系统的合规情况。

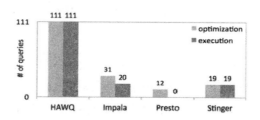

图 1-5 从 TPC-DS 套件返回的已完成查询个数

以 Greenplum 代码库提供的扩展 SQL 支持能力为基础，HAWQ 完成了全部 111 个查询。

3. 可实现灵活高效的连接

HAWQ 吸收了先进的基于成本的 SQL 查询优化器，自动生成执行计划，可优化使用 Hadoop 集群资源，还可以针对特定环境配置优化器内的成本函数，如版本、硬件、CPU、IOPS 等。HAWQ 声称，能够为涉及 50 多个关联表的查询快速找到理想的查询计划。这让用户能够以 HAWQ 提高用于大量数据分析的传统企业数据仓库工作负载的性能。

4. 利用线性可扩展加速 Hadoop 查询

HAWQ 为 PB 级数据操作专门设计。数据直接存储在 HDFS 上，并且其 SQL 查询优化器已经为基于 HDFS 的文件系统性能特征进行过细致的优化。

SQL-on-Hadoop 的主要设计目标之一是在 Hadoop 上执行 SQL 连接时最大限度地降低数据传输开销。HAWQ 采用 Dynamic pipelining 技术解决这一关键问题。Dynamic pipelining 是一种并行数据流框架，结合了以下技术：

- 适应性高速 UDP 互联。
- 针对大数据量调整操作运行时执行环境。
- 运行时资源管理，确保查询完整性。
- 无缝数据分配机制，将经常用于特定查询的部分数据集中处理。
- 使 IP 查找真正可扩展。

HAWQ 官方的性能分析显示，对于 Hadoop 上的分析与数据仓库工作场景，HAWQ 要比现有 Hive 查询引擎快一至两个数量级。

5. 一体化深度分析与机器学习

数据分析通常需要使用统计学、数学和机器学习算法，如聚类或主成分分析等，这正在成为 SQL-on-Hadoop 方案的基本要求。HAWQ 利用开源机器学习库 MADlib 提供这些功能，通过 UDF 扩展 SQL 能力。对于有此类需求的用户来说，将使其可以在通常的分析型工作中嵌入高级机器学习功能。

6. 外部数据处理能力

SQL-on-Hadoop 需要联合外部源数据，将各种来源的数据结合起来进行分析，提供更多灵活性。数据可以跨其他数据仓库、HDFS、HBase 以及 Hive 实例，且需要实施固有的并行性。HAWQ 通过名为 Pivotal eXtension Framework（PXF）的模块提供外部数据访问能力。PXF 提供的特色功能包括：

- PXF 使用智能抓取，其过滤器下推到 Hive 和 HBase。查询工作负载被下推到联合数据堆栈，从而尽可能减少数据移动，并改善延迟性能。
- PXF 提供框架 API，以便用户为其自有数据堆栈开发新的连接器，进而增强数据引擎的松耦合，避免数据重构操作。

- PXF 可利用 ANALYZE 收集外部数据的统计资料。这样就可以通过基于成本的优化器优化联合数据源统计信息，帮助构建更高效的查询。

7. 高可用性与容错

HAWQ 支持数据库事务，允许用户隔离 Hadoop 上的并行活动并在出错时进行回滚。HAWQ 的容错服务、可靠性和高可用三个特点能容忍磁盘级与节点级故障。这些能力可确保业务的连续性，同时增加了将更多关键业务分析迁移到 HAWQ 上运行的可能。

8. 原生 Hadoop 文件格式支持

HAWQ 支持 AVRO、Parquet 和原生的 HDFS 文件格式，在很大程度上降低了数据摄取期间 ETL 的复杂性。对 ETL 和数据移动需求的减少直接降低了分析解决方案的成本。

9. 通过 Apache Ambari 进行原生的 Hadoop 管理

HAWQ 使用 Apache Ambari 作为管理和配置的基础。合适的 Ambari 插件可以使得 HAWQ 像其他通用 Hadoop 服务一样被 Ambari 管理，IT 管理团队不再需要 Hadoop 与 HAWQ 两套管理界面。这使得用户专注于功能实现，最小化配置和管理等技术支持所需的工作量。同时，Ambari 是完全开源的 Hadoop 管理和配置工具，消除了供应商绑定风险。

10. Hortonworks Hadoop 兼容

HAWQ 可以与 Hortonworks HDP 大数据体系无缝兼容，使用户在已经投资的 Hortonworks 大数据平台上利用 HAWQ 提供的所有功能。

11. HAWQ 的其他主要特性

- 弹性执行引擎：可根据查询大小动态决定执行查询使用的节点个数。
- 支持多种分区方法及多级分区：如 List 分区和 Range 分区。
- 动态扩容：动态按需扩容，按照存储大小或者计算需求，快速添加节点。
- 多级资源管理：可与外部资源管理器 YARN 集成，也可以自己管理 CPU、Memory 等资源，支持多级资源队列。
- 支持多种第三方工具：如 Tableau、SAS 等。

1.3 HAWQ 系统架构

HAWQ 结合了 MPP 数据库的关键技术和 Hadoop 的可扩展性，在原生的 HDFS 上读写数据。MPP 架构使 HAWQ 表现出超越其他 SQL-on-Hadoop 解决方案的查询性能，Hadoop 又为 HAWQ 提供了传统数据库所不具备的线性扩展能力。

1.3.1 系统架构

图 1-6 给出了一个典型的 HAWQ 集群系统架构。通常 HAWQ 集群中包含一个 Master 节点和多个 Slave 节点。在 Master 节点上部署 HAWQ Master、HDFS NameNode、YARN ResourceManager。HAWQ 的元数据服务也在 Master 节点中。每个 Slave 节点上部署有 HDFS DataNode、YARN NodeManager 以及一个 HAWQ Segment。HAWQ Segment 在执行查询的时候会启动多个查询执行器（Query Executor，QE）。查询执行器运行在资源容器中。

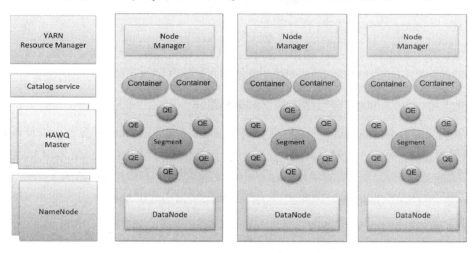

图 1-6　典型的 HAWQ 集群部署架构

1. HAWQ Master

HAWQ Master 是系统的入口，其上运行处理 SQL 命令的数据库进程。它负责的主要工作是：接受客户端连接；对连接请求进行鉴权；处理输入的 SQL 命令，解析 SQL 并生成执行计划；向 Segment 分发查询任务；协调每个 Segment 执行查询返回结果；向客户端程序输出最终结果。HAWQ Master 在本地存储的全局系统目录是一组系统表的集合，包含 HAWQ 系统自身的元数据。HAWQ Master 本地不存储任何用户数据，用户数据只存储在 HDFS 上。最终用户通过 Master 与 HAWQ 进行交互。可以使用如 psql 等客户端程序，或者类似 JDBC、ODBC 的应用程序接口（APIs）连接到数据库。

2. 物理 Segment 与虚拟 Segment

在 HAWQ 中，物理 Segment 是并行数据处理单元，每个主机上只有一个物理 Segment。每个物理 Segment 可以为一个查询启动多个 QE，这使得单一物理 Segment 看起来就像多个虚拟 Segment，从而使 HAWQ 能够更好地利用所有可用资源。虚拟 Segment 是内存、CPU 等资源的容器，每个虚拟 Segment 含有为查询启动的一个 QE，查询就是在虚拟 Segment 中被 QE 所执行。若没有特殊说明，则本书中所提及的 Segment 指的是物理 Segment。

与 Master 不同，Segment 是无状态的，并且 Segment 中不存储数据库元数据和本地文件系统中的数据。Master 节点将 SQL 请求连同相关的元数据信息分发给 Segment 进行处理。元数据中包含所请求表的 HDFS URL 地址，Segment 使用该 URL 访问相应的数据。

1.3.2 内部架构

HAWQ 可与 Hadoop 的资源管理框架 YARN 紧密结合，为查询提供资源管理。HAWQ 在一个资源池中缓存 YARN 容器，然后利用 HAWQ 自身的细粒度资源管理，为用户或组在本地管理这些资源。当执行一个查询时，HAWQ 根据查询成本、资源队列定义、数据局部化和当前系统中的资源使用情况，为查询规划资源分配。之后查询被分发到 Segment 所在的物理主机，可能是节点子集或整个集群。每个 Segment 节点上的资源实施器监控着查询对资源的实时使用情况，避免异常资源占用。图 1-7 是 HAWQ 内部架构图。

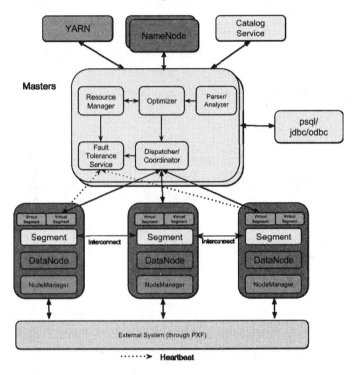

图 1-7　HAWQ 内部架构

可以看到在 Master 节点内部有如下几个重要组件：查询解析器（Parser/Analyzer），优化器，资源管理器，资源代理，HDFS 元数据缓存，容错服务，查询派遣器，元数据服务。在 Slave 节点上安装有一个 Segment，查询执行时，针对一条查询，弹性执行引擎会启动多个虚拟 Segment 同时执行该查询，节点间数据交换通过 Interconnect（高速互联网络）进行。如果一个查询启动了 1000 个虚拟 Segment，就意味着这个查询被均匀地分成了 1000 份任务，这些任务会并行执行。所以说虚拟 Segment 数其实表明了查询执行的并行度。查询并行度是由弹性执行引擎根据查询大小以及当前资源使用情况动态确定的。下面解释这些组件的作用以及它们之间的关系。

（1）查询解析器：解析查询，检查语法及语义，最终生成查询树并将其传递给优化器。

（2）优化器：负责接收查询树，生成查询计划。针对一个查询，可能有很多等价的查询计划，但执行性能差别很大。优化器的作用是找出最优的查询计划。

（3）资源管理器：通过资源代理向全局资源管理器（比如 YARN）动态申请资源，并缓存资源，在不需要的时候返回资源。缓存资源的主要目的是减少 HAWQ 与全局资源管理器之间的交互代价。如果每一个查询都去向资源管理器申请资源，那么性能会严重受到影响。资源管理器同时需要保证查询不使用超过分配给该查询的资源，否则查询之间会相互影响，严重时可能导致系统整体不可用。

（4）HDFS 元数据缓存：用于 HAWQ 确定哪些 Segment 扫描表的哪些部分。HAWQ 会把计算派遣到数据所在的地方，所以要匹配计算和数据的局部性，这需要 HDFS 块的位置信息。位置信息存储在 HDFS NameNode 上。每个查询都访问 NameNode 会造成瓶颈，因此在 HAWQ Master 节点上建立了 HDFS 元数据缓存。

（5）容错服务：负责检测集群中哪些节点可用、哪些节点不可用，不可用的节点会被排除出资源池。

（6）查询派遣器：优化器优化完查询后，查询派遣器将执行计划分发到各个 Segment 节点上执行，并协调查询执行的整个过程。查询派遣器是整个并行系统的粘合剂。

（7）元数据服务：负责存储 HAWQ 的各种元数据，包括数据库和表信息，以及访问权限信息等。另外，元数据服务也是实现分布式事务的关键。

（8）高速互联网络：负责在节点之间传输数据，由基于 UDP 协议的软件实现。

1.4 为什么选择 HAWQ

前面已经介绍了几种常用 SQL-on-Hadoop 的实现方式，也了解了 HAWQ 的功能特性与系统架构。那么站在用户的角度，我们为什么要选择 HAWQ？近年来我尝试过几种 SQL-on-Hadoop 产品，从最初的 Hive，到 Spark SQL，再到 Impala，在这些产品上进行了一系列 ETL、CDC、多维数据仓库、OLAP 实验。从数据库的角度看，这些产品与传统的 DBMS 相比，功能不够完善，性能差距很大，甚至很难找到一个相对完备的 Hadoop 数据仓库解决方案。这里就以个人的实践体验来简述这些产品的不足以及 HAWQ 的可行性。

1.4.1 常用 SQL-on-Hadoop 产品的不足

1. Hive

Hive 是一款老牌的 Hadoop 数据仓库产品，能够部署在所有 Hadoop 发行版本上。它在 MapReduce 计算框架上封装一个 SQL 语义层，极大简化了 MR 程序的开发。直到现在，Hive 依然以其稳定性赢得了大量用户。

Hive 的缺点也很明显——速度太慢。随着技术的不断进步，Hive 的执行引擎从 MapReduce 发展出 Hive on Spark、Hive on Tez 等。特别是运行在 Tez 框架上的 Hive，其性能有了很大改进。即便如此，Hive 的速度还是比较适合后台批处理应用场景，而不适合交互式即时查询和联机分析。

2. Spark SQL

Spark SQL 是 Hadoop 中另一个著名的 SQL 引擎，正如名字所表示的，它以 Spark 作为底层计算框架，实际上是一个 Scala 程序语言的子集。Spark 基本的数据结构是 RDD，一个分布于集群节点的只读数据集合。传统的 MapReduce 框架强制在分布式编程中使用一种特定的线性数据流处理方式。MapReduce 程序从磁盘读取输入数据，把数据分解成键/值对，经过混洗、排序、归并等数据处理后产生输出，并将最终结果保存在磁盘。Map 阶段和 Reduce 阶段的结果均要写磁盘，这大大降低了系统性能。也是由于这个原因，MapReduce 大都被用于执行批处理任务。

为了解决 MapReduce 的性能问题，Spark 使用 RDD 共享内存结构。这种内存操作减少了磁盘 IO，大大提高了计算速度。开发 Spark 的初衷是用于机器学习系统的培训算法，而不是 SQL 查询。Spark 宣称其应用的延迟可以比 MapReduce 降低几个数量级，但是在我们的实际使用中，20TB 的数据集合上用 Spark SQL 查询要 10 分钟左右出结果，这个速度纵然是比 Hive 快了 4 倍，但显然不能支撑交互查询和 OLAP 应用。Spark 还有一个问题，即需要占用大量内存，当内存不足时，很容易出现 OOM 错误。

3. Impala

Impala 的最大优势在于执行速度。官方宣称大多数情况下它能在几秒或几分钟内返回查询结果，而相同的 Hive 查询通常需要几十分钟甚至几小时完成，因此 Impala 适合对 Hadoop 文件系统上的数据进行分析式查询。Impala 默认使用 Parquet 文件格式，这种列式存储方式对于典型数据仓库场景下的大查询是较为高效的。

Impala 的问题主要体现在功能上的欠缺。例如，不支持 Date 数据类型，不支持 XML 和 JSON 相关函数，不支持 covar_pop、covar_samp、corr、percentile、percentile_approx、histogram_numeric、collect_set 等聚合函数，不支持 rollup、cube、grouping set 等操作，不支持数据抽样（Sampling），不支持 ORC 文件格式，等等。其中，分组聚合、取中位数等是数据分析中的常用操作，当前的 Impala 存在如此多的局限，使它在可用性上大打折扣，实际使用时要格外注意。

1.4.2 HAWQ 的可行性

介绍了几种 SQL-on-Hadoop 产品的主要问题后，现在看 HAWQ 是否有能力取而代之。作为用户，我们从功能与性能两方面简单讨论一下使用 HAWQ 在 Hadoop 上构建分析型数据仓库应用的可行性。

1. 功能

（1）兼容 SQL 标准

HAWQ 从代码级别上可以简单理解成是数据存储在 HDFS 上的 Greenplum 数据库，全面兼容 SQL 标准。它支持内连接、外连接、全连接、笛卡儿连接、相关子查询等所有表连接方式，支持并集、交集、差集等集合操作，并支持递归函数调用。作为一个数据库系统，提供这些功能很好理解。

（2）丰富的函数

除了包含诸多字符串、数字、日期时间、类型转换等常规标量函数以外，HAWQ 还包含丰富的窗口函数和高级聚合函数，这些函数经常被用于分析型数据查询。窗口函数包括 cume_dist、dense_rank、first_value、lag、last_valueexpr、lead、ntile、percent_rank、rank、row_number 等。高级聚合函数包括 median、percentile_cont (expr) within group (order by expr [desc/asc])、percentile_disc (expr) within group (order by expr [desc/asc])、sum(array[])、pivot_sum (label[], label, expr)等。

（3）过程化编程

HAWQ 支持内建的 SQL、C、Java、Perl、pgSQL、Python、R 等多种语言的过程化编程。

（4）原生 Hadoop 文件格式支持

HAWQ 支持 HDFS 上的 AVRO、Parquet、平面文本等多种文件格式，支持 snappy、gzip、quicklz、RLE 等多种数据压缩方法。与 Hive 不同，HAWQ 实现了 schema-on-write（写时模式）数据验证处理，不符合表定义或存储格式的数据是不允许进入到表中的，这点与传统数据库管理系统保持一致。

（5）外部数据整合

HAWQ 通过 PXF 模块提供访问 HDFS 上的 JSON 文件、Hive、HBase 等外部数据的能力。除了用于访问 HDFS 文件的 PXF 协议，HAWQ 还提供了 gpfdist 文件服务器，它利用 HAWQ 系统并行读写本地文件系统中的文件。

2. 性能

（1）基于成本的 SQL 查询优化器

HAWQ 采用基于成本的 SQL 查询优化器。该查询优化器以针对大数据模块化查询优化器架构的研究成果为基础而设计，能够生成高效的执行计划。

（2）与 Impala 的性能比较

同样采用 MPP 架构。图 1-8 是 HAWQ 提供的 TPC-DS 性能比较图，从中可以看出 HAWQ 平均比 Impala 快 4.55 倍。

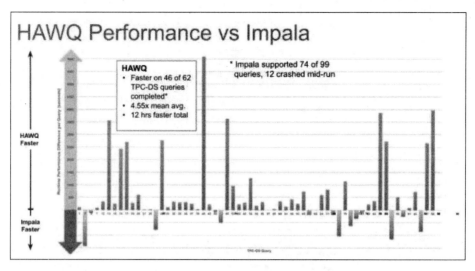

图 1-8　HAWQ 与 Impala 性能比较

（3）与 Hive 的性能比较

为了取得第一手数据，我们做了以下 HAWQ 与 Hive 查询的性能对比测试。

- 硬件环境

4 台 VMware 虚机组成的 Hadoop 集群，每台机器配置如下：

> 15K RPM SAS 100GB
> Intel(R) Xeon(R) E5-2620 v2 @ 2.10GHz，双核双 CPU
> 8GB 内存，8GB Swap
> 10000Mb/s 虚拟网卡

- 软件环境

> Linux：CentOS release 6.4，核心 2.6.32-358.el6.x86_64
> Ambari：2.4.1
> Hadoop：HDP 2.5.0
> Hive（Hive on Tez）：2.1.0
> HAWQ：2.1.1.0
> HAWQ PXF：3.1.1

- 数据模型

实验模拟一个记录页面单击数据的分析型应用。数据模型中包含日期、页面、浏览器、引用、状态 5 个维度表，1 个页面单击事实表。表结构和关系如图 1-9 所示。

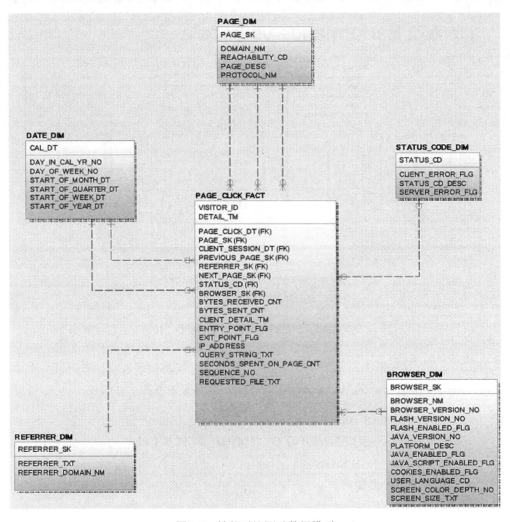

图 1-9 性能对比测试数据模型

- 记录数

各表的记录数如表 1-2 所示。

表 1-2 各表记录数

表名	行数	表名	行数
page_click_fact	1 亿	browser_dim	2 万
page_dim	20 万	status_code	70
referrer_dim	100 万	date_dim	366

- 查询

分别用 Hive 和 HAWQ 执行以下 5 个典型查询，记录执行时间。

➢ 查询给定周中 support.sas.com 站点上访问最多的目录。

> 查询各月从 www.google.com 访问的页面。
> 给定年份 support.sas.com 站点上的搜索字符串计数。
> 查询使用 Safari 浏览器访问页面的人数。
> 查询给定周中 support.sas.com 站点上浏览超过 10 秒的页面。

建表和查询语句参见 http://blog.csdn.net/wzy0623/article/details/71479539。

- 测试结果

Hive、HAWQ 外部表、HAWQ 内部表查询时间对比如表 1-3 所示。每种查询情况执行三次取平均值。

表 1-3　查询执行时间

查询	Hive（秒）	HAWQ 外部表（秒）	HAWQ 内部表（秒）
1	74.337	304.134	19.232
2	169.521	150.882	3.446
3	73.482	101.216	18.565
4	66.367	359.778	1.217
5	60.341	118.329	2.789

从图 1-10 中的对比可以看到，HAWQ 内部表比 Hive on Tez 快得多（4~50 倍）。同样的查询，在 HAWQ 的 Hive 外部表上执行却很慢。由此可见，在执行分析型查询时最好使用 HAWQ 内部表。

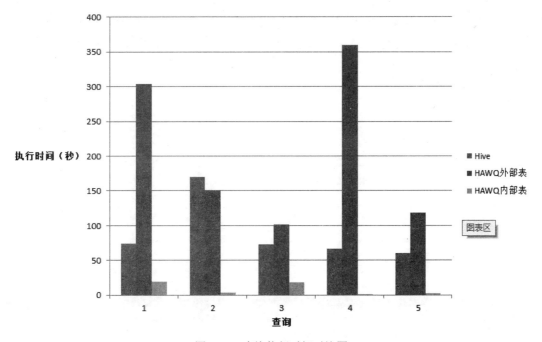

图 1-10　查询执行时间对比图

1.4.3 适合DBA的解决方案

HAWQ最吸引人的地方是它支持SQL过程化编程,这是通过用户自定义函数(user-defined functions,UDF)实现的。编写UDF的语言可以是SQL、C、Java、Perl、Python、R和pgSQL。数据库应用开发人员常用的自然是SQL和pgSQL,PL/pgSQL函数可以为SQL语言增加控制结构,执行复杂计算任务,并继承所有PostgreSQL的数据类型(包括用户自定义类型)、函数和操作符。

HAWQ是我所使用过的SQL-on-Hadoop解决方案中唯一支持SQL过程化编程的,Hive、Spark SQL、Impala、Kylin都没有此功能。对于习惯了编写存储过程的DBA来说,这无疑大大提高了HAWQ的易用性。HAWQ的UDF提供以下特性:

- 给HAWQ内部函数起别名。
- 返回结果集的表函数。
- 参数个数可变的函数。
- 多态数据类型。

1.5 小结

HAWQ是一个Hadoop上的SQL引擎,是以Greenplum Database为代码基础逐渐发展起来的。HAWQ采用MPP架构,改进了针对Hadoop的基于成本的查询优化器。除了能高效处理本身的内部数据,还可通过PXF访问HDFS、Hive、HBase、JSON等外部数据源。HAWQ全面兼容SQL标准,能编写SQL UDF,还可用SQL完成简单的数据挖掘和机器学习。无论是功能特性,还是性能表现,HAWQ都比较适用于构建Hadoop分析型数据仓库应用。

第 2 章

HAWQ安装部署

本章详细说明 HAWQ 的安装部署过程，以及如何启动和停止 HAWQ 服务。后面章节的实践部分都是在本章完成的安装环境中进行的。安装环境的主机、软硬件信息如下：

- 主机信息如表 2-1 所示，所有主机都能连接互联网。

表 2-1　主机信息

主机名	IP 地址	主机名	IP 地址
hdp1	172.16.1.124	hdp3	172.16.1.126
hdp2	172.16.1.125	hdp4	172.16.1.127

- 硬件配置：每台主机 CPU 4 核、内存 8GB、硬盘 100GB。
- 软件版本如表 2-2 所示。

表 2-2　系统软件版本

名称	版本
操作系统	CentOS release 6.4 (Final) 64 位
JDK	OpenJDK 64-Bit version "1.7.0_09-icedtea"
数据库	MySQL 5.6.14
JDBC	MySQL Connector Java 5.1.38
HDP	2.5.0
Ambari	2.4.1

2.1 安装规划

2.1.1 选择安装介质

HAWQ 的安装介质有两种选择，一是下载源码手工编译，二是使用官方提供的编译好的安装包。HAWQ 2.0.0 版本的源码下载地址为：http://apache.org/dyn/closer.cgi/incubator/hawq

/2.0.0.0-incubating/apache-hawq-src-2.0.0.0-incubating.tar.gz。源码编译和安装的 Apache 官方文档地址为：https://cwiki.apache.org/confluence/display/HAWQ/Build+and+Install。网上也有一些资料可供参考。

建议初学者不要使用源码编译方式，这种方法需要的依赖很多，对操作系统、Hadoop 的版本、安装与配置的要求都较高。推荐使用编译好的安装包，主要原因是过程相对简单、安装成功率高。

2.1.2 选择 HAWQ 版本

这里安装的是 HAWQ 2.1.1。该版本最主要的变化是实现了对 ORC 文件格式的支持，包含了所有 Apache HAWQ 孵化项目的功能特性，并修复了一些之前的 bug。

在选择 HAWQ 版本时，需要考虑它与所支持操作系统、Hadoop 平台和安装工具 Ambari 四者之间的版本匹配关系。表 2-3 显示了 HAWQ 2.1.1 版本的产品支持。

表 2-3　HAWQ 2.1.1 产品支持

HAWQ 版本	PXF 版本	Hortonworks HDP 版本	Ambari 版本	HAWQ Ambari Plug-inban 版本	MADlib 版本	RHEL/CentOS 版本
2.1.1.0	3.1.1	2.5	2.4.1	2.1.1	1.9, 1.9.1	6.4+ (64-bit)

注意：

- HAWQ 目前仅兼容 Hortonworks HDP 一种 Hadoop 发行版本。
- HAWQ 2.1.1 不支持 RHEL/CentOS 7。

实际上 HAWQ 只在 HDP 上经过了严格的测试。细心的读者也许已经注意到，前一章介绍 HAWQ 功能特性时，仅提到与 Hortonworks Hadoop 兼容，也就是这个原因。希望 HAWQ 能提高对 HDP 以外其他 Hadoop 发行版本的支持与普适度，以便在其他 Hadoop 平台上安装使用更为容易。

2.1.3 确认 Ambari 与 HDP 的版本兼容性

安装 HAWQ 之前首先需要安装 Ambari 和 Hortonworks Data Platform（HDP）。从表 2-3 看到，与 HAWQ 2.1.1 兼容的 Ambari 版本是 2.4.1、HDP 版本是 2.5。再次从 Hortonworks 官方的安装文档中确认版本兼容性，兼容矩阵如表 2-4 所示。

表 2-4　Ambari 与 HDP 的版本兼容性

Ambari*	HDP 2.5	HDP 2.4	HDP 2.3 (deprecated)	HDP 2.2 (deprecated)	HDP 2.1
2.4.1	✓	✓	✓	✓	
2.2.2		✓	✓	✓	
2.2.1		✓	✓	✓	✓
2.2.0			✓	✓	✓
2.1			✓**	✓	✓
2.0				✓	✓

* Ambari does not install Hue or HDP Search (Solr).

** If you plan to install and manage HDP 2.3.4 (or later), **you must use Ambari 2.2.0 (or later)**. Do **not** use Ambari 2.1x with HDP 2.3.4 (or later).

2.2　安装前准备

整个 HAWQ 的安装部署过程包括安装 Ambari、安装 HDP、安装 HAWQ 三个依次进行的步骤，在实施这些步骤前需要做一些准备工作。如果没有做特殊说明，那么所有配置或命令都用 root 用户执行。

2.2.1　确认最小系统需求

HAWQ 2.1.1 官方文档说明的最小系统需求如下：

- 操作系统：CentOS v6.x。
- 浏览器：Google Chrome 26 及以上。
- 依赖软件包：yum、rpm、scp、curl、unzip、tar、wget、OpenSSL (v1.01, build 16 or later)、Python 2.6.x、OpenJDK 7/8 64-bit。
- 系统内存与磁盘：Ambari 主机至少应该有 1GB 内存和 500MB 剩余磁盘空间。如果要使用 Ambari Metrics，所需内存和磁盘大小依据集群规模如表 2-5 所示。

表 2-5　资源需求与集群规模

主机数量	可用内存	磁盘空间
1	1024MB	10GB
10	1024MB	20GB

(续表)

主机数量	可用内存	磁盘空间
50	2048MB	50GB
100	4096MB	100GB
300	4096MB	100GB
500	8192MB	200GB
1000	12288MB	200GB

- 最大打开文件描述符：推荐值大于 10000。使用下面的命令检查每个主机的当前值：

```
ulimit -Sn
ulimit -Hn
```

- 如果小于 10000，使用下面的命令设置成 10000：

```
ulimit -n 10000
```

2.2.2　准备系统安装环境

（1）禁用防火墙

在安装期间 Ambari 需要与部署集群主机通信，因此特定的端口必须打开。最简单的实现方式是执行下面的命令禁用防火墙，所有主机都要执行：

```
/etc/init.d/iptables stop
chkconfig iptables off
```

（2）禁用 SELinux

Ambari 安装需要禁用 SELinux，所有主机都要执行：

```
setenforce 0
# 编辑/etc/selinux/config 文件，设置
SELINUX=disabled
```

（3）配置域名解析

编辑/etc/hosts 文件，添加如下四行，所有主机都要执行：

```
172.16.1.124 hdp1
172.16.1.125 hdp2
172.16.1.126 hdp3
172.16.1.127 hdp4
```

注意，不要删除文件中原有的如下两行，否则可能引起网络问题：

```
127.0.0.1       localhost localhost.localdomain localhost4 localhost4.localdomain4
::1             localhost localhost.localdomain localhost6 localhost6.localdomain6
```

在 hdp1 上执行：

```
hostname hdp1
# 编辑/etc/sysconfig/network 文件，设置如下两行：
NETWORKING=yes
HOSTNAME=hdp1
```

hdp2、hdp3、hdp4 上执行类似的配置。

（4）安装配置 NTP

安装 NTP 服务，所有主机都要执行：

```
yum install -y ntp
chkconfig ntpd on
service ntpd start
```

（5）配置 SSH 免密码

为了使 Ambari Server 在集群所有主机上自动安装 Ambari Agents，必须配置 Ambari Server 主机到集群其他主机的 SSH 免密码连接。以下配置用于在 hdp1 上运行 Ambari Server，在所有四台主机上运行 Ambari Agents 的情况。

在 hdp1 上执行：

```
ssh-keygen
... 一路回车 ...
ssh-copy-id hdp1
ssh-copy-id hdp2
ssh-copy-id hdp3
ssh-copy-id hdp4
```

在所有主机执行：

```
chmod 700 ~/.ssh
chmod 600 ~/.ssh/authorized_keys
```

（6）安装 MySQL JDBC 驱动

所有主机都执行：

```
tar -zxvf mysql-connector-java-5.1.38.tar.gz
cp ./mysql-connector-java-5.1.38/mysql-connector-java-5.1.38-bin.jar /usr/share/java/mysql-connector-java.jar
```

（7）安装 MySQL 数据库

在 hdp1、hdp2 上安装 MySQL，hdp1 上的 MySQL 用于 Ambari，hdp2 上的 MySQL 用于 Hive、Oozie 等 Hadoop 组件。在 hdp1、hdp2 上执行以下命令：

```
rpm -ivh MySQL-5.6.14-1.el6.x86_64.rpm
service mysql start
```

（8）在 MySQL 中建立数据库用户并授权

登录 hdp2 上的 MySQL，执行下面的 SQL 命令建立数据库用户并授权：

```
create database hive;
create database oozie;
create user 'hive'@'%' identified by 'hive';
grant all privileges on hive.* to 'hive'@'%';
create user 'oozie'@'%' identified by 'oozie';
grant all privileges on oozie.* to 'oozie'@'%';
flush privileges;
```

2.2.3 建立本地 Repository

联机安装过程中需要从远程的 Repository 中 yum 下载所需要的包。为了防止由于网络不稳定或远程 Repository 不可用等原因导致的安装失败，最好配置本地 Repository。在安装 HAWQ 时，本地和远程的 Repository 配合使用，既能加快安装进度，又能补全所需的包。

（1）下载以下两个文件到 hdp1：

```
wget http://public-repo-1.hortonworks.com/HDP/centos6/2.x/updates/2.5.0.0/HDP-2.5.0.0-centos6-rpm.tar.gz
wget http://public-repo-1.hortonworks.com/HDP-UTILS-1.1.0.21/repos/centos6/HDP-UTILS-1.1.0.21-centos6.tar.gz
```

（2）在 hdp1 上建立一个 HTTP 服务器：

```
yum install httpd
mkdir -p /var/www/html/
cd /var/www/html/
tar -zxvf ~/HDP-2.5.0.0-centos6-rpm.tar.gz
tar -zxvf ~/HDP-UTILS-1.1.0.21-centos6.tar.gz
service httpd start
```

（3）新建 /etc/yum.repos.d/hdp.repo 文件，添加如下行：

```
[hdp-2.5.0.0]
name=hdp-2.5.0.0
baseurl=http://172.16.1.124/HDP/centos6/
path=/
enabled=1
gpgcheck=0
priority=10
```

（4）新建 /etc/yum.repos.d/hdp-utils.repo 文件，添加如下行：

```
[HDP-UTILS-1.1.0.21]
name=HDP-UTILS-1.1.0.21
baseurl=http://172.16.1.124/HDP-UTILS-1.1.0.21/repos/centos6
path=/
enabled=1
gpgcheck=0
priority=10
```

(5)下载 CentOS6-Base-163.repo 到/etc/yum.repos.d 目录。

安装过程中发现本地仓库不全,还少 RPM 包,因此再加一个 163 的源。

```
cd /etc/yum.repos.d/
wget http://mirrors.163.com/.help/CentOS6-Base-163.repo
```

(6)新建/etc/yum.repos.d/fedora.repo 文件,添加如下行:

```
[epel]
name=epel
baseurl=http://dl.fedoraproject.org/pub/epel/6/x86_64/
enabled=1
gpgcheck=0
```

HAWQ 的 Repository 中缺少 libgsasl 库,因此再加一个包含 libgsasl 库的源。

(7)将新建的 Repository 配置文件复制到其他主机:

```
scp /etc/yum.repos.d/* root@hdp2:/etc/yum.repos.d/
scp /etc/yum.repos.d/* root@hdp3:/etc/yum.repos.d/
scp /etc/yum.repos.d/* root@hdp4:/etc/yum.repos.d/
```

2.3 安装 Ambari

Ambari 是 Apache Software Foundation 中的一个顶级项目。从 Ambari 的作用来说,就是创建、管理、监视 Hadoop 的集群。这里所说的 Hadoop 是广义的,指的是 Hadoop 整个生态圈(例如 Hive、HBase、Sqoop、ZooKeeper 等),而并不仅是特指 Hadoop。一言以蔽之,Ambari 就是让 Hadoop 以及相关的大数据软件更容易使用的一个工具。

Ambari 主要具有以下功能特性:

- 通过一步一步地安装向导简化了集群部署。
- 预先配置好关键的运维指标(metrics),可以直接查看 Hadoop Core(HDFS 和 MapReduce)及相关项目(如 HBase、Hive 和 HCatalog)是否健康。
- 支持作业与任务执行的可视化与分析,能够更好地查看依赖和性能。

- 通过一个完整的 RESTful API 把监控信息暴露出来，集成了现有的运维工具。
- 用户界面非常直观，用户可以轻松有效地查看信息并控制集群。

Ambari 使用 Ganglia 收集度量指标，用 Nagios 支持系统报警，当需要引起管理员的关注时（比如，节点停机或磁盘剩余空间不足等问题），系统将向其发送邮件。此外，Ambari 能够部署安全的（基于 Kerberos）Hadoop 集群，以此实现对 Hadoop 安全的支持，提供基于角色的用户认证、授权和审计功能，并为用户管理集成了 LDAP 和 Active Directory。

Ambari 自身也是一个分布式架构的软件，主要由两部分组成：Ambari Server 和 Ambari Agent。简单来说，用户通过 Ambari Server 通知 Ambari Agent 安装对应的软件，Agent 会定时发送各个机器上每个软件模块的状态给 Ambari Server，最终这些状态信息将呈现在 Ambari 的 GUI 中，方便用户了解集群的状态，并进行相应的维护。下面说明 Ambari 的安装步骤。

1. 下载 Ambari repository 到 hdp1

（1）下载 Ambari repository 文件

```
wget -nv
http://public-repo-1.hortonworks.com/ambari/centos6/2.x/updates/2.4.1.0/ambari.repo -O /etc/yum.repos.d/ambari.repo
```

注意，这里的文件名必须是 ambari.repo，当 Ambari Agent 注册到 Ambari Server 时需要此文件。

（2）确认 repository 配置

```
yum repolist | grep amba
```

应该看到类似下面的信息：

```
[root@hdp1 ~]# yum repolist | grep amba
Repository base is listed more than once in the configuration
Updates-ambari-2.4.2.0           ambari-2.4.2.0 - Updates           12
```

（3）安装 Ambari Server

```
yum install ambari-server
```

这一步也会安装 Ambari 默认使用的 PostgreSQL 数据库。出现提示符时输入 y，确认事务和依赖检查。

2. 为 Ambari 配置 MySQL 数据库

（1）在 hdp1 上的 MySQL 中建立 Ambari 数据库用户并授权：

```
create user 'ambari'@'%' identified by 'ambari';
grant all privileges on *.* to 'ambari'@'%';
flush privileges;
```

（2）建立 Ambari Server 数据库模式：

```
create database ambari;
use ambari;
source /var/lib/ambari-server/resources/Ambari-DDL-MySQL-CREATE.sql;
```

3. 配置 Ambari Server

启动 Ambari Server 前必须进行配置，指定 Ambari 使用的数据库、安装 JDK、指定运行 Ambari Server 守护进程的用户等。在 hdp1 上执行下面的命令管理配置过程。

```
ambari-server setup
```

- 出现 Customize user account for ambari-server daemon 提示时输入 n，使用 root 用户运行 Ambari Server。
- 选择 JDK 1.7。
- 出现 Enter advanced database configuration 提示时输入 y，选择 Option [3] MySQL/MariaDB，然后根据提示输入连接 MySQL 的用户名、密码和数据库名，就是上一步配置的信息，这里均为 ambari。

4. 启动 Ambari Server

在 hdp1 上执行下面的命令启动 Ambari Server：

```
ambari-server start
# 查看 Ambari Server 进程状态
ambari-server status
```

至此，Ambari 安装完成。

2.4 安装 HDP 集群

Hortonworks Data Platform 是 Hortonworks 公司开发的 Hadoop 数据平台。Hortonworks 由 Yahoo 的工程师创建，它为 Hadoop 提供了一种"service only"的分发模型。有别于其他商业化的 Hadoop 版本，Hortonworks 是一个可以自由使用的开放式企业级数据平台。其 Hadoop 发行版本即 HDP，可以被自由下载并整合到各种应用当中。

Hortonworks 是第一个提供基于 Hadoop 2.0 版产品的厂商，也是目前唯一支持 Windows 平台的 Hadoop 分发版本。用户可以通过 HDInsight 服务，在 Windows Azure 上部署 Hadoop 集群。

下面说明如何在浏览器中使用 Ambari 的安装向导交互式安装、配置、部署 HDP。

1. 登录 Ambari

在浏览器中打开 http://172.16.1.124:8080，初始的用户名/密码为 admin/admin。在欢迎页面单击 Launch Install Wizard，如图 2-1 所示。

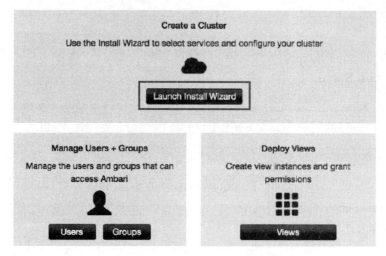

图 2-1　启动 Ambari 安装向导

2．给集群命名

集群名称中不要有空格和特殊字符，然后单击 Next。

3．选择 HDP 版本

选择 2.5.0.0，如图 2-2 所示。

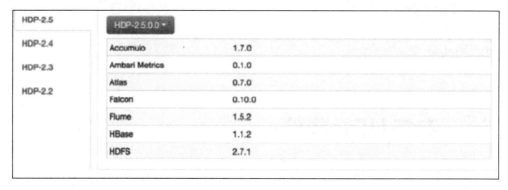

图 2-2　选择 HDP 版本

4．选择 Repositories

选择 Use Public Repositories，然后单击 Next。

5．安装选项

- 在 Target Hosts 编辑框中输入四个主机名，每个一行。
- 单击 Choose File 按钮，选择 2.2.2 "准备系统安装环境"第（5）步中 hdp1 上生成的私钥文件 id_rsa。

- 设置用户名 root、端口 22。
- 选择 Register and Confirm。

6. 确认主机

选中 4 台主机，单击 Next。

7. 选择服务

根据需要选择服务，或者接受默认配置，单击 Next。

8. 标识 Masters

根据需要选择 Masters，或者接受默认配置，单击 Next。

9. 标识 Slaves 和 Clients

根据需要选择 Slaves 和 Clients，或者接受默认配置，单击 Next。

10. 定制服务

- 为 hive 和 oozie 配置 MySQL 数据库连接。
- 设置所需的密码。
- 其他保持默认。

11. 复查确认

确认之前的配置无误后，单击 Deploy。

12. 安装、启动与测试

此时显示安装进度页面。Ambari 对 HDP 每个安装的组件执行安装、启动和测试。此时不要刷新浏览器，等待部署过程完全执行成功。当出现"Successfully installed and started the services"时，单击 Next。

13. 完成

汇总页面显示完成的任务列表。单击 Complete，显示 Ambari Web GUI 主页面。至此，HDP 安装完成。

2.5 安装 HAWQ

1. 选择 HAWQ 主机

在安装 HAWQ 之前，使用下面的步骤选择和准备所需主机。

（1）选择 HAWQ 主机。记住有以下限制：

- 每台主机都必须满足安装相应版本 HAWQ 的系统要求。
- 每个 HAWQ Segment 所在主机必须和其上运行的 HDFS DataNode 协同工作。
- HAWQ Master 和 Standby 必须部署在不同的主机上。

在本实验环境中，集群中的四台主机均作为 HAWQ Segment，其中两台分别作为 HAWQ Master 和 Standby，在安装时 Ambari 会自动部署主机。

（2）选择运行 PXF 主机。记住有以下限制：

- PXF 必须安装在 HDFS NameNode 和所有 HDFS DataNodes 主机上。
- 如果配置了 Hadoop HA，PXF 必须安装在包括所有 NameNode 和所有 HDFS Node 的主机上。
- 如果想通过 PXF 访问 HBase 和 Hive，必须在将要安装 PXF 的主机上首先安装 HBase 和 Hive 的客户端。

在本实验环境中，集群中的四台主机均安装 PXF，在安装时 Ambari 会自动部署主机。（在前面部署 HDP 时，已经在所有四台机器上安装了客户端程序。）

（3）确认所有主机上所需的端口没有被占用。

HAWQ Master 和 Standby 服务默认使用 5432 端口。前面安装 Ambari 时使用的是 MySQL 数据库存储元数据，而不是默认的 PostgreSQL，所以本次安装中不存在端口冲突问题。

2. 建立 HAWQ 的 Repositories

在安装 HAWQ 前需要建立两个本地 yum repositories。在运行 Ambari Server 的主机上（hdp1）以 root 用户执行下面的步骤。这台主机（称为 repo-node）必须能够访问 HAWQ 集群的所有节点。

（1）重启 httpd 服务器。

```
service httpd [re]start
```

（2）从 https://network.pivotal.io/products/pivotal-hdb 下载名为 hdb-2.1.1.0-7.tar 的 HAWQ 安装文件。

（3）建立一个临时目录存储解压后的 HAWQ 安装包。运行 httpd 进程的操作系统用户（本次安装中是 root）必须对该目录及其所有上级目录具有可读可执行的权限。

```
mkdir /staging
chmod a+rx /staging
```

注意，不要使用/tmp 目录，/tmp 下的文件可能在任意时间被删除。

（4）HAWQ 安装文件中包含一个 yum repository。解压安装文件后，运行 setup_repo.sh 脚本，将 HAWQ 软件的发布包添加到本地 yum 包的 repository 中。

```
cd /staging
```

```
tar -zxvf hdb-2.1.1.0-7.tar
cd hdb-2.1.1.0
./setup_repo.sh
```

setup_repo.sh 在本地建立一个名为 hdb-2.1.1.0.repo 的 HDB repository，并且在 httpd 服务器的根目录（默认为/var/www/html）下建立一个符号链接，指向 hdb-2.1.1.0-7.tar 的解压缩目录。在本次安装中为/var/www/html/hdb-2.1.1.0 → /staging/hdb-2.1.1.0。

（5）在 HAWQ 集群的所有节点上安装 epel-release 包：

```
yum install -y epel-release
```

3. 使用 Ambari 安装 HAWQ

（1）用 root 用户登录 Ambari Server 主机（hdp1）。

（2）从 HDB repository 安装 HAWQ Ambari 插件。

```
yum install -y hawq-ambari-plugin
```

以上命令会建立/var/lib/hawq 目录，并将所需的脚本和模板文件安装到该目录中。

（3）重启 Ambari 服务器。

```
ambari-server restart
```

（4）执行 add-hawq.py 脚本将 HDB repository 添加到 Ambari 服务器中。

```
cd /var/lib/hawq
./add-hawq.py --user admin --password admin --stack HDP-2.5
```

需要提供正确的 Ambari 管理员用户名和密码，默认都是 admin。

（5）重启 Ambari 服务器。

```
ambari-server restart
```

（6）登录 Ambari Web 控制台。

在浏览器中打开 http://172.16.1.124:8080，默认的用户名和密码都是 admin，确认 HAWQ 服务已经可用。

（7）选择 HDFS → Configs 标签。

（8）配置 HDFS。

- 选择 Settings 标签，修改 DataNode max data transfer threads 为 40960。
- 选择 Advanced 标签，点开 DataNode，设置 DataNode directories permission 为 750。
- 单击 General，设置 Access time precision 为 0。
- 单击 Advanced hdfs-site，设置表 2-6 所示属性的值。如果属性不存在，就选择 Custom hdfs-site，单击 Add property… 添加属性并设置表 2-6 中所示的值。

表2-6 hdfs-site 属性

Property	Setting
dfs.allow.truncate	true
dfs.block.access.token.enable	false for an unsecured HDFS cluster, or true for a secure cluster
dfs.block.local-path-access.user	gpadmin
HDFS Short-circuit read	true
dfs.client.socket-timeout	300000000
dfs.client.use.legacy.blockreader.local	false
dfs.datanode.handler.count	60
dfs.datanode.socket.write.timeout	7200000
dfs.namenode.handler.count	600
dfs.support.append	true

（9）点开 Advanced core-site，设置表 2-7 所示属性的值。如果属性不存在，就选择 Custom core-site，单击 Add property… 添加属性并设置表 2-7 中所示的值。

表2-7 core-site 属性

Property	Setting
ipc.client.connection.maxidletime	3600000
ipc.client.connect.timeout	300000
ipc.server.listen.queue.size	3300

（10）单击 Save 保存配置。

（11）在继续后面的步骤前先单击 Restart 重启服务。

（12）在主页选择 Actions→Add Service。

（13）从服务列表选择 HAWQ 和 PXF，单击 Next，显示 Assign Masters 页。

（14）选择 HAWQ Master 和 HAWQ Standby 的主机，或接受默认值，单击 Next 显示 Assign Slaves and Clients 页。

（15）选择运行 HAWQ Segments 和 PXF 的主机，或接受默认值，单击 Next。Add Service 助手会基于可用的 Hadoop 服务自动为 HAWQ 选择主机。注意，PXF 必须安装在 NameNode、Standby NameNode 和每一个 DataNode 节点上，而 HAWQ Segment 必须安装在每个 DataNode 节点上。

（16）在 Customize Services 页面接受默认设置。

（17）单击 Advanced 标签，输入 HAWQ 系统用户口令，单击 Next。

（18）在后面的页面均接受默认值，连续单击 Next。

（19）最后单击 Complete。如果 Ambari 提示集群上的组件需要重启，选择 Restart→Restart

All Affected 重启所有受影响的服务。

（20）验证 HAWQ 安装。

用 gpadmin 用户登录 HAWQ Master 所在主机，执行下面的命令：

```
source /usr/local/hawq/greenplum_path.sh    # 设置 HAWQ 环境变量
psql -d postgres
create database test;
\c test
create table t (i int);
insert into t select generate_series(1,100);
\timing
select count(*) from t;
```

结果如下所示。

```
[root@hdp3 ~]# su - gpadmin
[gpadmin@hdp3 ~]$ source /usr/local/hawq/greenplum_path.sh
[gpadmin@hdp3 ~]$ psql -d postgres
psql (8.2.15)
Type "help" for help.

postgres=# create database test;
CREATE DATABASE
postgres=# \c test
You are now connected to database "test" as user "gpadmin".
test=# create table t (i int);
CREATE TABLE
test=# insert into t select generate_series(1,100);
INSERT 0 100
test=# \timing
Timing is on.
test=# select count(*) from t;
 count
-------
   100
(1 row)

Time: 135.211 ms
test=#
```

至此，HAWQ 2.1.1 集群安装部署完成。

2.6 启动与停止 HAWQ

HAWQ 作为 Hadoop 上的一个服务提供给用户，与其他所有服务一样，最基本的操作就是启动、停止、重启服务。要完成这些操作，需要适当的环境设置。下面就 HAWQ 管理的一些基础概念、操作环境、启动停止及其推荐的操作进行讨论。

2.6.1 基本概念

如果将系统管理与开发分离，那这部分内容严格说应该是 HAWQ 系统管理员所关心的。要利用好 HAWQ 集群，应该有一些 Linux/UNIX 系统管理、数据库管理系统、DBA 和 SQL 等必备知识和经验。HAWQ 服务器实际上是一个以 HDFS 作为物理存储的分布式数据库系统，像 Oracle、MySQL 等软件一样，是一个真正的数据库。HAWQ 代码源自 Greenplum，而 Greenplum 是在 PostgreSQL 基础上开发的，这种沿袭关系使得 HAWQ 天生具有完整的数据库特性，就连 HAWQ 的官方文档也始终与 PostgreSQL 文档保持一致。这点与其他 SQL-on-Hadoop 方案显著不同。

1. HAWQ 用户

HAWQ 支持对用户及其操作权限的管理。HAWQ 系统安装后，数据库中包含一个预定义的超级用户，该用户与安装 HAWQ 的操作系统用户同名，叫做 gpadmin。gpadmin 作为操作系统用户，可以使用 HAWQ 的命令行工具执行管理任务，如启动或停止 HAWQ、扩展集群、删除集群中的节点等。而作为数据库用户，gpadmin 相当于 Oracle 的 sys 或 MySQL 的 root，具有数据库的最大权限。HAWQ 管理员用户可以创建其他数据库用户，并向他们赋予管理或操作数据库对象的权限。

可以选择使用 Ambari 或命令行管理 HAWQ 集群。当使用 Ambari 管理 HAWQ 时，用 Ambari 的管理员用户登录 Web 控制台页面即可，不需要使用 gpadmin。

2. HAWQ 系统部署

从前面的安装过程中看到，一个典型的 HAWQ 部署包括一个 HDFS NameNode、一个 HAWQ Master、一个 HAWQ Standby，以及多个 HAWQ Segment 与 HDFS DataNode。HAWQ 集群中还可能包括 PXF 或其他 Hadoop 服务。

使用 Ambari 在 HDP 上安装 HAWQ 时，会为 HAWQ 节点自动选择 HDP 集群中的主机，只要求 HAWQ Master 和 Standby 运行在不同主机上，Segment 可以和 Master、Standby 运行在相同主机上，通常在每个 DataNode 上运行一个 Segment。在我们的实验环境中，Ambari 选择 hdp3 作为 Master、hdp2 作为 Standby，HDP 集群中的所有 4 台主机，每个上面运行一个 Segment。

2.6.2 操作环境

在操作 HAWQ 集群前，必须设置 HAWQ 所需的环境。

1. 设置 HAWQ 操作环境

HAWQ 提供了一个名为 greenplum_path.sh 的 shell 脚本文件，位于 HAWQ 安装的根目录下，用于设置 HAWQ 所需的环境变量。这些环境变量中最重要的是$GPHOME，它指定了 HAWQ 安装的根目录，典型的 HAWQ 根目录是/usr/local/hawq。其他环境变量包括用于查找 HAWQ 相关文件的$PATH、动态链接库路径$LD_LIBRARY_PATH、python 路径$PYTHONPATH、openssl 配置文件$OPENSSL_CONF、HDFS3 客户端配置文件$LIBHDFS3_CONF、YARN 客户端配置文件$LIBYARN_CONF、HAWQ 的配置文件$HAWQSITE_CONF 等。默认设置可以满足大多数需求。如果环境有特殊要求，可以将相关环境变量添加到 greenplum_path.sh 文件中。

执行以下步骤设置 HAWQ 操作环境：

（1）用 gpadmin 操作系统用户登录 HAWQ 节点，或者切换到 gpadmin：

```
[root@hdp1 ~]# su - gpadmin
[gpadmin@hdp1 ~]$
```

（2）通过执行 greenplum_path.sh 文件设置 HAWQ 操作环境：

```
[gpadmin@hdp1 ~]$ source /usr/local/hawq/greenplum_path.sh
```

（3）编辑.bash_profile 或其他 shell 资源文件在登录时自动执行 greenplum_path.sh：

```
[gpadmin@hdp1 ~]$ echo "source /usr/local/hawq/greenplum_path.sh" >> ~/.bash_profile
```

（4）根据需要在 shell 初始化文件中设置与具体部署相关的 HAWQ 特定环境变量，包括 PGAPPNAME、PGDATABASE、PGHOST、PGPORT 和 PGUSER 等。设置这些环境变量可简化 psql 命令行，通过提供默认值省去在命令行中输入相关选项。例如：

- 如果定制了 HAWQ 主节点的端口号，在 shell 初始化文件中添加如下一行，设置 PGPORT 环境变量使该端口号成为默认值：export PGPORT=10432。
- 如果经常操作一个特定数据库，在 shell 资源文件中添加如下一行，设置 PGDATABASE 环境变量使其成为默认值：export PGDATABASE=<database-name>。

2. HAWQ 文件与目录

表 2-8 说明 HAWQ 默认安装的一些文件和目录。

表 2-8　HAWQ 文件与目录

文件/目录	内容
$HOME/hawqAdminLogs/	默认的 HAWQ 管理应用程序日志文件目录
$GPHOME/greenplum_path.sh	HAWQ 环境设置脚本
$GPHOME/bin/	HAWQ 客户端、数据库和管理应用程序

（续表）

文件/目录	内容
$GPHOME/etc/	HAWQ 配置文件，包括 hawq-site.xml
$GPHOME/include/	HDFS、PostgreSQL、libpq 的头文件
$GPHOME/lib/	HAWQ 库文件
$GPHOME/lib/postgresql/	PostgreSQL 共享库和 JAR 文件
$GPHOME/share/postgresql/	PostgreSQL 及其过程化语言的示例与脚本
/data/hawq/[master|segment]/	HAWQ Master 和 Segment 的默认数据目录
/data/hawq/[master|segment]/pg_log/	HAWQ Master 和 Segment 的默认日志目录
/etc/pxf/conf/	PXF 服务的配置文件
/usr/lib/pxf/	PXF 服务插件共享库
/var/log/pxf/	PXF 日志目录
/usr/hdp/current/	HDP 运行时配置文件

2.6.3 基本操作

在 HAWQ 系统中的 Master 节点和所有的 Segment 节点上都运行一个 PostgreSQL 数据库服务器实例。例如，在 hdp3 上可以看到如下两个 postgres 进程：

```
/usr/local/hawq_2_1_1_0/bin/postgres -D /data/hawq/master -i -M master -p 5432 --silent-mode=true
/usr/local/hawq_2_1_1_0/bin/postgres -D /data/hawq/segment -i -M segment -p 40000 --silent-mode=true
```

所有这些 DBMS 一起被当作单一 DBMS 启动和停止，通过这种方式能够统一启动、停止所有实例。因为 HAWQ 系统被分布于多个机器上，所以启动与停止 HAWQ 系统的过程不同于标准的 PostgreSQL DBMS 的启动停止过程。

启动和停止 HAWQ 的命令分别是 hawq start 和 hawq stop。hawq 命令行工具是一个 python 脚本，位于 $GPHOME/bin 目录下。可以在命令行输入 hawq -h、hawq start -h 或 hawq stop -h 等获得相关命令的联机帮助。启动停止 HAWQ 集群的命令都以 gpadmin 操作系统用户执行。

需要注意的是，不要使用操作系统的 kill 命令终止任何 postgres 进程。和其他所有数据库管理系统一样，强杀极有可能引起数据不一致的问题。每个客户端连接到 HAWQ 时，都会在 Master 节点上产生一个 postgres 进程，这与 Oracle 的专用服务器类似。终止用户会话 postgres 进程的正确方法是使用 pg_cancel_backend()数据库命令。下面是一个例子：

```
select datname,procpid,current_query from pg_stat_activity;
```

其中，datname 是会话连接的数据库名，procpid 是会话对应的操作系统进程号，current_query 是会话当前执行的 SQL 语句，查询结果如下：

```
gpadmin=# select datname,procpid,current_query from pg_stat_activity;
 datname | procpid |                    current_query
---------+---------+------------------------------------------------------
 mytest  | 354310  | <IDLE>
 gpadmin | 351905  | select datname,procpid,current_query from
pg_stat_activity;
(2 rows)
```

执行下面的语句取消 354310 进程。

```
select pg_cancel_backend(354310);
```

不能取消自己本身的会话：

```
gpadmin=# select pg_cancel_backend(351905);
ERROR:  canceling statement due to user request
gpadmin=#
```

1. 启动 HAWQ

初始安装或执行 hawq init cluster 命令后，HAWQ 集群会自动启动。该命令将初始化 HAWQ 的 Master 实例和每一个 Segment 实例。执行这条命令要求 HAWQ 在 HDFS 上的数据目录为空，也就是说先要清除所有用户数据，因此一般不要手动执行。

在 Master 实例上运行 hawq start 命令，启动 HAWQ 系统。下面的命令将启动 HAWQ 系统的 Master、Standby 和所有 Segment，并行执行且协调这个过程。该命令只能在 Master 节点上执行。

```
hawq start cluster
```

只启动 HAWQ 的 Master 节点，而不启动 Segment 节点，只能在 Master 节点上执行：

```
hawq start segment
```

启动本地 Segment 节点：

```
hawq start segment
```

启动 Standby 节点：

```
hawq start standby
```

一次启动所有 Segment 节点：

```
hawq start allsegments
```

如果希望忽略 ssh 无法连接的主机，可以使用 hawq start --ignore-bad-hosts 命令。

2. 重启 HAWQ

hawq restart 命令后跟适当的集群或节点类型，将停止 HAWQ，然后在完全终止后重启 HAWQ。如果 Master 或 Segment 已经停止，重启不受影响。

重启整个 HAWQ 集群，只能在 Master 节点上执行：

```
hawq restart cluster
```

只重启 Master 节点，只能在 Master 节点上执行：

```
hawq restart master
```

重启本地 Segment 节点：

```
hawq restart segment
```

重启 Standby：

```
hawq restart standby
```

一次重启所有 Segments：

```
hawq restart allsegments
```

3. 只重新加载配置文件

hawq stop 命令能够在不中断服务的情况下，重载 pg_hba.conf 配置文件（连接认证文件），以及 hawq-site.xml 和 pg_hba.conf 文件中的运行时参数，配置在新连接中生效。但许多服务器配置参数需要系统完全重启（hawq restart cluster）才能生效。

使用 hawq stop 命令重载配置文件而不停止系统：

```
hawq stop cluster --reload
```

或者

```
hawq stop cluster -u
```

4. 以维护模式启动 Master

可以只启动 Master 节点执行维护或管理任务，而不影响 Segment 节点上的数据。维护模式是一个超级用户模式，应该只在实施维护任务时使用。例如，在维护模式下，允许连接到 Master 节点实例上的数据库并编辑系统目录设置。

（1）在主节点上使用-m 选项运行 hawq start，启动维护模式：

```
hawq start master -m
```

（2）为维护系统目录，连接到维护模式下的 Master 节点：

```
PGOPTIONS='-c gp_session_role=utility' psql template1
```

（3）完成管理任务后，以生产模式重启 Master 节点：

```
hawq restart master
```

错误地使用维护模式连接，可能造成 HAWQ 系统状态不一致，因此应该只有专家级用户执行这个操作。

5. 停止 HAWQ

hawq stop cluster 命令停止 HAWQ 系统。当此命令执行时，会停止所有系统中的 postgres 进程，包括 Master、Standby 和所有 Segment 实例。hawq stop cluster 命令使用默认的最多 64 个并行线程停止所有构成 HAWQ 集群的 Segment。在停止前，系统会等待任何活动的事务结束。为了立即停止 HAWQ，可以使用 fast 停止方式。命令 hawq stop master、hawq stop segment、hawq stop standby 和 hawq stop allsegments 分别用于停止 Master 节点、本地 Segment 节点、Standby 节点和集群中的所有 Segment。只停止 Master 节点不会终止整个集群。下面是两个停止 HAWQ 集群的例子。

正常停止 HAWQ 集群：

```
hawq stop cluster
```

以快速模式停止 HAWQ：

```
hawq stop cluster -M fast
```

-M 选项提供了 smart、fast、immediate 三种停止方式，类似于 Oracle 中 shutdown 命令的 normal、immediate 和 abort。

- smart 是默认值，如果发现数据库中有活动的连接，停止失败，会发出一个错误消息：

```
20170302:15:37:00:376986 hawq_stop:hdp3:gpadmin-[INFO]:-Stop hawq cluster
...
20170302:15:37:00:376986 hawq_stop:hdp3:gpadmin-[ERROR]:-Active connections.
Aborting shutdown...
[gpadmin@hdp3 ~]$
```

- fast 方式中断并回滚当前处理的任何事务，此选项是安全的。
- immediate 方式终止正在处理的事务，并立即关掉所有相关 postgres 进程。数据库服务器不会完成事务处理，也不会清除任何临时数据或使用中的工作文件。工作文件的概念与 MySQL 的临时文件类似。查询执行过程中，如果不能在内存进行，就会在磁盘创建工作文件。不推荐使用 immediate 停止方式。在某些情况下，immediate 可能造成数据库损坏，并需要手动恢复。

6. 启动/停止 HAWQ 集群最佳实践

为了更好地使用 hawq start 和 hawq stop 管理系统，HAWQ 推荐使用下面的最佳实践。

（1）停止集群前执行 checkpoint SQL 命令，将所有数据文件中更新的数据刷新回磁盘，并更新日志文件。与其他数据库中检查点的概念相同，检查点确保在系统崩溃时，文件可以从检查点快照中被还原。

（2）在 Master 节点所在主机上执行 hawq stop cluster 命令停止整个 HAWQ 系统。

（3）要停止 Segment，并关掉任何运行的查询，而不造成数据丢失或不一致的问题，在 Master 上使用 fast 停止模式：hawq stop cluster -M fast。

（4）使用 hawq stop master 只停止 Master 节点。如果因为存在运行着的事务而不能停止 Master 节点，尝试使用 fast 方式。如果 fast 无法工作，再使用 immediate 方式。使用 immediate 会引发警告，因为在系统重新启动时，会导致执行崩溃恢复：hawq stop master -M fast 或 hawq stop master -M immediate。

（5）如果已经修改并希望重载服务器参数设置，并且 HAWQ 数据库上有活动连接，使用命令：hawq stop master -u -M fast。

（6）当停止本地 Segment 或所有 Segment 时，使用 smart 模式，这也是默认值。在 Segment 上使用 fast 或 immediate 模式是无效的，因为 Segment 是无状态的：hawq stop segment 或 hawq stop allsegments。

（7）典型地，应该总是使用 hawq start cluster 或 hawq restart cluster 启动集群。如果使用 hawq start standby|master|segment 的方式分别启动节点，确保总是在启动 Master 节点之前启动 Standby 节点，否则 Standby 可能与 Master 数据不同步。

2.7 小结

建议下载 HAWQ 编译后的安装包，用 Ambari 安装部署 HAWQ 集群。HAWQ 当前仅兼容 HDP 一种 Hadoop 发行版本，其安装过程通常按照 Ambari、HDP、HAWQ 的顺序依次进行。安装前需要确认各软件之间的版本兼容性。使用 hawq start 和 hawq stop 启停 HAWQ 时，建议参考其最佳实践。

第 3 章

◀ 连接管理 ▶

HAWQ 服务启动后，还要经过一系列配置，才能被客户端程序所连接。本章说明如何配置客户端身份认证，以及 HAWQ 的权限管理机制。我们还将演示 psql 和 Kettle 连接 HAWQ 的示例。psql 作为 HAWQ 最常用的命令行客户端工具，可类比为 mysql 命令之于 MySQL 数据库服务器的关系，因此我们还将 psql 与 mysql 命令常用的相似功能做一比较。本章最后还列举一些客户端连接 HAWQ 数据库的常见问题排查。

3.1 配置客户端身份认证

上一章曾经提到，HAWQ 系统初始安装后，数据库包含一个预定义的超级用户。这个用户和安装 HAWQ 的操作系统用户具有相同的名字，叫做 gpadmin。默认系统只允许使用 gpadmin 用户从本地连接至数据库。为了允许任何其他用户从本地或远程主机连接数据库，需要配置 HAWQ 允许此类连接。

1. 配置允许连接至 HAWQ

HAWQ 从代码级可以追溯到 PostgreSQL。它的客户端访问与认证是由标准的 PostgreSQL 基于主机的认证文件 pg_hba.conf 所控制的。Master 和每个 Segment 的数据目录下都存在一个 pg_hba.conf 文件。在 HAWQ 中，Master 实例的 pg_hba.conf 文件控制客户端对 HAWQ 系统的访问和认证。Segment 中 pg_hba.conf 文件的作用只是允许每个 Segment 作为 Master 节点主机的客户端连接数据库，而 Segment 本身并不接受其他客户端的连接。正因如此，不要修改 Segment 实例的 pg_hba.conf 文件。

pg_hba.conf 的格式是普通文本，其中每行一条记录，表示一个认证条目，HAWQ 忽略空行和任何#注释字符后面的文本。一行记录由四个或五个以空格或 tab 符分隔的字段构成。如果字段值中包含空格，则需要用双引号引起来，并且记录不能跨行。与 MySQL 类似，HAWQ 也接受 TCP 连接和本地的 UNIX 套接字连接。

每个 TCP 连接客户端的访问认证记录具有以下格式：

```
host|hostssl|hostnossl    <database>    <role>
<CIDR-address>|<IP-address>,<IP-mask>    <authentication-method>
```

本地 UNIX 域套接字的访问记录具有下面的格式：

```
local    <database>    <role>    <authentication-method>
```

表 3-1 描述了每个字段的含义。

表 3-1 pg_hba.conf 文件中的字段含义

字段	描述
local	匹配使用 UNIX 域套接字的连接请求。如果没有此种类型的记录，则不允许 UNIX 域套接字连接
host	匹配使用 TCP/IP 的连接请求。除非在服务器启动时使用了适当的 listen_addresses 服务器配置参数（默认值为"*"，允许所有 IP 连接），否则不能远程 TCP/IP 连接
hostssl	匹配使用 SSL 加密的 TCP/IP 连接请求。服务器启动时必须通过设置 ssl 配置参数启用 SSL
hostnossl	匹配不使用 SSL 的 TCP/IP 的连接请求
<database>	指定匹配此行记录的数据库名。值 "all" 表示允许连接所有数据库。多个数据库名用逗号分隔。也可以指定一个包含数据库名的文件，在文件名前加 "@"
<role>	指定匹配此行记录的数据库角色名。值 "all" 表示所有角色。如果指定的角色是一个组并且想包含所有的组成员，在角色名前面加一个 "+"。多个角色名可以通过逗号分隔。也可以指定一个包含角色名的文件，在文件名前加 "@"
<CIDR-address>	指定此行记录匹配的客户端主机的 IP 地址范围。它包含一个以标准点分十进制记法表示的 IP 地址，以及一个 CIDR 掩码长度。IP 地址只能用数字表示，不能是域名或主机名。掩码长度标识客户端 IP 地址必须匹配的高位数。在 IP 地址、斜杠和 CIDR 掩码长度之间不能有空格。CIDR 地址典型的例子有单一主机（如 192.0.2.2/32）、小型网络（如 192.0.2.0/24）、大型网络（如 192.0.0.0/16）。指定单一主机时，IPv4 的 CIDR 掩码是 32，Ipv6 的是 128。网络地址不要省略尾部的零
<IP-address>, <IP-mask>	这个字段是另一种 IP 地址表示方法，用掩码地址替换掩码长度。例如，255.255.255.255 对应的 CIDR 掩码长度是 32。此字段用于 host、hostssl 和 hostnossl 记录
<authentication-method>	指定连接认证时使用的方法。HAWQ 支持 PostgreSQL 9.0 所支持的认证方法，如信任认证、口令认证、Kerberos 认证、基于 Ident 的认证、PAM 认证等

2. 配置 pg_hba.conf 文件

这个例子显示如何编辑 Master 的 pg_hba.conf 文件，以允许远程客户端使用加密口令认证，用所有角色访问所有数据库。对于更高安全要求的系统，应考虑从 Master 的 pg_hba.conf 文件中删除所有信任认证方式（trust）的连接。信任方式意味着角色被授予访问权限而不需要任何认证，因此会绕过所有安全检查。

步骤 01 从 hawq-site.xml 文件的 hawq_master_directory 属性获取 Master 数据目录的位置，并

使用文本编辑器打开此目录下的 pg_hba.conf 文件。

步骤02 在该文件中，为允许的每个连接增加一行。记录逐行读取，因此记录的顺序至关重要。

```
local   all     gpadmin     ident
host    all     gpadmin     127.0.0.1/28                    trust
host    all     gpadmin     ::1/128                         trust
host    all     gpadmin     172.16.1.126/32                 trust
host    all     gpadmin     fe80::250:56ff:fea5:526f/128    trust
host    all     gpadmin     172.16.1.125/32                 trust
host    all     gpadmin     172.16.1.127/32                 trust
host    all     gpadmin     172.16.1.0/24                   trust
host    all     jsmith      172.16.1.0/24                   md5
host    all     testdb      0.0.0.0/0                       password
```

步骤03 保存并关闭文件。

步骤04 重载 pg_hba.conf 文件的配置，使修改生效：

```
hawq stop cluster -u -M fast
```

3. 限制并发连接数

HAWQ 的某些资源分配是以连接为基础的，因此最好配置允许的最大连接数。为了限制 HAWQ 系统的并发会话数量，设置 Master 的 max_connections 服务器配置参数，以及 Segment 的 seg_max_connections 服务器配置参数。这些参数是本地参数，因此必须在所有 HAWQ 实例的 hawq-site.xml 文件中进行设置。

如果设置了 max_connections，必须也要设置其依赖参数 max_prepared_transactions。该参数值必须大于等于 max_connections 的值，并且所有 HAWQ 实例要配置相同的值。下面是一个 $GPHOME/etc/hawq-site.xml 配置示例：

```xml
<property>
    <name>max_connections</name>
    <value>500</value>
</property>
<property>
    <name>max_prepared_transactions</name>
    <value>1000</value>
</property>
<property>
    <name>seg_max_connections</name>
    <value>3000</value>
</property>
```

以下是对这些参数具体含义的说明。

- max_connections：限制 Master 允许的最大客户端并发连接数，默认值是 200。在 HAWQ

系统中，用户客户端只能通过 Master 实例连接到系统。此参数的值越大，HAWQ 需要的共享内存越多。shared_buffers 参数设置一个 HAWQ Segment 实例使用的共享内存缓冲区大小，默认值是 125MB，最小值是 128KB 与 16KB * max_connections 的较大者。如果连接 HAWQ 时发生共享内存分配错误，可以尝试增加 SHMMAX 或 SHMALL 操作系统参数的值，或者降低 shared_buffers 或者 max_connections 参数的值解决此类问题。

- seg_max_connections：限制 Segment 允许对 Master 发起的最大并发连接数，默认值是 1280。该参数应该设置为 max_connections 的 5~10 倍。增加此参数时必须同时增加 max_prepared_transactions 参数的值。与 Master 类似，此参数的值越大，HAWQ 需要的共享内存越多。
- max_prepared_transactions：设置同时处于准备状态的事务数。HAWQ 内部使用准备事务保证跨 Segment 的数据完整性。该参数值必须大于等于 max_connections，并且在 Master 和 Segment 上应该设置成相同的值。

增加这些值会引起 HAWQ 需要更多的共享内存。为了缓解内存使用压力，可以考虑降低其他内存相关的服务器配置参数的值，如 gp_cached_segworkers_threshold 等。相对于手动编辑每个节点的 hawq-site.xml 文件，使用 Ambari 或命令行配置这些参数更为简单。

（1）使用 Ambari

步骤 01 通过 HAWQ service Configs→Advanced→Custom hawq-site→Add Property ... 配置 max_connections、seg_max_connections 和 max_prepared_transactions 属性。

步骤 02 选择 Service Actions→Restart All 重启所有相关服务使配置生效。

（2）使用命令行

步骤 01 作为管理员登录 HAWQ 的 Master 节点并设置环境：

```
source /usr/local/hawq/greenplum_path.sh
```

步骤 02 使用 hawq config 应用程序设置 max_connections、seg_max_connections 和 max_prepared_transactions 参数值，例如：

```
$ hawq config -c max_connections -v 100
$ hawq config -c seg_max_connections -v 640
$ hawq config -c max_prepared_transactions -v 200
```

步骤 03 重启 HAWQ 集群重载新的配置值。

```
$ hawq restart cluster
```

步骤 04 使用 hawq config 的 -s 选项显示服务器配置参数的值，确认配置生效：

```
$ hawq config -s max_connections
$ hawq config -s seg_max_connections
$ hawq config -s max_prepared_transactions
```

3.2 管理角色与权限

pg_hba.conf 文件限定了允许连接 HAWQ 的客户端主机、用户名、访问的数据库、认证方式等。用户名、口令以及用户对数据库对象的使用权限保存在 HAWQ 的元数据表中（pg_authid、pg_roles、pg_class 等）。

3.2.1 HAWQ 中的角色与权限

HAWQ 采用基于角色的访问控制机制。通过角色机制，简化了用户和权限的关联性。HAWQ 系统中的权限分为两种：系统权限和对象权限。系统权限是指系统规定用户使用数据库的权限，如连接数据库、创建数据库、创建用户等。对象权限是指在表、序列、函数等数据库对象上执行特殊动作的权限，其权限类型有 select、insert、update、delete、references、trigger、create、connect、temporary、execute 和 usage 等。

HAWQ 的角色与 Oracle、SQL Server 等数据库中的角色概念有所不同。这些系统中的所谓角色是权限的组合和抽象，创建角色最主要的目的是简化对用户的授权。举一个简单的例子，假设需要给五个用户授予相同的五种权限，如果没有角色，需要授权二十五次，而如果把五种权限定义成一种角色，只需要先进行一次角色定义，再授权五次即可。

HAWQ 中的角色既可以代表一个数据库用户，又可以代表一组权限。角色所拥有的预定义的系统权限是通过角色属性实现的。角色可以是数据库对象的属主，也可以给其他角色赋予访问对象的权限。角色可以是其他角色的成员，成员角色可以从父角色继承对象权限。

HAWQ 系统可能包含多个数据库角色（用户或组）。这些角色并不是运行服务器上操作系统的用户和组。但是为方便起见，可能希望维护操作系统用户名和 HAWQ 角色名的关系，因为很多客户端应用程序，如 psql，使用当前操作系统用户名作为默认的角色，gpadmin 就是最典型的例子。

用户通过 Master 实例连接 HAWQ。Master 使用 pg_hba.conf 文件里的条目验证用户的角色和访问权限。之后 Master 以当前登录的角色，从后台向 Segment 实例发布 SQL 命令。系统级定义的角色对所有数据库都是有效的。为了创建更多角色，首先需要使用超级用户 gpadmin 连接 HAWQ。

配置角色与权限时，应该注意以下问题：

- 保证 gpadmin 系统用户安全。HAWQ 需要一个 UNIX 用户 ID 安装和初始化 HAWQ 系统。这个系统用户 ID 就是 gpadmin。gpadmin 用户是 HAWQ 中默认的数据库超级用户，也是 HAWQ 安装及其底层数据文件的文件系统属主。这个默认的管理员账号是 HAWQ 的基础设计，缺少这个用户系统无法运行。并且，没有方法能够限制 gpadmin 用户对数据库的访问。应该只使用 gpadmin 账号执行诸如扩容和升级之类的系统维护任务。任何以这个用户登录 HAWQ 主机的人，都可以读取、修改和删除任何数据，尤其是系统目录相关的数据库访问权力。因此，gpadmin 用户的安全非常重要，仅应该提供给关键的系统管理员使用。应用的数据库用户应该永不作为 gpadmin 登录。
- 赋予每个登录用户不同的角色。出于记录和审核目的，每个登录 HAWQ 的用户都应该被赋予相应的数据库角色。对于应用程序或者 Web 服务，最好为每个应用或服务创建不同的角色。
- 使用组管理访问权限。
- 限制具有超级用户角色属性的用户。超级用户角色绕过 HAWQ 中所有的访问权限检查和资源队列，所以只应该将超级用户权限授予系统管理员。

3.2.2 管理角色及其成员

这里的角色指的是一个可以登录到数据库，并开启一个数据库会话的用户。使用 CREATE ROLE 命令创建一个角色时，必须授予 login 系统属性（功能类似于 Oracle 的 connect 角色），使得该角色可以连接数据库。例如：

```
create role jsmith with login;
```

一个数据库角色有很多属性，用以定义该角色可以在数据库中执行的任务，或者具有的系统权限。表 3-2 描述了有效的角色属性。

表 3-2 角色属性

属性	描述
SUPERUSER | NOSUPERUSER	确定一个角色是否是超级用户。只有超级用户才能创建新的超级用户。默认值为 NOSUPERUSER
CREATEDB | NOCREATEDB	确定角色是否被允许创建数据库。默认值为 NOCREATEDB
CREATEROLE | NOCREATEROLE	确定角色是否被允许创建和管理其他角色。默认值为 NOCREATEROLE
INHERIT | NOINHERIT	确定角色是否是从其所在的组继承权限。具有 INHERIT 属性的角色可以自动使用所属组已经被授予的数据库权限，无论角色是组的直接成员还是间接成员。默认值为 INHERIT
LOGIN | NOLOGIN	确定角色是否可以登录。具有 LOGIN 属性的角色可以将角色作为用户登录。没有此属性的角色被用于管理数据库权限（用户组）。默认值为 NOLOGIN
CONNECTION LIMIT *connlimit*	如果角色能够登录，此属性指定角色可以建立多少个并发连接。默认值为 -1，表示没有限制
PASSWORD *'password'*	设置角色的口令。如果不使用口令认证，可以忽略此选项。如果没有指定口令，口令将被设置为 null，此时该用户的口令认证总是失败。一个 null 口令也可以显式地写成 PASSWORD NULL

（续表）

属性	描述
ENCRYPTED | UNENCRYPTED	控制口令是否加密存储在系统目录中。默认行为由 password_encryption 配置参数所决定，当前设置是 MD5，如果要改为 SHA-256 加密，设置此参数为 password。如果给出的口令字符串已经是加密格式，那么它被原样存储，而不管指定 ENCRYPTED 还是 UNENCRYPTED。这种设计允许在 dump/restore 时重新导入加密的口令
VALID UNTIL '*timestamp*'	设置一个日期和时间，在该时间点后角色的口令失效。如果忽略此选项，口令将永久有效
RESOURCE QUEUE queue_name	赋予角色一个命名的资源队列用于负载管理。角色发出的任何语句都受到该资源队列的限制。注意，这个 RESOURCE QUEUE 属性不会被继承，必须在每个用户级（登录）角色设置
DENY {deny_interval | deny_point}	在此时间区间内禁止访问

可以在创建角色时，或者创建角色后使用 ALTER ROLE 命令指定这些属性：

```
alter role jsmith with password 'passwd123';
alter role admin valid until 'infinity';
alter role jsmith login;
alter role jsmith resource queue adhoc;
alter role jsmith deny day 'sunday';
```

使用 drop role 或 drop user 命令删除角色（用户）。在删除角色前，先要收回角色所拥有的全部权限，或者先删除与角色相关联的所有对象，否则删除角色时会提示"cannot be dropped because some objects depend on it"错误。

通常将多个权限合成一组，能够简化对权限的管理。使用这种方法，对于一个组中的用户，其权限可以被整体授予和回收。在 HAWQ 中的实现方式为先创建一个表示组的角色，然后将用户角色授予组角色的成员。下面的 SQL 命令使用 CREATE ROLE 创建一个名为 admin 组角色，该组角色具有 CREATEROLE 和 CREATEDB 系统权限。

```
create role admin createrole createdb;
```

一旦组角色存在，就可以使用 GRANT 和 REVOKE 命令添加或删除组成员（用户角色）：

```
grant admin to john, sally;
revoke admin from bob;
```

为简化对象权限的管理，应当只为组级别的角色授予适当的权限。成员用户角色继承组角色的对象权限：

```
grant all on table mytable to admin;
grant all on schema myschema to admin;
grant all on database mydb to admin;
```

角色属性 LOGIN、SUPERUSER、CREATEDB 和 CREATEROLE 不会当作普通的数据库对象权限被继承。为了让用户成员使用这些属性，必须执行 SET ROLE 设置一个角色具有这些属性。在上面的例子中，我们已经为 admin 指定了 CREATEDB 和 CREATEROLE 属性。sally 是 admin 的成员，当以 sally 连接到数据库后，执行以下命令，使 sally 可以拥有父角色的 CREATEDB 和 CREATEROLE 属性。

```
set role admin;
```

有关角色属性信息可以在系统表 pg_authid 中找到，pg_roles 是基于系统表 pg_authid 的视图。系统表 pg_auth_members 存储了角色与其成员的关系。

3.2.3 管理对象权限

当一个对象（表、视图、序列、数据库、函数、语言、模式或表空间）被创建时，它的权限被赋予属主。属主通常是执行 CREATE 语句的角色。对于大多数类型的对象，其初始状态是只允许属主或超级用户在对象上做任何操作。为了允许其他角色使用对象，必须授予适当的权限。HAWQ 对每种对象类型支持的权限如表 3-3 所示。

表 3-3 对象权限

对象类型	权限
Tables、Views、Sequences	SELECT、INSERT、RULE、ALL
External Tables	SELECT、RULE、ALL
Databases	CONNECT、CREATE、TEMPORARY | TEMP、ALL
Functions	EXECUTE
Procedural Languages	USAGE
Schemas	CREATE、USAGE、ALL
Custom Protocol	SELECT、INSERT、RULE、ALL

必须为每个对象单独授权。例如，授予数据库上的 ALL 权限，并不会授予数据库中全部对象的访问权限，而只是授予了该数据库自身的数据库级别的全部权限（CONNECT、CREATE、TEMPORARY 等）。

使用标准的 GRANT 和 REVOKE SQL 语句为角色授予或回收一个对象权限：

```
grant insert on mytable to jsmith;
revoke all privileges on mytable from jsmith;
```

可以使用 DROP OWNED 和 REASSIGN OWNED 命令为一个角色删除或重新赋予对象属主权限。只有对象的属主或超级用户能够执行此操作：

```
reassign owned by sally to bob;
drop owned by visitor;
```

HAWQ 不支持行级和列级的访问控制，但是可以通过视图来模拟，限制查询的行或列。此时角色被授予对视图而不是基表的访问权限。对象权限存储在 pg_class.relacl 列中。Relacl 是 PostgreSQL 支持的数组属性，该数组成员是抽象的数据类型 aclitem。每个 ACL 实际上是一个由多个 aclitem 构成的链表。

3.2.4 口令加密

HAWQ 默认使用 MD5 为用户口令加密，通过适当配置服务器参数，也能实现口令的 SHA-256 加密存储。为了使用 SHA-256 加密，客户端认证方法必须设置为 PASSWORD 而不是默认的 MD5。口令虽然以加密形式存储在系统表中，但仍然以明文在网络间传递。为了避免这种情况，应该建立客户端与服务器之间的 SSL 加密通道。

1. 系统级启用 SHA-256 加密

（1）使用 Ambari

步骤01 通过 HAWQ service Configs→Advanced→Custom hawq-site 下拉列表设置 password_hash_algorithm 配置属性，有效值为 SHA-256。

步骤02 选择 Service Actions→Restart All 使配置生效。

（2）使用命令行

步骤01 作为管理员登录 HAWQ Master 并设置路径：

```
$ source /usr/local/hawq/greenplum_path.sh
```

步骤02 使用 hawq config 应用程序设置 password_hash_algorithm 为 SHA-256：

```
$ hawq config -c password_hash_algorithm -v 'SHA-256'
```

步骤03 重载 HAWQ 配置：

```
$ hawq stop cluster -u
```

步骤04 验证设置：

```
$ hawq config -s password_hash_algorithm
```

2. 会话级启用 SHA-256 加密

为单个数据库会话设置 password_hash_algorithm 服务器参数：

（1）以超级用户登录 HAWQ 实例。

（2）设置 password_hash_algorithm 参数为 SHA-256：

```
set password_hash_algorithm = 'SHA-256';
```

（3）验证参数设置：

```
show password_hash_algorithm;
```

3. 验证口令加密方式生效

（1）建立一个具有 login 权限的新角色，并设置口令：

```
create role testdb with password 'testdb12345#' login;
```

（2）修改客户端认证方法，允许存储 SHA-256 加密的口令，打开 Master 的 pg_hba.conf 文件并添加下面一行：

```
host all testdb 0.0.0.0/0 password
```

（3）重启集群：

```
hawq restart cluster
```

（4）以刚创建的 testdb 用户登录数据库，在提示时输入正确的口令：

```
psql -d postgres -h hdp3 -U testdb
```

验证口令被以 SHA-256 哈希方式存储，加密后的口令存储在 pg_authid.rolpassword 字段中。作为超级用户登录，执行下面的查询：

```
gpadmin=# select rolpassword from pg_authid where rolname = 'testdb';
                          rolpassword
-----------------------------------------------------------------
 sha25650c2445bab257f4ea94ee12e5a6bf1400b00a2c317fc06b6ff9b57975bd1cde1
(1 row)
```

3.3 psql 连接 HAWQ

用户可以使用与一个 PostgreSQL 兼容的客户端程序连接到 HAWQ，最常用的客户端工具就是 psql。再次强调，用户和管理员总是通过 Master 连接到 HAWQ，Segment 不能接受客户端连接。为了建立一个到 Master 的连接，需要了解表 3-4 所示的连接信息，并在 psql 命令行给出相应参数或配置相关的环境变量。

表 3-4　psql 主要连接参数

连接参数	描述	环境变量
dbname	连接的数据库名称。对于一个新初始化的系统，首次连接使用 template1 数据库	$PGDATABASE
host	HAWQ Master 节点的主机名。默认主机为 localhost	$PGHOST
port	HAWQ Master 节点实例运行的端口号。默认是 5432	$PGPORT
username	连接数据库的用户（角色）名称。与操作系统用户名相同的用户名不需要此参数。注意，每个 HAWQ 系统都有一个在初始化时自动创建的超级用户账号。这个账号与初始化 HAWQ 系统的操作系统用户同名，典型的是 gpadmin	$PGUSER

下面的例子显示如何通过 psql 访问一个 HAWQ 数据库，没有指定的连接参数依赖于设置的环境变量或使用默认值。

```
psql -d mytest -h hdp3 -p 5432 -U gpadmin
psql mytest
psql
```

当用户定义的数据库还没有创建时，可以通过连接 template1 数据库访问系统：

```
psql template1
```

连接数据库后，psql 提供一个由当前连接的数据库名后跟=>构成的提示符（超级用户是=#）：

```
mytest=>
```

在提示符下，可以输入 SQL 命令。每个 SQL 命令必须以;（分号）结束，以发送到服务器执行：

```
select * from mytable;
```

psql 常用命令与 mysql 命令行的比较如表 3-5 所示。

表 3-5　psql 常用功能与 mysql 比较

功能描述	psql	mysql
联机帮助	help：简要帮助 \?：psql 命令帮助 \h：SQL 命令帮助	help、?、\?、\h：都是等价的简要帮助。后面可以跟 SQL 命令，显示详细的命令语法
执行 SQL	分号或\g	分号、\g 或\G
退出	\q	\q、exit 或 quit
列出所有数据库	\l	show databases;
改变当前连接的数据库	\c DBNAME	use db_name;
列出内部表	\dt	show tables;
列出外部表	\dx	无
表的描述	\d TABLENAME	desc tbl_name;
列出索引	无	show index from tbl_name;
列出视图	\dv	show tables;
列出序列	\ds	无
列出系统表	\dtS+	show tables from mysql; show tables from information_schema; show tables from performance_schema;

3.4 Kettle 连接 HAWQ

Kettle 是当前流行的开源 ETL 工具之一，是用 Java 语言开发的。它最初的作者 Matt Casters 原是一名 C 语言程序员，在着手开发 Kettle 时还是一名 Java 小白，但是他仅用了一年时间就开发出了 Kettle 的第一个版本。虽然有很多不足，但这个版本毕竟是可用的。使用自己并不熟悉的语言，仅凭一己之力在很短的时间里就开发出了复杂的 ETL 系统工具，作者的开发能力和实践精神令人十分佩服。后来 Pentaho 公司获得了 Kettle 源代码的版权，Kettle 也随之更名为 Pentaho Data Integration，简称 PDI。Kettle 的设计原则之一，就是尽量减少编程，几乎所有工作都可以通过简单拖拽来完成。它通过工作流和数据转换两种不同的模式进行数据操作，分别被称为作业和转换。

Kettle 里的转换和作业使用"数据库连接"主对象来连接关系型数据库。Kettle 数据库连接实际是数据库连接的描述，也就是建立实际连接需要的参数。实际连接只是在运行时才建立，定义一个 Kettle 数据库连接并不真正打开一个到数据库的连接。各个数据库的行为都是完全不同的，Kettle 7.0 可以连接的数据库多达 51 种，几乎覆盖了所有常用数据库，而且支持的数据库种类还在增多。下面就以 Kettle 7.0 为例，说明 Kettle 连接 HAWQ 的配置步骤。

（1）在 pg_hba.conf 文件中添加客户端连接。pg_hba.conf 文件作用在 3.1 节已经详细说明。这里连接 HAWQ 的用户名为 kettle，192.168.8.187 是 Kettle 所在主机的 IP 地址。

```
echo "host all kettle 192.168.8.187/32 md5" >> /data/hawq/master/pg_hba.conf
```

（2）重载 pg_hba.conf 文件使修改生效：

```
hawq stop cluster -u -M fast
```

（3）在 psql 中建立用户并授权。这里授予 kettle 用户对 public 模式下所有表的查询权限。

```
create role kettle with login;
alter role kettle with password '123456';
\t on
\o /tmp/grant.sql
select 'grant select on '||tablename || ' to kettle;' from pg_tables where schemaname='public';
\o
\i /tmp/grant.sql
```

（4）在 Kettle 中建立 DB 连接。

① 新建转换。
② 选中"主对象树"→转换→转换 1→DB 连接，右击"新建"。
③ 如图 3-1 所示配置数据库连接。

第 3 章 连接管理

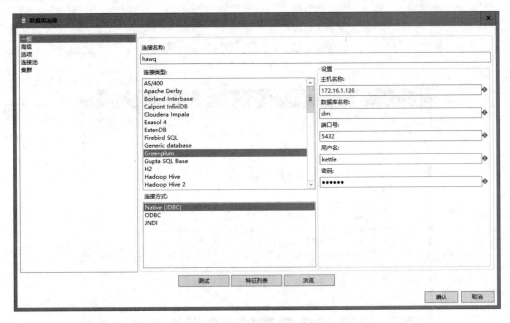

图 3-1 新建数据库连接

在数据库连接窗口中主要设置下面三个选项。

- 连接名称：设定一个作业或转换范围内唯一的数据库连接名称。
- 连接类型：从数据库列表中选择要连接的数据库类型。根据选中数据库的类型不同，要设置的访问方式和连接参数设置不尽相同，某些 Kettle 步骤或作业生成 SQL 语句时使用的方言也会有所不同。
- 连接方式：在列表里选择可用的连接方式，一般都使用 JDBC 连接。

如图 3-1 所示，这里的三个选项分别设置为 hawq、Greenplum、JDBC。右侧面板的连接参数如下：

- 主机名称：数据库服务器的主机名或 IP 地址，这里是 HAWQ Master 节点地址。
- 数据库名称：要访问的数据库名。
- 端口号：选中数据库服务器的端口号，HAWQ 默认使用 5432 端口。
- 用户名和密码：数据库服务器的用户名和密码，这里就是上一步建立的用户。

（5）单击"测试"按钮，弹出图 3-2 所示的测试成功页面。

图 3-2 测试数据库连接

53

（6）新建一个"表输入"步骤，在编辑窗口中，"数据库连接"选择"hawq"，然后单击"获取 SQL 查询语句"按钮，在图 3-3 所示的弹出窗口中选择一个表并确定，结果如图 3-4 所示。

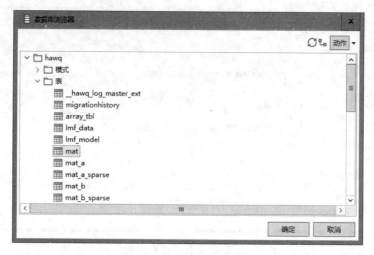

图 3-3　选择表

图 3-4　表输入步骤

（7）单击"预览"按钮，结果如图 3-5 所示。

图 3-5 预览表数据

可以看到，表输入步骤建立了到 HAWQ 的连接，SQL 查询成功执行，正确返回了数据。同样，利用"表输出"步骤，可以将其他源数据写入 HAWQ 表中，这里就不做演示了。

3.5 连接常见问题

很多问题会引起客户端连接 HAWQ 失败。表 3-6 提供了造成连接问题的常见原因及其排查方法。

表 3-6 常见连接问题排查

问题	排查
No pg_hba.conf entry for host or user	为了让 HAWQ 接受远程客户端连接，必须配置 HAWQ Master 节点实例上的 pg_hba.conf 文件，在该文件中增加适当的条目，允许客户端主机和数据库用户连接到 HAWQ
HAWQ is not running	如果 HAWQ Master 节点实例宕机，用户将不能连接。可以在 HAWQ Master 节点上运行 hawq state 应用程序，验证 HAWQ 系统正在运行
Network problem Interconnect timeouts	当用户从远程客户端连接到 HAWQ Master 节点时，网络问题可能阻止连接（例如，DNS 主机名解析问题、主机系统断网等）。为了确认不是网络问题，从远程客户端所在主机连接 HAWQ Master 节点所在主机。例如：ping hostname。 如果系统不能解析 HAWQ 主机 IP 地址所涉及的主机名，查询和连接将失败。有些操作使用 localhost 进行连接，而另一些操作使用实际的主机名，所以两种情况都必须能正确解析。如果碰到连接错误，首先核实能够从 HAWQ Master 节点所在主机连接到集群主机。在 Master 节点和所有 Segment 的/etc/hosts 文件中，确认有 HAWQ 集群中所有主机的正确主机名和 IP 地址。127.0.0.1 必须被解析成 localhost

（续表）

问题	排查
Too many clients already	默认 HAWQ Master 节点和 Segment 允许的最大并发连接数分别是 200 和 1280。超出此限制的连接将被拒绝。这个限制由 Master 节点实例的 max_connections 参数和 Segment 实例的 seg_max_connections 参数所控制。如果修改了 Master 节点的设置，也必须在 Segment 节点上做适当的修改
Query failure	HAWQ 集群网络中必须配置 DNS 反向解析。如果 DNS 反向解析没有配置，失败的查询将在 HAWQ Master 节点的日志文件中产生"Failed to reverse DNS lookup for ip <ip-address>"警告消息

3.6 小结

HAWQ 使用 Master 节点上的 pg_hba.conf 文件控制客户端访问与认证，这点继承自 PostgreSQL。该文件通常位于 Master 数据目录中，默认的文件位置是 /data/hawq/master/pg_hba.conf。与其他关系数据库不同，HAWQ 中的角色可以是用户或组。组角色主要用于简化权限管理，组中的成员默认会继承赋予组的权限。数据库对象的属主拥有对象上的所有权限，属主或超级用户（gpadmin）可以将对象权限授予其他用户。用户口令以加密形式存储于 pg_authid.rolpassword 列，默认使用 MD5，也可以配置成 SHA-256 加密。HAWQ 最常用的命令行工具是 psql。Kettle 使用 Greenplum 数据库类型连接 HAWQ，通过表输入和表输出步骤读写 HAWQ 表。

第 4 章 数据库对象管理

HAWQ 本质上是一个数据库系统。和其他关系数据库类似，HAWQ 中有数据库、表空间、表、视图、自定义数据类型、自定义函数、序列等对象。本章将简述这些对象的创建与管理。对 HAWQ 中表的存储方式与分布策略等特性的选择，会对应用性能产生极大影响，同时这也是一个复杂的话题，将在第 6 章单独讨论。

4.1 创建和管理数据库

HAWQ 中数据库的概念与 MySQL 类似，一个 HAWQ 实例中通常会建立多个数据库，这和 Oracle 中数据库的概念不同。在 Oracle 体系结构中，数据库是一个"最大"的概念，大多数情况下一个 Oracle 数据库对应一个实例，RAC 是一个数据库对应多个实例。尽管可以在一个 HAWQ 系统中创建很多数据库，但是客户端程序在某一时刻只能连接到一个数据库，这也决定了 HAWQ 不能执行跨库的查询。

1. 模板数据库

HAWQ 初始化后，就有了 template0、template1 和 postgres 等模板数据库。模板数据库是不能删除的：

```
gpadmin=# drop database template0;
ERROR:  cannot drop a template database
gpadmin=# drop database template1;
ERROR:  cannot drop a template database
gpadmin=# drop database postgres;
ERROR:  cannot drop a template database
```

开始时 template0 和 template1 两个库的内容是一样的。两者最主要的区别是，默认可以连接 template1 并在其中创建对象，但不能连接 template0。

```
gpadmin=# \c template0
```

```
FATAL:  database "template0" is not currently accepting connections
Previous connection kept
gpadmin=# \c template1
You are now connected to database "template1" as user "gpadmin".
```

每个新创建的数据库都基于一个模板,建库时如果不指定 TEMPLATE 属性,默认用的是 template1 模板库。除非希望某些对象在每一个新创建的数据库中都存在,否则不要在 template1 中创建任何对象。template1 是默认模板,并且其中的对象和数据会被克隆到每个以它为模板的新建数据库中。

```
template1=# create table t1 (a int);
CREATE TABLE
template1=# insert into t1 values (1);
INSERT 0 1
template1=# create database db1;
CREATE DATABASE
template1=# \c db1
You are now connected to database "db1" as user "gpadmin".
db1=# \dt
          List of relations
 Schema | Name | Type  |  Owner  |   Storage
--------+------+-------+---------+-------------
 public | t1   | table | gpadmin | append only
(1 row)

db1=# select * from t1;
 a
---
 1
(1 row)
```

HAWQ 还有一个模板库 postgres。不要修改 template0 或 postgres,HAWQ 内部需要使用它们。以 template0 为模板可以创建一个完全干净的数据库,其中只包含 HAWQ 在初始化时预定义的标准对象。如果修改了 template1,可能就需要这么做。指定以 template0 为模板创建数据库:

```
template1=# create database db2 with template template0;
CREATE DATABASE
```

通过适当配置,其实也可以连接 template0:

```
template1=# set allow_system_table_mods='DML';
SET
template1=# update pg_database set datallowconn='t' where datname='template0';
UPDATE 1
```

```
template1=# \c template0
You are now connected to database "template0" as user "gpadmin".
template0=# update pg_database set datallowconn='f' where datname='template0';
ERROR:  permission denied: "pg_database" is a system catalog
template0=# set allow_system_table_mods='DML';
SET
template0=# update pg_database set datallowconn='f' where datname='template0';
UPDATE 1
template0=# \q
[gpadmin@hdp3 ~]$ psql -d template0
psql: FATAL:  database "template0" is not currently accepting connections
```

2. 创建数据库

创建数据库的用户必须拥有适当的权限，比如超级用户，或者被设置了 CREATEDB 角色属性。除了像前面例子中使用 CREATE DATABASE 命令创建数据库，还可以使用客户端程序 createdb 创建一个数据库。例如，运行下面的命令将连接 HAWQ 主机并创建名为 db3 的数据库，主机名和端口号必须与 HAWQ 的 Master 节点相匹配。

```
[gpadmin@hdp4 ~]$ createdb -h hdp3 -p 5432 db3
[gpadmin@hdp4 ~]$ psql -h hdp3
psql (8.2.15)
Type "help" for help.

gpadmin=# \l
             List of databases
   Name    |  Owner  | Encoding | Access privileges
-----------+---------+----------+-------------------
 ...
 db3       | gpadmin | UTF8     |
 ...
(7 rows)
```

某些对象，如角色（用户），是被 HAWQ 中的所有数据库所共享的。而另外一些对象，如表，则只有它所在的数据库能感知它的存在。

3. 查看数据库列表

psql 客户端程序的 \l 元命令显示数据库列表。如果是数据库超级用户，还可以从 pg_database 系统目录表中查询数据库列表。

```
gpadmin=# \l
             List of databases
   Name    |  Owner  | Encoding | Access privileges
-----------+---------+----------+-------------------
 db1       | gpadmin | UTF8     |
```

```
 db2       | gpadmin | UTF8 |
 db3       | gpadmin | UTF8 |
 gpadmin   | gpadmin | UTF8 |
 postgres  | gpadmin | UTF8 |
 template0 | gpadmin | UTF8 |
 template1 | gpadmin | UTF8 |
(7 rows)

gpadmin=# select datname from pg_database;
  datname
-----------
 hcatalog
 template1
 postgres
 gpadmin
 template0
 db1
 db2
 db3
(8 rows)
```

可以看到，从 pg_database 查询出的结果比 \l 命令多返回一个库名为 hcatalog。此库仅 HAWQ 系统使用，并且不允许连接。

```
gpadmin=# \c hcatalog
FATAL:  "hcatalog" database is only for system use
Previous connection kept
```

4. 修改数据库

ALTER DATABASE 命令可以用于修改数据库的默认配置，如下面的命令修改 search_path 服务器配置参数，改变数据库 db1 默认的模式查找路径。

```
gpadmin=# alter database db1 set search_path to myschema, public, pg_catalog;
NOTICE:  schema "myschema" does not exist
ALTER DATABASE
```

HAWQ 不支持修改数据库名。

```
gpadmin=# alter database db1 rename to db11;
ERROR:  Cannot support rename database statement yet
```

5. 删除数据库

DROP DATABASE 命令删除一个数据库，除了删除数据库在系统目录中的条目外，同时会删除磁盘上的数据。只有数据库属主或超级用户才能删除数据库。并且，不能删除一个还有连接的数据库，包括不能删除自己当前会话连接的数据库。在删除一个数据库前，可先连接到 template1 或其他数据库。

```
gpadmin=# \c template1
You are now connected to database "template1" as user "gpadmin".
template1=# drop database db1;
DROP DATABASE
```

也可以使用客户端程序 dropdb 删除一个数据库。

```
[gpadmin@hdp4 ~]$ dropdb -h hdp3 -p 5432 db2
```

一个数据库有连接时是不允许删除的,必须先终止所有连接,在没有连接之后再删除。

```
gpadmin=# drop database db3;
ERROR:  database "db3" is being accessed by other users
gpadmin=# select procpid,current_query from pg_stat_activity where
datname='db3';
 procpid | current_query
---------+---------------
  790583 | <IDLE>
(1 row)

gpadmin=# select pg_terminate_backend(790583);
 pg_terminate_backend
----------------------
 t
(1 row)

gpadmin=# drop database db3;
DROP DATABASE
```

4.2 创建和管理表空间

很多数据库系统,如 Oracle 和 MySQL 等,都有表空间的概念。HAWQ 的表存储在 HDFS 上,其表空间管理有自己的特点。HAWQ 在表空间之上有一个文件空间的概念,系统中所有组件的文件系统位置的集合构成一个文件空间。文件空间可以被一个或多个表空间所使用。一个文件空间物理上实际就是一个 HDFS 的目录及其子目录。在表空间定义中需要指定所属的文件空间。一个文件空间下的所有表空间文件都存储在该文件空间所对应的 HDFS 目录下。

表空间允许为经常使用和不经常使用的数据库对象赋予不同的存储,或控制特定数据库对象的 I/O 性能。例如,将经常使用的表放在高性能文件系统(如 SSD)上,而将其他表放在普通标准硬盘上。通过这种方式,DBA 可以在 HAWQ 集群中使用多个 HDFS 目录,灵活规划数据库对象的物理存储。

1. 创建文件空间

文件空间是一个符号存储标识符,映射为一组 HAWQ 主机文件系统的位置,指示 HAWQ 系统的存储空间。为了创建一个文件空间,需要在 HAWQ 集群上准备 HDFS 文件系统目录,然后使用 hawq filespace 应用程序定义文件空间。必须以数据库超级用户创建文件空间。需要注意的是,HAWQ 并不直接感知底层的文件系统边界,它将文件存储在所指定的目录中,但不能人为控制单个文件的磁盘位置。下面创建一个文件空间。

(1)用 hdfs 用户为文件空间准备 HDFS 目录

```
[root@hdp4 ~]# su - hdfs
[hdfs@hdp4 ~]$ hdfs dfs -mkdir /hawq_data1
[hdfs@hdp4 ~]$ hdfs dfs -chown -R gpadmin:gpadmin /hawq_data1
```

(2)用 gpadmin 用户登录 HAWQ Master

```
$ su - gpadmin
```

(3)创建文件空间配置文件

在提示符下,输入文件空间的名字、文件空间的 HDFS 位置,副本集个数采用默认的 3。

```
[gpadmin@hdp3 ~]$ hawq filespace -o hawqfilespace_config
Enter a name for this filespace
> testfs
Enter replica num for filespace. If 0, default replica num is used (default=3)
>

Please specify the DFS location for the filespace (for example: localhost:9000/fs)
location> hdp1:8020/hawq_data1
20170306:11:24:52:352152 hawqfilespace:hdp3:gpadmin-[INFO]:-[created]
20170306:11:24:52:352152 hawqfilespace:hdp3:gpadmin-[INFO]:-
To add this filespace to the database please run the command:
   hawqfilespace --config /home/gpadmin/hawqfilespace_config

[gpadmin@hdp3 ~]$ more /home/gpadmin/hawqfilespace_config
filespace:testfs
fsreplica:3
dfs_url:: hdp1:8020/hawq_data1
[gpadmin@hdp3 ~]$
```

(4)用生成的配置文件创建文件空间

```
[gpadmin@hdp3 ~]$ hawq filespace --config /home/gpadmin/hawqfilespace_config
Reading Configuration file: '/home/gpadmin/hawqfilespace_config'
```

```
CREATE FILESPACE testfs ON hdfs
 ('hdp1:8020/hawq_data1/testfs') WITH (NUMREPLICA = 3);
 20170306:11:25:50:352658 hawqfilespace:hdp3:gpadmin-[INFO]:-Connecting to
database
 20170306:11:25:50:352658 hawqfilespace:hdp3:gpadmin-[INFO]:-Filespace
"testfs" successfully created
```

此时 HDFS 上会看到建立了 /hawq_data1/testfs 目录。

```
[hdfs@hdp2 ~]$ hdfs dfs -ls /hawq_data1
Found 1 items
drwx------   - gpadmin gpadmin          0 2017-03-07 14:32 /hawq_data1/testfs
```

2. 创建表空间

创建完文件空间，使用 CREATE TABLESPACE 命令建立一个使用该文件空间的表空间。

```
gpadmin=# create tablespace testts filespace testfs;
CREATE TABLESPACE
```

目前 HAWQ 只允许数据库超级用户定义表空间，并且不支持向其他用户 GRANT/REVOKE 表空间上的 CREATION 权限。

```
gpadmin=# create user wxy with superuser login password 'mypassword';
CREATE ROLE
gpadmin=# grant create on tablespace testts to wxy;
ERROR:  Cannot support GRANT/REVOKE on TABLESPACE statement
```

相关信息参见 https://issues.apache.org/jira/browse/HAWQ-24。

3. 使用表空间存储数据库对象

拥有表空间上 CREATE 权限的用户能够在此表空间中创建数据库对象，如数据库、表等。没有指定表空间的 CREATE TABLE 语句使用由 default_tablespace 参数指定的默认表空间。

与一个数据库关联的表空间存储数据库目录、数据库服务器进程创建的临时文件、数据库中在创建时没有指定 TABLESPACE 的表。如果创建数据库时不指定表空间，数据库使用其模板数据库相同的表空间。如果有适当的权限，可以在任意数据库中使用一个表空间。

```
[gpadmin@hdp3 ~]$ psql -d template1 -U wxy -h hdp3
template1=# create database db1 tablespace testts;
CREATE DATABASE
template1=# \c db1
You are now connected to database "db1" as user "wxy".
db1=# create table t1 (a int);
CREATE TABLE
db1=# create table t2 (a int) tablespace testts;
CREATE TABLE
db1=# set default_tablespace = testts;
```

```
SET
db1=# create table t3 (a int);
CREATE TABLE
db1=# set default_tablespace = dfs_default;
SET
db1=# create table t4 (a int);
CREATE TABLE
db1=# select relname,reltablespace from pg_catalog.pg_class where relname in
('t1','t2','t3','t4');
 relname | reltablespace
---------+---------------
 t1      |             0
 t2      |             0
 t3      |             0
 t4      |         16385
(4 rows)
```

pg_class.reltablespace 为 0，说明表保存在从数据库继承的默认表空间 testts 里。特别要指出的是，所有非共享的系统表（pg_class.relisshared 为 false）也都存放在这里。

4. 查看表空间和文件空间

每个 HAWQ 系统都有以下默认表空间：

- pg_global：共享系统目录表空间。
- pg_default：默认表空间，template1 和 template0 数据库使用。

这些表空间使用系统默认的文件空间 pg_system，指示系统初始化时创建的数据目录位置。pg_filespace 和 pg_filespace_entry 目录表存储文件空间信息。可以将这些表与 pg_tablespace 关联查看完整的表空间的定义。

```
db1=# select spcname as tblspc, fsname as filespc,
db1-#        fsedbid as seg_dbid, fselocation as datadir
db1-#   from  pg_tablespace pgts, pg_filespace pgfs,
db1-#        pg_filespace_entry pgfse
db1-#  where pgts.spcfsoid=pgfse.fsefsoid
db1-#    and pgfse.fsefsoid=pgfs.oid
db1-#  order by tblspc, seg_dbid;
   tblspc    |  filespc   | seg_dbid |                 datadir
-------------+------------+----------+------------------------------------------
 dfs_default | dfs_system |        0 | hdfs://hdp1:8020/hawq_data
 testts      | testfs     |        0 |
hdfs://{replica=3}hdp1:8020/hawq_data1/testfs
(2 rows)
```

5. 删除表空间和文件空间

只有表空间的属主或超级用户可以删除表空间。直到表空间所有的数据库对象都被删除后，才能删除表空间。

```
postgres=# drop tablespace testts;
ERROR:  tablespace "testts" is not empty: existing database.
postgres=# drop filespace testfs;
ERROR:  filespace "testfs" is not empty
postgres=# drop database db1;
DROP DATABASE
postgres=# drop filespace testfs;
ERROR:  filespace "testfs" is not empty
postgres=# drop tablespace testts;
DROP TABLESPACE
postgres=# drop filespace testfs;
DROP FILESPACE
postgres=#
```

此时 HDFS 上的/hawq_data1/testfs 目录已经删除。

```
[hdfs@hdp2 ~]$ hdfs dfs -ls /hawq_data1/testfs
ls: `/hawq_data1/testfs': No such file or directory
[hdfs@hdp2 ~]$
```

4.3 创建和管理模式

模式（schema）是一个有趣的概念，不同数据库系统中的模式代表完全不同的内容。如 Oracle 中，默认在创建用户的时候，就建立了一个和用户同名的模式，并且互相绑定，因此很多情况下 Oracle 的用户和模式可以通用。MySQL 中的 schema 是 database 的同义词。而 HAWQ 中的模式是从 PostgreSQL 继承来的，其概念与 SQL Server 的模式更为类似，是数据库中的逻辑对象。

HAWQ 的模式是数据库中对象和数据的逻辑组织。模式允许在一个数据库中存在多个同名的对象。如果对象属于不同的模式，同名对象之间不会冲突。使用 schema 有如下好处：

- 方便管理多个用户共享一个数据库，但是又可以互相独立。
- 方便管理众多对象，更有逻辑性。
- 方便兼容某些第三方应用程序，如果创建对象时是带 schema 的。

比如要设计一个复杂系统，由众多模块构成，有时候模块间又需要具有独立性。各模块存放单独的数据库显然是不合适的。此时就可使用 schema 来划分各模块间的对象，再对用户进行适当的权限控制，这样逻辑也非常清晰。

1. 默认的"Public"模式

每个数据库都有一个默认的名为 public 的模式。如果不建立任何模式，对象则被创建在 public 模式中。所有数据库角色（用户）都具有 public 模式中的 CREATE 和 USAGE 权限。创建了一个模式时，需要给用户授予访问该模式的权限。

2. 创建模式

使用 CREATE SCHEMA 命令创建一个新模式。为了在模式中建立和访问对象，完整的对象名称由模式名+对象名组成，对象名和模式名用点号分隔。可以创建一个属于其他人的模式，语法是：

```
CREATE SCHEMA <schemaname> AUTHORIZATION <username>;
```

3. 模式查找路径

可以设置 search_path 参数指定数据库对象有效模式的查找顺序。查找路径列表中的第一个存在的模式为默认模式。如果没有指定模式，对象在默认模式中创建。

（1）设置模式查找路径

search_path 参数设置模式查找顺序。

```
-- 设置数据库级的模式查找路径：
alter database db1 set search_path to u1, public, pg_catalog;
-- 设置当前会话的模式查找路径：
set search_path to u1, public, pg_catalog;
-- 不能为用户指定模式查找路径：
alter role etl set search_path=trade;
ERROR:  Cannot support alter role set statement yet
```

（2）查看当前模式

使用 current_schema() 函数查看当前模式。

```
select current_schema();
```

使用 SHOW 命令查看当前查找路径。

```
show search_path;
```

HAWQ 对于模式的使用建议是：在管理员创建一个具体数据库后，应该为所有可以连接到该数据库的用户分别创建一个与用户名相同的模式，然后，将 search_path 设置为"$user"，即默认的模式是与用户名相同的模式。

4. 删除模式

使用 DROP SCHEMA 命令删除一个模式。

```
drop schema myschema;
```

默认模式必须为空后才能删除。为了删除一个非空的模式，可以使用 DROP SCHEMA <schemaname> CASCADE 命令删除模式及模式下的所有对象。

5. 系统模式

使用 psql 的\dn 元命令查看当前连接数据库的所有模式。

```
gpadmin=# \dn
       List of schemas
       Name        |  Owner
--------------------+---------
 hawq_toolkit       | gpadmin
 information_schema | gpadmin
 pg_aoseg           | gpadmin
 pg_bitmapindex     | gpadmin
 pg_catalog         | gpadmin
 pg_toast           | gpadmin
 public             | gpadmin
(7 rows)
```

以下是每个数据库中的系统模式：

- pg_catalog：包含系统目录表、内建数据类型、函数和操作符等。它总是模式查找路径的一部分，即使在查找路径中没有显式指定。
- information_schema：由一系列标准视图构成的数据库对象信息。可用 \dv information_schema.*元命令列出该模式下的视图。这些视图以标准方式从系统目录表获取系统信息。
- pg_toast：存储大小超过页尺寸的大对象。该模式被 HAWQ 系统内部使用。
- pg_bitmapindex：存储位图索引对象，如值列表。该模式被 HAWQ 系统内部使用。
- hawq_toolkit：管理模式，包含可以从 SQL 命令访问的外部表、视图和函数。所有数据库用户可以访问 hawq_toolkit 查询系统日志文件或系统指标。
- pg_aoseg：存储 AO（Append-optimized）类型表对象的信息。该模式被 HAWQ 系统内部使用。

6. 模式示例

修改 Master 的 pg_hba.conf 文件，增加三个用户 u1、u2、u3 的认证：

```
[gpadmin@hdp3 ~]$ more /data/hawq/master/pg_hba.conf

...

host    all    u1    172.16.1.0/24    md5
host    all    u2    172.16.1.0/24    md5
host    all    u3    172.16.1.0/24    md5
```

使认证文件生效：

```
[gpadmin@hdp3 ~]$ hawq stop cluster -u -M fast
```

创建数据库 db1：

```
[gpadmin@hdp3 ~]$ createdb db1
```

使用 gpadmin 创建两个用户 u1、u2，授予超级用户权限：

```
[gpadmin@hdp3 ~]$ psql -c "create role u1 with superuser password 'mypassword' login;create role u2 with superuser password 'mypassword' login;"
```

使用 gpadmin 在 db1 数据库中创建两个与用户 u1、u2 同名的 schema，并指定对应的属主（此情况模拟 Oracle 的用户模式）：

```
[gpadmin@hdp3 ~]$ psql -d db1 -c "create schema u1 authorization u1; create schema u2 authorization u2;"
```

用 u1 用户执行：

```
[gpadmin@hdp3 ~]$ psql -d db1 -U u1 -h hdp3 -c "create table t1 (a int); insert into t1 values(1);"
```

用 u2 用户执行：

```
[gpadmin@hdp3 ~]$ psql -d db1 -U u2 -h hdp3 -c "create table t1 (a int); insert into t1 values(2);"
```

用 u1 用户执行：

```
[gpadmin@hdp3 ~]$ psql -d db1 -U u1 -h hdp3 -c "select *,current_schema() from t1;"
Password for user u1:
 a | current_schema
---+----------------
 1 | u1
(1 row)
```

用 u2 用户执行：

```
[gpadmin@hdp3 ~]$ psql -d db1 -U u2 -h hdp3 -c "select *,current_schema() from t1;"
Password for user u2:
 a | current_schema
---+----------------
 2 | u2
(1 row)
```

用 gpadmin 用户执行：

```
[gpadmin@hdp3 ~]$ psql -d db1 -h hdp3 -c "create table t1(a int);insert into t1 values(3);"
INSERT 0 1
[gpadmin@hdp3 ~]$ psql -d db1 -h hdp3 -c "select schemaname, tablename, tableowner from pg_tables where tablename='t1';"
 schemaname | tablename | tableowner
------------+-----------+------------
 u1         | t1        | u1
 u2         | t1        | u2
 public     | t1        | gpadmin
(3 rows)

[gpadmin@hdp3 ~]$ psql -d db1
psql (8.2.15)
Type "help" for help.

db1=# show search_path;
  search_path
----------------
 "$user",public
(1 row)

db1=# select * from t1;
 a
---
 3
(1 row)

db1=# set search_path='u1';
SET
db1=# select * from t1;
 a
---
 1
(1 row)

db1=# set search_path='u2';
SET
db1=# select * from t1;
 a
---
 2
```

```
(1 row)
```

建立只有 login 权限的用户 u3：

```
[gpadmin@hdp3 ~]$ psql -c "create role u3 with password 'mypassword' login;"
NOTICE:  resource queue required -- using default resource queue "pg_default"
CREATE ROLE
```

用 u3 用户执行：

```
[gpadmin@hdp3 ~]$ psql -d db1 -U u3 -h hdp3
Password for user u3:
psql (8.2.15)
Type "help" for help.

db1=> set search_path='u1';
SET
db1=> \dt
No relations found.
db1->
```

可以看到，u3 看不到表 u1.t1。

给 u3 赋予 usage 权限：

```
[gpadmin@hdp3 ~]$ psql -d db1 -c "grant usage on schema u1 to u3;"
GRANT
```

用 u3 用户执行：

```
[gpadmin@hdp3 ~]$ psql -d db1 -U u3 -h hdp3
Password for user u3:
psql (8.2.15)
Type "help" for help.

db1=> set search_path='u1';
SET
db1=> \dt
         List of relations
 Schema | Name | Type  | Owner | Storage
--------+------+-------+-------+-------------
 u1     | t1   | table | u1    | append only
(1 row)

db1=> select * from t1;
ERROR:  permission denied for relation t1
db1=>
```

可以看到，u3 可以看到表 u1.t1，但不能查询。

给 u3 赋予 select 权限：

```
[gpadmin@hdp3 ~]$ psql -d db1 -c "grant select on u1.t1 to u3;"
GRANT
```

用 u3 用户执行：

```
[gpadmin@hdp3 ~]$ psql -d db1 -U u3 -h hdp3 -c "set search_path='u1';select *,current_schema(),current_schemas(true) from t1;"
Password for user u3:
 a | current_schema | current_schemas
---+----------------+------------------
 1 | u1             | {pg_catalog,u1}
(1 row)
```

u3 现在可以查询 u1.t1。

用 u3 用户执行：

```
[gpadmin@hdp3 ~]$ psql -d db1 -U u3 -h hdp3 -c "create table t2(a int);"
Password for user u3:
CREATE TABLE
```

删除模式：

```
[gpadmin@hdp4 ~]$ psql -h hdp3 -d db1
psql (8.2.15)
Type "help" for help.

db1=# drop schema u1;
NOTICE:  append only table u1.t1 depends on schema u1
ERROR:  cannot drop schema u1 because other objects depend on it
HINT:  Use DROP ... CASCADE to drop the dependent objects too.
db1=# drop schema u1 cascade;
NOTICE:  drop cascades to append only table u1.t1
DROP SCHEMA
db1=# drop schema u2 cascade;
NOTICE:  drop cascades to append only table u2.t1
DROP SCHEMA
```

上面一系列示例验证了以下结论：

- 搜索路径参数 search_path 控制查询表时所属 schema 的搜索顺序。
- 创建的表存放在哪个 schema 跟 search_path 有关。
- 系统默认将 PUBLIC 模式的 usage、create 权限授予所有用户。
- usage 权限的含义是，可以"看到"模式中的对象，但是没有对象上的任何权限。

- pg_catalog 存放了各系统表、内置函数等。它总是在搜索路径中，可以通过 current_schemas 看到。

4.4 创建和管理表

这里所说的表是 HAWQ 数据库内部存储的表。除了表行分布在系统中的各个 Segment 上，HAWQ 中的表与其他关系数据库中的表类似。关于外部表，将在第 8 章"数据管理"中讨论。

4.4.1 创建表

CREATE TABLE 命令创建表并定义表结构，当建立一个表时，可以定义：

- 表列及其数据类型。
- 表或列包含的限定数据的约束。
- 表的分布策略，决定 HAWQ 如何在 Segment 中分布数据。
- 表在磁盘上的存储方式。
- 表分区策略，指定数据如何划分。

1. 选择列的数据类型

列的数据类型决定了列中可以包含何种数据。选择数据类型时应遵循以下通用原则：

- 选择可以容纳数据的最小可能空间，并能最好约束数据的数据类型。例如，能使用 INT 或 SMALLINT 表示数据时，就不要使用 BIGINT，因为这会浪费存储空间。
- 在 HAWQ 中，字符类型 CHAR、VARCHAR 和 TEXT 除了使用空间不同，它们在性能上并无太大差异。在大多数情况下，应该使用 TEXT 或 VARCHAR 而不是 CHAR。
- 考虑数据扩展。数据会随着时间的推移而不断扩展。在已经装载大量数据后，从小类型变为大类型的操作代价是很昂贵的。因此，如果当前的数据值可以用 SMALLINT，但是考虑到数据扩展性，那么出于长期需要，INT 可能是更好的选择。
- 为表连接的列使用相同的数据类型。如果数据类型不同，为了正确比较数据值，数据库必须进行隐式数据类型转换，这将增加不必要的系统消耗。

2. 设置约束

可以定义约束限制表中的数据。HAWQ 支持与 PostgreSQL 相同的约束，但是有一些限制：

- CHECK 约束只能引用它定义所属的表。
- 外键约束允许，但不起作用。
- 分区表上的约束作用于整个表，不能在一个表的单独部分上定义约束。

（1）Check 约束

Check 约束允许指定特定列中存储的数据值必须满足一个布尔表达式。例如，产品价格必须为正值：

```
db1=# create table products
       ( product_no integer,
         name text,
         price numeric check (price > 0) );
db1=# insert into products values (1,'a',10);
INSERT 0 1
db1=# insert into products values (1,'a',10.5);
INSERT 0 1
db1=# insert into products values (1,'a',10.5111);
INSERT 0 1
db1=# insert into products values (1,'a',-10.5111);
ERROR: One or more assertions failed  (seg0 hdp3:40000 pid=731975)
DETAIL: Check constraint products_price_check for table products was violated
db1=# insert into products values (1,'a',0);
ERROR: One or more assertions failed  (seg0 hdp3:40000 pid=731988)
DETAIL: Check constraint products_price_check for table products was violated
db1=# select * from products;
 product_no | name | price
------------+------+---------
          1 | a    |      10
          1 | a    |    10.5
          1 | a    | 10.5111
(3 rows)
```

（2）非空约束

非空约束指定一个列不能有空值。

```
db1=# create table products
       ( product_no integer not null,
         name text not null,
         price numeric );
db1=# insert into products values(1,'a',10.51);
INSERT 0 1
db1=# insert into products (price) values(10.51);
ERROR: null value in column "product_no" violates not-null constraint (CTranslatorUtils.cpp:2726)
db1=#
```

(3) 主键与外键

HAWQ 不支持主键与外键约束。因为主键是用唯一索引实现，而 HAWQ 不支持索引，因此不支持主键。根据外键的定义，既然没有主键，也就谈不上外键了。

```
db1=# create table t2(a int);
CREATE TABLE
db1=# create table t3(a int primary key);
ERROR: Cannot support create index statement yet
```

4.4.2 删除表

DROP TABLE 命令从数据库中删除表。DROP TABLE 总是删除表上的约束。指定 CASCADE 将删除引用表的视图。如果要清空表中的数据，但保留表定义，使用 TRUNCATE <tablename>。

```
db1=# create table t1 (a int);
CREATE TABLE
db1=# insert into t1 values (1);
INSERT 0 1
db1=# create view v1 as select * from t1;
CREATE VIEW
db1=# select * from v1;
 a
---
 1
(1 row)

db1=# drop table t1;
NOTICE:  rule _RETURN on view v1 depends on append only table t1
NOTICE:  view v1 depends on rule _RETURN on view v1
ERROR:  cannot drop append only table t1 because other objects depend on it
HINT:  Use DROP ... CASCADE to drop the dependent objects too.
db1=# drop table t1 cascade;
NOTICE:  drop cascades to rule _RETURN on view v1
NOTICE:  drop cascades to view v1
DROP TABLE
```

4.4.3 查看表对应的 HDFS 文件

假设在数据库 db1 中建立了表 public.t2，使用以下步骤查看 t2 所在的 HDFS 文件。

（1）确定 HAWQ 在 HDFS 上的根目录

```
db1=# select * from pg_filespace_entry;
 fsefsoid | fsedbid |       fselocation
```

```
----------+---------+------------------------------
    16384 |       0 | hdfs://mycluster/hawq_data
(1 row)
```

可以看到，HAWQ 在 HDFS 上的根目录是/hawq_data。实验环境的 Hadoop 集群配置了 HA，所以文件位置字段中的值使用 Nameservice ID（mycluster）代替了 NameNode FQDN（Fully Qualified Domain Name）。HDP HA 配置将在第 11 章"高可用性"中说明。

（2）检查 HAWQ 系统目录表中 t1 的相关信息

```
db1=# select d.dat2tablespace tablespace_id,
db1-#        d.oid database_id,
db1-#        c.relfilenode table_id
db1-#   from pg_database d, pg_class c, pg_namespace n
db1-#  where c.relnamespace = n.oid
db1-#    and d.datname = current_database()
db1-#    and n.nspname = 'public'
db1-#    and c.relname = 't2';
 tablespace_id | database_id | table_id
---------------+-------------+----------
         16385 |       25270 |   156634
(1 row)
```

一个数据库中不同 schema 下的表可能重名，但对应的表 ID 不同，因此需要关联 pg_namespace 系统表。d.oid 是一个系统的隐藏列，表示行的对象标识符，即对象 ID。该列只有在创建表的时候使用了 WITH OIDS，或者设置了 default_with_oids 配置参数时出现。用\d pg_database 命令是看不到 oid 列的。系统表 pg_class 的 relhasoids 列是布尔类型，true 表示对象具有 OID。

为了简化对表的管理，每个表中的数据都被保存在一个 HDFS 目录中。HAWQ 数据库表在 HDFS 上的目录结构为"文件空间根目录/表空间 ID/数据库 ID/表对象（分区表对象）ID"，例如表 public.t2 所对应的 HDFS 目录为/hawq_data/16385/25270/156634，该目录下是实际存储表数据的 HDFS 文件。

（3）查看表对应的 HDFS 文件

```
[gpadmin@hdp3 ~]$ hdfs dfs -ls /hawq_data/16385/25270/156634
Found 1 items
-rw-------   3 gpadmin gpadmin          0 2017-03-30 11:05 /hawq_data/16385/25270/156634/1
```

4.5 创建和管理视图

视图能够保存经常使用的或者复杂的查询，然后将它们看作表，在 SELECT 语句中进行访问。视图里的数据并不独立于表存储到磁盘。当访问视图时，查询作为一个子查询运行。HAWQ 不支持物化视图。

1. 创建视图

```
db1=# create table t1 (a int);
CREATE TABLE
db1=# insert into t1 values (10);
INSERT 0 1
db1=# insert into t1 values (1);
INSERT 0 1
db1=# select * from t1;
 a
----
 10
  1
(2 rows)

db1=# create view v1 as select * from t1 order by a;
CREATE VIEW
db1=# select * from v1;
 a
----
  1
 10
(2 rows)

db1=# drop view v1;
DROP VIEW
db1=# create view v1 as select * from t1 order by a desc;
CREATE VIEW
db1=# select * from v1;
 a
----
 10
  1
(2 rows)

db1=# select * from v1 order by a;
```

```
    a
 ----
    1
   10
(2 rows)
```

2. 查看视图定义

```
db1=# \d v1
      View "public.v1"
 Column |  Type   | Modifiers
--------+---------+-----------
 a      | integer |
View definition:
 SELECT t1.a
   FROM t1
  ORDER BY t1.a DESC;
```

3. 删除视图

```
db1=# drop view v1;
```

4.6 管理其他对象

HAWQ 还支持自定义数据类型、自定义函数、序列等对象。如果使用过 Oracle 数据库，对这些对象一定不会陌生。

自定义数据类型的例子：

```
gpadmin=# \c db1
You are now connected to database "db1" as user "gpadmin".
db1=# create type compfoo as (f1 int, f2 text);
CREATE TYPE
db1=# create table big_objs (
db1(#     id integer,
db1(#     obj compfoo
db1(# );
CREATE TABLE
db1=# insert into big_objs values (1,(1,'a'));
INSERT 0 1
```

序列的例子：

```
db1=# create sequence myseq start 101;
CREATE SEQUENCE
```

```
db1=# select currval('myseq'), nextval('myseq');
ERROR:  currval() not supported
db1=# select nextval('myseq');
 nextval
---------
     101
(1 row)

db1=# select nextval('myseq');
 nextval
---------
     102
(1 row)
```

自定义函数将在第 9 章 "过程语言" 中详细阐述。

4.7 小结

从逻辑上看，HAWQ 的文件空间是表空间的集合，而在物理上，它们都对应 HDFS 目录。表空间目录是文件空间目录的子目录。HAWQ 所有的数据（除 Master 上的全局系统表外）都存储在文件空间目录下。HAWQ 系统中可以创建多个数据库，每个数据库中可以定义多个模式。search_path 参数指定数据库对象所属模式的查找顺序。和其他关系数据库类似，HAWQ 中可以定义表、视图、函数、数据类型、序列等对象。但 HAWQ 仅支持表列上的非空与 CHECK 约束，不支持主外键和索引。

第 5 章

◀ 分区表 ▶

分区表功能通过改善可管理性、性能和可用性,为各式应用带来了便利。通常,分区可以使某些查询以及维护操作的性能得到提高。此外,分区还可以简化常见的管理任务,是构建高可用性系统的关键手段。本章说明 HAWQ 支持的分区类型、如何确定分区策略、分区表维护等相关问题。

5.1 HAWQ 中的分区表

与大多数关系数据库一样,HAWQ 也支持分区表。这里所说的分区表是指 HAWQ 的内部分区表,外部分区表在第 8 章"数据管理"中讨论。在数据仓库应用中,事实表通常有非常多的记录,分区可以将这样的大表在逻辑上分为小的、更易管理的数据片段。HAWQ 的查询优化器支持分区消除以提高性能。只要查询使用分区键作为过滤条件,那么 HAWQ 只需要扫描满足查询条件的分区,而不必进行全表扫描。

分区并不改变表数据在 Segment 间的物理分布。表的分布是物理的,无论是分区表还是非分区表,HAWQ 都会在 Segment 上物理地分布数据,并且并行处理查询。而表的分区是逻辑上的,HAWQ 逻辑分割大表以提高查询性能和数据仓库应用的可维护性。例如,将老的分区数据从数据仓库转储或移除,并建立新的数据分区等。HAWQ 支持以下分区类型:

- 范围分区:基于数字范围分区,如日期、价格等。
- 列表分区:基于列表值分区,如销售区域、产品分类等。
- 两者混合的分区类型。

图 5-1 是一个混合类型分区表的例子,sales 表以销售日期范围作为主分区,而以销售区域作为一个日期分区中的列表子分区键。注意,HAWQ 并没提供类似 Oracle 的在线重定义功能,它只能使用 CREATE TABLE 命令创建分区表,而没有简单的命令能够将一个非分区表转化成分区表。最好在建表之前就规划好分区方式和维护方法,因为当一个非分区表已经存在大量数据后再改作分区表的操作,时间和空间消耗都是很棘手的问题。

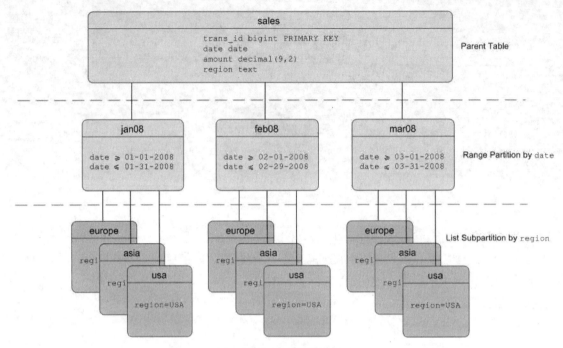

图 5-1 范围列表混合分区

在 CREATE TABLE 命令中使用 PARTITION BY 或可选的 SUBPARTITION BY 子句建立分区。上级分区可以包含一个或多个下级分区。HAWQ 内部创建上下级分区之间的层次关系。分区条件定义一个分区内可以包含的数据。在建立分区表时,HAWQ 为每个分区条件创建一个唯一的 CHECK 约束,限制一个分区所能含有的数据,保证各个分区中数据的互斥性。查询优化器利用该 CHECK 约束,决定扫描哪些分区以满足查询谓词条件。

HAWQ 在系统目录中存储分区的层次信息,因此插入到分区表中的行可以正确传递到子分区中。ALTER TABLE 命令的 PARTITION 子句用于修改分区表结构。在向分区表插入数据时,可以在 INSERT 命令中指定表的根分区或叶分区。如果数据对于指定的叶分区无效,将返回错误。INSERT 命令不支持向非叶分区的子分区中插入数据。

5.2 确定分区策略

并不是所有表都适合分区,需要进行实测以保证所期望的性能提升。下面是一些通用的分区指南,如果对以下问题的大部分答案是肯定的,那么分区表对于提高性能来说是可行的数据库设计方法。否则,表不适合分区。

- 表是否足够大?按照一般的经验,至少千万记录以上的表才算大表。数据仓库中的事实表适合作为分区表。对于小于这个数量级的表通常不需要分区,因为系统管理与维护分区的开销会抵消分区带来的可见的性能优势。

- 性能是否不可接受？只有当实施了其他优化手段后，响应时间仍然不可接受时，再考虑使用分区。
- 查询谓词条件中是否包含适合的分区键？检查查询的 WHERE 子句中是否包含适合作为分区的条件。例如，大部分查询都通过日期检索数据，那么按照月或周做范围分区可能是有益的。
- 是否需要维护一个数据仓库的历史数据窗口？例如，组织中的数据仓库只需要保持过去 12 个月的数据，那么按月分区，就可以很容易地删除最老的月份分区，并向最新的月份分区中装载当前数据。
- 根据分区定义条件，是否每个分区的数据量比较平均？分区条件应尽可能使数据平均划分。如果每个分区包含基本相同的记录数，性能会有所提升。例如，将一个大表分成 10 个相等的分区，如果查询条件中带有分区键，那么理论上查询应该比非分区表快将近 10 倍。

使用分区还要注意以下问题。首先，不要创建多余的分区。太多的分区将会减慢管理和维护任务，如检查磁盘使用、集群扩展、释放剩余空间等。其次，只有在查询条件可以利用分区消除时，性能才会得到提升。否则，一个需要扫描所有分区的查询会比非分区表还慢。可以通过查看一个查询的执行计划（explain plan）确认是否用到分区消除。最后是关于多级分区的问题。多级分区会使分区文件的数量快速增长。例如，如果一个表按日期和城市做分区，1000 天的 1000 个城市的数据，就会形成 100 万个分区。假设表有 100 列，并且假设表使用面向列的物理存储格式，那么系统需要为此表管理 1 亿个文件。

5.3 创建分区表

如前所述，创建分区表需要定义分区键、分区类型、分区层次。

5.3.1 范围分区与列表分区

1. 定义日期范围分区表

在定义日期分区表时，需要考虑以可接受的细节粒度做分区。例如，相对于以月份做主分区、日期做子分区的分区策略，每个日期一个分区，一年 365 个分区的方案可能更好。多级分区可以降低生成查询计划的时间，但平面化的分区设计运行得更快。

```
create table sales (id int, date date, amt decimal(10,2))
distributed by (id)
partition by range (date)
( start (date '2017-01-01') inclusive
  end (date '2017-02-01') exclusive
  every (interval '1 day') );
```

上面的语句以 date 列作为分区键，从 2017 年 1 月 1 月到 2017 年 2 月 1 日，每天一个分区，将建立 31 个分区。分区对应表对象的名称分别是 sales_1_prt_1、...、sales_1_prt_31。注意 inclusive 表示分区中包含定义的分区键值，exclusive 表示不包含。例如，sales_1_prt_1 包含 date >= (date '2017-01-01') and date < (date '2017-01-02')的数据，sales_1_prt_31 包含 date >= (date '2017-01-31') and date < (date '2017-02-01')的数据，即这个语句定义的分区是左闭右开的数据区间。

```
db1=# insert into sales values (1, (date '2016-12-31'),100);
ERROR:  no partition for partitioning key  (seg21 hdp4:40000 pid=60186)
db1=# insert into sales values (1, (date '2017-01-01'),100);
INSERT 0 1
db1=# insert into sales values (1, (date '2017-01-31'),100);
INSERT 0 1
db1=# insert into sales values (1, (date '2017-02-01'),100);
ERROR:  no partition for partitioning key  (seg23 hdp4:40000 pid=60190)
```

同样可以定义左开右闭的分区。

```
create table sales (id int, date date, amt decimal(10,2))
distributed by (id)
partition by range (date)
( start (date '2017-01-01') exclusive
  end (date '2017-02-01') inclusive
  every (interval '1 day') );

db1=# insert into sales values (1, (date '2017-01-01'),100);
ERROR:  no partition for partitioning key  (seg19 hdp4:40000 pid=60182)
db1=# insert into sales values (1, (date '2017-01-02'),100);
INSERT 0 1
db1=# insert into sales values (1, (date '2017-01-31'),100);
INSERT 0 1
db1=# insert into sales values (1, (date '2017-02-01'),100);
INSERT 0 1
```

也可以显式定义每个分区。

```
create table sales (id int, date date, amt decimal(10,2))
distributed by (id)
partition by range (date)
( partition p201701 start (date '2017-01-01') inclusive ,
  partition p201702 start (date '2017-02-01') inclusive ,
  partition p201703 start (date '2017-03-01') inclusive ,
  partition p201704 start (date '2017-04-01') inclusive ,
  partition p201705 start (date '2017-05-01') inclusive ,
  partition p201706 start (date '2017-06-01') inclusive ,
```

```
            partition p201707 start (date '2017-07-01') inclusive ,
            partition p201708 start (date '2017-08-01') inclusive ,
            partition p201709 start (date '2017-09-01') inclusive ,
            partition p201710 start (date '2017-10-01') inclusive ,
            partition p201711 start (date '2017-11-01') inclusive ,
            partition p201712 start (date '2017-12-01') inclusive
                     end (date '2018-01-01') exclusive );
```

上面的语句为 2017 年每个月建立一个分区。注意，不需要为每个分区指定 END 值，只要在最后一个分区（本例中的 p201712）指定 END 值即可。

2. 定义数字范围分区表

```
db1=# create table rank (id int, rank int, year int, gender
db1(# char(1), count int)
db1-# distributed by (id)
db1-# partition by range (year)
db1-# ( start (2017) end (2018) every (1),
db1(#   default partition extra );
NOTICE:  CREATE TABLE will create partition "rank_1_prt_extra" for table "rank"
NOTICE:  CREATE TABLE will create partition "rank_1_prt_2" for table "rank"
CREATE TABLE
db1=# \dt
              List of relations
 Schema |       Name        | Type  | Owner   | Storage
--------+-------------------+-------+---------+-------------
 public | rank              | table | gpadmin | append only
 public | rank_1_prt_2      | table | gpadmin | append only
 public | rank_1_prt_extra  | table | gpadmin | append only
(3 rows)

db1=# insert into rank values (1,1,2016,'M',100);
INSERT 0 1
db1=# insert into rank values (1,1,2017,'M',100);
INSERT 0 1
db1=# insert into rank values (1,1,2018,'M',100);
INSERT 0 1
db1=# insert into rank values (1,1,2019,'M',100);
INSERT 0 1
db1=# select * from rank;
 id | rank | year | gender | count
----+------+------+--------+-------
  1 |    1 | 2016 | M      |   100
  1 |    1 | 2018 | M      |   100
  1 |    1 | 2019 | M      |   100
```

```
    1 |    1 | 2017 | M        |   100
(4 rows)

db1=# select * from rank_1_prt_2;
 id | rank | year | gender | count
----+------+------+--------+-------
  1 |    1 | 2017 | M      |   100
(1 row)

db1=# select * from rank_1_prt_extra;
 id | rank | year | gender | count
----+------+------+--------+-------
  1 |    1 | 2016 | M      |   100
  1 |    1 | 2018 | M      |   100
  1 |    1 | 2019 | M      |   100
(3 rows)

db1=# drop table rank;
DROP TABLE
db1=# \dt
No relations found.
```

从上面的例子看到:

- HAWQ 默认的分区范围是左闭右开。
- 可以使用 default partition 子句增加一个默认分区,当数据不包含在任何明确定义的分区时,被包含在默认分区中。
- HAWQ 在查询时可以将分区当作表看待,但删除主表后,分区被一并删除。

3. 定义列表分区表

列表分区可以使用任何允许等值比较数据类型的列作为分区键。列表分区表必须显式定义每个分区。注意列表中的字符比较区分大小写。

```
db1=# create table rank (id int, rank int, year int, gender
db1(# char(1), count int )
db1-# distributed by (id)
db1-# partition by list (gender)
db1-# ( partition girls values ('f'),
db1(#   partition boys values ('m'),
db1(#   default partition other );
NOTICE:  CREATE TABLE will create partition "rank_1_prt_girls" for table "rank"
NOTICE:  CREATE TABLE will create partition "rank_1_prt_boys" for table "rank"
NOTICE:  CREATE TABLE will create partition "rank_1_prt_other" for table "rank"
CREATE TABLE
```

```
db1=# \dt
                List of relations
 Schema |       Name        | Type  | Owner   | Storage
--------+-------------------+-------+---------+-------------
 public | rank              | table | gpadmin | append only
 public | rank_1_prt_boys   | table | gpadmin | append only
 public | rank_1_prt_girls  | table | gpadmin | append only
 public | rank_1_prt_other  | table | gpadmin | append only
(4 rows)

db1=# insert into rank values (1,1,2016,'M',100);
INSERT 0 1
db1=# insert into rank values (1,1,2016,'m',100);
INSERT 0 1
db1=# insert into rank values (1,1,2016,'f',100);
INSERT 0 1
db1=# insert into rank values (1,1,2016,'F',100);
INSERT 0 1
db1=# insert into rank values (1,1,2016,'A',100);
INSERT 0 1
db1=# select * from rank;
 id | rank | year | gender | count
----+------+------+--------+-------
  1 |    1 | 2016 | f      |   100
  1 |    1 | 2016 | m      |   100
  1 |    1 | 2016 | M      |   100
  1 |    1 | 2016 | F      |   100
  1 |    1 | 2016 | A      |   100
(5 rows)

db1=# select * from rank_1_prt_boys;
 id | rank | year | gender | count
----+------+------+--------+-------
  1 |    1 | 2016 | m      |   100
(1 row)

db1=# select * from rank_1_prt_girls;
 id | rank | year | gender | count
----+------+------+--------+-------
  1 |    1 | 2016 | f      |   100
(1 row)

db1=# select * from rank_1_prt_other;
```

```
id | rank | year | gender | count
----+------+------+--------+-------
 1 |   1  | 2016 | M      |  100
 1 |   1  | 2016 | F      |  100
 1 |   1  | 2016 | A      |  100
(3 rows)
```

HAWQ 不支持多分区键列复合比较,分区键只能是单列。

```
db1=# create table rank (id int, rank int, year int, gender
db1(# char(1), count int )
db1-# distributed by (id)
db1-# partition by list (gender,year)
db1-# ( partition girls values ('f',2017),
db1(#   partition boys values ('m',2018),
db1(#   default partition other );
ERROR: Composite partition keys are not allowed
```

5.3.2 多级分区

可以在分区中定义子分区。使用 subpartition template 子句保证每个分区都有相同的子分区定义,包括以后添加的分区。

```
create table sales (trans_id int, date date, amount  decimal(9,2), region text)
distributed by (trans_id)
partition by range (date)
subpartition by list (region)
subpartition template
( subpartition usa values ('usa'),
  subpartition asia values ('asia'),
  subpartition europe values ('europe'),
  default subpartition other_regions)
  (start (date '2017-01-01') inclusive
   end (date '2018-01-01') exclusive
   every (interval '1 month'),
   default partition outlying_dates );
```

上面这条语句一共建立了 65 个分区。一级分区 13 个,每个一级分区包含 4 个子分区。范围上的多级分区很容易建立大量分区,其中有些分区可能只有很少的数据,甚至没有数据。随着分区数量的增加,系统表的记录不断增长,查询优化和执行时所需的内存也会增加。加大范围分区的粒度或者选择不同的分区策略有助于减少分区数量。

5.3.3 对已存在的非分区表进行分区

正如本章开始所提到的,HAWQ 只能使用 CREATE TABLE 命令创建分区表。如果想对一个已经存在的表进行分区，只能这样做：新建分区表→将原表数据导入分区表→删除原表→分区表改名→分析分区表→对新建的分区表重新授权。例如：

```
create table sales2 (like sales)
partition by range (date)
( start (date '2017-01-01') inclusive
  end (date '2018-01-01') exclusive
  every (interval '1 month') );
insert into sales2 select * from sales;
drop table sales;
alter table sales2 rename to sales;
analyze sales;
grant all privileges on sales to admin;
grant select on sales to guest;
```

查询 pg_partitions 视图可以获取分区定义。

```
select partitionboundary,
       partitiontablename,
       partitionname,
       partitionlevel,
       partitionrank
  from pg_partitions
 where tablename='sales';
```

以下表和视图提供了分区表的相关信息：

- pg_partition: 分区表及其层级关系。
- pg_partition_templates: 子分区使用的模板。
- pg_partition_columns: 分区键列。

5.4 分区消除

使用 EXPLAIN 可以检查查询的执行计划，验证查询优化器是否只扫描了相关分区的数据。下面以 sales 表上的年、月、地区三级分区为例进行说明。该 sales 表最底层存储数据的分区共有 4 * 13 * 4 = 208 个。

```
create table sales (id int, year int, month int, day int, region text)
distributed by (id)
partition by range (year)
   subpartition by range (month)
     subpartition template (
       start (1) end (13) every (1),
```

```
            default subpartition other_months )
        subpartition by list (region)
            subpartition template (
                subpartition usa values ('北京'),
                subpartition europe values ('上海'),
                subpartition asia values ('广州'),
                default subpartition other_regions )
( start (2017) end (2020) every (1),
  default partition outlying_years );
```

1. 插入一条数据

```
db1=# select * from sales;
 id | year | month | day | region
----+------+-------+-----+--------
(0 rows)


db1=# insert into sales values (1,2017,1,1,'北京');
INSERT 0 1
db1=# select * from sales;
 id | year | month | day | region
----+------+-------+-----+--------
  1 | 2017 |     1 |   1 | 北京
(1 row)
```

2. 无条件查询

查询计划如下：

```
db1=# explain select * from sales;
                              QUERY PLAN
-------------------------------------------------------------------------
 Gather Motion 24:1  (slice1; segments: 24)  (cost=0.00..431.00 rows=1 width=24)
   -> Sequence  (cost=0.00..431.00 rows=1 width=24)
         -> Partition Selector for sales (dynamic scan id: 1)  (cost=10.00..100.00 rows=5 width=4)
               Partitions selected:  208 (out of 208)
         -> Dynamic Table Scan on sales (dynamic scan id: 1)  (cost=0.00..431.00 rows=1 width=24)
 Settings:  default_hash_table_bucket_number=24
 Optimizer status: PQO version 1.684
(7 rows)
```

可以看到，该查询扫描了全部 208 个分区，没有分区消除。

3. 以年为条件查询

查询计划如下：

```
db1=# explain select * from sales where year=2017;
                                    QUERY PLAN
--------------------------------------------------------------------------------
Gather Motion 24:1  (slice1; segments: 24)  (cost=0.00..431.00 rows=1 width=24)
    -> Sequence  (cost=0.00..431.00 rows=1 width=24)
        -> Partition Selector for sales (dynamic scan id: 1)
(cost=10.00..100.00 rows=5 width=4)
             Filter: year = 2017
             Partitions selected:  104 (out of 208)
        -> Dynamic Table Scan on sales (dynamic scan id: 1)  (cost=0.00..431.00
rows=1 width=24)
             Filter: year = 2017
 Settings:  default_hash_table_bucket_number=24
 Optimizer status: PQO version 1.684
(9 rows)
```

可以看到，该查询扫描了全部 208 个分区的一半，104 个分区。顶级年份分区有 4 个，为什么 where year=2017 要扫描 104 而不是 52 个分区呢？在运行时,查询优化器会扫描这个表的层级关系，并使用 CHECK 表约束确定扫描哪些满足查询条件的分区。如果存在 DEFAULT 分区，则它总是被扫描，因此该查询扫描 year=2017 和 default 两个分区，这就是扫描的分区数是 104 而不是 52 的原因。可见,包含 DEFAULT 分区会增加整体扫描时间。按理说 DEFAULT 与其他所有分区的数据都是互斥的，完全不必在可以确定分区的条件下再去扫描它，这是不是 HAWQ 查询优化器的一个 bug 也未可知。

4. 以年、月为条件查询

查询计划如下：

```
db1=# explain select * from sales where year=2017 and month=1;
                                    QUERY PLAN
--------------------------------------------------------------------------------
 Gather Motion 24:1  (slice1; segments: 24)  (cost=0.00..431.00 rows=1
width=24)
    -> Sequence  (cost=0.00..431.00 rows=1 width=24)
        -> Partition Selector for sales (dynamic scan id: 1)
(cost=10.00..100.00 rows=5 width=4)
             Filter: year = 2017 AND month = 1
             Partitions selected:  16 (out of 208)
        -> Dynamic Table Scan on sales (dynamic scan id: 1)  (cost=0.00..431.00
rows=1 width=24)
             Filter: year = 2017 AND month = 1
```

```
Settings:  default_hash_table_bucket_number=24
Optimizer status: PQO version 1.684
(9 rows)
```

这次只扫描了 16 个分区。同样道理本应只扫描 4 个底层分区，因为 DEFAULT 的存在，需要扫描 16 个分区。

5. 以年、月、地区为条件查询

查询计划如下：

```
db1=# explain select * from sales where year=2017 and month=1 and region='北京';
                                      QUERY PLAN
--------------------------------------------------------------------------------
Gather Motion 24:1  (slice1; segments: 24)  (cost=0.00..431.00 rows=1 width=24)
   -> Sequence  (cost=0.00..431.00 rows=1 width=24)
         -> Partition Selector for sales (dynamic scan id: 1) (cost=10.00..100.00 rows=5 width=4)
               Filter: year = 2017 AND month = 1 AND region = '北京'::text
               Partitions selected:  1 (out of 208)
         -> Dynamic Table Scan on sales (dynamic scan id: 1)  (cost=0.00..431.00 rows=1 width=24)
               Filter: year = 2017 AND month = 1 AND region = '北京'::text
Settings:  default_hash_table_bucket_number=24
Optimizer status: PQO version 1.684
(9 rows)
```

这次只需扫描一个分区。当查询中包含所有层级的谓词条件时，没有扫描 DEFAULT，而是唯一确定了一个分区。

6. 以 DEFAULT 条件查询

查询计划如下：

```
db1=# explain select * from sales where year=2016;
                                      QUERY PLAN
--------------------------------------------------------------------------------
Gather Motion 24:1  (slice1; segments: 24)  (cost=0.00..431.00 rows=1 width=24)
   -> Sequence  (cost=0.00..431.00 rows=1 width=24)
         -> Partition Selector for sales (dynamic scan id: 1) (cost=10.00..100.00 rows=5 width=4)
               Filter: year = 2016
               Partitions selected:  52 (out of 208)
         -> Dynamic Table Scan on sales (dynamic scan id: 1)  (cost=0.00..431.00 rows=1 width=24)
               Filter: year = 2016
```

```
Settings:  default_hash_table_bucket_number=24
Optimizer status: PQO version 1.684
(9 rows)
```

这次只要扫描年份 DEFAULT 分区下的 52 个子分区。

HAWQ 的分区消除有以下限制：

- 查询优化器只有在查询条件中包含=、<、<=、>、>=、<>等比较运算符时才可能应用分区消除。
- 对于稳定的函数会应用分区消除，对于易变函数不会应用分区消除。例如，WHERE date→CURRENT_DATE 会应用分区消除，而 time→TIMEOFDAY 则不会。

5.5 分区表维护

ALTER TABLE 命令维护分区表。尽管可以通过引用分区对应的表对象的名字进行查询和装载数据，但修改分区表结构时，只能使用 ALTER TABLE...PARTITION 引用分区的名字。可以使用 PARTITION FOR (value)或 PARTITION FOR(RANK(number))指示分区。HAWQ 不支持在多级分区上的如下操作：

- 增加默认分区
- 增加分区
- 删除默认分区
- 删除分区
- 分裂分区
- 所有修改子分区的操作

1. 增加分区

```
-- 给 sales 表增加 2016 年的分区
alter table sales add partition start (2016) inclusive end (2017) exclusive;
```

使用 add partition 增加分区时不能存在 DEFAULT 分区，否则会报类似下面的错误，这时需要使用 split partition 增加分区：

```
ERROR:  cannot add RANGE partition to relation "sales" with DEFAULT partition
"outlying_years"
  HINT:  need to SPLIT partition "outlying_years"
```

为一个分区表增加子分区时，可以指定需要修改的分区。

```
alter table sales alter partition for (rank(12))
    add partition shenzhen values ('深圳');
```

```
alter table sales alter partition for (rank(1))
    add partition shenzhen values ('深圳');
```

2. 增加默认分区

```
alter table sales add default partition other;
```

如果没有 DEFAULT 分区，不能匹配分区 CHECK 约束的数据行将被拒绝入库，并且数据装载失败。通常为了避免出现这种情况而指定 DEFAULT 分区，任何不能与其他分区匹配的行都被装载进 DEFAULT 分区。

3. 分区改名

每个子分区对应一个表对象，可以用 \dt 元命令查看到。如果是自动生成的范围分区，没有指定名称的分区被赋予一个数字。分区对应表对象的命名规则如下：

```
<parentname>_<level>_prt_<partition_name>
```

例如：

```
sales_1_prt_1_2_prt_11_3_prt_other_regions
```

上面的名称表示该分区名为'other_regions'，是 sales 表的一个第三级分区，隶属第一级的 1 号分区下的第二级的 11 号分区下。

修改顶级父表的名称，会重命名所有分区子表名。

```
alter table sales rename to globalsales;
```

相关的分区子表名变为：

```
globalsales_1_prt_1_2_prt_11_3_prt_other_regions
```

也可以将顶级分区名改为自定义的名称：

```
alter table sales rename partition for (2017) to y2017;
```

表对象名的最大长度为 64 字节，超长会报错：

```
db1=# alter table globalsales rename partition for (2017) to year2017;
ERROR:  relation name
"globalsales_1_prt_year2017_2_prt_other_months_3_prt_other_regions" for child
partition is too long
```

使用 ALTER TABLE...PARTITION 命令修改分区表时，总是用分区名称（如 y2017）而不是分区对应的表对象全名（globalsales_1_prt_y2017）。

4. 删除分区

ALTER TABLE 命令也可用来删除分区，如果被删除的分区有子分区，则这些子分区及其数据也都被一起删除。

```
alter table globalsales drop partition for (2017);
```

```
alter table globalsales drop partition for (2018);
```

不能删除最后一个分区：

```
db1=# alter table globalsales drop partition for (2019);
ERROR:  cannot drop partition for value (2019) of relation "globalsales" -- only one remains
HINT:  Use DROP TABLE "globalsales" to remove the table and the final partition
```

5. 清空分区

使用 ALTER TABLE 命令清空一个分区及其所有子分区的数据。不能单独清空一个子分区。

```
alter table globalsales truncate partition for (2018);
```

6. 分区交换

分区交换指的是将一个表的数据与一个分区的数据互换。HAWQ 只支持单级分区表的分区交换。

```
db1=# alter table sales exchange partition for (2017)
db1-# with table stage_sales;
ERROR: cannot EXCHANGE PARTITION for relation "sales"-- partition has children
```

分区交换经常被用来向分区表装载数据。当然也能使用 COPY 或 INSERT 命令向分区表装载数据，此时数据被自动路由到正确的底层分区，就像普通表一样。但是，这种装载数据的方法会根据数据遍历整个分区层次结构，因此数据装载的性能很差。在前面 208 个分区的例子中，插入一条记录竟然用时 16 秒多：

```
db1=# \timing
Timing is on.
db1=# insert into sales values (2,2017,2,2,'上海');
INSERT 0 1
Time: 16512.156 ms
```

向分区表装载数据的推荐方法是：创建一个中间过渡表，装载过渡表，然后用过渡表与分区做交换。

```
db1=# -- 创建分区表
db1=# create table sales (id int, year int, month int, day int, region varchar(10))
db1-# distributed by (id)
db1-# partition by range (year)
db1-# ( start (2017) end (2020) every (1));
NOTICE:  CREATE TABLE will create partition "sales_1_prt_1" for table "sales"
NOTICE:  CREATE TABLE will create partition "sales_1_prt_2" for table "sales"
NOTICE:  CREATE TABLE will create partition "sales_1_prt_3" for table "sales"
CREATE TABLE
```

```
db1=# -- 添加记录
db1=# insert into sales values
db1-# (1,2017,1,1,'北京'), (2,2018,2,2,'上海'), (3,2019,3,3,'广州');
INSERT 0 3
db1=# -- 增加分区
db1=# alter table sales add partition start (2020) inclusive end (2021) exclusive;
NOTICE:  CREATE TABLE will create partition "sales_1_prt_r1873705512" for table "sales"
ALTER TABLE
db1=# -- 创建过渡表
db1=# create table stage_sales (like sales);
NOTICE:  Table doesn't have 'distributed by' clause, defaulting to distribution columns from LIKE table
CREATE TABLE
db1=# -- 添加新数据
db1=# insert into stage_sales values (4,2020,4,4,'深圳');
INSERT 0 1
db1=# -- 交换分区
db1=# \timing
Timing is on.
db1=# alter table sales exchange partition for (2020) with table stage_sales;
ALTER TABLE
Time: 61.744 ms
```

使用这种交换分区的方法会快得多，同样是添加一行，这次只用了 61 毫秒。此时分区表中有 4 条数据，而过渡表没有数据。

```
db1=# \dt
                List of relations
 Schema |          Name           | Type  | Owner   |   Storage
--------+-------------------------+-------+---------+-------------
 public | sales                   | table | gpadmin | append only
 public | sales_1_prt_1           | table | gpadmin | append only
 public | sales_1_prt_2           | table | gpadmin | append only
 public | sales_1_prt_3           | table | gpadmin | append only
 public | sales_1_prt_r1524203752 | table | gpadmin | append only
 public | stage_sales             | table | gpadmin | append only
(6 rows)

db1=# select * from sales;
 id | year | month | day | region
----+------+-------+-----+--------
  3 | 2019 |     3 |   3 | 广州
```

```
  4 | 2020 |    4  |  4  | 深圳
  1 | 2017 |    1  |  1  | 北京
  2 | 2018 |    2  |  2  | 上海
(4 rows)

db1=# select * from stage_sales;
 id | year | month | day | region
----+------+-------+-----+--------
(0 rows)
```

7. 分裂分区

分裂分区指的是将一个分区分裂成两个，HAWQ 只能分裂单级分区表。

```
db1=# alter table sales split partition for (2017)
db1-# at (2016)
db1-# into (partition y016, partition y2017);
ERROR:  cannot split partition with child partitions
HINT:  Try splitting the child partitions.
```

下面的例子将 2017 年 1 月的分区分割成 2017 年 1 月 1 日到 2017 年 1 月 15 日、2017 年 1 月 16 日到 2017 年 1 月 31 日两个分区，分割值包含在后一个分区中。

```
db1=# create table sales (id int, date date, amt decimal(10,2))
db1-# distributed by (id)
db1-# partition by range (date)
db1-# ( partition p201701 start (date '2017-01-01') inclusive ,
db1(#   partition p201702 start (date '2017-02-01') inclusive
db1(#                     end (date '2017-03-01') exclusive );
NOTICE:  CREATE TABLE will create partition "sales_1_prt_p201701" for table "sales"
NOTICE:  CREATE TABLE will create partition "sales_1_prt_p201702" for table "sales"
CREATE TABLE
db1=# insert into sales values (1, date '2017-01-15', 100);
INSERT 0 1
db1=# insert into sales values (1, date '2017-01-16', 100);
INSERT 0 1
db1=# select * from sales_1_prt_p201701;
 id |    date    |  amt
----+------------+--------
  1 | 2017-01-15 | 100.00
  1 | 2017-01-16 | 100.00
(2 rows)

db1=# alter table sales split partition for ('2017-01-01') at ('2017-01-16')
```

```
db1-# into (partition p20170101to0115, partition p20170116to0131);
NOTICE:  exchanged partition "p201701" of relation "sales" with relation "pg_temp_68011"
NOTICE:  dropped partition "p201701" for relation "sales"
NOTICE:  CREATE TABLE will create partition "sales_1_prt_p20170101to0115" for table "sales"
NOTICE:  CREATE TABLE will create partition "sales_1_prt_p20170116to0131" for table "sales"
ALTER TABLE
db1=# select * from sales_1_prt_p20170101to0115;
 id |    date    |  amt
----+------------+--------
  1 | 2017-01-15 | 100.00
(1 row)


db1=# select * from sales_1_prt_p20170116to0131;
 id |    date    |  amt
----+------------+--------
  1 | 2017-01-16 | 100.00
(1 row)
```

如果表有 DEFAULT 分区，必须使用分裂分区的方法添加分区。使用 INTO 子句的第二个分区为 DEFAULT 分区。

```
db1=# alter table sales add default partition other;
NOTICE:  CREATE TABLE will create partition "sales_1_prt_other" for table "sales"
ALTER TABLE
db1=# insert into sales values (3, date '2017-03-01', 100);
INSERT 0 1
db1=# insert into sales values (4, date '2017-04-01', 100);
INSERT 0 1
db1=# select * from sales_1_prt_other;
 id |    date    |  amt
----+------------+--------
  4 | 2017-04-01 | 100.00
  3 | 2017-03-01 | 100.00
(2 rows)

db1=# alter table sales split default partition
db1-# start ('2017-03-01') inclusive
db1-# end ('2017-04-01') exclusive
db1-# into (partition p201703, default partition);
NOTICE:  exchanged partition "other" of relation "sales" with relation
```

```
"pg_temp_68051"
NOTICE:  dropped partition "other" for relation "sales"
NOTICE:  CREATE TABLE will create partition "sales_1_prt_p201703" for table "sales"
NOTICE:  CREATE TABLE will create partition "sales_1_prt_other" for table "sales"
ALTER TABLE
db1=# select * from sales_1_prt_p201703;
 id |    date    |  amt
----+------------+--------
  3 | 2017-03-01 | 100.00
(1 row)

db1=# select * from sales_1_prt_other;
 id |    date    |  amt
----+------------+--------
  4 | 2017-04-01 | 100.00
(1 row)
```

8. 修改子分区模板

ALTER TABLE SET SUBPARTITION TEMPLATE 修改一个分区表的子分区模板。新模板只影响后面添加的数据，不修改现有的分区数据。

```
db1=# create table sales (trans_id int, date date, amount decimal(9,2), region text)
db1-#   distributed by (trans_id)
db1-#   partition by range (date)
db1-#   subpartition by list (region)
db1-#   subpartition template
db1-#   ( subpartition usa values ('usa'),
db1(#     subpartition asia values ('asia'),
db1(#     subpartition europe values ('europe'),
db1(#     default subpartition other_regions )
db1-#   ( start (date '2017-01-01') inclusive
db1(#     end (date '2017-04-01') exclusive
db1(#     every (interval '1 month') );
NOTICE:  CREATE TABLE will create partition "sales_1_prt_1" for table "sales"
...

CREATE TABLE
db1=# alter table sales set subpartition template
db1-# ( subpartition usa values ('usa'),
db1(#   subpartition asia values ('asia'),
db1(#   subpartition europe values ('europe'),
db1(#   subpartition africa values ('africa'),
db1(#   default subpartition regions );
```

```
NOTICE:  replacing level 1 subpartition template specification for relation
"sales"
ALTER TABLE
```

当添加一个分区时，使用新的子分区模板。

```
db1=# alter table sales add partition "4"
db1-#   start ('2017-04-01') inclusive
db1-#   end ('2017-05-01') exclusive ;
NOTICE:  CREATE TABLE will create partition "sales_1_prt_4" for table "sales"
...
ALTER TABLE
db1=# \dt sales*
                     List of relations
 Schema |            Name                | Type  |  Owner  |   Storage
--------+--------------------------------+-------+---------+-------------
...
 public | sales_1_prt_3                  | table | gpadmin | append only
 public | sales_1_prt_3_2_prt_asia       | table | gpadmin | append only
 public | sales_1_prt_3_2_prt_europe     | table | gpadmin | append only
 public | sales_1_prt_3_2_prt_other_regions| table | gpadmin | append only
 public | sales_1_prt_3_2_prt_usa        | table | gpadmin | append only
 public | sales_1_prt_4                  | table | gpadmin | append only
 public | sales_1_prt_4_2_prt_africa     | table | gpadmin | append only
 public | sales_1_prt_4_2_prt_asia       | table | gpadmin | append only
 public | sales_1_prt_4_2_prt_europe     | table | gpadmin | append only
 public | sales_1_prt_4_2_prt_regions    | table | gpadmin | append only
 public | sales_1_prt_4_2_prt_usa        | table | gpadmin | append only
(22 rows)
```

下面的命令移除子分区模板：

```
alter table sales set subpartition template ();
```

5.6 小结

和大多数数据库系统类似，HAWQ 也支持范围分区、列表分区和混合分区。分区主要起到两方面作用：利用分区消除提高查询性能，增强表的可维护性。每个分区对应一个 HAWQ 表对象。可以定义默认分区存储不属于其他任何分区的数据,但这样做可能对性能有负面影响。HAWQ 支持增加、删除、清空、分裂、交换、修改子分区模板等常规的分区维护操作，但有一些限制。HAWQ 不支持在线重定义，因此要将非分区表改成分区表，只能通过创建新的分区表并重新装载数据的方式实现。

第 6 章 存储管理

在 HAWQ 中创建一个表时，应该预先对数据如何分布、表的存储选项、数据导入导出方式和其他 HAWQ 特性做出选择，这些都将对查询性能产生极大影响。理解有效选项的含义以及如何在数据库中使用它们，有助于做出正确的选择。

6.1 数据存储选项

CREATE TABLE 的 WITH 子句用于设置表的存储选项：

```
create table t1 (a int) with
   (appendonly=true,
    blocksize=8192,
    orientation=row,
    compresstype=zlib,
    compresslevel=1,
    fillfactor=50,
    oids=false);
```

除了在表级别指定存储选项，HAWQ 还支持在一个特定分区或子分区上设置存储选项。以下语句在特定子分区上使用 WITH 子句，指定对应分区的存储属性。

```
create table sales
(id int, year int, month int, day int, region text)
   distributed by (id)
   partition by range (year)
     subpartition by range (month)
       subpartition template (
         start (1) end (13) every (1),
         default subpartition other_months )
     subpartition by list (region)
       subpartition template (
```

```
        subpartition usa values ('usa') with
            (appendonly=true,
            blocksize=8192,
            orientation=row,
            compresstype=zlib,
            compresslevel=1,
            fillfactor=50,
            oids=false),
        subpartition europe values ('europe'),
        subpartition asia values ('asia'),
        default subpartition other_regions)
    ( start (2002) end (2010) every (1),
    default partition outlying_years);
```

下面说明 HAWQ 所支持的存储选项。

1. APPENDONLY

指示是否只追加数据。因为目前 HDFS 文件中的数据只能追加，不允许修改或删除，所以该选项只能设置为 TRUE，否则会报错：

```
db1=# create table t1(a int) with (appendonly=true);
CREATE TABLE
db1=# create table t2(a int) with (appendonly=false);
ERROR:  tablespace "dfs_default" does not support heap relation
```

2. BLOCKSIZE

设置表中每个数据块的字节数，值在 8192～2097152 之间，而且必须是 8192 的倍数，默认值为 32768。该属性必须与 appendonly=true 一起使用，并且只支持行存储模型。

```
db1=# create table t1(a int) with (blocksize=8192);
ERROR:  invalid option 'blocksize' for base relation. Only valid for Append Only relations
db1=# create table t1(a int) with (appendonly=true,blocksize=8192);
CREATE TABLE
db1=# create table t2(a int) with (appendonly=true,blocksize=8192,orientation=parquet);
ERROR:  invalid option 'blocksize' for parquet table
db1=# create table t2(a int) with (appendonly=true,blocksize=8192,orientation=row);
CREATE TABLE
```

3. BUCKETNUM

设置一个哈希分布表使用的哈希桶数，有效值为大于 0 的整数，并且不要大于 default_hash_table_bucket_number 配置参数。默认值为"Segment 节点数 * 6"。推荐在创建哈希分布表时显式指定此值。该属性在建表时指定，表创建以后不能修改 bucketnum 的值。

```
db1=# create table t1(a int) with (bucketnum=1) distributed by (a);
CREATE TABLE
```

4. ORIENTATION

指定数据存储模型，有效值为 row（默认值）和 parquet，分别指的是面向行和列的存储格式。此选项只能与 appendonly=true 一起使用。

```
db1=# create table t1(a int) with (orientation=parquet);
ERROR:  invalid option "orientation" for base relation. Only valid for Append Only relations
db1=# create table t1(a int) with (orientation=parquet,appendonly=true);
CREATE TABLE
```

老版本的 HAWQ 还支持一种名为 column 的格式，但在 2.1.1 版本中已经过时而不再支持，应该用 parquet 存储格式代替 column 格式。

```
db1=# create table t1(a int) with (orientation=column,appendonly=true);
ERROR:  Column oriented tables are deprecated. Not support it any more.
```

row 格式对于全表扫描类型的读操作效率很高。适合行存储的情况主要有频繁插入，SELECT 或 WHERE 子句中包含表所有列或大部分列，并且一行中所有列的总长度相对较小时，这些是典型的 OLTP 应用特点。而 parquet 面向列的格式对于大型查询更高效，适合数据仓库应用。应该根据实际的数据和查询评估性能，选择最适当的存储类型。row 与 parquet 之间的格式转换工作由用户的应用程序完成，HAWQ 不会进行这种转换。

5. COMPRESSTYPE

指定使用的压缩算法，有效值为 ZLIB、SNAPPY 或 GZIP。默认值 ZLIB 的压缩率更高但速度更慢。parquet 表仅支持 SNAPPY 和 GZIP。该选项只能与 appendonly=true 一起使用。

```
db1=# create table t1(a int) with (compresstype=zlib);
ERROR:  invalid option 'compresstype' for base relation. Only valid for Append Only relations
db1=# create table t1(a int) with (compresstype=zlib,appendonly=true);
CREATE TABLE
db1=# create table t2(a int) with (compresstype=zlib,appendonly=true,orientation=parquet);
ERROR:  parquet table doesn't support compress type: 'zlib'
db1=# create table t2(a int) with (compresstype=snappy,appendonly=true,orientation=parquet);
CREATE TABLE
```

6. COMPRESSLEVEL

有效值为 1~9，数值越大压缩率越高。如果不指定，默认值为 1。该选项只对 zlib 和 gzip 有效，并且只能与 appendonly=true 一起使用。

```
db1=# create table t1(a int) with (compresstype=snappy,compresslevel=1);
ERROR:  invalid option 'compresslevel' for compresstype 'snappy'.
db1=# create table t1(a int) with (compresslevel=1);
ERROR:  invalid option 'compresslevel' for base relation. Only valid for Append Only relations
db1=# create table t1(a int) with (compresslevel=1,appendonly=true);
CREATE TABLE
```

7. OIDS

默认值为 FALSE，表示不给行赋予对象标识符。建议在创建表时不要启用 OIDS。首先，通常 OIDS 对用户应用没有用处。再者，典型 HAWQ 系统中的表行数都很大，如果为每行赋予一个 32 位的计数器，不但占用空间，而且可能给 HAWQ 系统的目录表造成问题。最后，每行节省 4 字节存储空间也能带来一定的查询性能提升。

8. FILLFACTOR

该选项控制插入数据时页存储空间的使用率，作用类似于 Oracle 的 PCTFREE，为后续的行更新预留空间。取值范围是 10~100，默认值为 100，即不为更新保留空间。HAWQ 表不支持 UPDATE 和 DELETE 操作，故保持默认值即可。该选项对 parquet 表无效。

```
db1=# create table t1(a int) with (fillfactor=100,orientation=parquet);
ERROR:  invalid option "orientation" for base relation. Only valid for Append Only relations
db1=# create table t1(a int) with (fillfactor=100);
CREATE TABLE
```

9. PAGESIZE 与 ROWGROUPSIZE

- PAGESIZE：描述 parquet 文件中每一列对应的 page 大小，可配置范围为[1KB,1GB)，默认为 1MB。
- ROWGROUPSIZE：描述 parquet 文件中 row group 的大小，可配置范围为 [1KB,1GB)，默认为 8MB。

这两个选项只对 parquet 表有效，并且只能与 appendonly=true 一起使用。PAGESIZE 的值应该小于 ROWGROUPSIZE，因为行组包含页的元信息。

```
db1=# create table t1(a int) with (pagesize=1024,rowgroupsize=1024,orientation=parquet);
ERROR:  row group size for parquet table must be larger than pagesize. Got rowgroupsize: 1024, pagesize 1024
db1=# create table t1(a int) with (pagesize=1024,rowgroupsize=8096,orientation=parquet);
ERROR:  invalid option "orientation" for base relation. Only valid for Append Only relations
db1=# create table t1(a int) with (pagesize=1024,rowgroupsize=8096,orientation=row);
ERROR:  invalid option 'pagesize' for non-parquet table
db1=# create table t1(a int) with (pagesize=1024,rowgroupsize=8096,orientation=parquet,appendonly=true);
CREATE TABLE
```

6.2 数据分布策略

必须要指出，这里所说的数据分布策略并不直接决定数据的物理存储位置，数据块的存储位置是由 HDFS 决定的。这里的数据分布策略概念是从 Greenplum 继承而来，存储移植到 HDFS 上后，数据分布决定了 HDFS 上数据文件的生成规则，以及在此基础上的资源分配策略。

6.2.1 数据分布策略概述

所有的 HAWQ 表（除 gpfdist 外部表）都是分布存储在 HDFS 上的。HAWQ 支持两种数据分布策略，即随机与哈希。在创建表时，DISTRIBUTED 子句声明 HAWQ 的数据分布策略。如果没有指定 DISTRIBUTED 子句，则 HAWQ 默认使用随机分布。当使用哈希分布时，bucketnum 属性设置哈希桶的数量。几何数据类型（Geometric Types）或用户定义数据类型的列不能作为 HAWQ 的哈希分布键列。哈希桶数影响处理查询时使用的虚拟段的数量。

默认哈希分布表使用的哈希桶数由 default_hash_table_bucket_number 服务器配置参数的值所指定。可以在会话级或使用建表 DDL 语句中的 bucketnum 存储参数覆盖默认值。

随机分布相对于哈希分布有一些益处。例如，集群扩容后，HAWQ 的弹性查询特性，使得在操作随机分布表时能够自动使用更多的资源，而不需要重新分布数据。重新分布大表数据时，资源与时间消耗都非常大。而且，随机分布表具有更好的数据本地化，这尤其表现在底层 HDFS 因为某个数据节点失效而执行 rebalance 操作重新分布数据后。在一个大规模 Hadoop 集群中，增删数据节点后 rebalance 的情况很常见。

然而，哈希分布表可能比随机分布表快。在 HAWQ 的 TPCH 测试中，哈希分布表在很多查询上具有更好的性能。图 6-1 是 HAWQ 提供的一个数据分布性能对比图，其中 CO 表示列存储格式，AO 表示行存储格式。

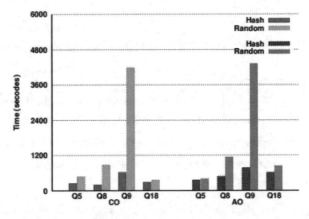

图 6-1　随机与哈希数据分布的性能对比

HAWQ 的文档中并没有说明图 6-1 中具体的测试环境，比如数据量和 Segment 节点数是多少，CPU 或内存等资源情况，default_hash_table_bucket_number、hawq_rm_nvseg_perquery_ perseg_limit、hawq_rm_nvseg_perquery_limit 等参数设置的是多少,具体查询语句是什么,等等,因此这个测试的结果也许并不适用于普遍情况。

HAWQ 的运行时弹性查询是以虚拟段为基础的,而虚拟段是基于查询成本按需分配的。每个节点使用一个物理段和一组动态分配的虚拟段。通常,为查询分配的虚拟段越多,查询执行得越快。可以通过设置 default_hash_table_bucket_number 和 hawq_rm_nvseg_perquery_limit 参数,控制一个查询使用的虚拟段数量,从而调整性能。如果 default_hash_table_bucket_number 的值改变了,哈希分布表的数据必须重新分布,这可能是一步成本很高的操作。因此,如果需要大量的虚拟段,最好在建表前预先设置好 default_hash_table_bucket_number。集群扩容后,可能需要调整 default_hash_table_bucket_number 的值。但要注意,该值不要超过 hawq_rm_nvseg_perquery_limit 参数的值。

表 6-1 是 HAWQ 给出的 Segment 节点数量与 default_hash_table_bucket_number 值的对应关系。不推荐将该参数设置为大于 1000 的值。

表 6-1 Segment 节点数与 default_hash_table_bucket_number 参数的对应关系

节点数	default_hash_table_bucket_number
<= 85	6 * #nodes
> 85 and <= 102	5 * #nodes
> 102 and <= 128	4 * #nodes
> 128 and <= 170	3 * #nodes
> 170 and <= 256	2 * #nodes
> 256 and <= 512	1 * #nodes
> 512	512

6.2.2 选择数据分布策略

在选择分布策略时,应该考虑具体数据和查询的情况,包括以下几点:

- 平均分布数据。为了达到更好的性能,所有 Segment 应该包含相似数据量。如果数据不平衡或存在"尖峰",拥有更多数据的 Segment 工作负载会比其他 Segment 高很多。
- 本地和分布式操作。本地操作比分布式操作更快。查询中有连接、排序或聚合等操作,如果能够在一个 Segment 上完成,那么这种本地处理查询是最快的。当多个表共享一个公共的哈希分布键,该列上的连接或排序操作是在本地进行的。对于随机分布策略,是否本地连接是不可控的。
- 平均处理查询。为了获得更好的性能,所有 Segment 应该处理基本等量的查询工作。如果表的数据分布策略和查询条件谓词匹配得不好,查询负载可能成为"尖峰"。例如,假设有一个销售事务表,以公司名称列作为分布键分布数据。如果查询中的一个谓词引用了单一的分布键,则查询可能只在一个 Segment 上进行处理。而如果查询谓词通常以公司名称外的其他条件选择数据，可能所有 Segment 共同处理查询。

HAWQ 利用运行时动态并行查询，这能显著提高查询性能。性能主要依赖于以下因素：

- 随机分布表的大小。
- 哈希分布表的 CREATE TABLE DDL 中指定的 bucketnum 存储参数。
- 数据本地化情况。
- default_hash_table_bucket_number。
- hawq_rm_nvseg_perquery_limit。

对随机分布表的查询资源分配与表的数据量有关，通常为每个 HDFS 块分配一个虚拟段，其结果是查询大表可能使用大量的资源。对于大的哈希分布表，为了在不同 Segment 节点上达到最好的负载均衡，bucketnum 应设置成 Segment 节点数量的倍数。运行时弹性查询将试图找到处理节点上最优的桶数量。大表需要更多的虚拟段，因此需要设置更大的 bucketnum。default_hash_table_bucket_number 是查询一个哈希分布表时使用的默认哈希数。由于资源是动态分配的，当查询实际执行时，分配的虚拟段数量可能与该值不同，但执行该查询虚拟段的总数永远不会超过 hawq_rm_nvseg_perquery_limit 的值。

对于任何一个特定的查询，前四个因素已经是固定值，只有最后一个配置参数 hawq_rm_nvseg_perquery_limit 可以被用于调整查询执行的性能。hawq_rm_nvseg_perquery_limit 指定集群范围内，一个查询语句在执行时可用的最大虚拟段数量，默认值为 512，取值范围是 1 ~ 65535。

除 hawq_rm_nvseg_perquery_limit 参数外，hawq_rm_nvseg_perquery_perseg_limit 也控制执行一个查询使用的虚拟段数量。该参数指示一个 Segment 在执行一个查询时可以使用的最大虚拟段数，默认值为 6，取值范围是 1 ~ 65535。它影响随机分布表和外部表，但不影响哈希分布表。减小 hawq_rm_nvseg_perquery_perseg_limit 的值可能提高并发查询数量，增加它的值可能提升单个查询执行的并行度。对于某些查询，如果已经达到硬件限制，提升并行度并不会提高性能，况且数据仓库应用的并发量通常也不会很高。因此，在绝大多数部署环境中，不应该修改此参数的默认值。

修改服务器配置参数最简便的方法是使用 Ambari 的 Web 界面交互式设置，如图 6-2 所示。大多数情况下，HAWQ 的运行时弹性查询将动态分配虚拟段以优化性能，因此通常不需要对相关参数做进一步的调优。

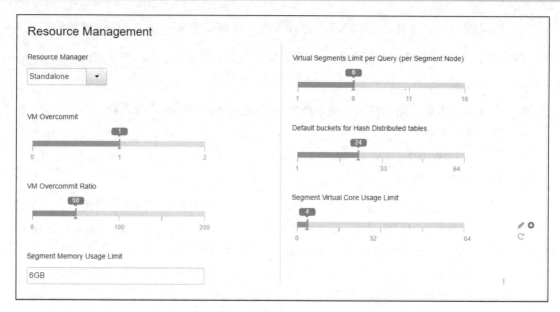

图 6-2　使用 Ambari 调整 HAWQ 虚拟段资源相关参数

下面用一个例子说明两种数据分布策略。建立三个表，t1 使用单列哈希分布，t2 使用随机分布，t3 使用多列哈希分布。

```
db1=# create table t1 (a int) distributed by (a);
CREATE TABLE
db1=# create table t2 (a int) distributed randomly;
CREATE TABLE
db1=# create table t3 (a int,b int,c int) distributed by (b,c);
CREATE TABLE
```

下面的语句能查询表的分布键列。

```
db1=# select c.relname, sub.attname
db1-#   from pg_namespace n
db1-#   join pg_class c on n.oid = c.relnamespace
db1-#   left join (select p.attrelid, p.attname
db1(#               from pg_attribute p
db1(#     join (select localoid, unnest(attrnums) as attnum
db1(#             from gp_distribution_policy) as g on g.localoid = p.attrelid
db1(#           and g.attnum = p.attnum) as sub on c.oid = sub.attrelid
db1-#   where n.nspname = 'public'
db1-#     and c.relname in ('t1', 't2', 't3')
db1-#     and c.relkind = 'r';
 relname | attname
---------+---------
 t1      | a
 t2      |
```

```
 t3       | c
 t3       | b
(4 rows)
```

前面已经提到哈希分布表中桶的概念。从根本上说，每个哈希桶对应一个 HDFS 文件。在数据库初始化时，default_hash_table_bucket_number 参数得到设置，默认值按表 6-1 所示的公式计算得到。我们的实验环境中有 4 个 Segment 节点，default_hash_table_bucket_number=24。现在表中没有数据，表目录下是空的。

```
db1=# select c.relname, d.dat2tablespace tablespace_id, d.oid database_id,
db1-#        c.relfilenode table_id
db1-#   from pg_database d, pg_class c, pg_namespace n
db1-#  where c.relnamespace = n.oid
db1-#    and d.datname = current_database()
db1-#    and n.nspname = 'public'
db1-#    and c.relname in ('t1', 't2');
 relname | tablespace_id | database_id | table_id
---------+---------------+-------------+----------
 t1      |         16385 |       25270 |   156897
 t2      |         16385 |       25270 |   156902
(2 rows)

[gpadmin@hdp3 ~]$ hdfs dfs -ls /hawq_data/16385/25270/156897
[gpadmin@hdp3 ~]$ hdfs dfs -ls /hawq_data/16385/25270/156902
[gpadmin@hdp3 ~]$
```

向表中插入数据后，哈希分布表 t1 对应的 HDFS 目录下有 24 个数据文件（每个哈希桶对应一个文件），而随机分布表 t2 只有一个数据文件。

```
db1=# insert into t1 values (1),(2),(3);
INSERT 0 3
db1=# insert into t2 values (1),(2),(3);
INSERT 0 3

[gpadmin@hdp3 ~]$ hdfs dfs -ls /hawq_data/16385/25270/156897
Found 24 items
-rw------ 3 gpadmin gpadmin 0 2017-04-01 14:40 /hawq_data/16385/25270/156897/1
...
-rw------ 3 gpadmin gpadmin 0 2017-04-01 14:40 /hawq_data/16385/25270/156897/9
[gpadmin@hdp3 ~]$ hdfs dfs -ls /hawq_data/16385/25270/156902
Found 1 items
-rw------ 3 gpadmin gpadmin 48 2017-04-01 14:40 /hawq_data/16385/25270/156902/1
[gpadmin@hdp3 ~]$
```

表一旦建立，哈希桶数就是固定不变且不能修改的。查询 t1 表时，将分配 24 个虚拟段，

每个文件一个。扩展集群时，查询 t1 依然分配 24 个虚拟段，但是这些虚拟段将在所有节点中分配。比如扩展到 8 节点，则 24 个虚拟段被分配到 8 个节点上。集群扩展后，应该根据集群中 Segment 节点的数量调整 default_hash_table_bucket_number 的值，并重建 t1 表，这样它才能获得正确的桶数。

相对于哈希分布策略，随机分布更具有弹性。可以看到，t2 表只在 HDFS 上创建了一个数据文件。查询 t2 表分配的虚拟段数量，由查询优化器在运行时决定。分配虚拟段数与表的数据量有关，对于小表的查询可能只分配一个虚拟段，而大表可能每个主机分配 6 个虚拟段。集群扩展时，不需要重新分布数据，如果需要数据库会自动增加查询一个随机分布表总的虚拟段数量。

HAWQ 推荐使用随机分布，这也是默认的分布策略，原因如下：

- 适合 HDFS。NameNode 只需要跟踪更少的文件。
- 更具弹性。集群增减节点时，不需要重新分配数据。
- 可以通过增加 hawq_rm_nvseg_perquery_perseg_limit 的值提高查询并行性。
- 优化器可以根据查询的需求动态分配虚拟段数量。

6.2.3 数据分布用法

数据分布的原理虽然有些复杂，但 DISTRIBUTED 子句的语法却很简单，DISTRIBUTED BY (<column>, […])用来声明一列或多列，作为哈希分布表的分布键。DISTRIBUTED RANDOMLY 显式指定表使用随机分布策略。

```
db1=# create table t1(id int) with (bucketnum=8) distributed by (id);
CREATE TABLE
db1=# create table t2(id int) with (bucketnum=8) distributed randomly;
CREATE TABLE
db1=# create table t3(id int) distributed randomly;
CREATE TABLE
```

注意 t2 表，虽然指定了 bucketnum=8，但其分布策略使用的是随机分布，bucketnum 参数不起作用。如果将 t2 的分布策略修改为哈希会报错：

```
db1=# \d t2
Append-Only Table "public.t2"
 Column |  Type   | Modifiers
--------+---------+-----------
 id     | integer |
Compression Type: None
Compression Level: 0
Block Size: 32768
Checksum: f
Distributed randomly
```

```
db1=# alter table t2 set distributed by (id);
ERROR: bucketnum requires a numeric value
```

查看相关系统表可以看到，虽然设置了 bucketnum=8，但 t2 的哈希分布键列为空，也说明是随机分布表。同时看到无论采用哪种分布策略，bucketnum 的默认值都是 default_hash_table_bucket_number 参数值，只是在随机分布表中不起作用。

```
db1=# select t1.*,t2.relname from gp_distribution_policy t1,pg_class t2
db1-# where t1.localoid=t2.oid;
 localoid | bucketnum | attrnums | relname
----------+-----------+----------+---------
    40651 |         8 | {1}      | t1
    40656 |         8 |          | t2
    40661 |        24 |          | t3
(3 rows)
```

可以在建表后修改它的分布策略。从随机分布修改为哈希分布，或者更改一个哈希分布表的分布键时，表数据会自动在所有 Segment 上重新分布。而从哈希分布修改为随机分布时，不会重新分布数据。

```
db1=# create table t1 (a int);
CREATE TABLE
db1=# \d t1
Append-Only Table "public.t1"
 Column |  Type   | Modifiers
--------+---------+-----------
 a      | integer |
Compression Type: None
Compression Level: 0
Block Size: 32768
Checksum: f
Distributed randomly

db1=# alter table t1 set distributed by (a);
ALTER TABLE
db1=# \d t1
Append-Only Table "public.t1"
 Column |  Type   | Modifiers
--------+---------+-----------
 a      | integer |
Compression Type: None
Compression Level: 0
Block Size: 32768
```

```
Checksum: f
Distributed by: (a)

db1=# alter table t1 set distributed randomly;
ALTER TABLE
```

为了重新分布随机分布表的数据，或者在没有改变哈希分布策略时需要重新分布数据，使用 reorganize=true。该命令使用当前分布策略在所有 Segment 中重新分布表数据。

```
db1=# alter table t1 set with (reorganize=true);
ALTER TABLE
```

这里有一个需要注意的细节，如果在建表时显式指定了 bucketnum，那么不能再使用 ALTER TABLE 语句修改表的分布策略，也不能重新分布数据。

```
db1=# create table t1(a int) with (bucketnum=10) distributed by (a);
CREATE TABLE
db1=# alter table t1 set distributed by (a);
ERROR:  bucketnum requires a numeric value
db1=# alter table t1 set distributed randomly;
ERROR:  bucketnum requires a numeric value
db1=# alter table t1 set with (reorganize=true);
ERROR:  bucketnum requires a numeric value
db1=# alter table t1 set with (bucketnum=10,reorganize=true);
ERROR:  option "bucketnum" not supported
```

如果在建表时需要使用不同于默认值的 bucketnum，可以在会话级设置 default_hash_table_bucket_number 系统参数，这样以后就可以使用 ALTER TABLE 语句修改表的分布策略或重新组织表数据了。

```
db1=# set default_hash_table_bucket_number=10;
SET
db1=# create table t1(a int) distributed by (a);
CREATE TABLE
db1=# alter table t1 set distributed randomly;
ALTER TABLE
db1=# alter table t1 set distributed by (a);
ALTER TABLE
db1=# alter table t1 set with (reorganize=true);
ALTER TABLE
```

推荐使用这种为表设置 bucketnum 的方法，而不要在 CREATE TABLE 中显式指定。

6.3 从已有的表创建新表

HAWQ 提供了四种方式从一个原始表创建新表，如表 6-2 所示。

表 6-2 从原始表创建新表的四种方式

方式	语法
INHERITS	CREATE TABLE new_table INHERITS (origintable) [WITH(bucketnum=x)] [DISTRIBUTED BY col]
LIKE	CREATE TABLE new_table (LIKE origintable) [WITH(bucketnum=x)] [DISTRIBUTED BY col]
AS	CREATE TABLE new_table [WITH(bucketnum=x)] AS SUBQUERY [DISTRIBUTED BY col]
SELECT INTO	CREATE TABLE origintable [WITH(bucketnum=x)] [DISTRIBUTED BY col] SELECT * INTO new_table FROM origintable

1. INHERITS

CREATE TABLE 语句的 INHERITS 子句指定一个或多个父表，新建的表作为子表，自动继承父表的所有列。INHERITS 在子表与父表之间建立了一种永久性关系。对父表结构的修改会传递到子表，默认时，子表中新增的数据也会在包含在父表中。

```
db1=# create table t1(a int);
CREATE TABLE
db1=# create table t2(b int);
CREATE TABLE
db1=# create table t3() inherits (t1,t2);
NOTICE:  Table has parent, setting distribution columns to match parent table
CREATE TABLE
db1=# \d t3
Append-Only Table "public.t3"
 Column |  Type   | Modifiers
--------+---------+-----------
 a      | integer |
 b      | integer |
Compression Type: None
Compression Level: 0
Block Size: 32768
Checksum: f
Inherits: t1,
          t2
Distributed randomly
```

```
db1=# alter table t1 alter a type text;
ALTER TABLE
db1=# \d t3
Append-Only Table "public.t3"
 Column |  Type   | Modifiers
--------+---------+-----------
 a      | text    |
 b      | integer |
Compression Type: None
Compression Level: 0
Block Size: 32768
Checksum: f
Inherits: t1,
         t2
Distributed randomly

db1=# insert into t3 values ('a',1);
INSERT 0 1
db1=# select * from t1;
 a
---
 a
(1 row)

db1=# select * from t2;
 b
---
 1
(1 row)

db1=# drop table t1;
NOTICE:  append only table t3 depends on append only table t1
ERROR:  cannot drop append only table t1 because other objects depend on it
HINT:  Use DROP ... CASCADE to drop the dependent objects too.
db1=# drop table t1 cascade;
NOTICE:  drop cascades to append only table t3
DROP TABLE
db1=# \dt
        List of relations
 Schema | Name | Type  | Owner   | Storage
--------+------+-------+---------+--------------
 public | t2   | table | gpadmin | append only
```

(1 row)

建立分区表时不能使用 INHERITS 子句。

```
db1=# create table sales (id int, date date, amt decimal(10,2)) inherits (t1)
db1-# distributed by (id)
db1-# partition by range (date)
db1-# ( partition jan08 start (date '2017-01-01') inclusive ,
db1(#   partition feb08 start (date '2017-02-01') inclusive
db1(#                   end (date '2018-01-01') exclusive );
error:  cannot mix inheritance with partitioning
```

如果存在多个父表中同名的列,当列的数据类型也相同时,在子表中会被合并为一个列,否则会报错。

```
db1=# create table t1(a int);
CREATE TABLE
db1=# create table t2(a smallint);
CREATE TABLE
db1=# create table t3 () inherits (t1,t2);
NOTICE:  Table has parent, setting distribution columns to match parent table
NOTICE:  merging multiple inherited definitions of column "a"
ERROR:  inherited column "a" has a type conflict
DETAIL:  integer versus smallint
db1=# alter table t2 alter a type int;
ALTER TABLE
db1=# create table t3 () inherits (t1,t2);
NOTICE:  Table has parent, setting distribution columns to match parent table
NOTICE:  merging multiple inherited definitions of column "a"
CREATE TABLE
```

如果新建表的列名也包含在父表中,处理方式类似,数据类型相同则合并成单列,否则报错。

```
db1=# create table t1(a int);
CREATE TABLE
db1=# create table t3 (a text) inherits (t1);
NOTICE:  Table has parent, setting distribution columns to match parent table
NOTICE:  merging column "a" with inherited definition
ERROR:  column "a" has a type conflict
DETAIL:  integer versus text
db1=# create table t3 (a int) inherits (t1);
NOTICE:  Table has parent, setting distribution columns to match parent table
NOTICE:  merging column "a" with inherited definition
CREATE TABLE
```

如果新建表指定了一个列的默认值,该默认值会覆盖从父表继承的列的默认值。

```
db1=# create table t1(a int default 1);
CREATE TABLE
db1=# create table t2(a int default 2) inherits (t1);
NOTICE:  Table has parent, setting distribution columns to match parent table
NOTICE:  merging column "a" with inherited definition
CREATE TABLE
db1-# \d t2
Append-Only Table "public.t2"
 Column |  Type   | Modifiers
--------+---------+-----------
 a      | integer | default 2
...
```

子表会自动从父表继承分布策略。

```
db1=# create table t1(a int) with (bucketnum=8) distributed by (a);
CREATE TABLE
db1=# create table t2 () inherits (t1);
NOTICE:  Table has parent, setting distribution columns to match parent table
CREATE TABLE
db1=# \d t2
Append-Only Table "public.t2"
 Column |  Type   | Modifiers
--------+---------+-----------
 a      | integer |
Compression Type: None
Compression Level: 0
Block Size: 32768
Checksum: f
Inherits: t1
Distributed by: (a)

db1=# create table t3 () inherits (t1) with (bucketnum=8) distributed by (a);
CREATE TABLE
db1=# create table t4 () inherits (t1) with (bucketnum=16) distributed by (a);
ERROR:  distribution policy for "t4" must be the same as that for "t1"
db1=# create table t4 (b int) inherits (t1) with (bucketnum=8) distributed by (b);
ERROR:  distribution policy for "t4" must be the same as that for "t1"
```

2. LIKE

LIKE 子句指示新建表从另一个已经存在的表中复制所有列的名称、数据类型、非空约束，以及表的数据分布策略。如果原表中指定了 bucketnum，而新表没有指定，则 bucketnum 将被复制，否则使用新表的 bucketnum。像 appendonly 这样的存储属性，或者分区结构不会被复制，默认值也不会被复制，新表中所有列的默认值都是 NULL。与 INHERITS 不同，新表与原始表是完全解耦的。

```
db1=# create table t1 (a int) with (bucketnum=8) distributed by (a);
CREATE TABLE
db1=# create table t2 (like t1);
NOTICE:  Table doesn't have 'distributed by' clause, defaulting to distribution columns from LIKE table
CREATE TABLE
db1=# create table t3 (like t1) with (bucketnum=16) distributed by (a);
CREATE TABLE
db1=# select t1.*,t2.relname from gp_distribution_policy t1,pg_class t2
db1-# where t1.localoid=t2.oid and t2.relname in ('t1','t2','t3');
 localoid | bucketnum | attrnums | relname
----------+-----------+----------+---------
    43738 |         8 | {1}      | t1
    43743 |        24 | {1}      | t2
    43748 |        16 | {1}      | t3
(3 rows)
```

非空约束总是被复制到新表。但对 CHECK 约束而言，只有指定了 INCLUDING CONSTRAINTS 子句时才会被复制到新表。

```
db1=# create table t1 (a int not null check (a > 0));
CREATE TABLE
db1=# create table t2 (like t1);
NOTICE:  Table doesn't have 'distributed by' clause, defaulting to distribution columns from LIKE table
CREATE TABLE
db1=# \d t2
Append-Only Table "public.t2"
 Column |  Type   | Modifiers
--------+---------+-----------
 a      | integer | not null
Compression Type: None
Compression Level: 0
Block Size: 32768
Checksum: f
Distributed randomly

db1=# create table t3 (like t1 including constraints);
NOTICE:  Table doesn't have 'distributed by' clause, defaulting to distribution columns from LIKE table
CREATE TABLE
```

```
db1=# \d t3
Append-Only Table "public.t3"
 Column |  Type   | Modifiers
--------+---------+-----------
 a      | integer | not null
Compression Type: None
Compression Level: 0
Block Size: 32768
Checksum: f
Check constraints:
    "t1_a_check" CHECK (a > 0)
Distributed randomly
```

LIKE 还有一点与 INHERITS 不同，它不会合并新表与原表的列。不能在新表或 LIKE 子句中显式定义列。

3. AS

CREATE TABLE AS 是很多数据库系统都提供的功能。它用一个 SELECT 查询命令的结果集创建新表并填充数据。新表的列就是 SELECT 返回的列，也可以显式定义新表的列名。新表的存储参数和分布策略与原表无关。

```
db1=# create table t1 (a int);
CREATE TABLE
db1=# insert into t1 values (100);
INSERT 0 1
db1=# create table t2 (b) with  --只定义列名，不能指定列的数据类型
db1-#     (bucketnum=8,
db1(#     appendonly=true,
db1(#     blocksize=8192,
db1(#     orientation=row,
db1(#     compresstype=zlib,
db1(#     compresslevel=1,
db1(#     fillfactor=50,
db1(#     oids=false)
db1-# as select * from t1
db1-# distributed by (b);
SELECT 1
db1=# select * from t2;
  b
-----
 100
(1 row)
```

4. SELECT INTO

SELECT INTO 在功能上与 AS 类似，也是从查询结果创建新表，但这种语法不能定义新表的存储选项和分布策略，而总是使用默认值。

```
db1=# create table t1 (a int) with
db1-#     (bucketnum=8,
db1(#     appendonly=true,
db1(#     blocksize=8192,
db1(#     orientation=row,
db1(#     compresstype=zlib,
db1(#     compresslevel=1,
db1(#     fillfactor=50,
db1(#     oids=false)
db1-# distributed by (a);
CREATE TABLE
db1=# insert into t1 values (1);
INSERT 0 1
db1=# select * into t2 from t1;
SELECT 1
db1=# \d t2
Append-Only Table "public.t2"
 Column |  Type   | Modifiers
--------+---------+-----------
 a      | integer |
Compression Type: None
Compression Level: 0
Block Size: 32768
Checksum: f
Distributed randomly

db1=# select * from t2;
 a
---
 1
(1 row)
```

6.4 小结

创建 HAWQ 表时，可以指定数据块大小、哈希桶数、行或列存储格式、数据压缩类型、压缩率、对象标识符、页大小、页空间的使用率等存储选项。HAWQ 的 DISTRIBUTED 子句为表指定随机或哈希两种数据分布策略之一，推荐使用随机分布，这也是默认值。数据分布决定了 HDFS 上数据文件的生成规则，以及在此基础上的资源分配策略，但并不直接决定数据的物理存储位置，数据块的存储位置由 HDFS 决定。存储选项与数据分布对查询性能有很大影响。HAWQ 提供了 INHERITS、LIKE、AS、SELECT INTO 四种方式从一个原始表创建新表。

第 7 章 资源管理

本章重点讨论 HAWQ 的资源管理器与资源队列。HAWQ 系统构建在 Hadoop 之上，能充分利用 Hadoop 集群中的 CPU 和内存等资源。HAWQ 支持两种资源管理方式：独立与全局。独立方式使用 HAWQ 自带的资源管理器，假设自己是 Hadoop 集群中的唯一应用而尽可能独占可用资源。全局方式将 HAWQ 作为 YARN 的一个应用，由 YARN 统一管理和分配集群资源。HAWQ 还支持定义多个级别的资源队列，通过它协调多用户多查询的并发资源使用。

7.1 HAWQ 资源管理概述

HAWQ 使用多种机制管理 CPU、内存、I/O、文件句柄等系统资源，包括全局资源管理、资源队列、强制资源使用限额等。

7.1.1 全局资源管理

Hadoop 通常使用 YARN 全局管理资源。YARN 是一个通用的资源管理框架，为 MapReduce 作业或其他配置了 YARN 的应用提供资源。在 YARN 环境中，资源分配的单位被称为容器（container）。YARN 还能强制限制每个集群节点上的可用资源。图 7-1 展示了 Hadoop YARN 环境下的 HAWQ 集群布局。

图 7-1　YARN 环境下的 HAWQ 集群架构

可以将 HAWQ 配置为一个在 YARN 中注册的应用，执行查询时，HAWQ 资源管理器与 YARN 通信以获取所需的资源。之后 HAWQ Master 主机上的资源管理器负责管理和分配这些从 YARN 获得的资源。当资源使用完毕后返还给 YARN。

7.1.2　HAWQ 资源队列

资源队列是 HAWQ 系统中并发管理的主要工具。它是一种数据库对象，可以使用 CREATE RESOURCE QUEUE 语句创建。资源队列控制可以并发执行的活跃查询数量，以及为每种查询类型分配的最大内存、CUP 数量。资源队列还可以限制单个查询消耗的资源总量，避免个别查询使用过多资源而影响系统的整体性能。HAWQ 内部基于资源队列层次系统动态管理资源。资源队列的数据结构为一棵 n 叉树，如图 7-2 所示。

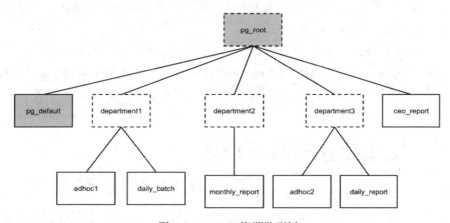

图 7-2　HAWQ 资源队列树

HAWQ 初始化后有一个根队列 pg_root 和一个默认的队列 pg_default。如果使用 YARN 模式，HAWQ 资源管理器自动从全局资源管理器获得根队列资源。当创建了一个新的资源队列时，必须指定其父队列，以这种方式将所有资源队列组织到一棵树中。

执行查询时，进行编译和语义分析后，优化器与 HAWQ 资源管理器协调查询的资源使用情况，得到基于可用资源的优化的查询计划。查询分发器将每个查询的资源分配与查询计划一同发送给 Segment。这样，Segment 上的查询执行器（QE）就知道当前查询的资源配额，并在整个执行过程中使用这些资源。查询结束或终止后，资源返还给 HAWQ 资源管理器。

资源队列树中，只有叶子队列可以被授予角色和接受查询。资源分配策略如下：

- 只为正在运行或排队的查询分配资源队列。
- 多个查询竞争时，HAWQ 资源管理器自动依据可用资源情况平衡队列间的资源。
- 在一个资源队列中，如果有多个查询等待资源，最终资源以 Best-Effort 方式分配给每个查询。

7.1.3　资源管理器配置原则

配置资源管理时应该遵循下面的实践原则，保证高效的资源管理和最佳系统性能：

- Segment 节点没有相同的 IP 地址。有些软件使用自动配置 IP 地址的虚拟网卡，这可能造成某些 HAWQ Segment 得到相同的 IP 地址。这种情况下，资源管理器的容错服务组件只能认到相同 IP 中的一个 Segment。
- 所有 Segment 配置相同的资源容量配额。
- 为避免资源碎片化，配置 Segment 资源配额为所有虚拟段资源限额的整数倍。
- 确保有足够已注册的 Segment 响应资源请求。如果失效的 Segment 超过了限额，资源请求被拒绝。
- Master 和 Segment 使用多个独立的大磁盘（2TB 或以上）的临时目录，如/disk1/tmp、/disk2/tmp，保证写临时文件的负载均衡。对于给定查询，HAWQ 为每个虚拟段使用一个单独的临时目录存储溢出文件。多个 HAWQ 会话也使用自己的临时目录避免磁盘竞争。如果临时目录过少，或者多个临时目录存储在同一个磁盘上，会增加磁盘竞争或磁盘空间用尽的风险。
- 最小化每个 Segment 的 YARN 容器数，并设置空闲资源返还 YARN 的超时时间。
- yarn.scheduler.minimum-allocation-mb 参数设置成可以被 1GB 整除，如 1024、512 等。

7.2 配置独立资源管理器

HAWQ 中的资源管理器配置主要涉及以下几方面：

- 确定使用哪种资源管理模式。HAWQ 支持两种管理模式：独立模式与外部全局资源管理器模式。独立模式也叫无全局资源管理模式。在该模式下，HAWQ 使用集群节点资源时，不考虑其他共存的应用，HAWQ 假设它能使用所有 Segment 节点的资源。对于专用 HAWQ 集群，独立模式是可选的方案。当前 HAWQ 支持 YARN 作为外部全局资源管理器。该模式下，HAWQ 作为一个 YARN 应用，使用 YARN 管理的集群资源。
- 如果选择独立资源管理模式，需要决定是否限制分配给每个 HAWQ Segment 使用的内存与 CPU。
- 如果使用 YARN 模式，需要在 YARN 中为 HAWQ 配置资源队列，还要在 HAWQ 中进行与 YARN 相关的配置。HAWQ 自动注册为 YARN 应用，使用 YARN 中配置的资源队列。
- 在 HAWQ 中创建资源队列。

1. 使用独立模式

为了配置 HAWQ 运行独立资源管理模式，在 Master 节点上的 hawq-site.xml 文件中设置以下属性：

```
<property>
```

```
        <name>hawq_global_rm_type</name>
        <value>none</value>
</property>
```

该属性为全局设置，需要重启 HAWQ 以生效。当然也可以使用 Ambari，在 HAWQ→Configs→Resource Manager 进行设置，然后重启 HAWQ 服务。

hawq_global_rm_type 参数代表 HAWQ 全局资源管理类型，有效值为 yarn 和 none。设置为 none 表示由 HAWQ 的资源管理器自己管理资源。设置为 yarn 意味着与 YARN 协调使用资源。默认安装时使用的是独立模式。

```
[gpadmin@hdp3 ~]$ hawq config -s hawq_global_rm_type
GUC     : hawq_global_rm_type
Value   : none
[gpadmin@hdp3 ~]$
```

2. 配置 Segment 资源限制

使用独立模式（hawq_global_rm_type=none）时，可以限制每个 HAWQ Segment 所能使用的资源。配置方法是在每个 Segment 节点上的 hawq-site.xml 文件中增加以下参数：

```
<property>
    <name>hawq_rm_memory_limit_perseg</name>
    <value>8GB</value>
</property>
<property>
    <name>hawq_rm_nvcore_limit_perseg</name>
    <value>4</value>
</property>
```

hawq_rm_memory_limit_perseg 参数设置独立资源管理模式下，每个 HAWQ Segment 使用的最大内存数。hawq_rm_nvcore_limit_perseg 参数设置每个 HAWQ Segment 使用的最大 CPU 虚拟核数。因为 XML 的配置验证，在 YARN 模式下也需要设置这些属性，即使该模式下不会使用这些参数。

所有 Segment 都配置成相同的值，内存应该设置成 1GB 的倍数。并且为了降低形成资源碎片的可能性，hawq_rm_memory_limit_perseg 应该配置成所有虚拟段资源限额（通过资源队列的 VSEG_RESOURCE_QUOTA 属性配置）的倍数。例如，hawq_rm_memory_limit_perseg 设置成 15GB，但是虚拟段资源限额设置成 2GB，那么一个 Segment 可以使用的最大内存只有 14GB。

3. 配置查询语句的资源限额

在某些场景下，可能需要在查询语句级增加资源限额。以下配置属性允许用户通过修改相应的资源队列控制限额：

- hawq_rm_stmt_vseg_memory

- hawq_rm_stmt_nvseg

hawq_rm_stmt_vseg_memory 定义每个虚拟段的内存限额，hawq_rm_stmt_nvseg 定义下个执行的查询所使用的虚拟段个数，默认值为 0，表示不能在语句级设置资源限额。下面的例子中，执行下个查询语句时，HAWQ 的资源管理器将分配 10 个虚拟段，每个虚拟段有 256MB 的内存限额。

```
db1=# set hawq_rm_stmt_vseg_memory='256mb';
SET
db1=# set hawq_rm_stmt_nvseg=10;
SET
db1=# create table t(i integer);
CREATE TABLE
db1=# insert into t values(1);
INSERT 0 1
db1=#
```

HAWQ 动态为给定查询语句分配资源，资源管理器只是分配 Segment 的资源，而不会为查询保留资源。并且，语句级设置的虚拟段数不要超过全局配置参数 hawq_rm_nvseg_perquery_limit 的值。

4. 配置最大虚拟段数

可以在服务器级限制语句执行时使用的虚拟段数，这能防止数据装载期间的资源瓶颈或过渡消耗资源带来的性能问题。在 Hadoop 集群中，NameNode 和 DataNode 可以并发打开的写文件数是有限制的，考虑下面的场景：需要向有 P 个分区的表导入数据；集群中有 N 个节点，为了导入数据，每个节点上启动了 V 个虚拟段，则每个 DataNode 要打开 $P×V$ 个文件，至少启动 $P×V$ 个线程。如果 P 和 V 的数量很大，DataNode 将成为瓶颈。而在 NameNode 上将有 $V×N$ 个连接，如果节点很多，那么 NameNode 可能成为瓶颈。

为缓解 NameNode 的负载，使用下面的服务器配置参数限制 V 的大小，即每个节点启动的最大虚拟段数。

- hawq_rm_nvseg_perquery_limit：在服务器级别上限制执行一条语句可以使用的最大虚拟段数量，默认值为 512。default_hash_table_bucket_number 定义的哈希桶数不能超过 hawq_rm_nvseg_perquery_limit。
- default_hash_table_bucket_number：定义哈希表使用的默认桶数。查询哈希表时，分配给查询的虚拟段数是固定的，等于表的桶数。一般集群扩容后应当调整此参数。

还可以在资源队列配置中限制查询使用的虚拟段数量。全局配置参数是"硬限制"，资源队列或语句级别的限额不能超过服务器级别的限额。

7.3 整合 YARN

HAWQ 支持 YARN 作为全局资源管理器。在 YARN 管理的环境中，HAWQ 动态向 YARN 请求资源容器，资源使用完返还 YARN。此特性让 HAWQ 有效利用 Hadoop 的资源管理能力，并使 HAWQ 成为 Hadoop 生态圈的一员。可以使用以下步骤整合 YARN 与 HAWQ：

（1）安装 YARN。如果使用 HDP 2.3 版本，必须设置 yarn.resourcemanager.system-metrics-publisher.enabled=false。

（2）配置 YARN 使用 CapacityScheduler 调度器，并为 HAWQ 保留一个单独的应用队列。

（3）启用 YARN 的高可用特性（可选）。

（4）配置 HAWQ 的 YARN 模式。

（5）根据需要调整 HAWQ 的资源使用：为 HAWQ 修改相应的 YARN 资源队列配额（刷新 YARN 资源队列不需要重启 HAWQ）；通过修改 HAWQ 资源队列进行资源消耗的细粒度控制；其他配置，如设置每个 HAWQ Segment 的最小 YARN 容器数，或修改 HAWQ 的空闲资源超时时间等。

1. 配置 YARN

查询请求资源时，HAWQ 资源管理器与 YARN 资源调度器协商资源分配。查询执行完毕后，HAWQ 资源管理器向 YARN 调度器返回占用的资源。为了使用 YARN，最好为 HAWQ 配置独立的应用资源队列。YARN 为特定的资源调度器配置资源队列，调度器根据资源队列的配置为应用分配资源。目前 HAWQ 仅支持 YARN 的 Capacity 调度器。

下面的例子说明如何使用 Ambari 配置 CapacityScheduler 作为 YARN 的调度器。

（1）登录 Ambari。

（2）选择 YARN → Configs → Advanced → Scheduler，如图 7-3 所示。

图 7-3　在 Ambari 中配置 CapacityScheduler

（3）在 yarn.resourcemanager.scheduler.class 输入框中输入以下值，该属性值设置 YARN 的调度器为 Capacity：

```
org.apache.hadoop.yarn.server.resourcemanager.scheduler.capacity.CapacityScheduler
```

（4）在 Capacity Scheduler 输入框中输入以下内容：

```
yarn.scheduler.capacity.maximum-am-resource-percent=0.2
yarn.scheduler.capacity.maximum-applications=10000
yarn.scheduler.capacity.node-locality-delay=40
yarn.scheduler.capacity.root.acl_administer_queue=*
yarn.scheduler.capacity.root.capacity=100
yarn.scheduler.capacity.root.queues=mrque1,mrque2,hawqque
yarn.scheduler.capacity.root.hawqque.capacity=20
yarn.scheduler.capacity.root.hawqque.maximum-capacity=80
yarn.scheduler.capacity.root.hawqque.state=RUNNING
yarn.scheduler.capacity.root.hawqque.user-limit-factor=1
yarn.scheduler.capacity.root.mrque1.capacity=30
yarn.scheduler.capacity.root.mrque1.maximum-capacity=50
yarn.scheduler.capacity.root.mrque1.state=RUNNING
yarn.scheduler.capacity.root.mrque1.user-limit-factor=1
yarn.scheduler.capacity.root.mrque2.capacity=50
yarn.scheduler.capacity.root.mrque2.maximum-capacity=50
yarn.scheduler.capacity.root.mrque2.state=RUNNING
yarn.scheduler.capacity.root.mrque2.user-limit-factor=1
```

默认的 Capacity 调度策略只有一个名为 default 的资源队列。以上属性在根队列下设置两个 MapReduce 队列和一个名为 hawqque 的 HAWQ 专用队列，三个队列并存，共享整个集群资源。hawqque 队列可以使用整个集群 20% 到 80% 的资源。表 7-1 说明了队列配置的主要属性及其含义。

表 7-1　YARN 配置参数

属性名称	描述
yarn.scheduler.capacity.maximum-am-resource-percent	集群中可用于运行 application master 的资源比例上限，通常用于限制并发运行的应用程序数目。示例中设置为 0.2，即 20%
yarn.scheduler.capacity.maximum-applications	集群中所有队列同时处于等待和运行状态的应用程序数目上限，这是一个强限制，一旦集群中应用程序数目超过该上限，后续提交的应用程序将被拒绝，默认值为 10000
yarn.scheduler.capacity.node-locality-delay	调度器尝试调度一个 rack-local container 之前，最多跳过的调度机会。默认值为 40，接近一个机架中的节点数目
yarn.scheduler.capacity.<queue_name>.queues	定义本队列下的资源队列，即资源队列树中某节点的直接下级节点

(续表)

属性名称	描述
yarn.scheduler.capacity.<queue_name>.capacity	一个百分比值,表示队列占用整个集群多少比例的资源
yarn.scheduler.capacity.<queue_name>.maximum-capacity	弹性设置,队列最大时占用多少比例的资源
yarn.scheduler.capacity.<queue_name>.state	队列状态,可以是 RUNNING 或 STOPPED
yarn.scheduler.capacity.<queue_name>.user-limit-factor	每个用户的低保百分比,比如设置为 1,则表示无论有多少用户在跑任务,每个用户最低占用不少于 1%的资源

（5）保存配置并重启 YARN 服务,之后可以在 YARN 资源管理器的用户界面看到配置的三个队列,如图 7-4 所示。

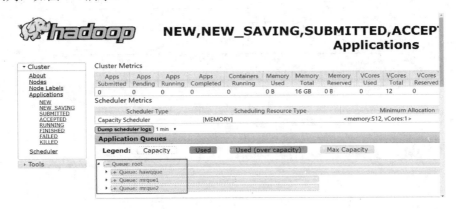

图 7-4　YARN 中的资源队列

2. 在 YARN 中设置 Segment 资源限制

与独立资源管理模式类似,也可以通过 YARN 管理 HAWQ Segment 的配额。HAWQ 推荐将所有 Segment 设置成相同的资源配额。在 yarn-site.xml 文件中设置如下属性:

```
<property>
  <name>yarn.nodemanager.resource.memory-mb</name>
  <value>4GB</value>
</property>
<property>
  <name>yarn.nodemanager.resource.cpu-vcores</name>
  <value>1</value>
</property>
```

或者使用 Ambari 的 YARN → Configs → Settings 进行配置,yarn.nodemanager.resource.memory-mb 和 yarn.nodemanager.resource.cpu-vcores 的设置分别如图 7-5、图 7-6 所示。

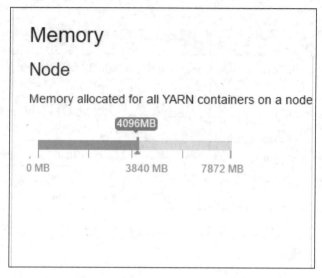

图 7-5 设置一个节点的内存配额

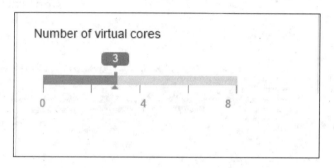

图 7-6 设置一个节点的 CPU 核数

- yarn.nodemanager.resource.memory-mb：表示该节点上 YARN 可使用的物理内存总量，默认是 8192MB。如果节点的内存资源不够 8GB，则需要调小这个值，YARN 不会智能地探测节点的物理内存总量。
- yarn.nodemanager.resource.cpu-vcores：表示该节点上 YARN 可使用的虚拟 CPU 个数，默认是 8。目前推荐将该值设置为与物理 CPU 核数相同。如果节点的 CPU 核数不够 8 个，则需要调小这个值，YARN 不会智能地探测节点的物理 CPU 总数。

与独立模式一样，HAWQ 推荐每核内存数是 1GB 的倍数，如每核 1GB、2GB、4GB 等。为了减少 YARN 模式下产生资源碎片的可能，应该遵从以下配置原则：

- yarn.nodemanager.resource.memory-mb 应该设置为虚拟段资源限额（在 HAWQ 资源队列中配置）的倍数。
- 每核内存应该设置为 yarn.scheduler.minimum-allocation-mb 的倍数。

比如 YARN 中的配置如下：

```
yarn.scheduler.minimum-allocation-mb=1024
yarn.nodemanager.resource.memory-mb=49152
```

```
yarn.nodemanager.resource.cpu-vcores=16
```

HAWQ 计算的每核内存为 3GB（48GB/16）。由于 yarn.scheduler.minimum-allocation-mb 设置是 1GB，每个 YARN 容器内存为 1GB。这样每核内存正好是 YARN 容器的 3 倍，不会形成资源碎片。但如果 yarn.scheduler.minimum-allocation-mb 设置为 4GB，则每个 YARN 容器形成 1GB（4GB-3GB）的内存碎片空间。为了避免这种情况，可以修改 yarn.nodemanager.resource.memory-mb 为 64GB，或者将 yarn.scheduler.minimum-allocation-mb 改为 1GB。注意，如果将 yarn.scheduler.minimum-allocation-mb 设置为 1GB 或更小的值，该值应该能被 1GB 整除，如 1024、512 等。

3. 启用 YARN 模式

配置完 YARN，就可以在 hawq-site.xml 文件中添加如下属性，将 YARN 启用为 HAWQ 的全局资源管理器。

```
<property>
    <name>hawq_global_rm_type</name>
    <value>yarn</value>
</property>
```

或者使用 Ambari 的 HAWQ → Configs → Settings → Resource Management 进行配置，如图 7-7 所示。

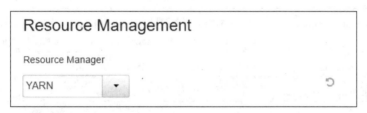

图 7-7　使用 Ambari 设置 HAWQ 的资源管理器

这样 HAWQ 资源管理器只会使用 YARN 分配的资源。YARN 模式的 HAWQ 需要在 hawq-site.xml 文件中配置以下属性：

```
<property>
    <name>hawq_rm_yarn_address</name>
    <value>hdp2:8050</value>
</property>
<property>
    <name>hawq_rm_yarn_scheduler_address</name>
    <value>hdp2:8030</value>
</property>
<property>
    <name>hawq_rm_yarn_queue_name</name>
    <value>hawqque</value></property>
<property>
```

```
<name>hawq_rm_yarn_app_name</name>
<value>hawq</value>
</property>
```

或者使用 Ambari 的 HAWQ → Configs → Advanced → Advanced hawq-site 进行配置，如图 7-8 所示。

hawq_rm_yarn_address	hdp2:8050
hawq_rm_yarn_app_name	hawq
hawq_rm_yarn_queue_name	hawqque
hawq_rm_yarn_scheduler_address	hdp2:8030

图 7-8　使用 Ambari 设置 HAWQ 的资源管理器参数

表 7-2 包含了相关属性的描述。

表 7-2　HAWQ 中的 YARN 资源管理器参数

属性名称	描述
hawq_rm_yarn_address	YARN 服务器的主机和端口号，与 yarn-site.xml 中的 yarn.resourcemanager.address 属性相同
hawq_rm_yarn_scheduler_address	YARN 调度器的主机和端口号，与 yarn-site.xml 中的 yarn.resourcemanager.scheduler.address 属性相同
hawq_rm_yarn_queue_name	YARN 中 HAWQ 使用的资源队列名称
hawq_rm_yarn_app_name	YARN 中 HAWQ 使用的应用名称

4. 用 YARN 协调 HAWQ 资源

为保证资源管理的效率与查询性能，应该做适当的配置，协调 YARN 为 HAWQ 资源管理器分配资源。

（1）调整每个 Segment 的最小 YARN 容器数

当 HAWQ 刚注册到 YARN，还没有工作负载时，HAWQ 不需要立即获得任何资源。只有在 HAWQ 接收到第一个查询请求时，HAWQ 资源管理器才开始向 YARN 请求资源。为保证为后续查询优化资源分配，同时避免与 YARN 协调过于频繁，可以调整 hawq_rm_min_resource_perseg 参数。无论初始查询的大小，每个 HAWQ Segment 都至少会被分配指定的 YARN 容器数。该参数的默认值为 2，这意味着即便查询所需的资源很少，HAWQ 资源管理器也至少为每个 Segment 获得两个 YARN 容器。

此属性配置不能超过 YARN 中 HAWQ 资源队列的配额。例如，如果 YARN 中 HAWQ 队列的配额不大于整个集群的 50%，并且每个 YARN 节点的最大内存与虚拟 CPU 核数分别为 64GB 和 16，那么 hawq_rm_min_resource_perseg 的设置不能大于 8。这是因为 HAWQ 资源管理器通过虚拟 CPU 核数获得 YARN 资源容器。

（2）设置 YARN 资源超时

如果 HAWQ 的工作负载很低，HAWQ 资源管理器已经获得的 YARN 资源有可能出现空闲的情况。可以调整 hawq_rm_resource_idle_timeout 参数，让 HAWQ 资源管理器更快或更慢地向 YARN 返还空闲资源。假设资源管理器获取资源的过程会造成查询延时。为了让 HAWQ 资源管理器可以更长时间地持有已经获得的资源，以备后面的查询工作使用，可增加 hawq_rm_resource_idle_timeout 的值。该参数的默认值为 300 秒。

7.4 管理资源队列

通过定义层次化的资源队列，系统管理员能够根据需要均衡分配系统资源。资源队列是使用 CREATE RESOURCE QUEUE 语句创建的数据库对象，是 HAWQ 系统中管理并发度的主要工具。可以使用资源队列控制并发执行的活跃查询数量，以及分配给每种查询类型使用的最大内存与 CPU 数量。资源队列还可以防止某些查询因消耗太多系统资源，影响系统整体性能的情况。HAWQ 内部将资源队列组织为一棵如图 7-2 所示的 n 叉树。HAWQ 初始化后，有一个名为 pg_root 的根队列，根下有一个名为 pg_default 的默认队列，在图 7-2 中用灰色节点表示。

1. 设置资源队列最大数

可以配置 HAWQ 集群中允许的资源队列最大数量，默认值为 128，值域范围是 3～1024。

```
[gpadmin@hdp3 ~]$ hawq config -s hawq_rm_nresqueue_limit
GUC     : hawq_rm_nresqueue_limit
Value   : 128
[gpadmin@hdp3 ~]$
```

也可执行以下 psql 命令检查运行时的参数设置：

```
db1=# show hawq_rm_nresqueue_limit;
 hawq_rm_nresqueue_limit
-------------------------
 128
(1 row)
```

手动编辑 hawq-site.xml 文件或使用 Ambari 界面配置该参数，新值在 HAWQ 重启后生效。下面的例子将最大资源队列数设置成 50。

```xml
<property>
   <name>hawq_rm_nresqueue_limit</name>
   <value>50</value>
</property>
```

该参数不能在服务器启动后动态改变：

```
db1=# set hawq_rm_nresqueue_limit=50;
ERROR:  attempted change of parameter "hawq_rm_nresqueue_limit" ignored
DETAIL:  This parameter cannot be changed after server start.
db1=#
```

2. 资源队列维护

只有管理员用户才能运行创建、修改、删除等维护资源队列的 DDL 语句。

（1）创建资源队列

使用 CREATE RESOURCE QUEUE 语句创建资源队列。创建资源队列时需要指定队列的名称、父队列名称、CPU 和内存限制等，还可限制队列中的活跃语句数量。下面的语句在 pg_root 下创建一个名为 myqueue 的资源队列，限制活跃查询数为 20，内存和 CPU 核数最多为集群总量的 10%：

```
db1=# create resource queue myqueue with (parent='pg_root',
active_statements=20,
  db1(# memory_limit_cluster=10%, core_limit_cluster=10%);
CREATE QUEUE
db1=#
```

一个父队列下所有直接子队列的资源限制总和不能超过 100%。例如，默认的 pg_default 队列的 memory_limit_cluster 和 core_limit_cluster 为 50%，刚刚建立的 myqueue 两者限制为 10%，两个队列的限制总和是 60%，此时无法再在 pg_root 下创建限制超过 40% 的资源队列。

```
db1=# create resource queue myqueue1 with (parent='pg_root',
active_statements=20,
  db1(# memory_limit_cluster=50%, core_limit_cluster=50%);
ERROR:  the value of core_limit_cluster and memory_limit_cluster exceeds parent
queue's limit, wrong value = 50%
db1=#
```

下面的语句在 pg_root 下创建一个名为 test_queue_1 的资源队列，内存和 CPU 核数最多为集群总量的 40%，并指定资源过度使用因子为 2：

```
db1=# create resource queue test_queue_1 with (parent='pg_root',
  db1(# memory_limit_cluster=30%, core_limit_cluster=30%,
resource_overcommit_factor=2);
CREATE QUEUE
db1=#
```

RESOURCE_OVERCOMMIT_FACTOR 是一个可选属性，定义有多少资源能够被过度使用，默认值为 2.0，最小值为 1.0。如果设置 RESOURCE_OVERCOMMIT_FACTOR 为 3.0，同时将 MEMORY_LIMIT_CLUSTER 设置为 30%，那么最大可能的资源分配为 90%（30% * 3）。

如果资源限制与 RESOURCE_OVERCOMMIT_FACTOR 相乘的结果值大于100%，采用100%。

（2）修改资源队列

使用 ALTER RESOURCE QUEUE 语句修改资源队列。该语句可以修改队列的资源限制和活跃语句数，但不能改变一个队列的父队列，创建资源队列时的约束同样适用于修改语句。

可以在队列正在被使用时进行修改，所有队列中排队的资源请求都调整为基于修改后的队列。在修改队列时，排队的资源请求可能遇到某些冲突，如可能发生资源死锁，或者修改后的资源限额不能满足排队的请求等。为防止冲突，默认情况下 HAWQ 会取消与新资源队列定义冲突的所有请求，该行为受服务器配置参数 hawq_rm_force_alterqueue_cancel_queued_request 控制，默认设置是 on。

```
[gpadmin@hdp3 ~]$ hawq config -s
hawq_rm_force_alterqueue_cancel_queued_request
   GUC     : hawq_rm_force_alterqueue_cancel_queued_request
   Value   : on
[gpadmin@hdp3 ~]$
```

该参数设置为 off 时，如果资源管理器发现新的队列定义与排队请求发生冲突，则取消 ALTER RESOURCE QUEUE 中指定的限制。

以下语句修改资源队列的内存和 CPU 核数限制：

```
db1=# alter resource queue test_queue_1 with
(memory_limit_cluster=40%,core_limit_cluster=40%);
ALTER QUEUE
db1=#
```

以下语句将资源队列的最大活跃语句数由默认的 20 改为 50：

```
db1=# alter resource queue test_queue_1 with (active_statements=50);
ALTER QUEUE
db1=#
```

（3）删除资源队列

使用 DROP RESOURCE QUEUE 语句删除一个资源队列。满足以下条件的资源队列才能被删除，并且不能删除系统资源队列 pg_root 和 pg_default：

- 没有查询正在使用该队列。
- 没有子队列。
- 没有被赋予角色。

从资源队列中删除角色（角色会被移到默认的 pg_default 资源队列中）：

```
db1=# alter role wxy resource queue none;
ALTER ROLE
db1=#
```

删除 myqueue 资源队列：

```
db1=# drop resource queue myqueue;
DROP QUEUE
db1=#
```

（4）检查现有资源队列

HAWQ 的系统表 pg_resqueue 保存资源队列数据。下面的查询语句显示 test_queue_1 的相关信息：

```
db1=# \x
Expanded display is on.
db1=# select rsqname,
db1-#        parentoid,
db1-#        activestats,
db1-#        memorylimit,
db1-#        corelimit,
db1-#        resovercommit,
db1-#        allocpolicy,
db1-#        vsegresourcequota,
db1-#        nvsegupperlimit,
db1-#        nvseglowerlimit,
db1-#        nvsegupperlimitperseg,
db1-#        nvseglowerlimitperseg
db1-#   from pg_resqueue
db1-#  where rsqname='test_queue_1';
-[ RECORD 1 ]---------+--------------
rsqname               | test_queue_1
parentoid             | 9800
activestats           | 50
memorylimit           | 40%
corelimit             | 40%
resovercommit         | 2
allocpolicy           | even
vsegresourcequota     | mem:256mb
nvsegupperlimit       | 0
nvseglowerlimit       | 0
nvsegupperlimitperseg | 0
nvseglowerlimitperseg | 0

db1=#
```

也可以从 pg_resqueue_status 系统视图检查资源队列的运行时状态：

```
db1=# select * from pg_resqueue_status;
```

```
     rsqname    | segmem  | segcore  | segsize  | segsizemax  |
inusemem   | inusecore   | rsqholders  | rsqwaiters  | paused
-----------+---------+----------+----------+-------------+----------+------------+-------------+-------------+--------
  pg_root      | 256     | 0.125000 | 72       | 72          | 0
           | 0.000000    | 0           | 0           | F
  pg_default   | 256     | 0.125000 | 36       | 72          | 0
           | 0.000000    | 0           | 0           | F
  test_queue_1 | 256     | 0.125000 | 28       | 57          | 0
           | 0.000000    | 0           | 0           | F
(3 rows)
```

表 7-3 描述了 pg_resqueue_status 视图中各字段的含义。

表 7-3　pg_resqueue_status 视图中字段的含义

字段名称	描述
rsqname	资源队列名称
segmem	每个虚拟段内存限额（MB）
segcore	每个虚拟段的 CPU 虚拟核数限额
segsize	资源队列能够为执行查询分配的虚拟段数
segsizemax	过度使用其他队列的资源时，资源队列可为执行查询分配的最大虚拟段数
inusemem	当前运行的语句使用的总内存（MB）
inusecore	当前运行的语句使用的总 CPU 虚拟核数
rsqholders	并发运行的语句数量
rsqwaiters	排队语句的数量
paused	指示在没有资源状态改变时，资源队列是否临时性暂停。'F'表示否，'T'表示是，'R'表示资源队列发生了资源碎片问题

下面的语句查询所有资源队列的当前活动 SQL：

```
  select usename, rsqname, locktype, objid, transaction, pid, mode, granted,
waiting
    from pg_stat_activity, pg_resqueue, pg_locks
    where pg_stat_activity.procpid=pg_locks.pid
      and pg_locks.objid=pg_resqueue.oid;
```

下面的语句查询所有资源队列的当前等待 SQL：

```
  select rolname, rsqname, pid, granted, current_query, datname
    from pg_roles, pg_resqueue, pg_locks, pg_stat_activity
    where pg_roles.rolresqueue=pg_locks.objid
      and pg_locks.objid=pg_resqueue.oid
```

```
and pg_stat_activity.procpid=pg_locks.pid;
```

3. 给角色赋予资源队列

资源队列对资源的管理是通过角色（用户）发挥作用的。角色被赋予某个资源队列，用该角色执行的查询受相应资源队列限制。只允许为角色赋予叶子队列。

创建或修改角色时赋予角色一个资源队列：

```
db1=# create role rmtest1 with login resource queue pg_default;
CREATE ROLE
db1=# alter role rmtest1 resource queue test_queue_1;
ALTER ROLE
db1=#
```

查看为角色分配的资源队列：

```
db1=# select rolname, rsqname from pg_roles, pg_resqueue
db1-#  where pg_roles.rolresqueue=pg_resqueue.oid;
 rolname  |   rsqname
----------+--------------
 gpadmin  | pg_default
 wxy      | pg_default
 rmtest1  | test_queue_1
(3 rows)
```

7.5 查询资源管理器状态

通过一些查询能够导出资源管理器的细节信息，如活跃资源上下文状态、当前资源队列状态、HAWQ Segment 状态等。

1. 连接状态跟踪

查询执行时从资源管理器请求分配资源，此时会有一个连接跟踪实例，记录查询资源使用的整个生命周期，从中可以找到所有的资源请求与分配情况。下面的查询获取保存连接跟踪状态信息的文件路径：

```
db1=# select * from dump_resource_manager_status(1);
                  dump_resource_manager_status
------------------------------------------------------------------
 Dump resource manager connection track status to
/tmp/resource_manager_conntrack_status
(1 row)
```

表7-4总结了该文件中的输出字段及其描述。

表7-4 连接状态跟踪文件中的字段

字段名称	描述
Number of free connection ids	空闲的连接跟踪实例数量。HAWQ 资源管理器支持最多 65536 个活动连接跟踪实例
Number of connection tracks having requests to handle	资源管理器已经接受但还没有处理的请求数量
Number of connection tracks having responses to send	资源管理器已经生成但还没有发送出去的响应数量
SOCK	请求的 socket 连接信息
CONN	请求的角色名称、目标队列和当前状态等信息： prog=1 表示连接建立 prog=2 表示连接由用户注册 prog=3 表示连接等待目标队列中的资源 prog=4 表示资源已经分配给连接 prog>5 表示失败或异常状态
ALLOC	会话相关信息，如请求的资源、会话级资源限制、语句级资源设置、按查询计划分片估计的工作量等
LOC	查询扫描 HDFS 的数据本地化信息
RESOURCE	已经分配的资源信息
MSG	最后收到的消息
COMMSTAT	当前 socket 通信缓冲区状态

2．资源队列状态

获取保存资源队列状态信息的文件路径：

```
db1=# select * from dump_resource_manager_status(2);
                 dump_resource_manager_status
-----------------------------------------------------------------------------------
 Dump resource manager resource queue status to /tmp/resource_manager_resqueue_status
(1 row)
```

表7-5总结了该文件中的输出字段及其描述。

表 7-5 资源队列状态文件中的字段

字段名称	描述
Maximum capacity of queue in global resource manager cluster	YARN 资源队列最大配额
Number of resource queues	HAWQ 资源队列总数量
QUEUE	资源队列的基本结构信息，以及是否忙于为查询调度资源
REQ	等待队列的计数和状态
SEGCAP	虚拟段资源限额和可派发的虚拟段数量
QUECAP	原始资源队列配额，以及一个队列可以使用的集群资源的实际百分比
QUEUSE	资源队列使用信息

3. HAWQ Segment 状态

获取保存 HAWQ Segment 状态信息的文件路径：

```
db1=# select * from dump_resource_manager_status(3);
                  dump_resource_manager_status
-----------------------------------------------------------------------------------
 Dump resource manager resource pool status to /tmp/resource_manager_respool_status
(1 row)
```

表 7-6 总结了该文件中的输出字段及其描述。

表 7-6 Segment 状态文件中的字段

字段名称	描述
HOST_ID	Segment 名称和内部 ID
HOST_INFO	Segment 配置的资源配额，GRMTotalMemoryMB 和 GRMTotalCore 显示 YARN 报告的限额，FTSTotalMemoryMB 和 FTSTotalCore 显示 HAWQ 中配置的限额
HOST_AVAILABILITY	Segment 对于 HAWQ 容错服务（fault tolerance service，FTS）和 YARN 是否可用
HOST_RESOURCE	当前分配的和可用的资源，以及估算的负载计数器
HOST_RESOURCE_CONTAINERSET	Segment 持有的资源容器

7.6 小结

HAWQ 支持独立与外部全局资源管理器两种资源管理模式。在独立模式下，HAWQ 使用集群节点资源时，假设它能使用所有 Segment 节点的资源，而不考虑其他共存的应用。如果使用 YARN 模式，需要在 YARN 中为 HAWQ 配置资源队列，还要在 HAWQ 中进行与 YARN 相关的配置。HAWQ 自动注册为 YARN 应用，使用 YARN 中配置的资源队列。目前 HAWQ 仅支持 YARN 的 Capacity 调度器。资源队列是 HAWQ 系统中并发管理的主要工具，是一种数据库对象。HAWQ 内部将资源队列组织为一棵 n 叉树。HAWQ 初始化后，有一个名为 pg_root 的根队列，根下有一个名为 pg_default 的默认队列，这两个队列不允许删除。资源队列对资源的管理是通过角色（用户）发挥作用的。可以通过检查相应的状态文件，跟踪连接状态、资源队列状态和 Segment 资源池状态。

第 8 章 数据管理

对数据仓库功能上的两个基本要求是外部数据访问和高效数据装载,HAWQ 在这两方面为用户提供了丰富的选项。本章着重介绍 HAWQ 的数据装载与外部表技术。除了基本的 INSERT 操作,使用 COPY 命令可以很方便地在 HAWQ 表与本地文件之间复制数据。对于大数据集的本地文件,gpfdist 提供了非常高效的数据装载方式。HAWQ 还支持基于 Web 的动态外部数据读取,以及对 HDFS 文件、Hive、HBase、JSON 等 Hadoop 上数据的访问。我们会用实例将这些技术一一展现。

8.1 基本数据操作

1. INSERT

在常用的增删改查数据库操作中,HAWQ 仅支持 INSERT 和 SELECT 两种,不支持 UPDATE 和 DELETE,这主要是因为 HDFS 是一个只能追加而不能更新数据的文件系统。开发人员应该对 SELECT 最熟悉不过,因为它是数据库中最常用的语句(在第 10 章"查询优化"时会进一步讨论)。INSERT 语句用于创建表行,该命令需要表名和表中每个列的值。在 HAWQ 中,该命令有四种用法,其中三种是 SQL 常规用法,另一种是对标准 SQL 的扩展。

(1) 指定列名与列值

可以按任何顺序指定列名,只需列值与列名一一对应。

```
insert into products (name, price, product_no) values ('Cheese', 9.99, 1);
```

(2) 仅指定列值

如果不指定列名,数据值列表的个数与顺序必须与表中定义的列保持一致。

```
insert into products values (1, 'Cheese', 9.99);
```

(3) 使用 SELECT 语句

通常数据值是常量,但是也可以使用标量表达式或查询语句。此时,利用 SELECT 语句选出的列表必须与插入表的列顺序一致、类型兼容。

```
insert into products select * from tmp_products where price < 100;
insert into products (name, price, product_no)
select * from tmp_products where price < 100;
```

如果列有默认值，可以在插入语句中省略该列名而使用默认值。

```
insert into products (name, product_no)
select name, product_no from tmp_products where price < 100;
```

（4）显式一次插入多行

这个 SQL 扩展与 MySQL 类似，一条 INSERT 语句中可以显式指定多条需要插入的记录。

```
db1=# create table t (a int);
CREATE TABLE
db1=# insert into t values (1),(2),(3);
INSERT 0 3
db1=# select * from t;
 a
---
 1
 2
 3
(3 rows)
```

如果需要快速插入大量数据，最好使用后面介绍的外部表或 COPY 命令，这些数据装载机制比 INSERT 更有效。

2. 整理系统目录表

（1）VACUUM

对于数据库中的对象，如表、视图、函数等，总是在不断地执行新增、删除、修改等操作，相应地会引起 HAWQ 系统目录表的增删改。因此对系统目录良好的空间管理非常重要，这能够给性能带来大幅提升。当删除一个数据库对象时，并不立即释放该条目在系统目录表中所占用的空间，而是在该条目上设置一个过期标志。只有当执行 VACUUM 命令时，才会物理删除那些已经标识为过期的数据条目并释放空间。应该定期执行 VACUUM 命令移除过期行，该命令也可收集表级的统计信息，如行数和页数等。

```
db1=# -- 整理 pg_class 系统表
db1=# vacuum pg_class;
VACUUM
db1=# -- 整理并分析 pg_class 系统表
db1=# vacuum analyze pg_class;
VACUUM
db1=# -- 整理并分析 pg_class 系统表的指定列
db1=# vacuum analyze pg_class
(relname,relam,relpages,reltuples,relkind,relnatts,relacl,reloptions);
```

VACUUM

（2）配置空余空间映射

过期行被保存在名为 free space map 的结构中。它必须足够大，能保存数据库中的所有过期行。VACUUM 命令不能回收超过 free space map 以外过期行占用的空间。HAWQ 中不推荐使用 VACUUM FULL，因为对于大表，该操作可能造成不可接受的执行时间。

```
db1=# vacuum full pg_class;
NOTICE:  'VACUUM FULL' is not safe for large tables and has been known to yield
unpredictable runtimes.
HINT:  Use 'VACUUM' instead.
VACUUM
```

free space map 的大小由以下服务器配置参数所控制：

- max_fsm_pages
- max_fsm_relations

max_fsm_pages（整数）设置 free space map 跟踪的最大页数。每个页槽位占用 6 字节的共享内存。至少要设置为 16 * max_fsm_relations。在初始安装时，系统依照可用内存的数量设置该参数的默认值。修改该参数后需要重启 HAWQ 使其生效。

```
[gpadmin@hdp3 ~]$ hawq config -s max_fsm_pages
GUC     : max_fsm_pages
Value   : 200000
```

max_fsm_relations（整数）设置 free space map 跟踪的最大表数。每个表槽位占用 7 字节左右的共享内存。默认值为 1000。设置该参数后需要重启 HAWQ 使其生效。

```
[gpadmin@hdp3 ~]$ hawq config -s max_fsm_relations
GUC     : max_fsm_relations
Value   : 1000
```

3. 其他操作

与 Oracle、MySQL 等常用数据库系统一样，HAWQ 支持 MVCC 并发访问控制和非锁定读，支持 ACCESS SHARE、ROW EXCLUSIVE、SHARE UPDATE EXCLUSIVE、SHARE、SHARE ROW EXCLUSIVE、ACCESS EXCLUSIVE 六种锁模式，以及读非提交、读提交、可重复读、串行化四种标准的事务隔离级别。因为数据仓库应用中的 ETL 操作通常为一个独立的后台程序，几乎没有并发调用，而前台的分析类应用大都是只读操作，所以这里不展开讨论并发控制与事务处理。

8.2 数据装载与卸载

HAWQ 既支持大数据量、多个文件的高性能并行数据装载、卸载操作，又支持小数据量、单个文件、非并发的数据导入导出。HAWQ 可读写多种外部数据源，包括本地文本文件、HDFS 或 Web 服务器。

在后面 8.4 节中会详细说明 PXF 外部表，这里先介绍使用 gpfdist 协议的外部表。PXF 外部表针对 HDFS 上的文件访问，而 gpfdist 用于对本地文件的并行访问。gpfdist 是一个 HAWQ 的并行文件分布程序。它是一个操作外部表的 HTTP 服务器，使 HAWQ 的 Segment 可以从多个文件系统的外部表并行装载数据。可以在多个不同的主机上运行 gpfdist 实例，并能够并行使用它们。

外部 Web 表提供了对动态数据的访问功能。它支持使用 HTTP 协议从 URL 访问数据，或者通过运行在 Segment 上的脚本输出数据。

hawq load 应用使用一个 YAML 格式的控制文件，自动完成数据装载任务。

HAWQ 中的 COPY SQL 命令可在 Master 主机上的文本文件与 HAWQ 数据库表之间转移数据。

所选择的数据装载方法依赖于数据源的特性，如位置、数据量、格式、需要的转换等。最简单的情况下，一条 COPY 命令就可将 HAWQ 主实例上的文本文件装载到表中。对于少量数据，这种方式不需要更多步骤，并提供了良好的性能。COPY 命令在 HAWQ Master 主机上的单个文件与数据库表之间复制数据。这种方式复制的数据量受限于文件所在系统所允许的单一文件最大字节数。对于大数据集，更为有效的数据装载方式是利用 HAWQ 的 MPP 架构，用多个 HAWQ Segments 并行装载数据。该方式允许同时从多个文件系统装载数据，实现很高的数据传输速率。用 gpfdist 创建的外部表会使用所有 HAWQ Segment 装载或卸载数据，并且完全并行操作。

无论使用哪种方法，装载完数据都应运行 ANALYZE。如果装载了大量表数据，运行 ANALYZE 或 VACUUM ANALYZE（只对系统目录表）为查询优化器更新表的统计信息，使当前统计信息保证优化器做出最好的查询计划，避免由于数据增长或缺失统计信息导致性能问题。

8.2.1 gpfdist 协议及其外部表

1. gpfdist

gpfdist 是 HAWQ 提供的一种文件服务器，利用 HAWQ 系统中的所有 Segment 读写外部表。它提供了良好的性能，并且非常容易运行。

（1）并行性

gp_external_max_segs 服务器配置参数控制可被单一 gpfdist 实例同时使用的虚拟段的数量，默认值为 64。在 Master 实例的 hawq-site.xml 文件中设置此参数。

```
[gpadmin@hdp3 ~]$ hawq config -s gp_external_max_segs
GUC       : gp_external_max_segs
Value     : 64
```

用户可能需要设置虚拟段的数量，例如一些用于处理外部表数据文件、一些执行其他的数据库操作。hawq_rm_nvseg_perquery_perseg_limit 和 hawq_rm_nvseg_perquery_limit 参数控制并行虚拟段的数量，它们限制集群中一个 gpfdist 外部表上执行查询时使用的最大虚拟段数。

（2）启动与停止

可以选择在 HAWQ Master 以外的其他机器上运行 gpfdist，例如一个专门用于 ETL 处理的主机。使用 gpfdist 命令启动 gpfdist。该命令位于 HAWQ Master 主机和每个 Segment 主机的 $GPHOME/bin 目录中。可以在当前目录位置或者指定任意目录启动 gpfdist，默认的端口是 8080。下面是一些启动 gpfdist 的例子。

处理当前目录中的文件，使用默认的 8080 端口。

```
[gpadmin@hdp4 ~]$ gpfdist &
```

/home/gpadmin/load_data/ 是要处理的文件目录，8081 是 HTTP 端口号，/home/gpadmin/log 是消息与错误日志文件，进程在后台运行。

```
[gpadmin@hdp4 ~]$ gpfdist -d /home/gpadmin/load_data/ -p 8081 -l /home/gpadmin/log &
```

在同一个 ETL 主机上运行多个 gpfdist 实例，每个实例使用不同的目录和端口。

```
[gpadmin@hdp4 ~]$ gpfdist -d /home/gpadmin/load_data1/ -p 8081 -l /home/gpadmin/log1 &
[gpadmin@hdp4 ~]$ gpfdist -d /home/gpadmin/load_data2/ -p 8082 -l /home/gpadmin/log2 &
```

HAWQ 没有提供停止 gpfdist 的特殊命令，直接使用操作系统的 kill 命令停止 gpfdist 进程。

```
[gpadmin@hdp4 ~]$ ps -ef |grep gpfdist |grep -v grep | awk '{print $2}'|xargs kill -9
```

（3）排错

如果 gpfdist 启动时报出类似没有 libapr-1.so.0 或 libyaml-0.so.2 文件的错误，则需要安装相应的包。

```
yum install apr
yum install libyaml
```

虚拟段在运行时访问 gpfdist，因此需要保证 HAWQ Segment 主机能访问 gpfdist 实例。gpfdist 实际上是一个 Web 服务器，可以在 HAWQ 的每个主机（Master 和 Segment）上执行下面的命令测试连通性：

```
[gpadmin@hdp3 ~]$ wget http://gpfdist_hostname:port/filename
```

CREATE EXTERNAL TABLE 定义必须为 gpfdist 提供正确的主机名、端口号和文件名。

2. gpfdist 外部表

（1）gpfdist 协议

在外部数据文件所在的主机上运行 gpfdist 命令，外部表定义中使用 gpfdist://协议引用一个运行的 gpfdist 实例。gpfdist 自动解压缩 gzip（.gz）和 bzip2（.bz2）文件。可以使用通配符（*）或其他 C 语言风格的模式匹配多个需要读取的文件。指定的文件应该位于启动 gpfdist 实例时指定的目录下。

所有虚拟段并行访问外部文件，虚拟段的数量受 gp_external_max_segments 参数、gpfdist 的位置列表长度，以及 hawq_rm_nvseg_perquery_limit 和 hawq_rm_nvseg_perquery_perseg_limit 参数的影响。在 CREATE EXTERNAL TABLE 语句中使用多个 gpfdist 数据源可扩展外部表扫描性能。

（2）创建 gpfdist 外部表

为了创建一个 gpfdist 外部表，需要指定输入文件的格式和外部数据源的位置。使用以下协议之一访问外部表数据源。一条 CREATE EXTERNAL TABLE 语句中使用的协议必须唯一，不能混用多个协议。

- gpfdist:// —— 指定主机上的一个目录，用于存储外部数据文件。HAWQ 的所有 Segment 可并行访问该目录下的文件。
- gpfdists:// —— gpfdist 的安全版本。

使用 gpfdist 外部表的步骤如下：

a. 启动 gpfdist 文件服务器。
b. 定义外部表。
c. 将数据文件放置于外部表定义中指定的位置。
d. 使用 SQL 命令查询外部表。

与 PXF 外部表一样，HAWQ 提供可读与可写两种 gpfdist 外部表，但一个外部表不能既可读又可写。

（3）gpfdist 外部表示例

例 1：单 gpfdist 实例外部表

启动 gpfdist。

```
[gpadmin@hdp4 ~]$ gpfdist -p 8081 -d /home/gpadmin/staging -l /home/gpadmin/log &
```

使用 gpfdist 协议创建只读外部表 example1，文件以管道符（|）作为列分隔符。

```
db1=# create external table example1
db1=#      ( name text, date date, amount float4, category text, desc1 text )
db1=#    location ('gpfdist://hdp4:8081/*')
db1=#    format 'text' (delimiter '|');
```

```
CREATE EXTERNAL TABLE
```

准备文本文件数据。

```
[gpadmin@hdp4 unload_data1]$ cd /home/gpadmin/staging
[gpadmin@hdp4 staging]$ more a.txt
aaa|2017-01-01|100.1|aaa|aaa
bbb|2017-01-02|100.2|bbb|bbb
[gpadmin@hdp4 staging]$ more b.txt
aaa|2017-03-01|200.1|aaa|aaa
bbb|2017-03-02|200.2|bbb|bbb
```

查询外部表。

```
db1=# select * from example1;
 name |    date    | amount | category | desc1
------+------------+--------+----------+-------
 aaa  | 2017-01-01 |  100.1 | aaa      | aaa
 bbb  | 2017-01-02 |  100.2 | bbb      | bbb
 aaa  | 2017-03-01 |  200.1 | aaa      | aaa
 bbb  | 2017-03-02 |  200.2 | bbb      | bbb
(4 rows)
```

例 2：多 gpfdist 实例外部表

在 hdp3 和 hdp4 上分别启动一个 gpfdist 实例。

```
[gpadmin@hdp3 ~]$ gpfdist -p 8081 -d /home/gpadmin/staging -l /home/gpadmin/log &
[gpadmin@hdp4 ~]$ gpfdist -p 8081 -d /home/gpadmin/staging -l /home/gpadmin/log &
```

使用 gpfdist 协议创建只读外部表 example2，文件以管道符(|)作为列分隔符，''表示 NULL。

```
db1=# create external table example2
db1-#     ( name text, date date, amount float4, category text, desc1 text )
db1-#     location ('gpfdist://hdp3:8081/*.txt', 'gpfdist://hdp4:8081/*.txt')
db1-#     format 'text' ( delimiter '|' null ' ') ;
CREATE EXTERNAL TABLE
```

查询外部表，因为 gpfdist://hdp3:8081/*.txt 不存在而报错。

```
db1=# select * from example2;
ERROR:  http response code 404 from gpfdist (gpfdist://hdp3:8081/*.txt):
HTTP/1.0 404 file not found (url.c:306)  (seg0 hdp4:40000 pid=53784)
(dispatcher.c:1801)
db1=#
```

将外部文件复制到 hdp3 的相关目录下。

```
[gpadmin@hdp4 staging]$ scp *.txt hdp3://home/gpadmin/staging/
```

再次查询外部表,可以正确读取全部外部文件数据。

```
db1=# select * from example2;
 name |    date    | amount | category | desc1
------+------------+--------+----------+-------
 aaa  | 2017-01-01 |  100.1 | aaa      | aaa
 bbb  | 2017-01-02 |  100.2 | bbb      | bbb
 aaa  | 2017-03-01 |  200.1 | aaa      | aaa
 bbb  | 2017-03-02 |  200.2 | bbb      | bbb
 aaa  | 2017-01-01 |  100.1 | aaa      | aaa
 bbb  | 2017-01-02 |  100.2 | bbb      | bbb
 aaa  | 2017-03-01 |  200.1 | aaa      | aaa
 bbb  | 2017-03-02 |  200.2 | bbb      | bbb
(8 rows)
```

例 3:带有错误日志的单 gpfdist 实例外部表

默认在访问外部表时只要遇到一行格式错误的数据,就立即返回错误,并导致查询失败。下面的语句设置了 SEGMENT REJECT LIMIT 的值,只有当一个 Segment 上的错误数大于等于 5 时,整个外部表操作才会失败,并且不处理任何行。而当错误数小于 5 时,会将被拒绝的行写入一个错误表 errs,其他数据行还可以正常返回。

```
db1=# create external table example3
db1-#         ( name text, date date, amount float4, category text, desc1 text )
db1-#      location ('gpfdist://hdp3:8081/*.txt', 'gpfdist://hdp4:8081/*.txt')
db1-#      format 'text' ( delimiter '|' null ' ')
db1-#      log errors into errs segment reject limit 5;
NOTICE:  Error table "errs" does not exist. Auto generating an error table with the same name
CREATE EXTERNAL TABLE
db1=# \d errs
       Append-Only Table "public.errs"
  Column   |           Type           | Modifiers
-----------+--------------------------+-----------
 cmdtime   | timestamp with time zone |
 relname   | text                     |
 filename  | text                     |
 linenum   | integer                  |
 bytenum   | integer                  |
 errmsg    | text                     |
 rawdata   | text                     |
 rawbytes  | bytea                    |
Compression Type: None
Compression Level: 0
```

```
Block Size: 32768
Checksum: f
Distributed randomly

db1=#
```

准备一条格式错误的数据。

```
[gpadmin@hdp4 staging]$ more a.txt
aaa|2017-01-01|100.1|aaa|aaa
bbb|2017-01-02|100.2|bbb|bbb
bbb,2017-01-02,100.2,bbb,bbb
```

查询外部表，返回 8 条数据，错误数据进入了 errs 表。

```
db1=# select * from example3;
NOTICE:  Found 1 data formatting errors (1 or more input rows). Rejected related input data.
 name |    date    | amount | category | desc1
------+------------+--------+----------+-------
 aaa  | 2017-01-01 |  100.1 | aaa      | aaa
 bbb  | 2017-01-02 |  100.2 | bbb      | bbb
 aaa  | 2017-03-01 |  200.1 | aaa      | aaa
 bbb  | 2017-03-02 |  200.2 | bbb      | bbb
 aaa  | 2017-01-01 |  100.1 | aaa      | aaa
 bbb  | 2017-01-02 |  100.2 | bbb      | bbb
 aaa  | 2017-03-01 |  200.1 | aaa      | aaa
 bbb  | 2017-03-02 |  200.2 | bbb      | bbb
(8 rows)

db1=# \x
Expanded display is on.
db1=# select * from errs;
-[ RECORD 1 ]----------------------------------------------------
cmdtime  | 2017-04-05 15:23:19.579421+08
relname  | example3
filename | gpfdist://hdp4:8081/*.txt [/home/gpadmin/staging/a.txt]
linenum  | 3
bytenum  |
errmsg   | missing data for column "date"
rawdata  | bbb,2017-01-02,100.2,bbb,bbb
rawbytes |

db1=#
```

准备 5 条错误数据。

```
[gpadmin@hdp4 staging]$ more a.txt
aaa|2017-01-01|100.1|aaa|aaa
bbb|2017-01-02|100.2|bbb|bbb
b1,2017-01-02,100.2,bbb,bbb
b2,2017-01-02,100.2,bbb,bbb
b3,2017-01-02,100.2,bbb,bbb
b4,2017-01-02,100.2,bbb,bbb
b5,2017-01-02,100.2,bbb,bbb
```

再次查询外部表,因为达到了错误上限,整条语句失败,没有数据被返回。

```
db1=# select * from example3;
ERROR:  Segment reject limit reached. Aborting operation. Last error was: missing data for column "date"  (seg16 hdp3:40000 pid=350431)
DETAIL:  External table example3, line 7 of gpfdist://hdp4:8081/*.txt: "b5,2017-01-02,100.2,bbb,bbb"
db1=#
```

例 4:gpfdist 可写外部表

建立可写外部表,并插入一条数据。

```
db1=# create writable external table example4 (name text, date date, amount float4, category text, desc1 text)
db1-#    location ('gpfdist://hdp4:8081/sales.out', 'gpfdist://hdp3:8081/sales.out')
db1-#    format 'text' ( delimiter '|' null ' ')
db1-#    distributed by (name);
CREATE EXTERNAL TABLE
db1=# insert into example4 values ('aaa','2017-01-01',100.1,'aaa','aaa');
INSERT 0 1
```

结果只在 hdp4 上建立了文件/home/gpadmin/staging/sales.out,而 hdp3 并没有建立输出文件。

```
[gpadmin@hdp4 staging]$ more sales.out
aaa|2017-01-01|100.1|aaa|aaa
```

再次建立可写外部表,将 gpfdist 位置调换,把 hdp3 放前面,并插入一条数据。

```
db1=# drop external table example4;
DROP EXTERNAL TABLE
db1=# create writable external table example4 (name text, date date, amount float4, category text, desc1 text)
db1-#    location ('gpfdist://hdp3:8081/sales.out', 'gpfdist://hdp4:8081/sales.out')
db1-#    format 'text' ( delimiter '|' null ' ')
db1-#    distributed by (name);
CREATE EXTERNAL TABLE
```

```
db1=# insert into example4 values ('aaa','2017-01-01',100.1,'aaa','aaa');
INSERT 0 1
```

这次只在 hdp3 上建立了文件/home/gpadmin/staging/sales.out。

```
[gpadmin@hdp3 staging]$ more sales.out
aaa|2017-01-01|100.1|aaa|aaa
```

在 LOCATION 子句中指定同一主机上的多个 gpfdist 实例，结果也是一样的。可见，在可写外部表上执行 INSERT 操作时，只在第一个 gpfdist 实例的位置上生成本地文件数据。

8.2.2 基于 Web 的外部表

外部表可以是基于文件的或基于 Web 的。基于文件的外部表访问静态平面文件。在查询运行时数据是静态的，数据可重复读。基于 Web 的外部表通过 Web 服务器的 http 协议或通过执行操作系统命令或脚本，访问动态数据源。数据不可重复读，因为在查询运行时数据可能改变。

CREATE EXTERNAL WEB TABLE 语句创建一个 Web 外部表。Web 外部表允许 HAWQ 将动态数据源视作一个常规数据库表。因为 Web 表数据可能在查询运行时改变，所以数据是不可重复读的。可以定义基于命令或基于 URL 的 Web 外部表，但不能在一条建表命令中混用两种定义。

1. 基于命令的 Web 外部表

用一个 shell 命令或脚本的输出定义基于命令的 Web 表数据。在 CREATE EXTERNAL WEB TABLE 语句的 EXECUTE 子句指定需要执行的命令。外部表中的数据是命令运行时的数据。EXECUTE 子句在特定 Master 或虚拟段上运行 shell 命令或脚本。脚本必须是 gpadmin 用户可执行的，并且位于所有 Master 和 Segment 主机的相同位置上，虚拟段并行运行命令。

外部表定义中指定的命令从数据库执行，数据库不能从.bashrc 或.profile 获取环境变量，因此需要在 EXECUTE 子句中设置环境变量。下面的外部表运行一个 HAWQ Master 主机上的命令：

```
create external web table output (output text)
execute 'PATH=/home/gpadmin/programs; export PATH; myprogram.sh'
    on master
format 'text';
```

下面的命令定义一个 Web 表，在 5 个虚拟段上运行一个名为 get_log_data.sh 脚本文件。

```
create external web table log_output (linenum int, message text)
execute '/home/gpadmin/get_log_data.sh' ON 5
format 'text' (delimiter '|');
```

资源管理器在运行时选取虚拟段。

2. 基于 URL 的 Web 外部表

基于 URL 的 Web 表使用 HTTP 协议从 Web 服务器访问数据，Web 表数据是动态的。在

LOCATION 子句中使用 http://指定文件在 Web 服务器上的位置。Web 数据文件必须在所有 Segment 主机能够访问的 Web 服务器上。URL 的数量对应访问该 Web 表时并行的最少虚拟段数量。下面的例子定义了一个从多个 URL 获取数据的 Web 表。

```
create external web table ext_expenses (
    name text, date date, amount float4, category text, description text)
location ('http://hdp1/sales/file.csv',
         'http://hdp1/exec/file.csv',
         'http://hdp1/finance/file.csv',
         'http://hdp1/ops/file.csv',
         'http://hdp1/marketing/file.csv',
         'http://hdp1/eng/file.csv'
)
format 'csv';
```

3. 基于 Web 的外部表示例

例 5：执行脚本的可读 Web 外部表

建立外部表。

```
db1=# create external web table example5 (linenum int, message text)
db1-# execute '/home/gpadmin/get_log_data.sh' on 5
db1-# format 'text' (delimiter '|');
CREATE EXTERNAL TABLE
```

HAWQ 集群中每台主机的相同位置上都必须有同一个可执行的脚本，否则查询会报错，如 hdp1 上没有/home/gpadmin/get_log_data.sh 文件：

```
db1=# select * from example5;
ERROR:  external table example5 command ended with error. sh: /home/gpadmin/get_log_data.sh: No such file or directory  (seg0 hdp1:40000 pid=360600)
  DETAIL:  Command: execute:/home/gpadmin/get_log_data.sh
```

对该外部表的查询会返回每个虚拟段输出的并集，如 get_log_data.sh 脚本内容如下：

```
#!/bin/bash
echo "1|aaa"
echo "2|bbb"
```

则该表将返回 10 条（每个虚拟段两条，5 个虚拟段）数据：

```
db1=# select * from example5;
 linenum | message
---------+---------
       1 | aaa
       2 | bbb
```

```
         1 | aaa
         2 | bbb
         1 | aaa
         2 | bbb
         1 | aaa
         2 | bbb
         1 | aaa
         2 | bbb
         1 | aaa
         2 | bbb
(10 rows)
```

执行查询时，资源管理器最少分配 5 个虚拟段。如果建表时指定的虚拟段数超过了允许的最大值，表仍然可以建立，但查询时会报错。

```
db1=# drop external web table example5;
DROP EXTERNAL TABLE
db1=# create external web table example5 (linenum int, message text)
db1-# execute '/home/gpadmin/get_log_data.sh' on 100
db1-# format 'text' (delimiter '|');
CREATE EXTERNAL TABLE
db1=# select * from example5;
ERROR:  failed to acquire resource from resource manager, minimum expected number of virtual segment 100 is more than maximum possible number 64 in queue pg_default (pquery.c:804)
```

例 6：执行脚本的可写 Web 外部表

创建外部表。

```
db1=# create writable external web table example6
db1-#       (name text, date date, amount float4, category text, desc1 text)
db1-#       execute 'PATH=/home/gpadmin/programs; export PATH; myprogram1.sh' on 6
db1-#       format 'text' (delimiter '|')
db1-#       distributed randomly;
CREATE EXTERNAL TABLE
```

myprogram1.sh 的内容如下：

```
#!/bin/bash
while read line
do
    echo "File:${line}" >> /home/gpadmin/programs/a.txt
done
```

向外部表中插入数据。

```
db1=# insert into example6 values ('aaa','2017-01-01',100.1,'aaa','aaa');
INSERT 0 1
```

```
db1=# insert into example6 values ('bbb','2017-02-01',200.1,'bbb','');
INSERT 0 1
```

插入的数据通过管道输出给 myprogram1.sh 并执行,可以看到插入的数据被写入了 a.txt 文件。与可读表不同,该文件只在一个 HAWQ 主机上生成,并且每次插入数据只生成一行。

```
[gpadmin@hdp4 programs]$ more /home/gpadmin/programs/a.txt
File:aaa|2017-01-01|100.1|aaa|aaa
File:bbb|2017-02-01|200.1|bbb|
```

8.2.3 使用外部表装载数据

使用 INSERT INTO target_table SELECT ... FROM source_external_table 命令向 HAWQ 表装载数据:

```
create table expenses_travel (like ext_expenses);
insert into expenses_travel
select * from ext_expenses where category='travel';
```

也可以在创建一个新表的同时装载数据:

```
create table expenses as select * from ext_expenses;
```

8.2.4 外部表错误处理

可读外部表通常被用于选择数据装载到普通的 HAWQ 数据库表中。使用 CREATE TABLE AS SELECT 或 INSERT INTO 命令查询外部表数据。默认,如果数据包含错误,则整条命令失败,没有数据装载到目标数据库表中。SEGMENT REJECT LIMIT 子句允许隔离外部表中格式错误的数据,并继续装载格式正确的行。使用 SEGMENT REJECT LIMIT 设置一个错误阈值,指定拒绝的数据行数(默认)或一个占总行数的百分比(1~100)。

如果错误行数达到了 SEGMENT REJECT LIMIT 的值,整个外部表操作失败,没有数据行被处理。注意,限制的错误行数是相对于一个虚拟段的,而不是整个操作的。如果错误行数没有达到 SEGMENT REJECT LIMIT 值,操作处理所有正确的行,丢弃错误行,或者可选地将格式错误的行写入日志表。LOG ERRORS 子句允许保存错误行以备后续检查。

设置 SEGMENT REJECT LIMIT 会使 HAWQ 以单行错误隔离模式扫描外部数据。当外部数据行出现多余属性、缺少属性、数据类型错误、无效的客户端编码序列等格式错误时,单行错误隔离模式将错误行丢弃或写入日志表。HAWQ 不检查约束错误,但可以在查询外部表时过滤约束错误。例如,消除重复键值错误:

```
insert into table_with_pkeys select distinct * from external_table;
```

1. 使用单行错误隔离定义外部表

下面的例子在 HAWQ 表中记录错误记录,并设置错误行阈值为 10。

```
db1=# create external table ext_expenses ( name text, date date, amount float4,
```

```
category text, desc1 text )
    db1-#    location ('gpfdist://hdp3:8081/*', 'gpfdist://hdp4:8081/*')
    db1-#    format 'text' (delimiter '|')
    db1-#    log errors into errs segment reject limit 10 rows;
CREATE EXTERNAL TABLE
```

2. 标识无效的 CSV 文件数据

如果一个 CSV 文件包含无效格式，错误日志表的 rawdata 字段可能包含多行。例如，某字段少了一个闭合的引号，后面所有的换行符都被认为是数据中内嵌的换行符。当这种情况发生时，HAWQ 在一行数据达到 64KB 时停止解析，并将此 64KB 数据作为单行写入错误日志表，然后重置引号标记，继续读取数据。如果这种情况在处理装载时发生三次，载入文件被认为是无效的，整个装载失败，错误信息为"rejected N or more rows"。

3. 表间迁移数据

可以使用 CREATE TABLE AS 或 INSERT...SELECT 语句将外部表的数据装载到其他非外部表中，数据将根据外部表的定义并行装载。如果一个外部表数据源有错误，依赖于使用的错误隔离模式，有以下两种处理方式：

- 表没有设置错误隔离模式：读取该表的任何操作都会失败。没有设置错误隔离模式的外部表上的操作将整体成功或失败。
- 表设置了错误隔离模式：除了发生错误的行，其他数据将被装载（依赖于 REJECT_LIMIT 的配置）。

8.2.5 使用 hawq load 装载数据

HAWQ 的 hawq load 应用程序使用可读外部表和 HAWQ 并行文件系统（gpfdist 或 gpfdists）装载数据。它并行处理基于文件创建的外部表，允许用户在单一配置文件中配置数据格式、外部表定义以及 gpfdist 或 gpfdists 的设置。

1. 确认建立了运行 hawq load 的环境

hawq load 需要依赖某些 HAWQ 安装中的文件，如 gpfdist 和 Python，还要能通过网络访问所有 HAWQ Segment 主机。

2. 创建控制文件

hawq load 的控制文件是一个 YAML（Yet Another Markup Language）格式的文件，在其中指定 HAWQ 连接信息、gpfdist 配置信息、外部表选项、数据格式等。下面是一个名为 my_load.yml 的控制文件内容：

```
---
VERSION: 1.0.0.1
DATABASE: db1
USER: gpadmin
```

```
    HOST: hdp3
    PORT: 5432
    GPLOAD:
       INPUT:
        - SOURCE:
            LOCAL_HOSTNAME:
              - hdp4
            PORT: 8081
            FILE:
              - /home/gpadmin/staging/*.txt
        - COLUMNS:
              - name: text
              - date: date
              - amount: float4
              - category: text
              - desc1: text
        - FORMAT: text
        - DELIMITER: '|'
        - ERROR_LIMIT: 25
        - ERROR_TABLE: errlog
       OUTPUT:
        - TABLE: t1
        - MODE: INSERT
       SQL:
        - BEFORE: "INSERT INTO audit VALUES('start', current_timestamp)"
        - AFTER: "INSERT INTO audit VALUES('end', current_timestamp)"
```

hawq load 控制文件使用 YAML 1.1 文档格式，为了定义 HAWQ 数据装载的各种步骤，它定义了自己的 schema。控制文件必须是一个有效的 YAML 文档。hawq load 程序按顺序处理控制文件文档，并使用空格识别文档中各段之间的层次关系，因此空格的使用非常重要。不要使用 TAB 符代替空格，YAML 文档中不要出现 TAB 符。

LOCAL_HOSTNAME 指定运行 hawq load 的本地主机名或 IP 地址。如果机器配置了多块网卡，可以为每块网卡指定一个主机名，允许同时使用多块网卡传输数据。比如 hdp4 上配置了两块网卡，可以如下配置 LOCAL_HOSTNAME：

```
LOCAL_HOSTNAME:
 - hdp4-1
 - hdp4-2
```

3. hawq load 示例

准备本地文件数据。

```
[gpadmin@hdp4 staging]$ more a.txt
aaa|2017-01-01|100.1|aaa|aaa
```

```
bbb|2017-01-02|100.2|bbb|bbb
[gpadmin@hdp4 staging]$ more b.txt
aaa|2017-03-01|200.1|aaa|aaa
bbb|2017-03-02|200.2|bbb|bbb
```

建立目标表和 audit 表。

```
db1=# create table t1 ( name text, date date, amount float4, category text, desc1 text );
CREATE TABLE
db1=# create table audit(flag varchar(10),st timestamp);
CREATE TABLE
```

执行 hawq load。

```
[gpadmin@hdp4 ~]$ hawq load -f my_load.yml
2017-04-05 16:41:44|INFO|gpload session started 2017-04-05 16:41:44
2017-04-05 16:41:44|INFO|setting schema 'public' for table 't1'
2017-04-05 16:41:44|INFO|started gpfdist -p 8081 -P 8082 -f "/home/gpadmin/staging/*.txt" -t 30
2017-04-05 16:41:49|INFO|running time: 5.63 seconds
2017-04-05 16:41:49|INFO|rows Inserted          = 4
2017-04-05 16:41:49|INFO|rows Updated           = 0
2017-04-05 16:41:49|INFO|data formatting errors = 0
2017-04-05 16:41:49|INFO|gpload succeeded
[gpadmin@hdp4 ~]$
```

查询目标表和 audit 表。

```
db1=# select * from t1;
 name |    date    | amount | category | desc1
------+------------+--------+----------+-------
 aaa  | 2017-01-01 |  100.1 | aaa      | aaa
 bbb  | 2017-01-02 |  100.2 | bbb      | bbb
 aaa  | 2017-03-01 |  200.1 | aaa      | aaa
 bbb  | 2017-03-02 |  200.2 | bbb      | bbb
(4 rows)

db1=# select * from audit;
 flag  |            st
-------+----------------------------
 start | 2017-04-05 16:41:44.736296
 end   | 2017-04-05 16:41:49.60153
(2 rows)
```

8.2.6 使用 COPY 复制数据

COPY 是 HAWQ 的 SQL 命令，它在标准输入和 HAWQ 表之间复制数据。COPY FROM 命令将本地文件追加到数据表中，而 COPY TO 命令将数据表中的数据覆盖写入本地文件。COPY 命令是非并行的，数据在 HAWQ Master 实例上以单进程处理，因此只推荐对非常小的数据文件使用 COPY 命令。本地文件必须在 Master 主机上，默认的文件格式是逗号分隔的 CSV 文本文件。HAWQ 使用客户端与 Master 服务器之间的连接，从 STDIN 或 STDOUT 复制数据。

```
[gpadmin@hdp4 ~]$ psql -h hdp3 -d db1
psql (8.2.15)
Type "help" for help.

db1=# create table t2 (like t1);
NOTICE:  Table doesn't have 'distributed by' clause, defaulting to distribution columns from LIKE table
CREATE TABLE
db1=# copy t2 from '/home/gpadmin/staging/a.txt' with delimiter '|';
COPY 2
db1=# select * from t2;
 name |    date    | amount | category | desc1
------+------------+--------+----------+-------
 aaa  | 2017-01-01 |  100.1 | aaa      | aaa
 bbb  | 2017-01-02 |  100.2 | bbb      | bbb
(2 rows)
```

将表数据卸载到 Master 的本地文件中，如果文件不存在则建立文件，否则会用卸载数据覆盖文件原来的内容。

```
db1=# copy (select * from t2) to '/home/gpadmin/staging/c.txt' with delimiter '|';
COPY 2

[gpadmin@hdp3 staging]$ more /home/gpadmin/staging/c.txt
bbb|2017-01-02|100.2|bbb|bbb
aaa|2017-01-01|100.1|aaa|aaa
```

默认，COPY 在遇到第一个错误时就会停止运行。如果数据含有错误，操作失败，没有数据被装载。如果以单行错误隔离模式运行 COPY，HAWQ 跳过含有错误格式的行，装载具有正确格式的行。如果数据违反了 NOT NULL 或 CHECK 等约束条件，操作仍然是 'all-or-nothing' 输入模式，整个操作失败，没有数据被装载。

```
[gpadmin@hdp3 staging]$ more a.txt
aaa|2017-01-01|100.1|aaa|aaa
bbb|2017-01-02|100.2|bbb|bbb
```

向表复制本地文件数据。

```
db1=# create table t3 ( name text not null, date date, amount float4, category text, desc1 text );
CREATE TABLE
db1=# copy t3 from '/home/gpadmin/staging/a.txt'
db1-#     with delimiter '|' log errors into errtable
db1-#     segment reject limit 5 rows;
...
COPY 2
db1=# select * from t3;
 name |    date    | amount | category | desc1
------+------------+--------+----------+-------
 bbb  | 2017-01-02 |  100.2 | bbb      | bbb
 aaa  | 2017-01-01 |  100.1 | aaa      | aaa
(2 rows)
```

修改文件,制造一行格式错误的数据。

```
[gpadmin@hdp3 staging]$ more a.txt
aaa,2017-01-01,100.1,aaa,aaa
bbb|2017-01-02|100.2|bbb|bbb
```

再次复制数据。与卸载不同,装载会向表中追加数据。

```
db1=# copy t3 from '/home/gpadmin/staging/a.txt'
db1-#     with delimiter '|' log errors into errtable
db1-#     segment reject limit 5 rows;
...
COPY 1
db1=# select * from t3;
 name |    date    | amount | category | desc1
------+------------+--------+----------+-------
 bbb  | 2017-01-02 |  100.2 | bbb      | bbb
 bbb  | 2017-01-02 |  100.2 | bbb      | bbb
 aaa  | 2017-01-01 |  100.1 | aaa      | aaa
(3 rows)

db1=# \x
Expanded display is on.
db1=# select * from errtable;
-[ RECORD 1 ]----------------------------
cmdtime    | 2017-04-05 16:56:02.402161+08
relname    | t3
filename   | /home/gpadmin/staging/a.txt
linenum    | 1
```

```
     bytenum    |
     errmsg     | missing data for column "date"
     rawdata    | aaa,2017-01-01,100.1,aaa,aaa
     rawbytes   |

db1=#
```

再次修改文件，将 name 字段对应的数据置空，因为该字段定义为 NOT NULL，所以违反约束，没有数据被复制。

```
[gpadmin@hdp3 staging]$ more a.txt
|2017-01-01|100.1|aaa|aaa
bbb|2017-01-02|100.2|bbb|bbb

db1=# truncate table t3;
TRUNCATE TABLE
db1=# copy t3 from '/home/gpadmin/staging/a.txt'
   with delimiter '|' null as '' log errors into errtable
   segment reject limit 5 rows;
ERROR:  null value in column "name" violates not-null constraint  (seg5 hdp1:40000 pid=370883)
CONTEXT:  COPY t3, line 1: "|2017-01-01|100.1|aaa|aaa"
db1=# select * from t3;
 name | date | amount | category | desc1
------+------+--------+----------+-------
(0 rows)
```

8.2.7 卸载数据

一个可写外部表允许用户从其他数据库表选择数据行并输出到文件、命名管道、应用或 MapReduce。如前面的例 4 和例 6 所示，可以定义基于 gpfdist 或 Web 的可写外部表。对于使用 gpfdist 协议的外部表，HAWQ Segment 将它们的数据发送给 gpfdist，gpfdist 将数据写入命名文件中。gpfdist 必须运行在 HAWQ Segment 能够在网络上访问的主机上。gpfdist 指向一个输出主机上的文件位置，将从 HAWQ Segment 接收到的数据写入文件。一个可写 Web 外部表的数据作为数据流发送给应用。例如，从 HAWQ 卸载数据并发送给一个连接其他数据库的应用或向别处装载数据的 ETL 工具。可写 Web 外部表使用 EXECUTE 子句指定一个运行在 Segment 主机上的 shell 命令、脚本或应用，接收输入数据流。

可以选择为可写外部表声明分布策略。默认，可写外部表使用随机分布。如果要导出的源表是哈希分布的，为外部表定义相同的分布键列会提升数据卸载性能，因为这消除了数据行在内部互联网络上的移动。如果卸载一个特定表的数据，可以使用 LIKE 子句复制源表的列定义与分布策略。

```
db1=# create writable external table unload_expenses
```

```
db1-# ( like t1 )
db1-# location ('gpfdist://hdp3:8081/expenses1.out',
db1(# 'gpfdist://hdp4:8081/expenses2.out')
db1-# format 'text' (delimiter ',');
NOTICE:  Table doesn't have 'distributed by' clause, defaulting to distribution
columns from LIKE table
CREATE EXTERNAL TABLE
```

可写外部表只允许 INSERT 操作。如果执行卸载的用户不是外部表的属主或超级用户,必须授予对外部表的 INSERT 权限。

```
grant insert on unload_expenses TO admin;
```

与例 4 不同,INSERT INTO 外部表 SELECT ... 语句中,外部表的输出文件只能在一个主机上,否则会报错。

```
db1=# insert into unload_expenses select * from t1;
ERROR:  External table has more URLs then available primary segments that can
write into them  (seg0 hdp1:40000 pid=387379)
db1=# drop external table unload_expenses;
DROP EXTERNAL TABLE
db1=# create writable external table unload_expenses
db1-# ( like t1 )
db1-# location ('gpfdist://hdp3:8081/expenses1.out')
db1-# format 'text' (delimiter ',');
NOTICE:  Table doesn't have 'distributed by' clause, defaulting to distribution
columns from LIKE table
CREATE EXTERNAL TABLE
db1=# insert into unload_expenses select * from t1;
INSERT 0 4
```

查看导出的数据。

```
[gpadmin@hdp3 staging]$ more expenses1.out
aaa,2017-01-01,100.1,aaa,aaa
bbb,2017-01-02,100.2,bbb,bbb
aaa,2017-03-01,200.1,aaa,aaa
bbb,2017-03-02,200.2,bbb,bbb
[gpadmin@hdp3 staging]$
```

如上面的例 6 所示,也可以定义一个可写的外部 Web 表,发送数据行到脚本或应用。脚本文件必须接收输入流,而且必须存在于所有 HAWQ Segment 主机的相同位置上,并可以被 gpadmin 用户执行。HAWQ 系统中的所有 Segment 都执行脚本,无论 Segment 是否有需要处理的输出行。

允许外部表执行操作系统命令或脚本会带来相应的安全风险。为了在可写外部 Web 表定义中禁用 EXECUTE，可在 HAWQ Master 的 hawq-site.xml 文件中设置 gp_external_enable_exec 服务器配置参数为 off。

```
gp_external_enable_exec = off
```

正如前面说明 COPY 命令时所看到的，COPY TO 命令也可以用来卸载数据。它使用 HAWQ Master 主机上的单一进程，将表中数据复制到 HAWQ Master 主机上的一个文件（或标准输入）中。COPY TO 命令重写整个文件，而不是追加记录。

8.2.8 hawq register

该命令将 HDFS 上的 Parquet 表数据装载并注册到对应的 HAWQ 表中。hawq register 的使用场景好像很有限，因为它只能注册 HAWQ 或 Hive 已经生成的 Parquet 表文件。关于该命令的使用可参考：http://hawq.incubator.apache.org/docs/userguide/2.1.0.0-incubating/datamgmt/load/g-register_files.html。

8.2.9 格式化数据文件

使用 HAWQ 工具装载或卸载数据时，必须指定数据的格式。CREATE EXTERNAL TABLE、hawq load 和 COPY 都包含指定数据格式的子句。数据可以是固定分隔符的文本或逗号分隔值（CSV）格式。外部数据必须是 HAWQ 可以正确读取的格式。

1. 行分隔符

HAWQ 需要数据行以换行符（LF，Line feed，ASCII 值 0x0A）、回车符（CR，Carriage return，ASCII 值 0x0D）或回车换行符（CR+LF，0x0D 0x0A）作为行分隔符。LF 是类 UNIX 操作系统中标准的换行符。而 Windows 或 Mac OS X 使用 CR 或 CR+LF。所有这些表示一个新行的特殊符号都被 HAWQ 作为行分隔符所支持。

2. 列分隔符

文本文件和 CSV 文件默认的列分隔符分别是 TAB（ASCII 值为 0x09）和逗号（ASCII 值为 0x2C）。在定义数据格式时，可以在 CREATE EXTERNAL TABLE 或 COPY 命令的 DELIMITER 子句，或者 hawq load 的控制文件中，声明一个单字符作为列分隔符。分隔符必须出现在字段值之间，不要在一行的开头或结尾放置分隔符。如使用管道符（|）作为列分隔符：

```
data value 1|data value 2|data value 3
```

下面的建表命令显示以管道符作为列分隔符：

```
create external table ext_table (name text, date date)
location ('gpfdist://host:port/filename.txt)
format 'text' (delimiter '|');
```

3. 空值

空值（NULL）表示一列中的未知数据。可以指定数据文件中的一个字符串表示空值。文本文件中表示空值的默认字符串为\N，CSV 文件中表示空值的默认字符串为不带引号的空串（两个连续的逗号）。定义数据格式时，可以在 CREATE EXTERNAL TABLE、COPY 命令的 NULL 子句，或者 hawq load 的控制文件中，声明其他字符串表示空值。例如，若不想区分空值与空串，就可以指定空串表示 NULL。使用 HAWQ 装载工具时，任何与声明代表 NULL 的字符串相匹配的数据项都被认为是空值。

4. 转义

列分隔符与行分隔符在数据文件中具有特殊含义。如果实际数据中也含有这个符号，必须对这些符号进行转义，以使 HAWQ 将它们作为普通数据而不是列或行的分隔符。文本文件默认的转义符为一个反斜杠（\），CSV 文件默认的转义符为一个双引号（"）。

（1）文本文件转义

可以在 CREATE EXTERNAL TABLE、COPY 的 ESCAPE 子句，或者 hawq load 的控制文件中指定转义符。假设有以下三个字段的数据：

```
backslash = \
vertical bar = |
exclamation point = !
```

指定管道符（|）为列分隔符、反斜杠（\）为转义符，则对应的数据行格式如下：

```
backslash = \\ | vertical bar = \| | exclamation point = !
```

可以对八进制或十六进制序列应用转义符。在装载进 HAWQ 时，转义后的值就是八进制或十六进制的 ASCII 码所表示的字符。例如，取址符（&）可以使用十六进制的（\0x26）或八进制的（\046）表示。

如果要在 CREATE EXTERNAL TABLE、COPY 命令的 ESCAPE 子句，或者 hawq load 的控制文件中禁用转义，可如下设置：

```
ESCAPE 'OFF'
```

该设置常用于输入数据中包含很多反斜杠（如 Web 日志数据）的情况。

（2）CSV 文件转义

可以在 CREATE EXTERNAL TABLE、COPY 的 ESCAPE 子句，或者 hawq load 的控制文件中指定转义符。假设有以下三个字段的数据：

```
Free trip to A,B
5.89
Special rate "1.79"
```

指定逗号（,）为列分隔符、一个双引号（"）为转义符，则数据行格式如下：

```
"Free trip to A,B","5.89","Special rate ""1.79"""
```

将字段值置于双引号中能保留字符串中头尾的空格。

5. 字符编码

在将一个 Windows 操作系统上生成的数据文件装载到 HAWQ 前，先使用 dos2unix 系统命令去除只有 Windows 使用的字符，如删除文件中的 CR（'\x0D'）。

6. 导入导出固定宽度数据

HAWQ 的函数 fixedwith_in 和 fixedwidth_out 支持固定宽度的数据格式。这些函数的定义保存在$GPHOME/share/postgresql/cdb_external_extensions.sql 文件中。下面的例子声明一个自定义格式，然后调用 fixedwidth_in 函数指定为固定宽度的数据格式。

```
db1=# create readable external table students (
db1(#   name varchar(5), address varchar(10), age int)
db1-# location ('gpfdist://hdp4:8081/students.txt')
db1-# format 'custom' (formatter=fixedwidth_in, name='5', address='10', age='4');
CREATE EXTERNAL TABLE
db1=# select * from students;
 name  | address    | age
-------+------------+-----
 abcde | 1234567890 |  40
(1 row)
```

students.txt 文件内容如下：

```
[gpadmin@hdp4 unload_data1]$ more students.txt
abcde12345678900040
```

文件中一行记录的字节必须与建表语句中字段字节数的和一致，如上例中一行必须严格为 19 字节，否则读取文件时会报错。再看一个含有中文的例子。

```
[gpadmin@hdp4 unload_data1]$ echo $LANG
zh_CN.UTF-8
[gpadmin@hdp4 unload_data1]$ more students.txt
中文中文中 0040
```

操作系统和数据库的字符集都是 UTF8，一个中文占用三字节，记录一共 19 字节，满足读取条件。

```
db1=# select * from students;
 name | address | age
------+---------+-----
 中   | 中文中  |  40
(1 row)
```

name 字段 5 字节，address 字段 10 字节，理论上这两个字段都应该含有不完整的字符，但从查询结果看到，HAWQ 在这里做了一些处理，name 字段读取了一个完整的中文，address 字段读取了三个完整的字符。而中间按字节分列的中文字符被不可见字符所取代。

```
db1=# select char_length(name),octet_length(name),substr(name,1,1),substr(name,2,1) from students;
 char_length | octet_length | substr | substr
-------------+--------------+--------+--------
           2 |            5 | 中     |
(1 row)


db1=# select char_length(address),octet_length(address),substr(address,1,1),substr(address,2,1) from students;
 char_length | octet_length | substr | substr
-------------+--------------+--------+--------
           4 |           10 |        | 中
(1 row)
```

以下选项指定如何读取固定宽度数据文件。

- 读取全部数据。装载固定宽度数据一行中的所有字段，并按它们的物理顺序进行装载。必须指定字段长度，不能指定起始与终止位置。固定宽度参数中字段名的顺序必须与 CREATE TABLE 命令中的顺序相匹配。
- 设置空格与 NULL 特性。默认尾部空格被截取。为了保留尾部空格，使用 preserve_blanks=on 选项。使用 null='null_string_value'选项指定代表 NULL 的字符串。
- 如果指定了 preserve_blanks=on，也必须定义代表 NULL 值的字符串，否则会报 ERROR: A null_value was not defined. When preserve_blanks is on, a null_value。
- 如果指定了 preserve_blanks=off，没有定义 NULL，并且一个字段只包含空格，HAWQ 向表中写一个 null。如果定义了 NULL，HAWQ 向表中写一个空串。
- 使用 line_delim='line_ending'参数指定行尾字符。下面的例子覆盖大多数情况。'E' 表示转义，就是说如果记录正文中含有 line_delim，需要进行转义。

```
line_delim=E'\n'
line_delim=E'\r'
line_delim=E'\r\n'
line_delim='abc'
```

8.3 数据库统计

统计信息指的是数据库中所存储数据的元信息描述。查询优化器需要依据最新的统计信息，为查询生成最佳执行计划。例如，查询连接了两个表，一个表必须被广播到所有 Segment，那么优化器会选择广播其中的小表，使网络流量最小化。

ANALYZE 命令计算优化器所需的统计信息，并将结果保存到系统目录中。有三种方式启动分析操作：

- 直接运行 ANALYZE 命令。
- 在数据库外运行 analyzedb 命令行应用程序。
- 执行 DML 操作的表上没有统计信息，或者 DML 操作影响的行数超过了指定的阈值时，系统自动执行分析操作。

计算统计信息会消耗时间和资源，因此 HAWQ 会在大表上进行采样，通过计算部分数据，产生统计信息的估算值。大多数情况下，默认设置能够提供生成正确查询执行计划的信息。如果产生的统计不能生成优化的查询执行计划，管理员可以调整配置参数，通过增加样本数据量，产生更加精确的统计。统计信息越精确，所消耗的 CPU 和内存资源越多，因此可能由于资源的限制，无法生成更好的计划。此时就需要查看执行计划并测试查询性能，目标是要通过增加的统计成本达到更好的查询性能。

8.3.1 系统统计

1. 表大小

查询优化器使用查询必须处理的数据行数和必须访问的磁盘页数等统计信息，寻找查询所需的最小磁盘 I/O 和网络流量的执行计划。用于估算行数和页数的数据分别保存在 pg_class 系统表的 reltuples 和 relpages 列中，其中的值是最后运行 VACUUM 或 ANALYZE 命令时生成的数据。对于默认的 AO（Append Only）表，系统目录中的 tuples 数是最近的值，因此 reltuples 统计是精确值而不是估算值，但 relpages 是 AO 数据块的估算值。如果 reltuples 列的值与 SELECT COUNT(*)的返回值差很多，应该执行分析更新统计信息。

2. pg_statistic 系统表与 pg_stats 视图

pg_statistic 系统表保存每个数据库表上最后执行 ANALYZE 操作的结果。每个表列有一行记录，具有以下字段：

- starelid: 列所属表的对象 ID。
- staatnum: 所描述列在表中的编号，从 1 开始。
- stanullfrac: 列中空值占比。
- stawidth: 非空数据项的平均宽度，单位是字节。

- stadistinct：列中不同非空数据值的个数。
- stakindN：表示后面 number、values 所示的数据用途，被用于生成 pg_stats。例如，1 表示是 MCV（Most Common Values）的值；2 表示直方图（histogram）的值；3 表示相关性（correlation）的值等。kind 的取值范围：1~99，内核占用；100~199，PostGIS 占用；200~299，ESRI ST_Geometry 几何系统占用；300~9999，公共占用。
- staopN：表示该统计值支持的操作，如 '=' 或 '<' 等。
- stanumbersN：如果是 MCV 类型（kind=1），那么这里就是下面对应的 stavaluesN 出现的概率值，即 MCF。
- stavaluesN：anyarray 类型的数据，内核特殊类型，不可更改，是统计信息的值部分，与 kind 对应。例如，kind=2 时，这里的值表示直方图。

pg_statistic 表将不同的统计类型分为四类，分别用四个字段表示。pg_stats 视图以一种更友好的方式表示 pg_statistic 的内容，其定义如下：

```
SELECT n.nspname AS schemaname, c.relname AS tablename, a.attname, s.stanullfrac 
AS null_frac, s.stawidth AS avg_width, s.stadistinct AS n_distinct,
    CASE 1
        WHEN s.stakind1 THEN s.stavalues1
        WHEN s.stakind2 THEN s.stavalues2
        WHEN s.stakind3 THEN s.stavalues3
        WHEN s.stakind4 THEN s.stavalues4
        ELSE NULL::anyarray
    END AS most_common_vals,
    CASE 1
        WHEN s.stakind1 THEN s.stanumbers1
        WHEN s.stakind2 THEN s.stanumbers2
        WHEN s.stakind3 THEN s.stanumbers3
        WHEN s.stakind4 THEN s.stanumbers4
        ELSE NULL::real[]
    END AS most_common_freqs,
    CASE 2
        WHEN s.stakind1 THEN s.stavalues1
        WHEN s.stakind2 THEN s.stavalues2
        WHEN s.stakind3 THEN s.stavalues3
        WHEN s.stakind4 THEN s.stavalues4
        ELSE NULL::anyarray
    END AS histogram_bounds,
    CASE 3
        WHEN s.stakind1 THEN s.stanumbers1[1]
        WHEN s.stakind2 THEN s.stanumbers2[1]
        WHEN s.stakind3 THEN s.stanumbers3[1]
        WHEN s.stakind4 THEN s.stanumbers4[1]
```

```
            ELSE NULL::real
        END AS correlation
 FROM pg_statistic s
 JOIN pg_class c ON c.oid = s.starelid
 JOIN pg_attribute a ON c.oid = a.attrelid AND a.attnum = s.staattnum
 LEFT JOIN pg_namespace n ON n.oid = c.relnamespace
 WHERE has_table_privilege(c.oid, 'select'::text);
```

新建的表没有统计信息。

3. 采样

为大表计算统计信息时，HAWQ 通过对基表采样数据的方式建立一个小表。如果基表是分区表，从全部分区中采样。样本表中的行数取决于由 gp_analyze_relative_error 系统配置参数指定的最大可接受错误数。该参数的默认值是 0.25（25%）。通常该值已经足够生成正确的查询计划。如果 ANALYZE 不能产生好的表列估算，可以通过调低该参数值，增加采样的数据量。需要注意的是，降低该值可能导致大量的采样数据，并明显增加分析时间。

```
[gpadmin@hdp3 ~]$ hawq config -s gp_analyze_relative_error
GUC      : gp_analyze_relative_error
Value    : 0.25
```

4. 统计更新

不带参数运行 ANALYZE 会更新当前数据库中所有表的统计信息，这可能需要执行很长时间。所以最好分析单个表，在一个表中的数据大量修改后分析该表。也可以选择分析一个表列的子集，例如只分析 join、where、order by、group by、having 等子句中用到的列。

```
db1=# analyze t1 (name,category);
ANALYZE
```

5. 分析分区和 AO 表

在分区表上运行 ANALYZE 命令时，逐个分析每个叶级别的子分区。也可以只在新增或修改的分区文件上运行 ANALYZE，避免分析没有变化的分区。analyzedb 命令行应用自动跳过无变化的分区，并且它是多会话并行的，可以同时分析几个分区。默认运行 5 个会话，会话数可以通过命令行的-p 选项设置值域为 1~10。每次运行 analyzedb，它都会将 AO 表和分区的状态信息保存在 Master 节点数据目录中的 db_analyze 目录下，如/data/hawq/master/db_analyze/。下次运行时，analyzedb 比较每个表的当前状态与上次保存的状态，不分析没有变化的表或分区。但是，系统表总是会被分析。

HAWQ 新的 GPORCA 查询优化器需要分区表根级别的统计信息，而老的优化器不使用该统计。通过设置 optimizer 和 optimizer_analyze_root_partition 系统配置参数启用新的查询优化器，默认是启用的。

```
[gpadmin@hdp3 ~]$ hawq config -s optimizer
GUC      : optimizer
```

```
    Value            : on
[gpadmin@hdp3 ~]$ hawq config -s optimizer_analyze_root_partition
    GUC              : optimizer_analyze_root_partition
    Value            : on
```

每次运行 ANALYZE 或 ANALYZE ROOTPARTITION 时，根级别的统计信息被更新。analyzedb 应用默认更新根分区统计。当在父表上使用 ANALYZE 收集统计信息时，既会收集每个叶子分区的统计信息，又会收集分区表的全局统计信息。生成分区表查询计划时两个统计信息都需要。如果所有子分区的统计信息都已经更新，ROOTPARTITION 选项可用于只收集分区表的全局状态信息，这可以节省分析时间。如果在一个非根分区或非分区表上使用 ROOTPARTITION 选项，ANALYZE 命令将跳过该选项并发出一个警告信息。

```
db1=# analyze rootpartition t1;
WARNING:  skipping "t1" --- cannot analyze a non-root partition using ANALYZE ROOTPARTITION
ANALYZE
db1=# analyze rootpartition sales;
ANALYZE
```

8.3.2 统计配置

1. 统计目标

统计目标指的是一个列的 most_common_vals、most_common_freqs 和 histogram_bounds 数组的大小。这些数组的含义可以从上面 pg_stats 视图的定义得到。默认目标值为 25。可以通过设置服务器配置参数修改全局目标值，也可以使用 ALTER TABLE 命令设置任何表列的目标值。目标值越大，优化器评估质量越高，但 ANALYZE 需要的时间也越长。

default_statistics_target 服务器配置参数设置系统默认的统计目标。默认值 25 通常已经足够，只有经过测试确定要定义一个新目标时，才考虑更改此参数的值。可以通过 Ambari Web UI 和命令行两种方法修改配置参数值。下面的例子使用 hawq config 命令行将统计目标从 25 改为 50。

以 HAWQ 管理员（默认为 gpadmin）登录 HAWQ Master 主机并设置环境。

```
$ source /usr/local/hawq/greenplum_path.sh
```

使用 hawq config 应用设置 default_statistics_target。

```
$ hawq config -c default_statistics_target -v 50
```

重载使配置生效。

```
$ hawq stop cluster -u
```

单个列的统计目标可以用 ALTER TABLE 命令设置。例如，某些查询可以通过为特定列，尤其是分布不规则的列增加目标值提高性能。如果将一列的目标值设置为 0，ANALYZE 忽略该列。下面的命令将 desc1 列的统计目标设置为 0，因为该列对于查询优化没有任何作用。

```
db1=# alter table t1 alter column desc1 set statistics 0;
ALTER TABLE
```

统计目标可以设置为 0～1000 之间的值，或者设置成-1，此时恢复使用系统默认的统计目标值。父分区表上设置的统计目标影响子分区。如果父表上某列的目标设置为 0，所有子分区上的该列统计目标也为 0。但是，如果以后增加或者交换了其他子分区，新增的子分区将使用默认目标值，交换的子分区使用以前的统计目标。因此如果增加或交换了子分区，应该在新的子分区上设置统计目标。

2. 自动收集统计信息

如果一个表没有统计信息，或者在表上执行的特定操作改变了大量数据时，HAWQ 可以在表上自动运行 ANALYZE。对于分区表，自动统计收集仅当直接操作叶表时被触发，它仅分析叶表。自动收集统计信息有三种模式：

- none：禁用自动收集。
- on_no_stats：在一个没有统计信息的表上执行 CREATE TABLE AS SELECT、INSERT、COPY 命令时触发分析操作。
- on_change：在表上执行 CREATE TABLE AS SELECT、INSERT、COPY 命令，并且影响的行数超过了 gp_autostats_on_change_threshold 配置参数设定的阈值时触发分析操作。

依据 CREATE TABLE AS SELECT、INSERT、COPY 这些命令是单独执行还是在函数中执行，自动收集统计信息模式的设置方法也不一样。如果是在函数外单独执行，gp_autostats_mode 配置参数控制统计模式，默认值为 on_no_stats。

```
[gpadmin@hdp3 ~]$ hawq config -s gp_autostats_mode
GUC     : gp_autostats_mode
Value   : ON_NO_STATS
```

on_change 模式仅当影响的行数超过 gp_autostats_on_change_threshold 配置参数设置的阈值时触发 ANALYZE，该参数的默认值为 2147483647。

```
[gpadmin@hdp3 ~]$ hawq config -s gp_autostats_on_change_threshold
GUC     : gp_autostats_on_change_threshold
Value   : 2147483647
```

on_change 模式可能触发不希望的、大的分析操作，严重时会使系统中断，因此不推荐在全局修改该参数，但可以在会话级设置，例如装载数据后自动分析。

为了禁用函数外部的自动统计收集，设置 gp_autostats_mode 参数为 none：

```
$ hawq configure -c gp_autostats_mode -v none
```

如果想记录自动统计收集操作的日志，可以设置 log_autostats 系统配置参数为 on。

8.4 PXF

HAWQ 不但可以读写自身系统中的表，而且能够访问 HDFS、Hive、HBase 等外部系统的数据。这是通过一个名为 PXF 的扩展框架实现的。大部分外部数据以 HAWQ 外部表形式进行访问，但对于 Hive，除外部表方式，PXF 还能够与 HCatalog 结合直接查询 Hive 表。PXF 内建多个连接器，用户也可以按照 PXF API 创建自己的连接器，访问其他并行数据存储或处理引擎。

8.4.1 安装配置 PXF

如果使用 Ambari 安装管理 HAWQ 集群，那么不需要执行任何手动命令行安装步骤，从 Ambari Web 接口就可以安装所有需要的 PXF 插件。详细安装步骤参考第 2 章 "HAWQ 安装部署"。如果使用命令行安装 PXF，参考 http://hawq.incubator.apache.org/docs/userguide/2.1.0.0-incubating/pxf/InstallPXFPlugins.html#install_pxf_plug_cmdline。PXF 相关的默认安装目录和文件如表 8-1 所示。

表 8-1 PXF 安装目录

目录	描述
/usr/lib/pxf	PXF 库目录
/etc/pxf/conf	PXF 配置目录。该目录下包含 pxf-public.classpath 和 pxf-private.classpath 及其他配置文件
/var/pxf/pxf-service	PXF 服务实例所在目录
/var/log/pxf	该目录包含 pxf-service.log 和所有 Tomcat 相关的日志文件（PXF 需要在主机上运行 Tomcat，用 Ambari 安装 PXF 时会自动安装 Tomcat），这些文件的属主是 pxf:pxf，对其他用户是只读的
/var/run/pxf/catalina.pid	PXF Tomcat 容器的 PID 文件，存储进程号

与安装一样，PXF 也可以使用 Ambari 的图形界面进行交互式配置，完成后重启 PXF 服务以使配置生效。手动配置步骤参考 http://hawq.incubator.apache.org/docs/userguide/2.1.0.0-incubating/pxf/ConfigurePXF.html，需要修改所有集群主机上的相关配置文件，然后重启所有节点上的 PXF 服务。

8.4.2 PXF profile

PXF profile 是一组通用元数据属性的集合，用于简化外部数据读写。PXF 自带多个内建的 profile，每个 profile 将一组元数据属性归于一类，使得对以下数据存储系统的访问更加容易：

- HDFS 文件（读写）
- Hive（只读）

- HBase（只读）
- JSON（只读）

表 8-2 说明了 PXF 的内建 profile 及其相关 Java 类。这些 profile 在 /etc/pxf/conf/pxf-profiles.xml 文件中定义。

表 8-2 PXF 内建 profile

Profile	描述	相关 Java 类
HdfsTextSimple	读写 HDFS 上平面文本文件，每条记录由固定分隔符的一行构成	org.apache.hawq.pxf.plugins.hdfs.HdfsDataFragmenter org.apache.hawq.pxf.plugins.hdfs.LineBreakAccessor org.apache.hawq.pxf.plugins.hdfs.StringPassResolver
HdfsTextMulti	从 HDFS 上的平面文件中读取具有固定分隔符的记录，每条记录由一行或多行（记录中包含换行符）构成。此 profile 是不可拆分的（非并行），比 HdfsTextSimple 读取慢	org.apache.hawq.pxf.plugins.hdfs.HdfsDataFragmenter org.apache.hawq.pxf.plugins.hdfs.QuotedLineBreakAccessor org.apache.hawq.pxf.plugins.hdfs.StringPassResolver
Hive	读取 Hive 表，支持 text、RC、ORC、Sequence 或 Parquet 存储格式	org.apache.hawq.pxf.plugins.hive.HiveDataFragmenter org.apache.hawq.pxf.plugins.hive.HiveAccessor org.apache.hawq.pxf.plugins.hive.HiveResolver org.apache.hawq.pxf.plugins.hive.HiveMetadataFetcher org.apache.hawq.pxf.service.io.GPDBWritable
HiveRC	优化读取 RCFile 存储格式的 Hive 表，必须指定 DELIMITER 参数	org.apache.hawq.pxf.plugins.hive.HiveInputFormatFragmenter org.apache.hawq.pxf.plugins.hive.HiveRCFileAccessor org.apache.hawq.pxf.plugins.hive.HiveColumnarSerdeResolver org.apache.hawq.pxf.plugins.hive.HiveMetadataFetcher org.apache.hawq.pxf.service.io.Text
HiveORC	优化读取 ORCFile 存储格式的 Hive 表	org.apache.hawq.pxf.plugins.hive.HiveInputFormatFragmenter org.apache.hawq.pxf.plugins.hive.HiveORCAccessor org.apache.hawq.pxf.plugins.hive.HiveORCSerdeResolver org.apache.hawq.pxf.plugins.hive.HiveMetadataFetcher org.apache.hawq.pxf.service.io.GPDBWritable

(续表)

Profile	描述	相关 Java 类
HiveText	优化读取 TextFile 存储格式的 Hive 表，必须指定 DELIMITER 参数	org.apache.hawq.pxf.plugins.hive.HiveInputFormatFragmenter org.apache.hawq.pxf.plugins.hive.HiveLineBreakAccessor org.apache.hawq.pxf.plugins.hive.HiveStringPassResolver org.apache.hawq.pxf.plugins.hive.HiveMetadataFetcher org.apache.hawq.pxf.service.io.Text
HBase	读取 HBase 存储引擎数据	org.apache.hawq.pxf.plugins.hbase.HBaseDataFragmenter org.apache.hawq.pxf.plugins.hbase.HBaseAccessor org.apache.hawq.pxf.plugins.hbase.HBaseResolver
Avro	读取 Avro 文件	org.apache.hawq.pxf.plugins.hdfs.HdfsDataFragmenter org.apache.hawq.pxf.plugins.hdfs.AvroFileAccessor org.apache.hawq.pxf.plugins.hdfs.AvroResolver
JSON	读取 HDFS 上的 JSON 文件	org.apache.hawq.pxf.plugins.hdfs.HdfsDataFragmenter org.apache.hawq.pxf.plugins.json.JsonAccessor org.apache.hawq.pxf.plugins.json.JsonResolver

8.4.3 访问 HDFS 文件

HDFS 是 Hadoop 应用的主要分布式存储机制。PXF 的 HDFS 插件用于读取存储在 HDFS 文件中的数据，支持具有固定分隔符的文本和 Avro 两种文件格式。在使用 PXF 访问 HDFS 文件前，确认已经在集群所有节点上安装了 PXF HDFS 插件（Ambari 会自动安装），并授予了 HAWQ 用户（典型的是 gpadmin）对 HDFS 文件相应的读写权限。

1. PXF 支持的 HDFS 文件格式

PXF HDFS 插件支持对以下两种文件格式的读取：

- comma-separated value（.csv）或其他固定分隔符的平面文本文件。
- 由 JSON 定义的、基于 Schema 的 Avro 文件格式。

PXF HDFS 插件包括以下 Profile 以支持上面的两类文件：

- HdfsTextSimple：单行文本文件。
- HdfsTextMulti：内嵌换行符的多行文本文件。
- Avro：Avro 文件。

2. 查询外部 HDFS 数据

HAWQ 通过外部表的形式访问 HDFS 文件。下面是创建一个 HDFS 外部表的语法。

```
CREATE EXTERNAL TABLE <table_name>
    ( <column_name> <data_type> [, ...] | LIKE <other_table> )
 LOCATION ('pxf://<host>[:<port>]/<path-to-hdfs-file>
    ?PROFILE=HdfsTextSimple|HdfsTextMulti|Avro[&<custom-option>=<value>[...]
]')
    FORMAT '[TEXT|CSV|CUSTOM]' (<formatting-properties>);
```

CREATE EXTERNAL TABLE 语句中使用的各个关键字和相应值的描述如表 8-3 所示。

表 8-3 HDFS 外部表建表语句说明

关键字	值
<host>[:<port>]	HDFS NameNode 主机名、端口
<path-to-hdfs-file>	HDFS 文件路径
PROFILE	PROFILE 关键字指定为 HdfsTextSimple、HdfsTextMulti 或 Avro 之一
<custom-option>	与特定 PROFILE 对应的用户自定义选项
FORMAT 'TEXT'	当<path-to-hdfs-file>指向一个单行固定分隔符的平面文件时，使用该关键字
FORMAT 'CSV'	当<path-to-hdfs-file>指向一个单行或多行的逗号分隔值（CSV）平面文件时，使用该关键字
FORMAT 'CUSTOM'	Avro 文件使用该关键字。Avro 'CUSTOM' 格式只支持内建的 (formatter='pxfwritable_import')格式属性
<formatting-properties>	与特定 PROFILE 对应的格式属性

下面是几个 HAWQ 访问 HDFS 文件的例子。

（1）使用 HdfsTextSimple Profile

HdfsTextSimple Profile 用于读取一行表示一条记录的平面文本文件或 CSV 文件，支持的<formatting-properties>是 delimiter，用来指定文件中每条记录的字段分隔符。

为 PXF 创建一个 HDFS 目录。

```
su - hdfs
hdfs dfs -mkdir -p /data/pxf_examples
hdfs dfs -chown -R gpadmin:gpadmin /data/pxf_examples
```

建立一个名为 pxf_hdfs_simple.txt 的平面文本文件，生成四条记录，使用逗号作为字段分隔符。

```
echo 'Prague,Jan,101,4875.33
Rome,Mar,87,1557.39
Bangalore,May,317,8936.99
Beijing,Jul,411,11600.67' > /tmp/pxf_hdfs_simple.txt
```

将文件传到 HDFS 上。

```
hdfs dfs -put /tmp/pxf_hdfs_simple.txt /data/pxf_examples/
```

显示 HDFS 上的 pxf_hdfs_simple.txt 文件内容。

```
hdfs dfs -cat /data/pxf_examples/pxf_hdfs_simple.txt
```

使用 HdfsTextSimple profile 创建一个可从 pxf_hdfs_simple.txt 文件查询数据的 HAWQ 外部表。delimiter=e','中的 e 表示转义,就是说如果记录正文中含有逗号,需要用\符号进行转义。

```
db1=# create external table pxf_hdfs_textsimple(location text, month text, num_orders int, total_sales float8)
db1-#             location 
('pxf://hdp1:51200/data/pxf_examples/pxf_hdfs_simple.txt?profile=hdfstextsimple')
db1-#             format 'text' (delimiter=e',');
CREATE EXTERNAL TABLE
db1=# select * from pxf_hdfs_textsimple;
 location   | month | num_orders | total_sales
------------+-------+------------+-------------
 Prague     | Jan   |        101 |     4875.33
 Rome       | Mar   |         87 |     1557.39
 Bangalore  | May   |        317 |     8936.99
 Beijing    | Jul   |        411 |    11600.67
(4 rows)
```

用 CSV 格式创建第二个外部表。当指定格式为'CSV'时,逗号是默认分隔符,不再需要使用 delimiter 说明。

```
db1=# create external table pxf_hdfs_textsimple_csv(location text, month text, num_orders int, total_sales float8)
db1-#             location 
('pxf://hdp1:51200/data/pxf_examples/pxf_hdfs_simple.txt?profile=hdfstextsimple')
db1-#             format 'csv';
CREATE EXTERNAL TABLE
db1=# select * from pxf_hdfs_textsimple_csv;
 location   | month | num_orders | total_sales
------------+-------+------------+-------------
 Prague     | Jan   |        101 |     4875.33
 Rome       | Mar   |         87 |     1557.39
 Bangalore  | May   |        317 |     8936.99
 Beijing    | Jul   |        411 |    11600.67
(4 rows)
```

(2) 使用 HdfsTextMulti Profile

HdfsTextMulti profile 用于读取一条记录中含有换行符的平面文本文件。因为 PXF 将换行符作为行分隔符,所以当数据中含有换行符时需要用 HdfsTextMulti 进行特殊处理。HdfsTextMulti Profile 支持的<formatting-properties>是 delimiter,用来指定文件中每条记录的字段分隔符。

创建一个平面文本文件。

```
vi /tmp/pxf_hdfs_multi.txt
```

输入以下记录,以冒号作为字段分隔符,第一个字段中含有换行符。

```
"4627 Star Rd.
San Francisco, CA 94107":Sept:2017
"113 Moon St.
San Diego, CA 92093":Jan:2018
"51 Belt Ct.
Denver, CO 90123":Dec:2016
"93114 Radial Rd.
Chicago, IL 60605":Jul:2017
"7301 Brookview Ave.
Columbus, OH 43213":Dec:2018
```

使用 HdfsTextMulti profile 创建一个可从 pxf_hdfs_multi.txt 文件查询数据的外部表,指定分隔符是冒号。

```
db1=# create external table pxf_hdfs_textmulti(address text, month text, year int)
db1-#           location
('pxf://hdp1:51200/data/pxf_examples/pxf_hdfs_multi.txt?profile=hdfstextmulti')
db1-#           format 'csv' (delimiter=e':');
CREATE EXTERNAL TABLE
db1=# select * from pxf_hdfs_textmulti;
         address              | month | year
------------------------------+-------+------
 4627 Star Rd.                | Sept  | 2017
San Francisco, CA  94107
 113 Moon St.                 | Jan   | 2018
San Diego, CA  92093
 51 Belt Ct.                  | Dec   | 2016
Denver, CO  90123
 93114 Radial Rd.             | Jul   | 2017
Chicago, IL  60605
 7301 Brookview Ave.          | Dec   | 2018
Columbus, OH  43213
(5 rows)
```

（3）访问 HDFS HA 集群中的文件

为了访问 HDFS HA 集群中的外部数据，将 CREATE EXTERNAL TABLE LOCATION 子句由<host>[:<port>]修改为<HA-nameservice>。

```
gpadmin=# create external table pxf_hdfs_textmulti_ha (address text, month text, year int)
          location
('pxf://mycluster/data/pxf_examples/pxf_hdfs_multi.txt?profile=hdfsextmulti')
          format 'csv' (delimiter=e':');
gpadmin=# select * from pxf_hdfs_textmulti_ha;
        address             | month | year
----------------------------+-------+------
 4627 Star Rd.              | Sept  | 2017
 San Francisco, CA  94107
 113 Moon St.               | Jan   | 2018
 San Diego, CA  92093
 51 Belt Ct.                | Dec   | 2016
 Denver, CO  90123
 93114 Radial Rd.           | Jul   | 2017
 Chicago, IL  60605
 7301 Brookview Ave.        | Dec   | 2018
 Columbus, OH  43213
(5 rows)
```

8.4.4　访问 Hive 数据

Hive 是 Hadoop 的分布式数据仓库框架，支持多种文件格式，如 CSV、RC、ORC、Parquet 等。PXF 的 Hive 插件用于读取存储在 Hive 表中的数据。PXF 提供两种方式查询 Hive 表：

- 通过整合 PXF 与 HCatalog 直接查询。
- 通过外部表查询。

在使用 PXF 访问 Hive 前，确认满足以下前提条件：

- 在 HAWQ 和 HDFS 集群的所有节点上（Master、Segment、NameNode、DataNode）安装了 PXF HDFS 插件。
- 在 HAWQ 和 HDFS 集群的所有节点上安装了 PXF Hive 插件。
- 如果配置了 Hadoop HA，PXF 也必须安装在所有运行 NameNode 服务的 HDFS 节点上。
- 所有 PXF 节点上都安装了 Hive 客户端。
- 集群所有节点上都安装了 Hive JAR 文件目录和 conf 目录。
- 已经测试了 PXF 访问 HDFS。
- 在集群中的一台主机上运行 Hive Metastore 服务。

- 在 NameNode 上的 hive-site.xml 文件中设置了 hive.metastore.uris 属性。

看似条件不少，但是如果使用 Ambari 安装管理 HAWQ 集群，并安装了 Hadoop 相关服务，则所有这些前置条件都已自动配置好，不需要任何手动操作。

1. PXF 支持的 Hive 文件格式

PXF Hive 插件支持的 Hive 文件格式及其访问这些格式对应的 profile 如表 8-4 所示。

表 8-4　PXF Hive 文件格式及 profile

文件格式	描述	Profile
TextFile	逗号、tab 或空格分隔的平面文件格式	Hive、HiveText
SequenceFile	二进制键值对组成的平面文件	Hive
RCFile	记录由键值对组成的列数据，具有行高压缩率	Hive、HiveRC
ORCFile	优化的列式存储，减小数据大小	Hive
Parquet	压缩的列式存储	Hive
Avro	基于 schema 的、由 JSON 所定义的序列化格式	Hive

2. 数据类型映射

为了在 HAWQ 中表示 Hive 数据，需要将使用 Hive 私有数据类型的数据值映射为等价的 HAWQ 类型值。表 8-5 是对 Hive 私有数据类型的映射规则汇总。

表 8-5　Hive 与 HAWQ 数据类型映射

Hive 数据类型	HAWQ 数据类型	Hive 数据类型	HAWQ 数据类型
boolean	Bool	float	float4
int	int4	double	float8
smallint	int2	string	Text
tinyint	int2	binary	Bytea
bigint	int8	timestamp	Timestamp

除简单类型外，Hive 还支持 array、struct、map 等复杂数据类型。由于 HAWQ 原生不支持这些类型，PXF 将它们统一映射为 text 类型。可以创建 HAWQ 函数或使用应用程序抽取复杂数据类型子元素的数据。

3. HAWQ 访问 Hive 表的示例

（1）准备示例数据

步骤 01　准备数据文件，添加如下记录，用逗号分隔字段。

```
vi /tmp/pxf_hive_datafile.txt
Prague,Jan,101,4875.33
```

```
Rome,Mar,87,1557.39
Bangalore,May,317,8936.99
Beijing,Jul,411,11600.67
San Francisco,Sept,156,6846.34
Paris,Nov,159,7134.56
San Francisco,Jan,113,5397.89
Prague,Dec,333,9894.77
Bangalore,Jul,271,8320.55
Beijing,Dec,100,4248.41
```

步骤02 创建文本格式的 Hive 表 sales_info。

```
create database test;
use test;
create table sales_info (location string, month string,
    number_of_orders int, total_sales double)
    row format delimited fields terminated by ','
    stored as textfile;
```

步骤03 向 sales_info 表装载数据。

```
load data local inpath '/tmp/pxf_hive_datafile.txt' into table sales_info;
```

步骤04 查询 sales_info 表数据，验证装载数据成功。

```
select * from sales_info;
```

步骤05 确认 sales_info 表在 HDFS 上的位置，在创建 HAWQ 外部表时需要用到该信息。

```
describe extended sales_info;
...
location:hdfs://mycluster/apps/hive/warehouse/test.db/sales_info
...
```

（2）使用 PXF 和 HCatalog 查询 Hive

HAWQ 可以获取存储在 HCatalog 中的元数据，通过 HCatalog 直接访问 Hive 表，而不用关心 Hive 表对应的底层文件存储格式。HCatalog 建立在 Hive metastore 之上，包含 Hive 的 DDL 语句。使用这种方式的好处是：

- 不需要知道 Hive 表结构。
- 不需要手动输入 Hive 表的位置与格式信息。
- 如果表的元数据改变，HCatalog 自动提供更新后的元数据。这是使用 PXF 静态外部表方式无法做到的。

HAWQ 使用 HCatalog 查询 Hive 表的示意图如图 8-1 所示。

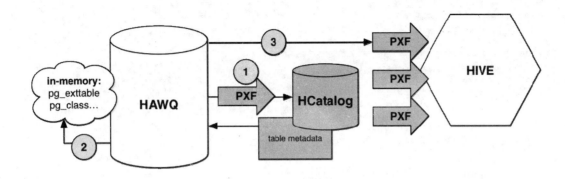

图 8-1　HAWQ 通过 HCatalog 查询 Hive 表

HAWQ 使用 PXF 从 HCatalog 查询表的元数据，然后用查询到的元数据创建一个内存目录表。如果一个查询中多次引用了同一个表，内存目录表可以减少对外部 HCatalog 的调用次数。PXF 使用内存目录表的元数据信息查询 Hive 表。查询结束后，内存目录表将被删除。

如果使用 Ambari 安装管理 HAWQ，并且已经启动了 Hive 服务，则不需要任何额外配置，就可以查询 Hive 表。

```
db1=# select * from hcatalog.test.sales_info;
   location     | month | number_of_orders | total_sales
----------------+-------+------------------+-------------
 Prague         | Jan   |              101 |     4875.33
 Rome           | Mar   |               87 |     1557.39
 Bangalore      | May   |              317 |     8936.99
 Beijing        | Jul   |              411 |    11600.67
 ...
(10 rows)
```

获取 Hive 表的字段和数据类型映射。

```
db1=# \d+ hcatalog.test.sales_info;
    PXF Hive Table "test.sales_info"
      Column       | Type   | Source type
-------------------+--------+-------------
 location          | text   | string
 month             | text   | string
 number_of_orders  | int4   | int
 total_sales       | float8 | double
```

可以使用通配符获取所有 Hive 库表的信息。

```
\d+ hcatalog.test.*;
\d+ hcatalog.*.*;
```

还可以使用 pxf_get_item_fields 函数获得 Hive 表的描述信息。该函数目前仅支持 Hive profile。

```
db1=# select * from pxf_get_item_fields('hive','test.sales_info');
 path  |  itemname   |    fieldname      | fieldtype | sourcefieldtype
-------+-------------+-------------------+-----------+-----------------
 test  | sales_info  | location          | text      | string
 test  | sales_info  | month             | text      | string
 test  | sales_info  | number_of_orders  | int4      | int
 test  | sales_info  | total_sales       | float8    | double
(4 rows)
```

pxf_get_item_fields 函数同样也支持通配符。

```
select * from pxf_get_item_fields('hive','test.*');
select * from pxf_get_item_fields('hive','*.*');
```

（3）查询 Hive 外部表

使用外部表方式需要标识适当的 profile。PXF Hive 插件支持三种 Hive 相关的 profile，Hive、HiveText 和 HiveRC。HiveText 和 HiveRC 分别针对 TEXT 和 RC 文件格式做了特别优化，而 Hive profile 可用于所有 PXF 支持的 Hive 文件存储类型。当底层 Hive 表由多个分区组成，并且分区使用了不同的文件格式时，需要使用 Hive profile。以下语法创建一个 HAWQ 的 Hive 外部表：

```
CREATE EXTERNAL TABLE <table_name>
    ( <column_name> <data_type> [, ...] | LIKE <other_table> )
LOCATION ('pxf://<host>[:<port>]/<hive-db-name>.<hive-table-name>
    ?PROFILE=Hive|HiveText|HiveRC[&DELIMITER=<delim>']')
FORMAT 'CUSTOM|TEXT' (formatter='pxfwritable_import' | delimiter='<delim>')
```

CREATE EXTERNAL TABLE 语句中 Hive 插件使用关键字和相应值的描述如表 8-6 所示。

表 8-6 Hive 外部表建表语句说明

关键字	值
<host>[:]	HDFS NameNode 主机名、端口号
<hive-db-name>	Hive 数据库名，如果忽略，默认是 defaults
<hive-table-name>	Hive 表名
PROFILE	必须是 Hive、HiveText 或 HiveRC 之一
DELIMITER	指定字段分隔符，必须是单个 ascii 字符或相应字符的十六进制表示
FORMAT (Hive profile)	必须指定为 CUSTOM，仅支持内建的 pxfwritable_import formatter
FORMAT (HiveText and HiveRC profiles)	必须指定为 TEXT，并再次指定字段分隔符

① Hive Profile

Hive profile 适用于任何 PXF 支持的 Hive 文件存储格式，它实际上是为底层文件存储类型选择最优的 Hive* profile。

```
db1=# create external table salesinfo_hiveprofile(location text, month text,
num_orders int, total_sales float8)
db1-#             location ('pxf://hdp1:51200/test.sales_info?profile=hive')
db1-#             format 'custom' (formatter='pxfwritable_import');
CREATE EXTERNAL TABLE
db1=#
db1=# select * from salesinfo_hiveprofile;
   location   | month | num_orders | total_sales
--------------+-------+------------+-------------
 Prague       | Jan   |        101 |     4875.33
 Rome         | Mar   |         87 |     1557.39
 Bangalore    | May   |        317 |     8936.99
 Beijing      | Jul   |        411 |    11600.67
 ...
(10 rows)
```

注意外部表和 Hcatalog 查询计划的区别：

```
gpadmin=# explain select * from salesinfo_hiveprofile;
                         QUERY PLAN
-------------------------------------------------------------------------------
---- Gather Motion 24:1  (slice1; segments: 24)  (cost=0.00..501.44 rows=1000000
width=28)
   -> External Scan on salesinfo_hiveprofile  (cost=0.00..432.33 rows=41667
width=28)
 Settings:  default_hash_table_bucket_number=24
 Optimizer status: PQO version 1.684
(4 rows)


gpadmin=# explain select * from hcatalog.test.sales_info;
                         QUERY PLAN
-------------------------------------------------------------------------------
----
 Gather Motion 1:1  (slice1; segments: 1)  (cost=0.00..431.00 rows=1 width=28)
   -> External Scan on sales_info  (cost=0.00..431.00 rows=1 width=28)
 Settings:  default_hash_table_bucket_number=24
 Optimizer status: PQO version 1.684
(4 rows)
```

外部表查询使用了全部 24 个虚拟段，而 Hcatalog 查询只使用了 1 个虚拟段，显然外部表更加有效地利用了资源。

② HiveText Profile

使用 HiveText profile 时，必须在 LOCATION 和 FORMAT 两个子句中都指定分隔符选项。

```
db1=# create external table salesinfo_hivetextprofile(location text, month text,
num_orders int, total_sales float8)
db1-# location ('pxf://hdp1:51200/test.sales_info?profile=hivetext&delimiter=,')
db1-# format 'text' (delimiter=e',');
CREATE EXTERNAL TABLE
db1=# select * from salesinfo_hivetextprofile where location='Beijing';
 location | month | num_orders | total_sales
----------+-------+------------+-------------
 Beijing  | Jul   |        411 |    11600.67
 Beijing  | Dec   |        100 |     4248.41
(2 rows)
```

③ HiveRC Profile

建立一个 rcfile 格式的 Hive 表，并插入数据。

```
create table sales_info_rcfile (location string, month string,
    number_of_orders int, total_sales double)
  row format delimited fields terminated by ','
  stored as rcfile;

insert into table sales_info_rcfile select * from sales_info;
```

在 HAWQ 中查询 Hive 表。

```
db1=# create external table salesinfo_hivercprofile(location text, month text,
num_orders int, total_sales float8)
db1-#          location ('pxf://hdp1:51200/test.sales_info_rcfile?profile=hiverc&delimiter=,')
db1-#          format 'text' (delimiter=e',');
CREATE EXTERNAL TABLE
db1=#
db1=# select location, total_sales from salesinfo_hivercprofile;
   location    | total_sales
---------------+-------------
 Prague        |     4875.33
 Rome          |     1557.39
 Bangalore     |     8936.99
 Beijing       |    11600.67
...
(10 rows)
```

④ 访问 Parquet 格式的 Hive 表

PXF Hive profile 支持分区或非分区的 Parquet 存储格式。建立一个 Parquet 格式的 Hive 表，并插入数据。

```
create table sales_info_parquet (location string, month string,
    number_of_orders int, total_sales double)
    stored as parquet;

insert into sales_info_parquet select * from sales_info;
```

在 HAWQ 中查询 Hive 表。

```
db1=# create external table salesinfo_parquet (location text, month text,
num_orders int, total_sales float8)
db1-#    location ('pxf://hdp1:51200/test.sales_info_parquet?profile=hive')
db1-#    format 'custom' (formatter='pxfwritable_import');
CREATE EXTERNAL TABLE
db1=#
db1=# select * from salesinfo_parquet;
   location    | month | num_orders | total_sales
---------------+-------+------------+-------------
 Prague        | Jan   |        101 |     4875.33
 Rome          | Mar   |         87 |     1557.39
 Bangalore     | May   |        317 |     8936.99
 Beijing       | Jul   |        411 |    11600.67
...
(10 rows)
```

（4）复杂数据类型

步骤 01 准备数据文件，添加如下记录，用逗号分隔字段，第三个字段是 array 类型，第四个字段是 map 类型。

```
vi /tmp/pxf_hive_complex.txt
3,Prague,1%2%3,zone:euro%status:up
89,Rome,4%5%6,zone:euro
400,Bangalore,7%8%9,zone:apac%status:pending
183,Beijing,0%1%2,zone:apac
94,Sacramento,3%4%5,zone:noam%status:down
101,Paris,6%7%8,zone:euro%status:up
56,Frankfurt,9%0%1,zone:euro
202,Jakarta,2%3%4,zone:apac%status:up
313,Sydney,5%6%7,zone:apac%status:pending
76,Atlanta,8%9%0,zone:noam%status:down
```

步骤 02 建立 Hive 表。

```
create table table_complextypes( index int, name string, intarray array<int>,
propmap map<string, string>)
        row format delimited fields terminated by ','
        collection items terminated by '%'
        map keys terminated by ':'
        stored as textfile;
```

步骤 03 向 Hive 表装载数据。

```
load data local inpath '/tmp/pxf_hive_complex.txt' into table
table_complextypes;
```

步骤 04 查询 Hive 表，验证数据正确导入。

```
select * from table_complextypes;
```

步骤 05 在 HAWQ 中建立 Hive 外部表并查询数据。

```
db1=# create external table complextypes_hiveprofile(index int, name text,
intarray text, propmap text)
db1-#             location
('pxf://hdp1:51200/test.table_complextypes?profile=hive')
db1-#             format 'custom' (formatter='pxfwritable_import');
CREATE EXTERNAL TABLE
db1=# select * from complextypes_hiveprofile;
 index |   name    | intarray |              propmap
-------+-----------+----------+-----------------------------------
     3 | Prague    | [1,2,3]  | {"zone":"euro","status":"up"}
    89 | Rome      | [4,5,6]  | {"zone":"euro"}
   400 | Bangalore | [7,8,9]  | {"zone":"apac","status":"pending"}
   183 | Beijing   | [0,1,2]  | {"zone":"apac"}
...
(10 rows)
```

可以看到，复杂数据类型都被简单地转化为 HAWQ 的 TEXT 类型。

（5）访问 Hive 分区表

PXF Hive 插件支持 Hive 的分区特性与目录结构，并且提供了所谓的分区过滤下推功能，可以利用 Hive 的分区消除特性，以降低网络流量和 I/O 负载。PXF 的分区过滤下推与 MySQL 的索引条件下推（Index Condition Pushdown，ICP）的概念类似，都是将过滤条件下推至更底层的存储以提高性能。为了利用 PXF 的分区过滤下推功能，查询的 where 子句中应该只使用分区字段。否则，PXF 忽略分区过滤，过滤将在 HAWQ 端执行，影响查询性能。PXF 的 Hive 插件只对分区键执行过滤下推。

分区过滤下推默认是启用的：

```
db1=# show pxf_enable_filter_pushdown;
```

```
pxf_enable_filter_pushdown
----------------------------
 on
(1 row)
```

步骤 01 使用 Hive Profile 访问同构分区数据。

创建 Hive 表并装载数据。

```
create table sales_part (name string, type string, supplier_key int, price double)
        partitioned by (delivery_state string, delivery_city string)
        row format delimited fields terminated by ',';
 insert into table sales_part partition(delivery_state = 'CALIFORNIA', delivery_city = 'Fresno')
  values ('block', 'widget', 33, 15.17);
 insert into table sales_part partition(delivery_state = 'CALIFORNIA', delivery_city = 'Sacramento')
  values ('cube', 'widget', 11, 1.17);
 insert into table sales_part partition(delivery_state = 'NEVADA', delivery_city = 'Reno')
  values ('dowel', 'widget', 51, 31.82);
 insert into table sales_part partition(delivery_state = 'NEVADA', delivery_city = 'Las Vegas')
  values ('px49', 'pipe', 52, 99.82);
```

查询 sales_part 表。

```
select * from sales_part;
```

检查 sales_part 表在 HDFS 上的目录结构。

```
sudo -u hdfs hdfs dfs -ls -R /apps/hive/warehouse/test.db/sales_part
```

建立 PXF 外部表并查询数据。

```
db1=# create external table pxf_sales_part(
db1(#            item_name text, item_type text,
db1(#            supplier_key integer, item_price double precision,
db1(#            delivery_state text, delivery_city text)
db1-#            location ('pxf://hdp1:51200/test.sales_part?profile=hive')
db1-#            format 'custom' (formatter='pxfwritable_import');
CREATE EXTERNAL TABLE
db1=# select * from pxf_sales_part;
 item_name | item_type | supplier_key | item_price | delivery_state | delivery_city
```

```
 ----------+----------+---------------+------------+----------------+------
---------
  block    | widget   |            33 |      15.17 | CALIFORNIA     |Fresno
  dowel    | widget   |            51 |      31.82 | NEVADA         |Reno
  cube     | widget   |            11 |       1.17 | CALIFORNIA     |Sacramento
  px49     | pipe     |            52 |      99.82 | NEVADA         |Las Vegas
(4 rows)
```

执行一个非过滤下推的查询。

```
  db1=# select * from pxf_sales_part where delivery_city = 'Sacramento' and
item_name = 'cube';
  item_name | item_type | supplier_key | item_price | delivery_state |
delivery_city
 ----------+----------+---------------+------------+----------------+------
---------
  cube     | widget   |           11 |       1.17 | CALIFORNIA     | Sacramento
(1 row)
```

该查询会利用 Hive 过滤 delivery_city='Sacramento'的分区，但 item_name 上的过滤条件不会下推至 Hive，因为它不是分区列。当所有 Sacramento 分区的数据传到 HAWQ 后，在 HAWQ 端执行 item_name 的过滤。

执行一个过滤下推的查询。

```
  db1=# select * from pxf_sales_part where delivery_state = 'CALIFORNIA';
  item_name | item_type | supplier_key | item_price | delivery_state |
delivery_city
 ----------+----------+---------------+------------+----------------+------
---------
  cube     | widget   |           11 |       1.17 | CALIFORNIA     | Sacramento
  block    | widget   |           33 |      15.17 | CALIFORNIA     | Fresno
(2 rows)
```

步骤 02 使用 Hive Profile 访问异构分区数据。

一个 Hive 表中的不同分区可能有不同的存储格式，PXF Hive profile 也支持这种情况。

建立 Hive 表。

```
create external table hive_multiformpart( location string, month string,
number_of_orders int, total_sales double)
        partitioned by( year string )
        row format delimited fields terminated by ',';
```

记下 sales_info 和 sales_info_rcfile 表在 HDFS 中的位置。

```
describe extended sales_info;
...
```

```
location:hdfs://mycluster/apps/hive/warehouse/test.db/sales_info
...
describe extended sales_info_rcfile;
...
location:hdfs://mycluster/apps/hive/warehouse/test.db/sales_info_rcfile
...
```

给 hive_multiformpart 表增加两个分区，位置分别指向 sales_info 和 sales_info_rcfile。

```
alter table hive_multiformpart add partition (year = '2013')
location 'hdfs://mycluster/apps/hive/warehouse/test.db/sales_info';
alter table hive_multiformpart add partition (year = '2016')
location 'hdfs://mycluster/apps/hive/warehouse/test.db/sales_info_rcfile';
```

显式标识与 sales_info_rcfile 表对应分区的文件格式。

```
alter table hive_multiformpart partition (year='2016') set fileformat rcfile;
```

此时查看两个分区的存储格式，sales_info 表对应的分区使用的是默认的 TEXTFILE 格式，而 sales_info_rcfile 表对应的分区是 RCFILE 格式。

```
hive> show partitions hive_multiformpart;
OK
year=2013
year=2016
Time taken: 0.553 seconds, Fetched: 2 row(s)
hive> desc formatted hive_multiformpart partition(year=2013);
...
InputFormat:         org.apache.hadoop.mapred.TextInputFormat
OutputFormat:
org.apache.hadoop.hive.ql.io.HiveIgnoreKeyTextOutputFormat
...
hive> desc formatted hive_multiformpart partition(year=2016);
...
InputFormat:         org.apache.hadoop.hive.ql.io.RCFileInputFormat
OutputFormat:        org.apache.hadoop.hive.ql.io.RCFileOutputFormat
...
```

使用 Hcatalog 方式查询 hive_multiformpart 表。

```
db1=# select * from hcatalog.test.hive_multiformpart;
   location    | month | number_of_orders | total_sales | year
---------------+-------+------------------+-------------+------
 ...
 Prague        | Dec   |              333 |     9894.77 | 2013
 Bangalore     | Jul   |              271 |     8320.55 | 2013
```

```
    Beijing         | Dec  |            100     |     4248.41   | 2013
...
(20 rows)
```

使用外部表方式查询 hive_multiformpart 表。

```
db1=# create external table pxf_multiformpart(location text, month text,
num_orders int, total_sales float8, year text)
db1-#       location
('pxf://hdp1:51200/test.hive_multiformpart?profile=hive')
db1-#       format 'custom' (formatter='pxfwritable_import');
CREATE EXTERNAL TABLE
db1=# select * from pxf_multiformpart;
    location    | month | num_orders | total_sales | year
----------------+-------+------------+-------------+------
...
 Prague         | Dec   |        333 |     9894.77 | 2013
 Bangalore      | Jul   |        271 |     8320.55 | 2013
 Beijing        | Dec   |        100 |     4248.41 | 2013
...
(20 rows)

db1=# select sum(num_orders) from pxf_multiformpart where month='Dec' and year='2013';
 sum
-----
 433
(1 row)
```

8.4.5 访问 JSON 数据

PXF 的 JSON 插件用于读取存储在 HDFS 上的 JSON 文件，支持 N 层嵌套。为了使用 HAWQ 访问 JSON 数据，必须将 JSON 文件存储在 HDFS 上，并从 HDFS 数据存储创建外部表。在使用 PXF 访问 JSON 文件前，确认满足以下前提条件：

- 已经在集群所有节点上安装了 HDFS 插件（Ambari 会自动安装）。
- 已经在集群所有节点上安装了 JSON 插件（Ambari 会自动安装）。
- 已经测试了 PXF 对 HDFS 的访问。

1. PXF 与 JSON 文件协同工作

JSON 是一种基于文本的数据交换格式，其数据通常存储在一个以.json 为后缀的文件中。一个.json 文件包含一组对象的集合，一个 JSON 对象是一组无序的名/值对，值可以是字符串、数字、true、false、null，或者是一个对象或数组。对象和数组可以嵌套。下面是一个 JSON 数据文件的内容：

```
{
  "created_at":"MonSep3004:04:53+00002013",
  "id_str":"3845292566681725952",
  "user": {
    "id":31424214,
    "location":"COLUMBUS"
  },
  "coordinates":null
}
```

（1）JSON 到 HAWQ 的数据类型映射

为了在 HAWQ 中表示 JSON 数据，需要将使用私有数据类型的 JSON 值映射为等价的 HAWQ 数据类型值。表 8-7 是对 JSON 数据类型映射规则的总结。

表 8-7　JSON 与 HAWQ 数据类型映射

JSON 数据类型	HAWQ 数据类型
私有类型（integer、float、string、boolean、null）	使用对应的 HAWQ 内建数据类型（integer、real、double precision、char、varchar、text、boolean）
Array	使用[]标识一个特定数组中具有私有数据类型成员的下标
Object	使用．点标识符指定每个级别的具有私有数据类型的嵌套成员

（2）JSON 文件读模式

PXF 的 JSON 插件用两个模式之一读取数据。默认模式是每行一个完整的 JSON 记录，同时也支持对多行构成的 JSON 记录的读操作。下面是每种读模式的例子。示例 schema 包含数据列的名称和数据类型如下：

- "created_at": text
- "id_str": text
- "user": object（"id": integer，"location": text）
- "coordinates": object（"type": text，"values": array（integer））

每行一条 JSON 记录的读模式：

```
{"created_at":"FriJun0722:45:03+00002013","id_str":"3431365513221365576","user":{"id":395504494,"location":"NearCornwall"},"coordinates":{"type":"Point","values": [ 6, 50 ]}},
{"created_at":"FriJun0722:45:02+00002013","id_str":"3431365471152537761","user":{"id":26643566,"location":"Austin,Texas"}, "coordinates": null},
{"created_at":"FriJun0722:45:02+00002013","id_str":"3431365471362334772","user":{"id":287819058,"location":""}, "coordinates": null}
```

多行 JSON 记录读模式：

```
{
  "root":[
    {
      "record_obj":{
        "created_at":"MonSep3004:04:53+00002013",
        "id_str":"3845292566681725952",
        "user":{
          "id":31424214,
          "location":"COLUMBUS"
        },
        "coordinates":null
      },
      "record_obj":{
        "created_at":"MonSep3004:04:54+00002013",
        "id_str":"3845292608722228864",
        "user":{
          "id":67600981,
          "location":"KryberWorld"
        },
        "coordinates":{
          "type":"Point",
          "values":[
            8,
            52
          ]
        }
      }
    }
  ]
}
```

下面演示如何从 PXF 的 JSON 外部表查询示例数据。

2. 将 JSON 数据装载到 HDFS

PXF 的 JSON 插件读取存储在 HDFS 中的 JSON 文件。因此在 HAWQ 查询 JSON 数据前，必须先将 JSON 文件传到 HDFS 上。将前面的单行和多行 JSON 记录分别保存到 singleline.json 和 multiline.json 文件中，并且确保 JSON 文件中没有空行，然后将文件传到 HDFS。

```
su - hdfs
hdfs dfs -mkdir /user/data
hdfs dfs -chown -R gpadmin:gpadmin /user/data
hdfs dfs -put singleline.json /user/data
hdfs dfs -put multiline.json /user/data
```

文件传到 HDFS 后，就可以通过 HAWQ 查询 JSON 数据。

3. 查询外部的 JSON 数据

使用下面的语法创建一个表示 JSON 数据的 HAWQ 外部表。

```
CREATE EXTERNAL TABLE <table_name>
    ( <column_name> <data_type> [, ...] | LIKE <other_table> )
LOCATION
( 'pxf://<host>[:<port>]/<path-to-data>?PROFILE=Json[&IDENTIFIER=<value>]' )
      FORMAT 'CUSTOM' ( FORMATTER='pxfwritable_import' );
```

CREATE EXTERNAL TABLE 语句中使用的各个关键字和相应值的描述如表 8-8 所示。

表 8-8　JSON 外部表建表语句说明

关键字	值
<host>[:<port>]	HDFS NameNode 主机名、端口
PROFILE	PROFILE 关键字必须指定为 JSON
IDENTIFIER	只有当 JSON 文件是多行记录格式时，LOCATION 字符串中才包含 IDENTIFIER 关键字及其对应的值。<value>应该标识用以确定一个返回的 JSON 对象成员名称，例如上面的示例中，应该指定&IDENTIFIER=created_at
FORMAT	FORMAT 子句必须指定为 CUSTOM
FORMATTER	JSON 'CUSTOM'格式只支持内建的'pxfwritable_import'格式属性

创建一个基于单行记录的 JSON 外部表并查询。

```
db1=# create external table sample_json_singleline_tbl(
db1(#   created_at text,
db1(#   id_str text,
db1(#   text text,
db1(#   "user.id" integer,
db1(#   "user.location" text,
db1(#   "coordinates.values[0]" integer,
db1(#   "coordinates.values[1]" integer
db1(# )
db1-# location('pxf://hdp1:51200/user/data/singleline.json?profile=json')
db1-# format 'custom' (formatter='pxfwritable_import');
CREATE EXTERNAL TABLE
db1=# select "user.id", "user.location", "coordinates.values[0]",
"coordinates.values[1]"
db1-#   from sample_json_singleline_tbl;
  user.id  | user.location | coordinates.values[0] | coordinates.values[1]
-----------+---------------+-----------------------+-----------------------
 395504494 | NearCornwall  |                     6 |                    50
```

```
 26643566 | Austin,Texas   |                      |
287819058 |                |                      |
(3 rows)
```

原来 JSON 中的嵌套数据都被平面化展开。在查询结果中，使用 . 访问嵌套 user 对象（user.id 和 user.location），使用 [] 访问 coordinates.values 数组的元素（coordinates.values[0] 和 coordinates.values[1]）。

多行记录的 JSON 外部表与单行的类似，只是需要指定 identifier，指定标识记录的键。

```
db1=# create external table sample_json_multiline_tbl(
db1(#    created_at text,
db1(#    id_str text,
db1(#    text text,
db1(#    "user.id" integer,
db1(#    "user.location" text,
db1(#    "coordinates.values[0]" integer,
db1(#    "coordinates.values[1]" integer
db1(# )
db1-# location('pxf://hdp1:51200/user/data/multiline.json?profile=json&identifier=created_at')
db1-# format 'custom' (formatter='pxfwritable_import');
CREATE EXTERNAL TABLE
db1=# select "user.id", "user.location", "coordinates.values[0]", "coordinates.values[1]"
db1-#   from sample_json_multiline_tbl;
 user.id  | user.location | coordinates.values[0] | coordinates.values[1]
----------+---------------+-----------------------+-----------------------
 31424214 | COLUMBUS      |                       |
 67600981 | KryberWorld   |                     8 |                    52
(2 rows)
```

8.4.6　向 HDFS 中写入数据

PXF 只能向 HDFS 文件中写入数据，而对 Hive、HBase 和 JSON 等外部数据都是只读的。在使用 PXF 向 HDFS 文件写数据前，确认已经在集群所有节点上安装了 PXF HDFS 插件（Ambari 会自动安装），并授予了 HAWQ 用户（典型的是 gpadmin）对 HDFS 文件相应的读写权限。

1. 写 PXF 外部表

PXF HDFS 插件支持两种可写的 profile：HdfsTextSimple 和 SequenceWritable。创建 HAWQ 可写外部表的语法如下：

```
CREATE WRITABLE EXTERNAL TABLE <table_name>
```

```
    ( <column_name> <data_type> [, ...] | LIKE <other_table> )
  LOCATION ('pxf://<host>[:<port>]/<path-to-hdfs-file>
    ?PROFILE=HdfsTextSimple|SequenceWritable[&<custom-option>=<value>[...]]'
)
  FORMAT '[TEXT|CSV|CUSTOM]' (<formatting-properties>);
```

CREATE EXTERNAL TABLE 语句中使用的各个关键字和相应值的描述如表 8-9 所示。

表 8-9 可写 HDFS 外部表建表语句说明

关键字	值
<host>[:<port>]	HDFS NameNode 主机名、端口
<path-to-hdfs-file>	HDFS 文件路径
PROFILE	PROFILE 关键字指定为 HdfsTextSimple 或 SequenceWritable
<custom-option>	与特定 PROFILE 对应的用户自定义选项
FORMAT 'TEXT'	当<path-to-hdfs-file>指向一个单行固定分隔符的平面文件时，使用该关键字
FORMAT 'CSV'	当<path-to-hdfs-file>指向一个单行或多行的逗号分隔值（CSV）平面文件时，使用该关键字
FORMAT 'CUSTOM'	SequenceWritable profile 使用该关键字。SequenceWritable 'CUSTOM'格式仅支持内建的 formatter='pxfwritable_export（写）和 formatter='pxfwritable_import（读）格式属性

2. 定制选项

HdfsTextSimple 和 SequenceWritable profile 支持表 8-10 所示的定制选项。

表 8-10 可写外部表建表定制选项

选项	值描述	Profile
COMPRESSION_CODEC	压缩编解码对应的 Java 类名。如果不提供，不会执行数据压缩。支持的压缩编解码包括 org.apache.hadoop.io.compress.DefaultCodec 和 org.apache.hadoop.io.compress.BZip2Codec	HdfsTextSimple、SequenceWritable
COMPRESSION_CODEC	org.apache.hadoop.io.compress.GzipCodec	HdfsTextSimple
COMPRESSION_TYPE	使用的压缩类型，支持的值为 RECORD（默认）或 BLOCK	HdfsTextSimple、SequenceWritable
DATA-SCHEMA	写入器的序列化/反序列化类名。类所在的 jar 文件必须在 PXF classpath 中。该选项被 SequenceWritable profile 使用，并且没有默认值	SequenceWritable
THREAD-SAFE	该 Boolean 值决定表查询是否运行在多线程模式，默认值为 TRUE	HdfsTextSimple、SequenceWritable

3. 使用 HdfsTextSimple Profile 写数据

HdfsTextSimple Profile 用于向单行每条记录（不含内嵌换行符）的固定分隔符平面文件写数据。使用 HdfsTextSimple Profile 建立可写表时，可以选择记录或块压缩，支持以下压缩编解码方法：

- org.apache.hadoop.io.compress.DefaultCodec
- org.apache.hadoop.io.compress.GzipCodec
- org.apache.hadoop.io.compress.BZip2Codec

HdfsTextSimple profile 支持的格式属性为'delimiter'，标识字段分隔符，默认值为逗号（,）。

（1）创建可写外部表，数据写到 HDFS 的/data/pxf_examples/pxfwritable_hdfs_textsimple1 目录中，字段分隔符为逗号。

```
create writable external table pxf_hdfs_writabletbl_1(location text, month text, num_orders int, total_sales float8)
    location('pxf://hdp1:51200/data/pxf_examples/pxfwritable_hdfs_textsimple1?profile=hdfstextsimple') format 'text' (delimiter=e',');
```

（2）向 pxf_hdfs_writabletbl_1 表插入数据。

```
insert into pxf_hdfs_writabletbl_1 values ( 'Frankfurt', 'Mar', 777, 3956.98 );
insert into pxf_hdfs_writabletbl_1 values ( 'Cleveland', 'Oct', 3812, 96645.37 );
insert into pxf_hdfs_writabletbl_1 select * from pxf_hdfs_textsimple;
```

（3）查看 HDFS 文件的内容。

```
[hdfs@hdp1 ~]$ hdfs dfs -cat /data/pxf_examples/pxfwritable_hdfs_textsimple1/*
Frankfurt,Mar,777,3956.98
Cleveland,Oct,3812,96645.37
Prague,Jan,101,4875.33
Rome,Mar,87,1557.39
Bangalore,May,317,8936.99
Beijing,Jul,411,11600.67
[hdfs@hdp1 ~]$ hdfs dfs -ls /data/pxf_examples/pxfwritable_hdfs_textsimple1
Found 3 items
-rw-r--r--   3 pxf gpadmin         26 2017-03-22 10:45 /data/pxf_examples/pxfwritable_hdfs_textsimple1/236002_0
-rw-r--r--   3 pxf gpadmin         28 2017-03-22 10:45 /data/pxf_examples/pxfwritable_hdfs_textsimple1/236003_0
-rw-r--r--   3 pxf gpadmin         94 2017-03-22 10:46 /data/pxf_examples/pxfwritable_hdfs_textsimple1/236004_15
[hdfs@hdp1 ~]$ hdfs dfs -cat /data/pxf_examples/pxfwritable_hdfs_textsimple1/236002_0
```

```
    Frankfurt,Mar,777,3956.98
    [hdfs@hdp1 ~]$ hdfs dfs -cat
/data/pxf_examples/pxfwritable_hdfs_textsimple1/236003_0
    Cleveland,Oct,3812,96645.37
    [hdfs@hdp1 ~]$ hdfs dfs -cat
/data/pxf_examples/pxfwritable_hdfs_textsimple1/236004_15
    Prague,Jan,101,4875.33
    Rome,Mar,87,1557.39
    Bangalore,May,317,8936.99
    Beijing,Jul,411,11600.67
    [hdfs@hdp1 ~]$
```

可以看到，一共写入了 6 条记录，生成了 3 个文件。其中两个文件各有 1 条记录，另外一个文件中有 4 条记录，记录以逗号作为字段分隔符。

（4）查询可写外部表。

HAWQ 不支持对可写外部表的查询。为了查询可写外部表的数据，需要建立一个可读外部表，指向 HDFS 的相应文件。

```
    db1=# select * from pxf_hdfs_writabletbl_1;
    ERROR:  External scan error: It is not possible to read from a WRITABLE external
table. Create the table as READABLE instead. (CTranslatorDXLToPlStmt.cpp:1041)
    db1=# create external table pxf_hdfs_textsimple_r1(location text, month text,
num_orders int, total_sales float8)
    db1-#             location
('pxf://hdp1:51200/data/pxf_examples/pxfwritable_hdfs_textsimple1?profile=hdfs
textsimple')
    db1-#             format 'csv';
CREATE EXTERNAL TABLE
    db1=# select * from pxf_hdfs_textsimple_r1;
  location  | month | num_orders | total_sales
------------+-------+------------+-------------
  Cleveland | Oct   |       3812 |    96645.37
  Frankfurt | Mar   |        777 |     3956.98
  Prague    | Jan   |        101 |     4875.33
  Rome      | Mar   |         87 |     1557.39
  Bangalore | May   |        317 |     8936.99
  Beijing   | Jul   |        411 |    11600.67
(6 rows)
```

（5）建立一个使用 Gzip 压缩，并用冒号（:）做字段分隔符的可写外部表，注意类名区分大小写。

```
    create writable external table pxf_hdfs_writabletbl_2 (location text, month text,
num_orders int, total_sales float8)
```

```
    location
('pxf://hdp1:51200/data/pxf_examples/pxfwritable_hdfs_textsimple2?profile=hdfs
textsimple&compression_codec=org.apache.hadoop.io.compress.GzipCodec')
    format 'text' (delimiter=e':');
```

（6）插入数据。

```
    insert into pxf_hdfs_writabletbl_2 values ( 'Frankfurt', 'Mar', 777, 3956.98 );
    insert into pxf_hdfs_writabletbl_2 values ( 'Cleveland', 'Oct', 3812,
96645.37 );
```

（7）使用-text参数查看压缩的数据。

```
[hdfs@hdp1 ~]$ hdfs dfs -text
/data/pxf_examples/pxfwritable_hdfs_textsimple2/*
    Frankfurt:Mar:777:3956.98
    Cleveland:Oct:3812:96645.37
[hdfs@hdp1 ~]$
```

可以看到刚插入的两条记录，记录以冒号作为字段分隔符。

8.5 小结

HAWQ 支持 SELECT、INSERT 语句，不支持 UPDATE、DELETE 语句。向 HAWQ 表装载数据的常用方法有 gpfdist 外部表、Web 外部表、hawq load 命令行工具、COPY SQL 命令等。其中，gpfdist 是 HAWQ 提供的一种文件服务器，它利用 HAWQ 系统中的所有 Segment 并行读写本地文件。向 HAWQ 表中导入大量数据后，应该执行 ANALYZE SQL 命令，为查询优化器更新系统统计信息。通过 PXF 扩展框架，HAWQ 可以访问 HDFS、Hive、JSON 等外部数据。使用 Ambari 安装 HAWQ 时，默认安装 PXF 服务，不需要额外手动配置。

第 9 章 过程语言

HAWQ 支持用户自定义函数（user-defined functions，UDF），还支持给 HAWQ 的内部函数起别名。编写 UDF 的语言可以是 SQL、C、Java、Perl、Python、R 和 pgSQL，其中除 SQL 和 C 是 HAWQ 的内建语言，其他语言通常被称为过程语言（PLs），支持过程语言编程是对 HAWQ 核心的功能性扩展。本章主要研究 HAWQ 内建的 SQL 语言函数和 PL/pgSQL 函数编程。为了便于说明，执行下面的 SQL 语句创建一个名为 channel 的示例表，并生成一些数据。后面定义的函数大都以操作 channel 表为例。

```
create table channel (
    id int not null,
    cname varchar(200) not null,
    parent_id int not null);

insert into channel values
(13,'首页',-1), (14,'tv580',-1), (15,'生活580',-1),
(16,'左上幻灯片',13), (17,'帮忙',14),  (18,'栏目简介',17);

analyze channel;
```

9.1 HAWQ 内建 SQL 语言

默认时，在 HAWQ 的所有数据库中都可以使用 SQL 和 C 语言编写用户自定义函数。SQL 函数中可执行任意条数的 SQL 语句。在 SQL 函数体中，每条 SQL 语句必须以分号（;）分隔。SQL 函数可以返回 void 或返回 return 语句指定类型的数据。由于 HAWQ 只有函数而没有存储过程的概念，returns void 可用来模拟没有返回值的存储过程。所有非 returns void 函数的最后一句 SQL 必须是返回指定类型的 SELECT 语句，函数返回最后一条查询语句的结果，可以是单行或多行结果集。下面是几个 SQL 函数的例子。

```
create function fn_count_channel() returns bigint as $$
  select count(*) from channel;
```

```
$$ language sql;
```

该函数没有参数，并返回 channel 表的记录数，函数调用结果如下：

```
db1=# select fn_count_channel();
 fn_count_channel
------------------
                6
(1 row)
```

修改上面定义的函数：

```
create or replace function fn_count_channel() returns bigint as $$
    select count(*) from channel;
    select count(*) from channel where parent_id=-1;
$$ language sql;
```

该函数体内执行了两条查询语句。在函数参数和返回值的定义没有变化时，可以使用 CREATE OR REPLAC 重新定义函数体，该语法与 Oracle 类似。如果函数参数或返回值的定义发生变化，必须先删除再重建函数。函数返回最后一条查询语句的结果，即 parent_id=-1 的记录数，调用结果如下：

```
db1=# select fn_count_channel();
 fn_count_channel
------------------
                3
(1 row)
```

再次修改 fn_count_channel()函数：

```
create or replace function fn_count_channel() returns bigint as $$
    select count(*) from channel;
    create table t1 (a int);
    drop table t1;
    select count(*) from channel where parent_id=-1;
$$ language sql;
```

函数体中也能执行 DDL 语句，调用结果和前一个的函数相同。

改变 fn_count_channel()函数的返回值类型，必须先删除再重建，不能使用 CREATE OR REPLACE 语法。

```
db1=# create or replace function fn_count_channel() returns void as $$
db1$# $$ language sql;
ERROR:  cannot change return type of existing function
HINT:  Use DROP FUNCTION first.
db1=# select fn_count_channel();
 fn_count_channel
```

```
                 3
(1 row)

db1=# drop function fn_count_channel();
DROP FUNCTION
db1=# create or replace function fn_count_channel() returns void as $$
db1$# $$ language sql;
CREATE FUNCTION
db1=# select fn_count_channel();
 fn_count_channel
------------------

(1 row)
```

该函数没有返回值,而且函数体内没有任何 SQL 语句。

9.2 PL/pgSQL 函数

SQL 是关系数据库使用的查询语言,其最大特点是简单易用,但主要问题是每条 SQL 语句必须由数据库服务器独立执行,而且缺少必要的变量定义、流程控制等编程手段。过程语言解决的就是这个问题。顾名思义,PL/pgSQL 以 PostgreSQL 作为编程语言,自动在所有 HAWQ 数据库中安装。它能实现以下功能:

- 建立 plpgsql 函数。
- 为 SQL 语言增加控制结构。
- 执行复杂计算。
- 继承所有 PostgreSQL 的数据类型(包括用户自定义类型)、函数和操作符。

每条 SQL 语句由数据库服务器独立执行的情况下,客户端应用向数据库服务器发送一个查询请求后,必须等待处理完毕,接收处理结果,做相应的计算,然后再向服务器发送后面的查询。通常客户端与数据库服务器不在同一物理主机上,这种频繁的进程间通信增加了网络开销。使用 PL/pgSQL 函数,可以将一系列查询和计算作为一组保存在数据库服务器中。它结合了过程语言的强大功能与 SQL 语言的易用性,并且显著降低了客户端/服务器的通信开销。正因如此,很多时候 UDF 的性能比不使用存储函数的情况会有很大提高:

- 消除了客户端与服务器之间的额外往复,只需要一次调用并接收结果即可。
- 客户端不需要中间处理结果,从而避免了它和服务器之间的数据传输或转换。
- 避免多次查询解析。

PL/pgSQL 函数参数接收任何 HAWQ 服务器所支持的标量数据类型或数组类型，也可以返回这些数据类型。除此之外，PL/pgSQL 还可以接收或返回任何自定义的复合数据类型，也支持返回单行记录（record 类型）或多行结果集（set of record 或 table 类型）。返回结果集的函数通过执行 RETURN NEXT 语句生成一条返回的记录（与 PostgreSQL 不同，HAWQ 函数不支持 RETURN QUERY 语法）。

PL/pgSQL 可以声明输出参数，这种方式可代替用 returns 语句显式指定返回数据类型的写法。当返回值是单行多列时，用输出参数的方式更方便。

9.3 给 HAWQ 内部函数起别名

许多 HAWQ 的内部函数是用 C 语言编写的。这些函数在 HAWQ 集群初始化时声明，并静态连接到 HAWQ 服务器。用户不能自己定义新的内部函数，但可以给已存在的内部函数起别名。下面的例子创建了一个新的函数 fn_all_caps，它是 HAWQ 内部函数 upper 的别名。

```
create function fn_all_caps (text) returns text as 'upper' language internal strict;
```

该函数的调用结果如下：

```
db1=# select fn_all_caps('change me') ;
 fn_all_caps
-------------
 CHANGE ME
(1 row)
```

9.4 表函数

表函数返回多行结果集，调用方法就像查询一个 FROM 子句中的表、视图或子查询。如果表函数返回单列，那么返回的列名就是函数名。下面是一个表函数的例子，该函数返回 channel 表中给定 ID 值的数据。

```
create function fn_getchannel(int) returns setof channel as $$
select * from channel where id = $1;
$$ language sql;
```

可以使用以下语句调用该函数：

```
select * from fn_getchannel(-1) as t1;
select * from fn_getchannel(13) as t1;
```

调用结果如下:

```
db1=# select * from fn_getchannel(-1) as t1;
 id | cname | parent_id
----+-------+-----------
(0 rows)

db1=# select * from fn_getchannel(13) as t1;
 id | cname | parent_id
----+-------+-----------
 13 | 首页  |        -1
(1 row)
```

与 PostgreSQL 不同,HAWQ 的表函数不能用于表连接。在 PostgreSQL 中以下查询可以正常执行:

```
db1=# create table t1 (a int);
CREATE TABLE
db1=# insert into t1 values (1);
INSERT 0 1
db1=# select * from t1,fn_getchannel(13);
 a | id | cname | parent_id
---+----+-------+-----------
 1 | 13 | 首页  |        -1
(1 行记录)
```

但是在 HAWQ 中,同样的查询会报错:

```
db1=# create table t1 (a int);
CREATE TABLE
db1=# insert into t1 values (1);
INSERT 0 1
db1=# select * from t1,fn_getchannel(13);
ERROR: function cannot execute on segment because it accesses relation "public.channel" (functions.c:152)  (seg0 hdp3:5432 pid=575910) (dispatcher.c:1801)
DETAIL: SQL function "fn_getchannel" during startup
db1=#
```

单独查询表函数是可以的:

```
db1=# create view vw_getchannel as select * from fn_getchannel(13);
CREATE VIEW
db1=# select * from vw_getchannel;
 id | cname | parent_id
----+-------+-----------
```

```
 13 | 首页    |       -1
(1 row)
```

在某些场景下，函数返回的结果依赖于调用它的参数。为了支持这种情况，表函数可以被声明为返回伪类型（pseudotype）的记录。当这种函数用于查询中时，必须由查询本身指定返回的行结构。下面的例子使用动态SQL，返回结果集依赖于作为入参的查询语句。

```
create or replace function fn_return_pseudotype (str_sql text)
 returns setof record as
$$
declare
   v_rec record;
begin
  for v_rec in execute str_sql loop
    return next v_rec;
  end loop;
  return;
end;
$$
language plpgsql;
```

调用函数时必须显式指定返回的字段名及其数据类型。

```
db1=# select * from fn_return_pseudotype('select 1') t (id int);
 id
----
  1
(1 row)


db1=# select * from fn_return_pseudotype('select * from channel') t (id int,cname varchar(200),parent_id int);
 id |   cname    | parent_id
----+------------+-----------
 13 | 首页       |       -1
 14 | tv580      |       -1
 15 | 生活580    |       -1
 16 | 左上幻灯片 |       13
 17 | 帮忙       |       14
 18 | 栏目简介   |       17
(6 rows)
```

https://www.postgresql.org/docs/8.2/static/datatype-pseudo.html 显示了 PostgreSQL 8.2 支持的伪类型。伪类型不能作为表列或变量的数据类型，但可以被用于函数的参数或返回值类型。

9.5 参数个数可变的函数

HAWQ 从 PostgreSQL 继承了一个非常好的特性，即函数参数的个数可变。在某些数据库系统中，想实现这个功能是很麻烦的。参数个数可变是通过一个动态数组实现的，因此所有参数都应该具有相同的数据类型。这种函数将最后一个参数标识为 VARIADIC，并且参数必须声明为数组类型。下面是一个例子，实现类似原生函数 greatest 的功能。

```
create or replace function fn_mgreatest(variadic numeric[]) returns numeric as
$$
declare
    l_i numeric:=-99999999999999;
    l_x numeric;
    array1 alias for $1;
begin
  for i in array_lower(array1, 1) .. array_upper(array1, 1)
  loop
    l_x:=array1[i];
    if l_x > l_i then
      l_i := l_x;
    end if;
  end loop;
  return l_i;
end;
$$ language 'plpgsql';
```

执行函数结果如下：

```
db1=# select fn_mgreatest(array[10, -1, 5, 4.4]);
 fn_mgreatest
--------------
       10
(1 row)

db1=# select fn_mgreatest(array[10, -1, 5, 4.4, 100]);
 fn_mgreatest
--------------
       100
(1 row)
```

9.6 多态类型

PostgreSQL 中的 anyelement、anyarray、anynonarray 和 anyenum 四种伪类型被称为多态类型。使用这些类型声明的函数叫做多态函数。多态函数的同一参数在每次调用函数时可以有不同数据类型，实际使用的数据类型由调用函数时传入的参数所确定。

当一个查询调用多态函数时，特定的数据类型在运行时解析。每个声明为 anyelement 的位置（参数或返回值）允许是任何实际的数据类型，但是在任何一次给定的函数调用中，anyelement 必须具有相同的实际数据类型。同样，每个声明为 anyarray 的位置允许是任何实际的数组数据类型，但是在任何一次给定的函数调用中，anyarray 也必须具有相同类型。如果某些位置声明为 anyarray，而另外一些位置声明为 anyelement，那么实际的数组元素类型必须与 anyelement 的实际数据类型相同。

anynonarray 在操作上与 anyelement 完全相同，它只是在 anyelement 的基础上增加了一个额外约束，即实际类型不能是数组。anyenum 在操作上也与 anyelement 完全相同，它只是在 anyelement 的基础上增加了一个额外约束，即实际类型必须是枚举（enum）类型。anynonarray 和 anyenum 并不是独立的多态类型，它们只是在 anyelement 上增加了约束而已。例如，f(anyelement, anyenum)与 f(anyenum, anyenum)是等价的，实际参数都必须是同样的枚举类型。

如果一个函数的返回值被声明为多态类型，那么它的参数中至少应该有一个是多态的，并且参数与返回结果的实际数据类型必须匹配。例如，函数声明为 assubscript(anyarray, integer) returns anyelement。此函数的第一个参数为数组类型，而且返回值必须是实际数组元素的数据类型。再比如一个函数的声明为 asf(anyarray) returns anyenum，那么参数只能是枚举类型的数组。

参数个数可变的函数也可以使用多态类型，实现方式是声明函数的最后一个参数为 VARIADIC anyarray。

下面看几个多态类型函数的列子。

例 1：判断两个入参是否相等，每次调用的参数类型可以不同，但两个入参的类型必须相同。

```
create or replace function fn_equal (anyelement,anyelement)
    returns boolean as $$
begin
    if $1 = $2 then return true;
    else return false;
    end if;
end; $$
language 'plpgsql';
```

函数调用：

```
db1=# select fn_equal(1,1);
 fn_equal
```

```
----------
 t
(1 row)

db1=# select fn_equal(1,'a');
ERROR:  invalid input syntax for integer: "a"
LINE 1: select fn_equal(1,'a');
                          ^
db1=# select fn_equal('a','A');
ERROR:  could not determine anyarray/anyelement type because input has type
"unknown"
LINE 1: select fn_equal('a','A');
                        ^
db1=# select fn_equal(text 'a',text 'A');
 fn_equal
----------
 f
(1 row)

db1=# select fn_equal(text 'a',text 'a');
 fn_equal
----------
 t
(1 row)
```

例 2：遍历任意类型的数组，数组元素以行的形式返回。

```
create or replace function fn_unnest(anyarray)
returns setof anyelement
language 'sql' as
$$
   select $1[i] from generate_series(array_lower($1,1),array_upper($1,1)) i;
$$;
```

下面是调用函数返回情况：

```
db1=# select fn_unnest(array[1,2,3,4]);
 fn_unnest
-----------
         1
         2
         3
         4
(4 rows)
```

```
db1=# select fn_unnest(array['a','b','c']);
 fn_unnest
-----------
 a
 b
 c
(3 rows)
```

例3：新建 fn_mgreatest1 函数，使它能返回任意数组类型中的最大元素。

```
create or replace function fn_mgreatest1(v anyelement, variadic anyarray)
returns anyelement as $$
declare
   l_i v%type;
   l_x v%type;
   array1 alias for $2;
begin
   l_i := array1[1];
   for i in array_lower(array1, 1) .. array_upper(array1, 1) loop
      l_x:=array1[i];
      if l_x > l_i then
         l_i := l_x;
      end if;
   end loop;
   return l_i;
end;
$$ language 'plpgsql';
```

说明：

- 变量不能定义成伪类型，但可以通过参数进行引用，如上面函数中的 l_i v%type。
- 动态数组必须是函数的最后一个参数。
- 第一个参数的作用仅是为变量定义数据类型，所以在调用函数时传空即可。

下列是调用函数返回情况：

```
db1=# select fn_mgreatest1(null, array[10, -1, 5, 4.4]);
 fn_mgreatest1
---------------
           10
(1 row)


db1=# select fn_mgreatest1(null, array['a', 'b', 'c']);
 fn_mgreatest1
---------------
 c
(1 row)
```

9.7 UDF 管理

1. 查看 UDF 定义

psql 的元命令\df 可以查看 UDF 的定义，返回函数的参数与返回值的类型。用 psql 命令行的-E 参数，还能够看到元命令对应的对系统表的查询语句。

```
[gpadmin@hdp3 ~]$ psql -d db1 -E
psql (8.2.15)
Type "help" for help.

db1=# \df
********* QUERY **********
SELECT n.nspname as "Schema",
  p.proname as "Name",
  CASE WHEN p.proretset THEN 'SETOF ' ELSE '' END ||
  pg_catalog.format_type(p.prorettype, NULL) as "Result data type",
  CASE WHEN proallargtypes IS NOT NULL THEN
    pg_catalog.array_to_string(ARRAY(
      SELECT
        CASE
          WHEN p.proargmodes[s.i] = 'i' THEN ''
          WHEN p.proargmodes[s.i] = 'o' THEN 'OUT '
          WHEN p.proargmodes[s.i] = 'b' THEN 'INOUT '
          WHEN p.proargmodes[s.i] = 'v' THEN 'VARIADIC '
        END ||
        CASE
          WHEN COALESCE(p.proargnames[s.i], '') = '' THEN ''
          ELSE p.proargnames[s.i] || ' '
        END ||
        pg_catalog.format_type(p.proallargtypes[s.i], NULL)
      FROM
        pg_catalog.generate_series(1, pg_catalog.array_upper(p.proallargtypes, 1)) AS s(i)
    ), ', ')
  ELSE
    pg_catalog.array_to_string(ARRAY(
      SELECT
        CASE
          WHEN COALESCE(p.proargnames[s.i+1], '') = '' THEN ''
          ELSE p.proargnames[s.i+1] || ' '
        END ||
```

```
              pg_catalog.format_type(p.proargtypes[s.i], NULL)
       FROM
         pg_catalog.generate_series(0, pg_catalog.array_upper(p.proargtypes,
1)) AS s(i)
       ), ', ')
     END AS "Argument data types",
   CASE
     WHEN p.proisagg THEN 'agg'
     WHEN p.prorettype = 'pg_catalog.trigger'::pg_catalog.regtype THEN
'trigger'
     ELSE 'normal'
   END AS "Type"
FROM pg_catalog.pg_proc p
     LEFT JOIN pg_catalog.pg_namespace n ON n.oid = p.pronamespace
WHERE pg_catalog.pg_function_is_visible(p.oid)
      AND n.nspname <> 'pg_catalog'
      AND n.nspname <> 'information_schema'
ORDER BY 1, 2, 4;
***************************

                              List of functions
 Schema |       Name         |Result data type |Argument data types   |Type
--------+--------------------+-----------------+----------------------+------
 public | fn_all_caps        | text            | text                 | normal
 ...
 public |fn_return_pseudotype| SETOF record    | str_sql text         | normal (8 rows)
```

可以看到，用户自定义函数包含在 **pg_proc** 系统表中。以下语句查看函数体：

```
db1=# select prosrc from pg_proc where proname='fn_return_pseudotype';
              prosrc
-----------------------------------------

 declare
    v_rec record;
 begin
   for v_rec in execute str_sql loop
      return next v_rec;
   end loop;
   return;
 end;
```

```
(1 row)
```

2. 删除 UDF

使用 DROP FUNCTION <function_name>命令删除函数。该命令需要加上函数定义的参数类型列表，但不须带参数名。

```
db1=# drop function fn_mgreatest1;
ERROR:  syntax error at or near ";"
LINE 1: drop function fn_mgreatest1;
                                   ^
db1=# drop function fn_mgreatest1();
ERROR:  function fn_mgreatest1() does not exist
db1=# drop function fn_mgreatest1(anyelement, variadic anyarray);
DROP FUNCTION
```

9.8 UDF 实例——递归树形遍历

经常在一个表中存在表示父子关系的两个字段，比如 empno 与 manager，本章开始建立的示例表 channel 也属于这种结构。Oracle 中可以使用 connect by 简单解决此类树的遍历问题，PostgreSQL 9 也有相似功能的 with recursive 语法。

```
db1=# with recursive t (id, cname, parent_id, path, depth) as (
db1(#     select id, cname, parent_id, array[id] as path, 1 as depth
db1(#       from channel
db1(#      where parent_id = -1
db1(#     union all
db1(#     select c.id, c.cname, c.parent_id, t.path || c.id, t.depth + 1 as depth
db1(#       from channel c
db1(#       join t on c.parent_id = t.id
db1(# )
db1-# select id, cname, parent_id, path, depth from t order by path;
 id |   cname    | parent_id |    path    | depth
----+------------+-----------+------------+-------
 13 | 首页       |        -1 | {13}       |     1
 16 | 左上幻灯片 |        13 | {13,16}    |     2
 14 | tv580      |        -1 | {14}       |     1
 17 | 帮忙       |        14 | {14,17}    |     2
 18 | 栏目简介   |        17 | {14,17,18} |     3
 15 | 生活580    |        -1 | {15}       |     1
(6 行记录)
```

但 HAWQ 不支持 with recursive 语法，同样的查询，会返回错误：

```
db1=# with recursive t (id, cname, parent_id, path, depth) as (
db1(#     select id, cname, parent_id, array[id] as path, 1 as depth
db1(#       from channel
db1(#      where parent_id = -1
db1(#      union all
db1(#     select c.id, c.cname, c.parent_id, t.path || c.id, t.depth + 1 as depth
db1(#       from channel c
db1(#       join t on c.parent_id = t.id )
db1-# select id, cname, parent_id, path, depth from t order by path;
ERROR:  RECURSIVE option in WITH clause is not supported
db1=#
```

我们可以使用 HAWQ 的递归函数功能，自己编写 UDF 来实现树的遍历。

1. 从某节点向下遍历子节点

（1）建立函数递归生成节点信息，函数返回以 '|' 作为字段分隔符的字符串。

```
create or replace function fn_ChildLst(int, int)
returns setof character varying
as
$$
declare
    v_rec character varying;
begin
    for v_rec in (select case when node = 1 then
                         q.id||'|'||q.cname||'|'||q.parent_id||'|'||$2
                         else fn_ChildLst(q.id, $2 + 1)
                         end
                    from (select id, cname, parent_id, node
                            from (select 1 as node
                                  union all
                                  select 2) nodes, channel
                           where parent_id = $1
                           order by id, node) q) loop
        return next v_rec;
    end loop;
    return;
end;
$$
language 'plpgsql';
```

（2）建立节点复合数据类型。

```
create type tp_depth as (rn int, id int, cname varchar(200), parent_id int, depth int);
```

（3）将 fn_ChildLst 函数的返回值转换为 tp_depth 类型。

```sql
create or replace function fn_ChildLst_split(int, int)
returns setof tp_depth
as
$$
    select cast(rownum as int) rn,
           cast(a[1] as int) id,
           a[2] cname,
           cast(a[3] as int) parent_id,
           cast(a[4] as int) depth
      from (select rownum,string_to_array(fn_ChildLst,'|') a
              from (select row_number() over() as rownum,*
                      from fn_ChildLst($1, $2)
                    union all
                    select 0,id||'|'||cname||'|'||parent_id||'|'||($2 -1) from channel where id = $1)
           t) t;
$$
language 'sql';
```

（4）建立查询结果复合数据类型。

```sql
create type tp_result as
(id int,
 name1 varchar(1000),
 parent_id int,
 depth int,path varchar(200),
 pathname varchar(1000));
```

（5）实现类似 Oracle SYS_CONNECT_BY_PATH 的功能，递归输出某节点 id 路径。

```sql
create or replace function fn_path(a_id integer)
returns character varying as $$
declare
    v_result character varying;
    v_parent_id int;
begin
    select t.parent_id into v_parent_id
      from channel as t where t.id = a_id;
    if found then
        v_result := fn_path(v_parent_id) || '/' || a_id;
    else
        return '';
    end if;
    return v_result;
```

```
end;
$$ language 'plpgsql';
```

(6) 递归输出某节点的 name 路径。

```
create or replace function fn_pathname(a_id integer)
returns character varying as $$
declare
   v_result character varying;
   v_parent_id int;
 v_cname varchar(200);
 begin
   select t.cname,t.parent_id into v_cname,v_parent_id
     from channel as t where t.id = a_id;
   if found then
      v_result := fn_pathname(v_parent_id) || '/' || v_cname;
   else
      return '';
   end if;
   return v_result;
end;
$$ language 'plpgsql';
```

(7) 建立输出子节点的函数。

```
create or replace function fn_showChildLst(int)
returns setof tp_result
as
$$
   select t1.id,
          repeat(' ', t1.depth)||'--'||t1.cname name1,
          t1.parent_id,
          t1.depth,
          fn_path(t1.id) path,
          fn_pathname(t1.id) pathname
     from fn_ChildLst_split($1,1) t1
    order by t1.rn;
$$
language 'sql';
```

调用函数结果如下：

```
db1=# select * from fn_showChildLst(-1);
 id |     name1        | parent_id | depth |   path    |  pathname
----+------------------+-----------+-------+-----------+-------------
 13 | --首页           |    -1     |   1   |   /13     |  /首页
```

```
 16 |     --左上幻灯片          |       13   |   2  | /13/16     | /首页/左上幻灯片
 14 |     --tv580             |       -1   |   1  | /14        | /tv580
 17 |       --帮忙            |       14   |   2  | /14/17     | /tv580/帮忙
 18 |         --栏目简介      |       17   |   3  | /14/17/18  | /tv580/帮忙/栏目简介
 15 |     --生活580           |       -1   |   1  | /15        | /生活580
(6 rows)

db1=# select * from fn_showChildLst(13);
 id  |     name1          | parent_id | depth |    path    |     pathname
-----+--------------------+-----------+-------+------------+---------------------
 13  | --首页             |     -1    |   0   | /13        | /首页
 16  |   --左上幻灯片     |     13    |   1   | /13/16     | /首页/左上幻灯片
(2 rows)

db1=# select * from fn_showChildLst(14);
 id  |     name1          | parent_id | depth |    path    |     pathname
-----+--------------------+-----------+-------+------------+---------------------
 14  | --tv580            |     -1    |   0   | /14        | /tv580
 17  |   --帮忙           |     14    |   1   | /14/17     | /tv580/帮忙
 18  |     --栏目简介     |     17    |   2   | /14/17/18  | /tv580/帮忙/栏目简介
(3 rows)

db1=# select * from fn_showChildLst(17);
 id  |     name1          | parent_id | depth |    path    |     pathname
-----+--------------------+-----------+-------+------------+---------------------
 17  | --帮忙             |     14    |   0   | /14/17     | /tv580/帮忙
 18  |   --栏目简介       |     17    |   1   | /14/17/18  | /tv580/帮忙/栏目简介
(2 rows)

db1=# select * from fn_showChildLst(18);
 id  |     name1          | parent_id | depth |    path    |     pathname
-----+--------------------+-----------+-------+------------+---------------------
 18  | --栏目简介         |     17    |   0   | /14/17/18  | /tv580/帮忙/栏目简介
(1 row)

db1=#
```

2. 从某节点向上追溯根节点

（1）从某节点向上递归生成节点信息，函数返回以'|'作为字段分隔符的字符串。

```
create or replace function fn_ParentLst(int, int)
returns setof character varying
as
$$
```

```
declare
    v_rec character varying;
begin
    for v_rec in (select case when node = 1 then
                            q.id||'|'||q.cname||'|'||q.parent_id||'|'||$2
                        else fn_ParentLst(q.parent_id, $2 + 1)
                        end
                    from (select id, cname, parent_id, node
                            from (select 1 as node
                                    union all
                                    select 2) nodes, channel
                            where id = $1
                            order by id, node) q) loop
        return next v_rec;
    end loop;
    return;
end;
$$
language 'plpgsql';
```

（2）将 fn_ParentLst 函数的返回值转换为 tp_depth 类型。

```
create or replace function fn_ParentLst_split(int, int)
returns setof tp_depth
as
$$
    select cast(rownum as int) rn,
           cast(a[1] as int) id,
           a[2] cname,
           cast(a[3] as int) parent_id,
           cast(a[4] as int) depth
      from (select rownum,string_to_array(fn_ParentLst,'|') a
              from (select row_number() over() as rownum,* from fn_ParentLst($1,$2)) t) t;
$$
language 'sql';
```

（3）建立输出父节点的函数。

```
create or replace function fn_showParentLst(int)
returns setof tp_result
as
$$
    select t1.id,
           repeat(' ', t1.depth)||'--'||t1.cname name1,
```

```
            t1.parent_id,
            t1.depth,
            fn_path(t1.id) path,
            fn_pathname(t1.id) pathname
   from fn_ParentLst_split($1,0) t1
   order by t1.rn;
$$
language 'sql';
```

调用函数结果如下:

```
db1=# select * from fn_showParentLst(-1);
 id | name1 | parent_id | depth | path | pathname
----+-------+-----------+-------+------+----------
(0 rows)

db1=# select * from fn_showParentLst(13);
 id | name1    | parent_id | depth | path | pathname
----+----------+-----------+-------+------+----------
 13 | --首页   |        -1 |     0 | /13  | /首页
(1 row)

db1=# select * from fn_showParentLst(14);
 id | name1    | parent_id | depth | path | pathname
----+----------+-----------+-------+------+----------
 14 | --tv580  |        -1 |     0 | /14  | /tv580
(1 row)

db1=# select * from fn_showParentLst(17);
 id | name1    | parent_id | depth | path   | pathname
----+----------+-----------+-------+--------+---------------
 17 | --帮忙   |        14 |     0 | /14/17 | /tv580/帮忙
 14 | --tv580  |        -1 |     1 | /14    | /tv580
(2 rows)

db1=# select * from fn_showParentLst(18);
 id |  name1    | parent_id | depth |   path    |    pathname
----+-----------+-----------+-------+-----------+---------------------
 18 | --栏目简介|        17 |     0 | /14/17/18 | /tv580/帮忙/栏目简介
 17 | --帮忙    |        14 |     1 | /14/17    | /tv580/帮忙
 14 | --tv580   |        -1 |     2 | /14       | /tv580
(3 rows)

db1=#
```

9.9 小结

HAWQ 支持用户自定义函数（user-defined functions，UDF），编写 UDF 的语言可以是 SQL、C、Java、Perl、Python、R 和 pgSQL，其中除 SQL 和 C 是 HAWQ 的内建语言，其他语言通常被称为过程语言。HAWQ 本身没有存储过程的概念，但可以通过 returns void 函数来模拟存储过程。HAWQ UDF 支持表函数、参数个数可变的函数、多态类型等有用特性，还可以为 HAWQ 内建函数起别名。虽然 HAWQ 不支持递归查询，但通过自定义递归函数，亦能实现树形遍历。从 pg_proc 系统表能查看函数定义。在删除函数时，必须加上函数的参数类型列表，但不用带参数名。

第 10 章 查询优化

HAWQ 的查询有其自己的特点,因此即便对 SELECT 等数据库查询语句已经很熟悉了,还是需要认真研究一下。我们就用单独的一章来说明。

10.1 HAWQ 的查询处理流程

理解 HAWQ 的查询处理过程有助于写出更加优化的查询。与任何其他数据库管理系统类似,HAWQ 有如下查询执行步骤:

(1)用户使用客户端程序(如 psql)连接到 HAWQ Master 主机上的数据库实例,并向系统提交 SQL 语句。

(2)Master 接收到查询后,由查询编译器解析提交的 SQL 语句,并将生成的查询解析树递交给查询优化器。

(3)查询优化器根据查询所需的磁盘 I/O、网络流量等成本信息,生成它认为最优的执行计划,并将查询计划交给查询分发器。

(4)查询分发器依照查询计划的成本信息,向 HAWQ 资源管理器请求所需的资源。

(5)获得资源后,查询分发器在 Segment 上启动虚拟段,并向虚拟段分发查询计划。

(6)查询执行器使用多个虚拟段并行执行查询,将结果传回至 Master,最后 Master 向客户端返回查询结果。

HAWQ 基本的查询处理流程如图 10-1 所示。

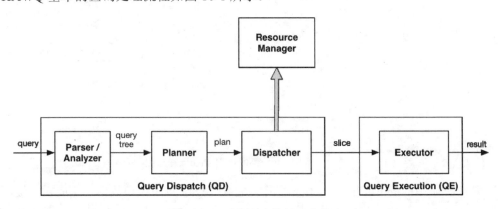

图 10-1　HAWQ 查询处理流程

1. 查询计划

一个查询计划是 HAWQ 为了产生查询结果而要执行的一系列操作。查询计划中的每个节点或步骤，表示一个数据库操作，如表扫描、连接、聚合、排序等等。查询计划被由底向上读取和执行。

除了通常的扫描、连接等数据库操作，HAWQ 还有一种叫做 motion 的操作类型。查询处理期间，motion 操作通过内部互联网络在节点间移动数据，并不是每个查询都需要 motion 操作。为了实现查询执行的最大并行度，HAWQ 将查询计划分成多个 slice，每个 slice 可以在 Segment 上独立执行。查询计划中的 motion 操作总是分片的，迁移数据的源和目标上各有一个 slice。

下面的查询连接两个数据库表：

```
select customer, amount
  from sales join customer using (cust_id)
 where datecol = '04-30-2016';
```

图 10-2 显示了为该查询生成的三个 slice。每个 Segment 接收一份查询计划的复制，查询计划在多个 Segment 上并行工作。

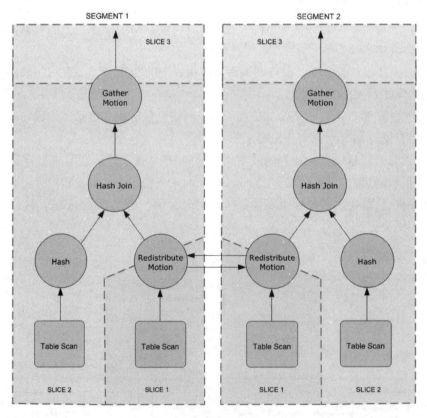

图 10-2　查询计划 slice

注意 slice 1 中的 redistribute motion 操作，它在 Segment 间移动数据以完成表连接。假设 customer 表通过 cust_id 字段在 Segment 上分布，而 sales 表通过 sale_id 字段分布。为了连接

两个表，sales 的数据必须通过 cust_id 重新分布。因此查询计划在每个分片上各有一个 redistribute motion 操作。

在这个执行计划中还有一种叫做 gather motion 的 motion 操作。当 Segment 将查询结果发送回 Master，用于向客户端展示时，会使用 gather motion。因为查询计划中发生 motion 的部分总是被分片，所以在图 10-2 的顶部还有一个隐含的 slice 3。并不是所有查询计划都包含 gather motion，例如，CREATE TABLE x AS SELECT ... 语句就没有 gather motion 操作，因为结果数据被发送到新表而不是 Master。

2. 并行执行

HAWQ 会创建许多数据库进程处理一个查询。Master 和 Segment 上的查询工作进程分别被称为查询分发器（query dispatcher，QD）和查询执行器（query executor，QE）。QD 负责创建和分发查询计划，并返回最终的查询结果。QE 在虚拟段中完成实际的查询工作，并与其他工作进程互通中间结果。

查询计划的每个 slice 至少需要一个工作进程。工作进程独立完成被赋予的部分查询计划。一个查询执行时，每个虚拟段中有多个并行执行的工作进程。工作在不同虚拟段中的相同 slice 构成一个 gang。查询计划被从下往上执行，一个 gang 的中间结果数据向上流向下一个 gang。不同虚拟段的进程间通信是由 HAWQ 的内部互联组件完成的。

图 10-3 显示了示例中 Master 和 Segment 上的工作进程，查询计划分成了三个 slice，两个 Segment 上的相同 slice 构成了 gang。

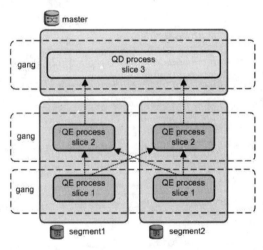

图 10-3　gang 与 slice

10.2 GPORCA 查询优化器

当前 HAWQ 默认使用的查询优化器是 GPORCA，但遗留的老优化器与 GPORCA 并存。

HAWQ 尽可能使用 GPORCA 生成查询的执行计划，当 GPORCA 没有启用或无法使用时，HAWQ 用老的查询优化器生成执行计划。可以通过 EXPLAIN 命令的输出确定查询使用的是哪种优化器。GPORCA 会忽略与老优化器相关的服务器配置参数，但当查询使用老优化器时，这些参数仍然影响查询计划的生成。相对于老优化器，GPORCA 在多核环境中的优化能力更强，并且在分区表查询、子查询、连接、排序等操作上提升了性能。图 10-4 显示了 HAWQ 查询优化器。

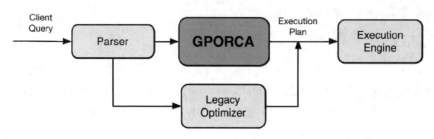

图 10-4　HAWQ 查询优化器

10.2.1　GPORCA 的改进

1. 分区表查询

GPORCA 在查询分区表时做了以下增强：

- 改进分区消除。
- 查询计划中包含了分区选择器操作符。
- 如果查询中分区键与常量进行比较，GPORCA 在 EXPLAIN 输出中的分区选择器操作符下列出需要扫描的分区数。如果查询中分区键与变量进行比较，只有在查询执行时才能知道需要扫描的分区数，因此 EXPLAIN 的输出中无法显示选择的分区。
- 查询计划的大小与分区数量无关。
- 减少了由于分区数量引起的内存溢出（Out of memory，OOM）错误。

下面看一个分区表查询的例子。

```
db1=# create table sales (order_id int, item_id int, amount numeric(15,2), date date, yr_qtr int)
db1-# partition by range (yr_qtr)
db1-# ( partition p201701 start (201701) inclusive ,
db1(#   partition p201702 start (201702) inclusive ,
db1(#   partition p201703 start (201703) inclusive ,
db1(#   partition p201704 start (201704) inclusive ,
db1(#   partition p201705 start (201705) inclusive ,
db1(#   partition p201706 start (201706) inclusive ,
db1(#   partition p201707 start (201707) inclusive ,
db1(#   partition p201708 start (201708) inclusive ,
```

```
db1(#     partition p201709 start (201709) inclusive ,
db1(#     partition p201710 start (201710) inclusive ,
db1(#     partition p201711 start (201711) inclusive ,
db1(#     partition p201712 start (201712) inclusive
db1(#                end (201801) exclusive );
...
CREATE TABLE
db1=#
```

GPORCA 改进了分区表上以下类型的查询：

- 全表扫描时，查询计划中不罗列分区，只显示分区数量。

```
db1=# explain select * from sales;
                                QUERY PLAN
------------------------------------------------------------------------
 Gather Motion 1:1  (slice1; segments: 1)  (cost=0.00..431.00 rows=1 width=24)
   -> Sequence  (cost=0.00..431.00 rows=1 width=24)
        -> Partition Selector for sales (dynamic scan id: 1)
(cost=10.00..100.00 rows=100 width=4)
             Partitions selected: 12 (out of 12)
        -> Dynamic Table Scan on sales (dynamic scan id: 1)  (cost=0.00..431.00
rows=1 width=24)
 Settings:  default_hash_table_bucket_number=24
 Optimizer status: PQO version 1.684
(7 rows)
```

- 查询中如果包含常量过滤谓词，执行分区消除。下面的查询只需要扫描 12 个分区中的 1 个。

```
db1=# explain select * from sales where yr_qtr = 201706;
                                QUERY PLAN
------------------------------------------------------------------------
 Gather Motion 1:1  (slice1; segments: 1)  (cost=0.00..431.00 rows=1 width=24)
   -> Sequence  (cost=0.00..431.00 rows=1 width=24)
        -> Partition Selector for sales (dynamic scan id: 1)
(cost=10.00..100.00 rows=100 width=4)
             Filter: yr_qtr = 201706
             Partitions selected: 1 (out of 12)
        -> Dynamic Table Scan on sales (dynamic scan id: 1)  (cost=0.00..431.00
rows=1 width=24)
             Filter: yr_qtr = 201706
 Settings:  default_hash_table_bucket_number=24
 Optimizer status: PQO version 1.684
(9 rows)
```

- 范围选择同样执行分区消除。下面的查询扫描 4 个分区。

```
db1=# explain select * from sales where yr_qtr between 201701 and 201704 ;
                                     QUERY PLAN
--------------------------------------------------------------------------------
 Gather Motion 1:1  (slice1; segments: 1)  (cost=0.00..431.00 rows=1 width=24)
   ->  Sequence  (cost=0.00..431.00 rows=1 width=24)
         ->  Partition Selector for sales (dynamic scan id: 1)  (cost=10.00..100.00 rows=100 width=4)
               Filter: yr_qtr >= 201701 AND yr_qtr <= 201704
               Partitions selected:  4 (out of 12)
         ->  Dynamic Table Scan on sales (dynamic scan id: 1)  (cost=0.00..431.00 rows=1 width=24)
               Filter: yr_qtr >= 201701 AND yr_qtr <= 201704
 Settings:  default_hash_table_bucket_number=24
 Optimizer status: PQO version 1.684
(9 rows)
```

- 查询中包含子查询过滤谓词，查询计划中显示扫描全部 12 个分区，但运行时可以进行动态分区消除。

```
db1=# explain select * from sales where yr_qtr = (select 201701);
                                     QUERY PLAN
--------------------------------------------------------------------------------
 Hash Join  (cost=0.00..431.00 rows=1 width=24)
   Hash Cond: "outer"."?column?" = sales.yr_qtr
   ->  Result  (cost=0.00..0.00 rows=1 width=4)
         ->  Result  (cost=0.00..0.00 rows=1 width=1)
   ->  Hash  (cost=431.00..431.00 rows=1 width=24)
         ->  Gather Motion 1:1  (slice1; segments: 1)  (cost=0.00..431.00 rows=1 width=24)
               ->  Sequence  (cost=0.00..431.00 rows=1 width=24)
                     ->  Partition Selector for sales (dynamic scan id: 1)  (cost=10.00..100.00 rows=100 width=4)
                           Partitions selected:  12 (out of 12)
                     ->  Dynamic Table Scan on sales (dynamic scan id: 1)  (cost=0.00..431.00 rows=1 width=24)
 Settings:  default_hash_table_bucket_number=24; optimizer=on
 Optimizer status: PQO version 1.684
(12 rows)
```

2. 子查询

GPORCA 能够更有效地处理子查询：

```
select * from part where price > (select avg(price) from part);
```

GPORCA 也能高效处理相关子查询（correlated subquery，CSQ）。相关子查询在子查询中引用了外层查询的值：

```
select * from part p1 where price > (select avg(price) from part p2 where  p2.brand = p1.brand);
```

GPORCA 为下面类型的相关子查询生成更有效的查询计划：

- 相关子查询出现在 SELECT 列表中。

```
select *,
 (select min(price) from part p2 where p1.brand = p2.brand)
 as foo
from part p1;
```

- 相关子查询出现在 OR 过滤中。

```
select * from part p1 where p_size > 40 or
    p_retailprice >
    (select avg(p_retailprice)
      from part p2
     where p2.p_brand = p1.p_brand);
```

- 多级嵌套相关子查询。

```
select * from part p1 where p1.p_partkey
in (select p_partkey from part p2 where p2.p_retailprice =
    (select min(p_retailprice)
      from part p3
     where p3.p_brand = p1.p_brand));
```

- 不等于条件的相关子查询。

```
select * from part p1 where p1.p_retailprice =
 (select min(p_retailprice) from part p2 where p2.p_brand <> p1.p_brand);
```

- 返回单行的相关子查询。

```
select p_partkey,
     (select p_retailprice from part p2 where p2.p_brand = p1.p_brand )
from part p1;
```

3. 共用表表达式

GPORCA 能处理包含 WITH 子句的查询。WITH 子句又被称为共用表表达式（common table expression，CTE），是在查询时系统自动生成的一个临时表。

```
db1=# create table t (a int,b int,c int);
CREATE TABLE
db1=# insert into t values (1,1,1), (2,2,2);
```

```
INSERT 0 2
db1=# with v as (select a, sum(b) as s from t where c < 10 group by a)
db1-#   select * from v as v1 , v as v2
db1-#   where v1.a <> v2.a and v1.s < v2.s;
 a | s | a | s
---+---+---+---
 1 | 1 | 2 | 2
(1 row)
```

作为查询优化的一部分，GPORCA 能将谓词过滤条件下推至 CTE，如下面的查询。

```
db1=# explain
db1-# with v as (select a, sum(b) as s from t group by a)
db1-#   select *
db1-#     from v as v1, v as v2, v as v3
db1-#     where v1.a < v2.a
db1-#     and v1.s < v3.s
db1-#     and v1.a = 10
db1-#     and v2.a = 20
db1-#     and v3.a = 30;
                    QUERY PLAN
--------------------------------------------------------------------------------
 ...
  -> Table Scan on t  (cost=0.00..431.00 rows=2 width=8)
       Filter: a = 10 OR a = 20 OR a = 30
 ...
 Settings: default_hash_table_bucket_number=24
 Optimizer status: PQO version 1.684
(34 rows)
```

GPORCA 可以处理以下类型的 CTE：

- 一条查询语句中定义多个 CTE。

```
db1=# with cte1 as (select a, sum(b) as s from t
db1(#                where c < 10 group by a),
db1-#      cte2 as (select a, s from cte1 where s > 1)
db1-#   select *
db1-#     from cte1 as v1, cte2 as v2, cte2 as v3
db1-#     where v1.a < v2.a and v1.s < v3.s;
 a | s | a | s | a | s
---+---+---+---+---+---
 1 | 1 | 2 | 2 | 2 | 2
(1 row)
```

- 嵌套 CTE。

```
db1=# with v as (with w as (select a, b from t
db1(#                       where b < 5)
db1(#           select w1.a, w2.b
db1(#             from w as w1, w as w2
db1(#             where w1.a = w2.a and w1.a > 1)
db1-#   select v1.a, v2.a, v2.b
db1-#     from v as v1, v as v2
db1-#     where v1.a <= v2.a;
 a | a | b
---+---+---
 2 | 2 | 2
(1 row)
```

4. INSERT 语句的提升

- 查询计划中增加 Insert 操作符。
- 引入 Assert 操作符用于约束检查。

```
db1=# drop table t;
DROP TABLE
db1=# create table t (a int not null, b int, c int);
CREATE TABLE
db1=# explain insert into t values (1,1,1);
                          QUERY PLAN
--------------------------------------------------------------------
 Insert  (cost=0.00..0.08 rows=1 width=12)
   -> Result  (cost=0.00..0.00 rows=1 width=20)
       -> Assert  (cost=0.00..0.00 rows=1 width=20)
            Assert Cond: NOT a IS NULL
            -> Result  (cost=0.00..0.00 rows=1 width=20)
                 -> Result  (cost=0.00..0.00 rows=1 width=1)
 Settings: default_hash_table_bucket_number=24
 Optimizer status: PQO version 1.684
(8 rows)
```

5. 去重聚合

GPORCA 提升了一类去重聚合查询的性能。当查询中包含有去重限定的聚合操作（distinct qualified aggregates，DQA），并且没有分组列，表也不是以聚合列做的分布，则 GPORCA 在三个阶段计算聚合函数，分别是本地、中间和全局聚合。

```
db1=# explain select count(distinct b) from t;
                          QUERY PLAN
--------------------------------------------------------------------
- Aggregate  (cost=0.00..431.00 rows=1 width=8)
    -> Gather Motion 1:1  (slice2; segments: 1)  (cost=0.00..431.00 rows=2
```

```
width=4)
            ->  GroupAggregate  (cost=0.00..431.00 rows=2 width=4)
                  Group By: b
                  ->  Sort  (cost=0.00..431.00 rows=2 width=4)
                        Sort Key: b
                        ->  Redistribute Motion 1:1  (slice1; segments: 1)
(cost=0.00..431.00 rows=2 width=4)
                              Hash Key: b
                              ->  GroupAggregate  (cost=0.00..431.00 rows=2 width=4)
                                    Group By: b
                                    ->  Sort  (cost=0.00..431.00 rows=2 width=4)
                                          Sort Key: b
                                          ->  Table Scan on t  (cost=0.00..431.00
rows=2 width=4)
 Settings:  default_hash_table_bucket_number=24
 Optimizer status: PQO version 1.684
(15 rows)
```

optimizer_prefer_scalar_dqa_multistage_agg 配置参数控制处理 DQA 的行为，该参数默认是启用的。

```
[gpadmin@hdp3 ~]$ hawq config -s optimizer_prefer_scalar_dqa_multistage_agg
GUC      : optimizer_prefer_scalar_dqa_multistage_agg
Value    : on
[gpadmin@hdp3 ~]$
```

启用该参数会强制 GPORCA 使用三阶段 DQA 计划，保证 DQA 查询具有可预测的性能。如果禁用该参数，则 GPORCA 使用基于成本的方法生成执行计划。

10.2.2 启用 GPORCA

预编译版本的 HAWQ 默认启用 GPORCA 查询优化器，不需要额外配置。当然也可以手工启用 GPORCA，这需要设置以下两个配置参数：

- 设置 optimizer_analyze_root_partition 参数收集分区表的根分区统计信息。
- 设置 optimizer 参数启用 GPORCA。

分区表上使用 GPORCA 时必须用 ANALYZE ROOTPARTITION 命令收集根分区的统计信息。该命令只收集根分区统计信息，而不收集叶分区。作为一项例行的数据库维护工作，应该在分区表数据大量改变（如装载了大量数据）后刷新根分区的统计。

1. 设置 optimizer_analyze_root_partition 参数

以 gpadmin 用户登录 HAWQ Master 主机设置环境。

```
[gpadmin@hdp3 ~]$ source /usr/local/hawq/greenplum_path.sh
```

使用 hawq config 应用程序设置 optimizer_analyze_root_partition 参数。

```
[gpadmin@hdp3 ~]$ hawq config -c optimizer_analyze_root_partition -v on
```

重载 HAWQ 配置。

```
[gpadmin@hdp3 ~]$ hawq stop cluster -u
```

2. 在系统级启用 GPORCA

以 gpadmin 用户登录 HAWQ Master 主机设置环境。

```
[gpadmin@hdp3 ~]$ source /usr/local/hawq/greenplum_path.sh
```

使用 hawq config 应用程序设置 optimizer 参数。

```
[gpadmin@hdp3 ~]$ hawq config -c optimizer -v on
```

重载 HAWQ 配置。

```
[gpadmin@hdp3 ~]$ hawq stop cluster -u
```

3. 在数据库级别启用 GPORCA

使用 ALTER DATABASE 命令设置一个数据库的优化器：

```
db1=# alter database db1 set optimizer = on ;
ALTER DATABASE
```

4. 在会话级启用 GPORCA

可以使用 SET 命令在会话级别设置优化器参数：

```
db1=# set optimizer = on ;
SET
```

为特定查询指定 GPORCA 优化器时，在运行查询前执行该 set 命令。

10.2.3 使用 GPORCA 需要考虑的问题

1. 使用 GPORCA 优化器的前提条件

为了使用 GPORCA 优化器执行查询，应该满足以下条件：

- 表不包含多列分区键。
- 表不包含多级分区。
- 不是查询仅存储在 Master 上的表，如全局系统表。

```
db1=# explain select * from pg_attribute;
                           QUERY PLAN
-----------------------------------------------------------------------
 Seq Scan on pg_attribute  (cost=0.00..62.70 rows=104880 width=103)
```

```
    Settings:  default_hash_table_bucket_number=24; optimizer=on
    Optimizer status: legacy query optimizer
    (3 rows)
```

- 已经收集了分区表的根分区统计信息。
- 表中的分区数不要太多，如果一个表的分区数超过了20000，应该重新设计表模式。

2. 确认查询使用的优化器

启用了 GPORCA 时，可以从 EXPLAIN 查询计划的输出中查看一个查询是使用了 GPORCA 还是老的优化器。如果使用的是 GPORCA，在查询计划的最后会显示 GPORCA 的版本：

```
db1=# explain select * from sales where yr_qtr = 201706;
                          QUERY PLAN
--------------------------------------------------------------------------------
...

    Settings:  default_hash_table_bucket_number=24; optimizer=on
    Optimizer status: PQO version 1.684
(9 rows)
```

如果查询使用了老的优化器生成执行计划，输出的最后会显示"legacy query optimizer"：

```
db1=# explain select 1;
                    QUERY PLAN
-----------------------------------------------------------
 Result  (cost=0.00..0.01 rows=1 width=0)
 Settings:  default_hash_table_bucket_number=24; optimizer=on
 Optimizer status: legacy query optimizer
(3 rows)
```

下面的操作只会出现在 GPORCA 生成的执行计划中，老优化器不支持这些操作。

- Assert operator
- Sequence operator
- DynamicIndexScan
- DynamicTableScan
- Table Scan

3. 生成查询优化上下文

GPORCA 可以生成 minidump 文件描述给定查询的优化细节,该文件可被用来分析 HAWQ 的问题。minidump 文件位于 Master 的数据目录下，文件名称的格式为：

```
Minidump_date_time.mdp
```

下面看一个生成 minidump 文件的例子。

(1) 运行一个 psql 会话，设置 optimizer_minidump 参数为 always。

```
[gpadmin@hdp3 ~]$ psql -d db1
psql (8.2.15)
Type "help" for help.

db1=# set optimizer_minidump=always;
SET
```

(2) 执行一个查询。

```
db1=# select * from t;
 a | b | c
---+---+---
 1 | 1 | 1
 1 | 2 | 2
(2 rows)
```

(3) 查看生成的 minidump 文件。

```
[gpadmin@hdp3 ~]$ ls -l /data/hawq/master/minidumps/
总用量 12
-rw------- 1 gpadmin gpadmin 8949 4月  11 17:07
Minidump_20170411_170712_72720_2.mdp
[gpadmin@hdp3 ~]$
```

(4) 运行 xmllint 将 minidump 文件格式化，并将格式化后内容输出到一个新文件。

```
[gpadmin@hdp3 ~]$ xmllint --format
/data/hawq/master/minidumps/Minidump_20170411_170712_72720_2.mdp >
/data/hawq/master/minidumps/MyTest.xml
```

(5) 查看良好格式的 minidump 文件。

```
[gpadmin@hdp3 ~]$ cat /data/hawq/master/minidumps/MyTest.xm
```

10.2.4 GPORCA 的限制

GPORCA 有一些限制，也正是因为 GPORCA 并不支持所有的 HAWQ 特性，GPORCA 与老优化器才会在 HAWQ 中并存。

1. 不支持的 SQL 特性

GPORCA 不支持以下 SQL：

- PERCENTILE_窗口函数。

```
db1=# explain select a, percentile_cont (0.5) within group (order by b desc)
db1-#          from t group by a;
                QUERY PLAN
```

```
---------------------------------------------------------------
...
 Settings:  default_hash_table_bucket_number=24; optimizer=on
 Optimizer status: legacy query optimizer
(24 rows)
```

- CUBE 和 GROUPING SETS 分析函数。

```
db1=# explain select count(*) from t group by cube(a,b);
                    QUERY PLAN
---------------------------------------------------------------
...
 Settings:  default_hash_table_bucket_number=24; optimizer=on
 Optimizer status: legacy query optimizer
(27 rows)
```

2. 性能衰退的情况

启用 GPORCA 时，以下是已知的性能衰减情况：

- 短查询。对于短查询来说，GPORCA 为了确定优化的查询执行计划，可能带来额外的开销。
- ANALYZE。启用 GPORCA 时，ANALYZE 命令生成分区表根分区的统计信息，而老的优化器不收集此统计。

10.3 性能优化

HAWQ 为查询动态分配资源，数据所在的位置、查询所使用的虚拟段数量、集群的总体健康状况等因素都会影响查询性能。

1. 常用优化手段

以下是 HAWQ 内部常用的优化手段，当进行了适当的服务器参数设置后，这些优化是系统自动实施的，理解它们对于开发高性能应用大有裨益。对用户来说，表设计与 SQL 语句的写法对性能的影响很大，然而这些技术对大部分数据库系统来说是通用的，如规范化设计、索引设计、连接时驱动表的选择、利用提示影响优化器等等。有很多这方面的资料，本书不展开讨论这些内容。

（1）动态分区消除

HAWQ 有两种分区消除：静态消除与动态消除。静态消除发生在编译期间，在执行计划生成的时候，已经知道哪些分区会被使用。而动态消除发生在运行时，也就是说在运行的时候，才会知道哪些分区会被用到。例如，WHERE 字句里面包含一个函数或者子查询用于返回分区键的值。查询过滤条件的值可用于动态分区消除时，查询处理速度将得到提升。该特性由服务器配置参数 gp_dynamic_partition_pruning 控制，默认是开启的。

```
[gpadmin@hdp3 ~]$ hawq config -s gp_dynamic_partition_pruning
GUC    : gp_dynamic_partition_pruning
Value  : on
[gpadmin@hdp3 ~]$
```

（2）内存优化

HAWQ 针对查询中的不同操作符分配最佳内存，并且在查询处理的各个阶段动态释放和重新分配内存。

（3）自动终止资源失控的查询

当服务器中所有查询占用的内存超过一定阈值，HAWQ 可以终止某些查询。HAWQ 的资源管理器会计算得到一个为 Segment 分配的虚拟内存限额，再结合可配的系统参数计算阈值。阈值计算公式为：vmem threshold = (资源管理器计算的虚拟内存限额 + hawq_re_memory_overcommit_max) * runaway_detector_activation_percent。

hawq_re_memory_overcommit_max 参数设置每个物理 Segment 可以超过资源管理器动态分配的内存限额的最大值，默认为 8192MB。当 HAWQ 使用 YARN 管理资源时，为了避免内存溢出错误，应该为该参数赋予一个较大值。runaway_detector_activation_percent 参数设置触发自动终止查询的虚拟内存限额百分比，默认值为 95，如果设置为 100，将禁用虚拟内存检测和自动查询终止。

```
[gpadmin@hdp3 ~]$ hawq config -s hawq_re_memory_overcommit_max
GUC    : hawq_re_memory_overcommit_max
Value  : 8192
[gpadmin@hdp3 ~]$ hawq config -s runaway_detector_activation_percent
GUC    : runaway_detector_activation_percent
Value  : 95
[gpadmin@hdp3 ~]$
```

当一个物理 Segment 使用的虚拟内存数量超过了该阈值，HAWQ 就从内存消耗最大的查询开始终止查询，直到虚拟内存的使用低于指定的百分比。假设 HAWQ 的资源管理器计算得到的一个物理 Segment 的虚拟内存限额为 9GB，hawq_re_memory_overcommit_max 设置为 1GB，runaway_detector_activation_percent 设置为 95，那么当虚拟内存使用超过 9.5GB 时，HAWQ 开始终止查询。

2. 查询性能问题排查

当一个查询没有达到希望的执行速度时，应该从以下方面检查造成查询缓慢的可能原因。

（1）检查集群健康状况，如是否有 DataNode 或 Segment 宕机，是否存在磁盘损坏等。

（2）检查表的统计信息，确认是否需要执行分析。

（3）检查查询的执行计划确定瓶颈。对于某些操作如 Hash Join，如果没有足够的内存，该操作会使用溢出文件（spill files）。相对于完全在内存中执行的操作，磁盘溢出文件会慢得多。

（4）检查查询计划中的数据本地化统计。

（5）检查资源队列状态。HAWQ 的 pg_resqueue 系统目录表保存资源队列信息。还可以查询 pg_resqueue_status 视图检查资源队列的运行时状态。

（6）分析资源管理器状态。这一点可以参考 http://hawq.incubator.apache.org/docs/userguide/2.1.0.0-incubating/resourcemgmt/ResourceManagerStatus.html。

3. 数据本地化统计

EXPLAIN ANALYZE 语句可以获得数据本地化统计：

```
db1=# explain analyze select * from t;
                          QUERY PLAN
-------------------------------------------------------------------------
...
 Data locality statistics:
   data locality ratio: 1.000; virtual segment number: 1; different host number:
1; virtual segment number per host(avg/min/max): (1/1/1); segment size(avg/min/max):
(56.000 B/56 B/56 B); segment size with penalty(avg/min/max): (56.000 B/56 B/56
B); continuity(avg/min/max): (1.000/1.000/1.000); DFS metadatacache: 0.138 ms;
resource allocation: 1.159 ms; datalocality calculation: 0.252 ms.
 Total runtime: 8.205 ms
(17 rows)
```

表 10-1 说明数据本地化相关度量值的含义，用这些信息可以检查潜在的查询性能问题。

表 10-1 数据本地化度量值含义

统计项	描述
data locality ratio	表示查询总的本地化读取比例。比例越低，从远程节点读取的数据越多。由于远程读取 HDFS 需要网络 IO，可能增加查询的执行时间。对于哈希分布表，一个文件中的所有数据块将由一个 Segment 处理，因此如果 HDFS 上的数据重新分布，比如做了 HDFS Rebalance，那么数据本地化比例将会降低。这种情况下，可以执行 CREATE TABLE AS SELECT 语句，通过重建表手工执行数据的重新分布
number of virtual segments	查询使用的虚拟段数量。通常虚拟段数越多，查询执行得越快。如果虚拟段太少，需要检查 default_hash_table_bucket_number、hawq_rm_nvseg_perquery_limit 或哈希分布表的桶数是否过小
different host number	表示有多少主机用于运行此查询。当虚拟段数量大于等于 HAWQ 集群主机总数时，所有主机都应该被使用。对于一个大查询，如果该度量值小于主机数，通常意味着有些主机宕机了。这种情况下，应该执行 "select * from gp_segment_configuration" 语句检查节点状态
segment size and segment size with penalty	"segment size" 表示一个虚拟段处理的数据量（平均/最小/最大），以字节为单位。"segment size with penalty" 表示一个虚拟段处理的包含了远程读取的数据量（平均/最小/最大），以字节为单位，远程读取量计算公式为 "net_disk_ratio" * block size。包含远程读取的虚拟段应该比只有本地读取的虚拟段处理更少的数据。"net_disk_ratio" 配置参数用于测量远程读取比本地读取慢多少，默认值为 1.01。可依据不同的网络环境调整该参数的值

(续表)

统计项	描述
continuity	间断地读取 HDFS 文件会引入额外的查找，减慢查询的表扫描，一个较低的 continuity 值说明文件在 DataNode 上的分布并不连续
DFS metadatacache	表示查询元数据缓存的时间。HDFS 块信息被 HAWQ 的 DFS Metadata Cache process 进程缓存。如果缓存没有命中，该时间会增加
resource allocation	表示从资源管理器获取资源所花的时间
datalocality calculation	表示运行将 HDFS 块分配给虚拟段的算法和计算数据本地化比例的时间

4．虚拟段数量

执行查询使用的虚拟段数量直接影响查询并行度，从而影响查询性能。

（1）影响虚拟段数量的因素

分配给查询的虚拟段数量受以下因素影响：

- 查询成本。大查询使用更多的虚拟段。
- 查询运行时的可用资源情况。如果资源队列中有更多的资源，查询就会使用它。
- 哈希分布表及其桶数。如果只查询一个哈希分布表，查询的并行度是固定的，等于创建哈希表时分配的桶数。如果查询中既有哈希分布表又有随机分布表，当所有哈希表都具有相同的桶数，并且随机表的大小不大于哈希表大小的 1.5 倍时，分配的虚拟段数等于桶数。否则，分配的虚拟段数依赖于查询成本，此时哈希表的虚拟段分配行为与随机表类似。
- 查询类型。对于包含外部表或用户定义函数（UDF）的查询，计算其查询成本比较困难。对于此类查询，分配的虚拟段数量由 hawq_rm_nvseg_perquery_limit 和 hawq_rm_nvseg_perquery_perseg_limit 参数，以及定义外部表时 ON 子句中的位置列表数量所控制。如果查询结果是装载一个哈希表（如 INSERT into hash_table），虚拟段的数量等于结果哈希表的桶数。COPY 或 ANALYZE 等 SQL 命令将使用不同的策略计算虚拟段数量。

（2）分配虚拟段的一般规则

如果有足够的可用资源，HAQW 使用以下一般规则确定为查询分配的虚拟段数量：

- SELECT 列表中仅包含随机分布表：虚拟段数量依赖于表大小。
- SELECT 列表中仅包含哈希分布表：虚拟段数量依赖于表的桶数。
- SELECT 列表中既有随机分布表，又有哈希分布表：如果所有哈希表都具有相同的桶数，并且随机表的大小不大于哈希表大小的 1.5 倍，分配的虚拟段数等于桶数。否则，分配的虚拟段数依赖于随机表的大小。
- 查询中存在用户定义函数：虚拟段数量依赖于 hawq_rm_nvseg_perquery_limit 和 hawq_rm_nvseg_perquery_perseg_limit 参数。

- 查询中存在 PXF 外部表：虚拟段数量依赖于 default_hash_table_bucket_number 参数。
- 查询中存在 gpfdist 外部表：虚拟段数量不少于 location 列表中的位置数。
- CREATE EXTERNAL TABLE 命令：虚拟段数量对应命令中 ON 子句的位置列表数量。
- 哈希分布表与本地文件互拷数据：虚拟段数量依赖于哈希表的桶数。
- 复制随机分布表数据到本地文件：虚拟段数量依赖于表大小。
- 将本地文件内容复制到随机分布表中：虚拟段数量是固定值，如果资源足够，为 6。
- ANALYZE 表：分析一个非分区表比等量的分区表使用更多的虚拟段。
- 哈希分布结果表：虚拟段数量等于结果哈希表的桶数。

10.4 查询剖析

遇到性能不良的查询时，最常用的调查手段就是查看执行计划。HAWQ 选择与每个查询相匹配的查询计划，查询计划定义了 HAWQ 在并行环境中如何运行查询。查询优化器根据数据库系统维护的统计信息选择成本最低的查询计划。成本以磁盘 I/O 作为考量，以查询需要读取的磁盘页数为测量单位。优化器的目标就是制定最小化执行成本的查询计划。

和其他 SQL 数据库一样，HAWQ 也是用 EXPLAIN 命令查看一个给定查询的执行计划。EXPLAIN 会显示查询优化器估计出的计划成本。EXPLAIN ANALYZE 命令会实际执行查询语句，它除了显示估算的查询成本，还会显示实际执行时间，从这些信息可以分析优化器所做的估算与实际之间的接近程度。

再次强调在 HAWQ 中老的优化器与 GPORCA 并存，默认的查询优化器为 GPORCA。HAWQ 尽可能使用 GPORCA 生成执行计划。GPORCA 和老优化器的 EXPLAIN 输出是不同的：

```
db1=# set optimizer=on;
SET
db1=# explain select * from t;
                          QUERY PLAN
-------------------------------------------------------------------------
 Gather Motion 1:1  (slice1; segments: 1)  (cost=0.00..431.00 rows=2 width=12)
   ->  Table Scan on t  (cost=0.00..431.00 rows=2 width=12)
 Settings:  default_hash_table_bucket_number=24; optimizer=on
 Optimizer status: PQO version 1.684
(4 rows)


db1=# set optimizer=off;
SET
db1=# explain select * from t;
                          QUERY PLAN
-------------------------------------------------------------------------
```

```
Gather Motion 1:1  (slice1; segments: 1)  (cost=0.00..1.02 rows=2 width=12)
  -> Append-only Scan on t  (cost=0.00..1.02 rows=2 width=12)
Settings:  default_hash_table_bucket_number=24; optimizer=off
Optimizer status: legacy query optimizer
(4 rows)
```

1. 读取 EXPLAIN 的输出

查询计划的输出是一个由节点构成的树形结构,每个节点表示一个单一操作,例如表扫描、连接、聚合、排序等等。查询计划应该由底向上进行读取,每个节点操作返回的行提供给直接上级节点。最底层的节点通常为一个表扫描操作,连接、聚合、排序等其他操作节点在表扫描节点之上。计划的顶层通常为 motion 节点,如 redistribute、broadcast 或 gather motions。在查询执行期间,这些操作将在节点间移动数据行。

计划树中的每个节点对应 EXPLAIN 输出中的一行,显示基本的节点类型和为该操作估算的执行成本。

- cost: 读取磁盘页的测量单位。1.0 表示一次顺序磁盘页读取。前一个值表示获取第一行的成本估算,后一个值表示获取全部行的总成本估算。总成本假定查询返回所有行,但当使用 LIMIT 时,并不返回全部的行,因此这种情况下的总成本是不对的。需要注意的是,节点成本包含了其子节点的成本,因此顶层节点的成本就是该计划执行的总成本估算,也就是优化器认为的最小成本。而且成本仅反映了查询优化器考虑的计划执行成本,不包括将结果行传送到客户端的开销。
- rows: 该节点输出的总行数。此行数通常会少于节点需要扫描或处理的行数,反映了对 WHERE 条件选择性的估算。理想情况下,顶层节点的估算值应该接近查询实际返回的行数。
- width: 该节点输出的所有行的总字节数。

EXPLAIN 输出读取示例。

```
db1=# explain select * from t where b=1;
                        QUERY PLAN
------------------------------------------------------------------------------
Gather Motion 1:1  (slice1; segments: 1)  (cost=0.00..431.00 rows=1 width=12)
  -> Table Scan on t  (cost=0.00..431.00 rows=1 width=12)
        Filter: b = 1
Settings:  default_hash_table_bucket_number=24; optimizer=on
Optimizer status: PQO version 1.684
(5 rows)
```

查询计划的 EXPLAIN 输出只有 5 行,其中最后一行表示生成该计划的优化器是 GPORCA,倒数第二行表示哈希桶数和优化器等基本参数的设置。这两行不属于查询计划树。

现在开始自底向上读取计划。底层是一个表扫描节点,顺序扫描 t 表。WHERE 子句表现为一个过滤条件,表示扫描操作会检查扫描到的每一行是否满足过滤条件,并且只向直接上级节点返回满足条件的行。

扫描操作的结果传给上级的 gather motion 操作。在 HAWQ 中，Segment 实例向 Master 实例发送数据即为 gather motion 操作。该操作在并行查询执行计划的 slice1 分片中完成，并且该分片只在一个 Segment 上执行。正如介绍优化器时所述，查询计划被分成 slice，因此 Segment 可以并行执行部分查询计划。

该计划估算的启动成本（返回首行的成本）为 0，总成本为 431 个磁盘页读取，优化器估算查询返回 1 行。这是一个最简单的示例，只有两步操作，实际的 EXPLAIN 可能复杂得多。

2. 读取 EXPLAIN ANALYZE 的输出

与 EXPLAIN 不同，EXPLAIN ANALYZE 命令不但生成执行计划，还会实际执行查询语句。

```
db1=# select * from t;
 a | b | c
---+---+---
(0 rows)

db1=# explain insert into t values (1,1,1);
                        QUERY PLAN
--------------------------------------------------------------------------------
Insert (slice0; segments: 1)  (rows=1 width=0)
   -> Redistribute Motion 1:1  (slice1; segments: 1)  (cost=0.00..0.01 rows=1 width=0)
         -> Result  (cost=0.00..0.01 rows=1 width=0)
 Settings:  default_hash_table_bucket_number=24; optimizer=off
 Optimizer status: legacy query optimizer
(5 rows)


db1=# select * from t;
 a | b | c
---+---+---
(0 rows)

db1=# explain analyze insert into t values (1,1,1);
                        QUERY PLAN
--------------------------------------------------------------------------------
 Insert (slice0; segments: 1)  (rows=1 width=0)
   -> Redistribute Motion 1:1  (slice1; segments: 1)  (cost=0.00..0.01 rows=1 width=0)
         Rows out:  Avg 1.0 rows x 1 workers at destination.
Max/Last(seg0:hdp3/seg0:hdp3) 1/1 rows with 14/14 ms to end
, start offset by 161/161 ms.
         -> Result  (cost=0.00..0.01 rows=1 width=0)
            Rows out:  Avg 1.0 rows x 1 workers.  Max/Last(seg0:hdp3/seg0:hdp3)
```

```
1/1 rows with 0.004/0.004 ms to first
   row, 0.005/0.005 ms to end, start offset by 176/176 ms.
 ...
   Total runtime: 210.536 ms
 (18 rows)

 db1=# select * from t;
  a | b | c
 ---+---+---
  1 | 1 | 1
 (1 row)
```

EXPLAIN ANALYZE 显示优化器的估算成本与查询的实际执行成本，因此可以分析估算与实际的接近程度。EXPLAIN ANALYZE 的输出还显示如下内容：

- 查询总的执行时间，单位是毫秒。
- 查询计划每个分片使用的内存，以及为整个查询语句估算的内存。
- 查询分发器的统计信息，包括当前查询使用的执行器数量（总数/缓存数/新连接数），分发时间（总时间/连接建立时间/分发数据时间），及其分发数据、执行器消耗、释放执行器的时间细节（最大/最小/平均）。
- 如表 10-1 所示的数据本地化统计信息。
- 节点操作涉及的 Segment（workers）数量，只对返回行的 Segment 计数。
- 节点操作输出的最多行数和用时最长的 Segment 统计。
- 一个操作中产生最多行的 Segment id。
- 操作返回首行和返回所有行所用的时间（毫秒），如果两个时间相同，输出中省略返回首行的时间。
- 连接操作使用的内存（work_mem）。如果内存不足，计划显示溢出到磁盘的数据量，及其受到影响的 Segment 数，例如：

```
Work_mem used: 64K bytes avg, 64K bytes max (seg0).
Work_mem wanted: 90K bytes avg, 90K byes max (seg0) to lessen
workfile I/O affecting 2 workers.
```

EXPLAIN ANALYZE 输出读取示例。为了方便与前面的 EXPLAIN 做对比，执行同样的查询语句。

```
db1=# explain analyze select * from t where b=1;
                         QUERY PLAN
-------------------------------------------------------------------
 Gather Motion 1:1  (slice1; segments: 1)  (cost=0.00..431.00 rows=1 width=12)
   Rows out:  Avg 1.0 rows x 1 workers at destination.  Max/Last(seg-1:hdp3/seg-1:hdp3)
 1/1 rows with 11/11 ms to end, start offset by 1.054/1.054 ms.
   -> Table Scan on t  (cost=0.00..431.00 rows=1 width=12)
```

```
            Filter: b = 1
            Rows out:  Avg 1.0 rows x 1 workers.  Max/Last(seg0:hdp3/seg0:hdp3) 1/1
rows with 2.892/2.892 ms to first row, 2.989/2.989 ms to end, start offset by
8.579/8.579 ms.
   Slice statistics:
     (slice0)    Executor memory: 163K bytes.
     (slice1)    Executor memory: 279K bytes (seg0:hdp3).
   Statement statistics:
     Memory used: 262144K bytes
   Settings:  default_hash_table_bucket_number=24; optimizer=on
   Optimizer status: PQO version 1.684
   Dispatcher statistics:
     executors used(total/cached/new connection): (1/1/0); dispatcher
time(total/connection/dispatch data): (0.342 ms/0.000 ms/0.095 ms).
     dispatch data time(max/min/avg): (0.095 ms/0.095 ms/0.095 ms); consume
executor data time(max/min/avg): (0.020 ms/0.020 ms/0.020 ms); free executor
time(max/min/avg): (0.000 ms/0.000 ms/0.000 ms).
   Data locality statistics:
     data locality ratio: 1.000; virtual segment number: 1; different host number:
1; virtual segment number per host(avg/min/max): (1/1/1); segment size(avg/min/max):
(24.000 B/24 B/24 B); segment size with penalty(avg/min/max): (24.000 B/24 B/24
B); continuity(avg/min/max): (1.000/1.000/1.000); DFS metadatacache: 0.092 ms;
resource allocation: 0.911 ms; datalocality calculation: 0.221 ms.
   Total runtime: 13.304 ms
   (18 rows)
```

与 EXPLAIN 不同相比，这次的输出长得多，有 18 行。第 11 行表示哈希桶数和优化器等基本参数的设置，第 12 行表示生成该计划的优化器为 GPORCA。这两行与 EXPLAIN 的输出相同。前 5 行是执行计划树，比 EXPLAIN 的输出多出第 2、5 两行，这两行是节点的实际执行情况，包括返回数据行数、首末行时间、最大最长 Segment 等。Table Scan 操作只有一个 Segment（seg0）返回行，并且只返回 1 行。Max/Last 统计是相同的，因为只有一个 Segment 返回行。找到首行使用的时间为 2.892 毫秒，返回所有行的时间为 2.989 毫秒。注意 start offset by，它表示的是从分发器开始执行操作到 Segment 返回首行经历的时间为 8.579 毫秒。查询实际返回行数与估算返回的行数相同。gather motion 操作接收 1 行，并传送到 Master。gather motion 节点的时间统计包含了其子节点 Table Scan 操作的时间。最后一行显示该查询总的执行时间为 13.304 毫秒。

输出中的其他行是各种统计信息，包括分片统计、语句统计、分发器统计、数据本地化统计等。

3. 分析查询计划中的问题

查询慢时，需要查看执行计划并考虑以下问题：

- 计划中的某些特定操作是否花费了很长时间？找到最消耗时间的操作并分析原因。例如，哈希表的扫描时间出乎意料的长，可能是由于数据本地化程度低，导致节点间的网络 IO 花费大量时间。此时重新装载数据可能提高查询速度。
- 查询优化器估算的行数是否与实际的相近？运行 EXPLAIN ANALYZE 检查实际与估算的返回行数是否接近。如果相差很多，收集相关表列的统计信息。
- 是否在计划的早期应用了过滤谓词？在计划早期应用选择过滤使得向上层节点传递的行更少。若查询计划错误地估计了查询谓词的选择性，收集相关表列的统计信息。也可以尝试改变 SQL 语句中 WHERE 子句中列的顺序（查看 Filter 显示的顺序）。
- 查询优化器是否选择了最好的表连接顺序？过滤行数越多的表越应该先处理。如果计划没有选择优化的连接顺序，可能需要收集关联列的统计信息，或者设置 join_collapse_limit 配置参数为 1。后者会导致查询按 SQL 语句中指定的连接顺序执行。
- 优化器是否使用了分区消除？确认分区策略和查询谓词中的过滤条件是否匹配。
- 优化器是否选择了适当的哈希聚合与哈希连接？哈希操作通常比其他的连接或聚合类型快，因为行的比较和排序在内存中完成，而不是读写磁盘。为了让优化器适当地选择哈希操作，必须有足够的可用内存，以存储估算的行数。如果可能，运行 EXPLAIN ANALYZE 显示查询需要的内存和溢出到磁盘的数据量。例如，以下输出显示查询使用了 23430KB 内存，还需要 33649KB，此时考虑调整内存相关配置优化查询。

```
Work_mem used: 23430K bytes avg, 23430K bytes max (seg0).
Work_mem wanted: 33649K bytes avg, 33649K bytes max (seg0) to lessen workfile I/O affecting 2 workers.
```

必须注意，不要在 HAWQ 中使用 PL/pgSQL 函数生成动态查询的执行计划，这可能引起服务器崩溃！下面的例子在 PostgreSQL 8.4.20 中可以正常执行，但在 HAWQ2.1.1 中数据库直接宕机。

```
db1=# create or replace function explain_plan_func() returns varchar as $$
declare

  a varchar;
  b int;
  c varchar;

  begin
    a = '';
    b = 1;
    for c in execute 'explain select * from t where b=' || cast(b as varchar) loop
      a = a || e'\n' || c;
    end loop;
    return a;
```

```
         end;
$$
language plpgsql
volatile;
CREATE FUNCTION
db1=# select explain_plan_func();
             explain_plan_func
----------------------------------------------------
 Seq Scan on t  (cost=0.00..34.25 rows=10 width=12)
   Filter: (b = 1)
(1 行记录)
```

10.5 小结

HAWQ 中新旧查询优化器并存，优先选择新的 GPORCA 优化器，它针对分区表、子查询、WITH、INSERT、去重聚合等查询类型有所改进。查询计划是在 Segment 上分片并行执行的。数据本地化情况、为查询分配的虚拟段数量对查询性能具有直接影响。同很多数据库系统类似，EXPLAIN 用于语句输出查询执行计划。学会读懂 EXPLAIN 的信息，对于排查性能问题十分有用。EXPLAIN ANALYZE 会实际执行 SQL 语句，并且比单纯的 EXPLAIN 输出更多的信息。

第 11 章 高可用性

HAWQ 的高可用性体现在三个不同层次：（1）数据库备份与恢复机制；（2）HAWQ Master 高可用；（3）HAWQ 文件空间高可用。其中第三个依赖于 Hadoop NameNode HA。虽然 HDFS 的副本集特性提供了一定的数据高可用性，但默认安装的 Hadoop 集群，NameNode 仍然存在单点故障风险。本章首先讨论 HAWQ 数据库的备份与恢复方法，然后介绍 HAWQ Master 的配置过程，最后说明如何在 HDP HA 基础之上配置 HAWQ 文件空间高可用。

11.1 备份与恢复

HAWQ 作为一个数据库管理系统，备份与恢复是其必备功能之一。HAWQ 的用户数据存储在 HDFS 上，系统主目录存储在 Master 节点本地主机。HDFS 上的每个数据块默认自带三份副本，而且一个数据块的三份副本不会存储在同一个 DataNode 上，因此一个 DataNode 节点失效不会造成数据丢失。而配置了 HDFS NameNode HA 与 HAWQ Master HA 后，NameNode 和 Master 的单点故障问题也得到了解决。似乎 HAWQ 没有提供额外备份功能的必要。

事实上，Hadoop 集群上存储和处理的数据量通常非常大，大到要想做全备份，在时间与空间消耗上都是不可接受的。这也就是 HDFS 的数据块自带副本容错的主要原因。那么说到 HAWQ 数据库提供的数据备份功能，作为高可用性的补充，主要体现在三个方面：一是自然地从 PostgreSQL 继承，本身就带备份功能；二是提供了一种少量数据迁移的简便方法，比如把一个小表从生产环境迁移到到测试环境；三是处理人为误操作引起的数据问题，例如误删除一个表时，就可以使用备份进行恢复，将数据丢失最小化。

11.1.1 备份方法

HAWQ 提供以下三个应用程序帮助用户备份数据：

- gpfdist
- PXF
- pg_dump

gpfdist 与 PXF 是并行数据装载/卸载工具，能提供最佳性能。pg_dump 是一个从 PostgreSQL 继承的非并行应用程序。除此之外，有些情况下还需要从 ETL 过程备份原始数据，比如将抽取到的源数据装载到多份目标中。用户可以根据自己的实际场景选择适当的备份/恢复方法。

1. gpfdist 和 PXF

用户可以在 HAWQ 中使用 gpfdist 或 PXF 执行并行备份,将数据卸载到外部表中。备份文件可以存储在本地文件系统或 HDFS 上。恢复表的过程就是简单将数据从外部表装载回数据库。

(1)备份步骤

执行以下步骤并行备份:

① 检查数据库大小,确认文件系统有足够的空间保存备份文件。
② 使用 pg_dump 应用程序导出源数据库的 schema。
③ 在目标数据库中,为每个需要备份的表创建一个可写外部表。
④ 向新创建的外部表中装载表数据。

需要注意的是,需将所有表的 INSERT 语句放在一个单独的事务中,以避免因在备份期间执行任何数据操作而产生问题。

(2)恢复步骤

执行以下步骤从备份还原:

① 创建一个数据库用于恢复。
② 从 schema 文件(在 pg_dump 过程中被创建)重建 schema。
③ 为数据库中的每个表建立一个可读外部表。
④ 从外部表向实际的表中导入数据。
⑤ 装载完成后,运行 ANALYZE 命令,保证基于最新表统计信息生成优化的查询计划。

(3)gpfdist 与 PXF 备份的区别

gpfdist 与 PXF 备份的区别体现在以下方面:

- gpfdist 在本地文件系统存储备份文件,PXF 将文件存储在 HDFS 上。
- gpfdist 只支持平面文本格式,PXF 还支持如 AVRO 的二进制格式,以及用户自定义的格式。
- gpfdist 不支持生成压缩文件,PXF 支持压缩,用户可以在 Hadoop 中指定使用的压缩算法,如 org.apache.hadoop.io.compress.GzipCodec。
- gpfdist 和 PXF 都提供快速装载性能,但 gpfdist 要比 PXF 快得多。

2. pg_dump 与 pg_restore

HAWQ 支持 PostgreSQL 的备份与还原应用程序,pg_dump 和 pg_restore。pg_dump 应用在 Master 节点所在主机上创建一个 dump 文件,其中包含所有注册 Segment 中的数据。pg_restore 从 pg_dump 创建的备份中还原一个 HAWQ 数据库。大多数情况下,整库备份/还原是不切实际的,因为一般在 Master 节点上没有足够的磁盘空间存储整个分布式数据库的单个备份文件。HAWQ 支持这些应用的主要目的是用于从 PostgreSQL 向 HAWQ 迁移数据。下面是一些 pg_dump 用法的简单示例。

为数据库 mytest 创建一个备份，导出数据文件格式为 tar：

```
$ pg_dump -Ft -f mytest.tar mytest
```

使用自定义格式创建一个压缩的备份，压缩级别为 3：

```
$ pg_dump -Fc -Z3 -f mytest.dump mytest
```

使用 pg_restore 从备份还原：

```
$ pg_restore -d new_db mytest.dump
```

3. 原始数据备份

大多数情况使用 gpfdist 或 PXF 并行备份能够很好地工作，但以下两种情况不能执行并行备份与还原操作：

- 周期性增量备份。
- 导出大量数据到外部表，原因是此过程花费的时间太长。

在这些情况下，用户可以在 ETL 处理期间生成原始数据的备份，并装载到 HAWQ。ETL 程序提供了选择在本地还是 HDFS 存储备份文件的灵活性。

4. 备份方法对比

表 11-1 汇总了上面讨论的 4 种备份方法的区别。

表 11-1　备份方法比较

	gpfdist	PXF	pg_dump	原始数据备份
并行执行	Yes	Yes	No	No
增量备份	No	No	No	Yes
备份文件存储位置	本地文件系统	HDFS	本地文件系统	本地文件系统，HDFS
备份文件格式	Text，CSV	Text，CSV，自定义格式	Text，Tar，自定义格式	依赖原始数据的格式
压缩	No	Yes	只支持自定义格式	可选
可伸缩性	好	好	—	好
性能	读取快速，写入快速	读取快速，写入一般	—	快（只复制文件）

5. 估计空间需求

备份数据库前，需要确认有足够的空间存储备份文件。下面说明如何获取数据库大小和估算备份文件所需空间。

（1）从 hawq_toolkit 查询需要备份的数据库大小

hawq_toolkit 是 HAWQ 的一个管理模式，使用与下面类似的命令在模式查找路径中增加 hawq_toolkit 模式：

```
db1=# set role 'gpadmin';
SET
db1=# set search_path to public, hawq_toolkit;
SET
db1=#
```

该模式中包含若干可以使用 SQL 命令访问的视图，这些视图提供了系统目录表、日志文件、操作环境，以及系统状态的相关信息。hawq_toolkit 可被所有数据库用户访问，hawq_log_command_timings、hawq_log_master_concise、hawq_size_of_table_and_indexes_licensing、hawq_size_of_table_uncompressed 视图的查询需要超级用户权限。下面语句查询数据库大小。

```
gpadmin=# select sodddatsize from hawq_toolkit.hawq_size_of_database where sodddatname='db1';
 sodddatsize
--------------
    16063436
(1 row)
```

输出结果以字节为单位，如果数据库中的表是压缩的，此查询显示压缩后的数据库大小。

（2）估算备份文件总大小

- 如果数据库表和备份文件都是压缩的，可以使用 sodddatsize 作为估算值。
- 如果数据库表是压缩的，备份文件是非压缩的，需要用 sodddatsize 乘以压缩率。尽管压缩率依赖于压缩算法，但通常可以使用经验值如三倍进行估算。
- 如果备份文件是压缩的，数据库表是非压缩的，需要用 sodddatsize 除以压缩率。

（3）得出空间需求

- 如果使用 PXF 与 HDFS，所需空间为：备份文件大小 * 复制因子。
- 如果使用 gpfdist，每个 gpfdist 实例的所需空间为：备份文件大小 / gpfdist 实例个数，这是因为表数据将最终分布到所有 gpfdist 实例。

11.1.2 备份与恢复示例

1. gpfdist 示例

为使用 gpfdist，在要还原备份文件的主机上启动 gpfdist 服务器程序，可以在同一个主机或不同主机上启动多个 gpfdist 实例。每个 gpfdist 实例需要指定一个对应目录，gpfdist 从该目录向可读外部表提供文件，或者创建可写外部表的输出文件。例如，如果用户有一台两块磁盘的专门用于备份的机器，则可以启动两个 gpfdist 实例，每个实例使用一块磁盘，如图 11-1 所示。

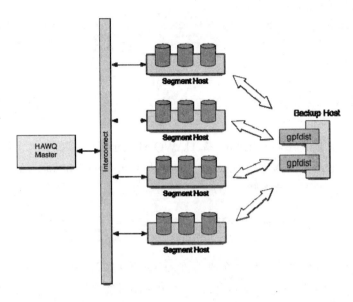

图 11-1　专用备份主机上的 gpfdist 实例

也可以在每个 Segment 主机上运行 gpfdist 实例，如图 11-2 所示。备份期间，表数据将最终分布于所有 CREATE EXTERNAL TABLE 定义的 LOCATION 子句中指定的 gpfdist 实例上。

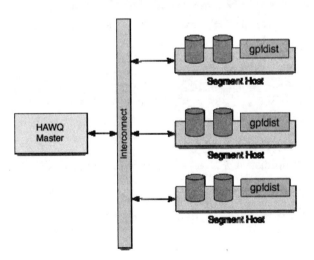

图 11-2　每个 Segment 主机上运行 gpfdist 实例

（1）使用 gpfdist 备份

使用 gpfdist 备份 mytest 数据库：

步骤 01　创建备份位置，并启动 gpfdist 实例。在 hdp4 上启动一个 gpfdist 实例，端口号为 8081，外部数据目录是 /home/gpadmin/mytest_20170223。

```
gpfdist -d /home/gpadmin/mytest_20170223 -p 8081 &
# 如果碰到 "cannot open shared object file: No such file or directory" 类似的错
误，需要先安装相应的依赖包，如：
gpfdist: error while loading shared libraries: libapr-1.so.0: cannot open shared
object file: No such file or directory
yum install apr
gpfdist: error while loading shared libraries: libyaml-0.so.2: cannot open
shared object file: No such file or directory
yum install libyaml
```

步骤02 保存 mytest 数据库的 schema。在 HAWQ Master 节点所在主机，使用 pg_dump 应用程序，将 mytest 数据库的 schema 保存到文件 mytest.schema。将 schema 文件复制到备份目录，用于以后还原数据库 schema。

```
pg_dump --schema-only -f mytest.schema mytest
scp mytest.schema hdp4:/home/gpadmin/mytest_20170223
```

步骤03 为数据库中的每个表创建一个可写外部表。在 LOCATION 子句中指定 gpfdist 实例。本例使用 CSV 文本格式，但也可以选择其他固定分隔符的文本格式。

```
psql mytest
mytest=# create writable external table wext_base_table (like base_table)
mytest=# location('gpfdist://hdp4:8081/base_table.csv') format 'csv';
mytest=# create writable external table wext_t (like t)
mytest=# location('gpfdist://hdp4:8081/t.csv') format 'csv';
```

步骤04 向外部表卸载数据。所有外部表的数据插入在一个事务中进行。

```
mytest=# begin;
mytest=# insert into wext_base_table select * from base_table;
mytest=# insert into wext_t select * from t;
mytest=# commit;
```

步骤05 （可选）停止 gpfdist 服务，为其他进程释放端口。下面的命令查找 gpfdist 进程号并杀掉这些进程。

```
ps -efww|grep gpfdist|grep -v grep|cut -c 9-15|xargs kill -9
```

（2）从 gpfdist 备份还原

步骤01 如果 gpfdist 没运行则启动 gpfdist 实例。下面的命令在 hdp4 上启动一个 gpfdist 实例。

```
gpfdist -d /home/gpadmin/mytest_20170223 -p 8081 &
```

步骤02 创建一个新的数据库 mytest2，并将 mytest 的 schema 还原到新库中。

```
[gpadmin@hdp3 ~]$ createdb mytest2
[gpadmin@hdp3 ~]$ scp hdp4:/home/gpadmin/mytest_20170223/mytest.schema .
[gpadmin@hdp3 ~]$ psql -f mytest.schema -d mytest2
```

步骤 03 为每个表创建可读外部表。

```
[gpadmin@hdp3 ~]$ psql mytest2
mytest2=# create external table rext_base_table (like base_table) location('gpfdist://hdp4:8081/base_table.csv') format 'csv';
mytest2=# create external table rext_t (like t) location('gpfdist://hdp4:8081/t.csv') format 'csv';
```

这里的 location 子句与前面备份时创建的可写外部表相同。

步骤 04 从外部表导回数据。

```
mytest2=# insert into base_table select * from rext_base_table;
mytest2=# insert into t select * from rext_t;
```

步骤 05 装载数据后运行 ANALYZE 命令生成表的统计信息。

```
mytest2=# analyze base_table;
mytest2=# analyze t;
```

2. PXF 示例

（1）使用 PXF 备份

使用 PXF 备份 mytest 数据库：

步骤 01 在 HDFS 上建立一个用作备份的文件夹。

```
[root@hdp3 ~]# su - hdfs
[hdfs@hdp3 ~]$ hdfs dfs -mkdir -p /backup/mytest-2017-02-23
[hdfs@hdp3 ~]$ hdfs dfs -chown -R gpadmin:gpadmin /backup
```

步骤 02 使用 pg_dump 导出数据库 schema，并将 schema 文件存储到备份文件夹中。

```
[gpadmin@hdp3 ~]$ pg_dump --schema-only -f mytest.schema mytest
[gpadmin@hdp3 ~]$ hdfs dfs -copyFromLocal mytest.schema /backup/mytest-2017-02-23
```

步骤 03 为数据库中每个表创建一个可写外部表。

```
[gpadmin@hdp3 ~]$ psql mytest
mytest=# create writable external table wext_base_table (like base_table)
mytest=# location('pxf://hdp1:51200/backup/mytest-2017-02-23/base_table?Profile=HdfsTextSimple&COMPRESSION_CODEC=org.apache.hadoop.io.compress.SnappyCodec')
mytest=# format 'text';

mytest=# create writable external table wext_t (like t)
mytest=# location('pxf://hdp1:51200/backup/mytest-2017-02-23/t?Profile=HdfsTextSimple&COMPRESSION_CODEC=org.apache.hadoop.io.compress.SnappyCodec')
```

```
mytest=# format 'text';
```

这里,所有 base_table 表的备份文件存储在/backup/mytest-2017-02-23/base_table 文件夹中,所有 t 表的备份文件存储在/backup/mytest-2017-02-23/t 文件夹中。外部数据文件使用 snappy 压缩保存到磁盘。

步骤04 向外部表导出数据。

```
mytest=# begin;
mytest=# insert into wext_base_table select * from base_table;
mytest=# insert into wext_t select * from t;
mytest=# commit;
```

外部表使用 snappy 压缩时可能遇到如下错误:

```
mytest=# insert into wext_t select * from t;
ERROR:  remote component error (500) from '172.16.1.125:51200': type
Exception report   message   native snappy
  library not available: this version of libhadoop was built without snappy support.
description   The server
  encountered an internal error that prevented it from fulfilling this request.
exception
  java.lang.RuntimeException: native snappy library not available: this version
of libhadoop was built without
  snappy support. (libchurl.c:897)    (seg0 hdp4:40000 pid=8565)
(dispatcher.c:1801)
```

使用下面的方法解决此问题(参考 https://issues.apache.org/jira/browse/HAWQ-951):

```
# 创建目录和软连接
mkdir -p /usr/lib/hadoop/lib && cd /usr/lib/hadoop/lib && ln -s
/usr/hdp/current/hadoop-client/lib/native native
# 添加 pxf-public-classpath 属性
登录 ambari,在 Services -> PXF -> Configs -> Advanced pxf-public-classpath 中
添加一行:/usr/hdp/current/hadoop-client/lib/snappy*.jar
```

步骤05 (可选)改变备份文件夹的 HDFS 复制因子。默认 HDFS 每个数据块复制三份以提供可靠性。根据需要,可以为备份文件降低这个数,以下命令将复制因子设置为 2:

```
su - pxf
-bash-4.1$ hdfs dfs -setrep 2 /backup/mytest-2017-02-23
```

该命令只改变已经存在文件的备份因子,新文件仍然使用默认备份因子值。

(2)从 PXF 备份还原

步骤01 创建一个新的数据库并还原 schema。

```
[gpadmin@hdp3 ~]$ createdb mytest3
```

```
[gpadmin@hdp3 ~]$ hdfs dfs -copyToLocal
/backup/mytest-2017-02-23/mytest.schema .
[gpadmin@hdp3 ~]$ psql -f mytest.schema -d mytest3
```

步骤02 为每个需要还原的表创建一个可读外部表。

```
[gpadmin@hdp3 ~]$ psql mytest3
mytest3=# create external table rext_base_table (like base_table)
mytest3=#
location('pxf://hdp1:51200/backup/mytest-2017-02-23/base_table?Profile=HdfsTextSimple')
mytest3=# format 'text';

mytest3=# create external table rext_t (like t)
mytest3=#
location('pxf://hdp1:51200/backup/mytest-2017-02-23/t?Profile=HdfsTextSimple')
mytest3=# format 'text';
```

除了不需要指定 COMPRESSION_CODEC，LOCATION 子句与前面创建可写外部表的相同。PXF 会自动检测压缩算法。

步骤03 从外部表导入数据。

```
mytest3=# insert into base_table select * from rext_base_table;
mytest3=# insert into t select * from rext_t;
```

步骤04 导入数据后运行 ANALYZE。

```
mytest3=# analyze base_table;
mytest3=# analyze t;
```

11.2 高可用性

11.2.1 HAWQ 高可用简介

HAWQ 作为一个传统数据仓库在 Hadoop 上的替代品，其高可用性至关重要。通常需要考虑并实施硬件容错、HAWQ HA、HDFS HA 等多种措施以保持系统高可用。另外实时监控和定期维护，也是保证集群所有组件健康的必不可少的工作。总的来说，HAWQ 容错高可用的实现方式包括：

- 硬件冗余
- Master 镜像
- 双集群

1. 硬件级别的冗余（RAID 和 JBOD）

硬件组件的正常磨损或意外情况最终会导致损坏，因此有必要提供备用的冗余硬件，当一个组件发生损坏时，不中断服务。某些情况下，冗余的成本高于用户所能容忍的服务中断。此时，目标是保证所有服务能够在一个预期的时间范围内被还原。

虽然 Hadoop 集群本身是硬件容错的，但 HAWQ 有其特殊性。HAWQ Master 的数据存储在主机本地硬盘上，是一个单点。作为最佳实践，HAWQ 建议在部署时，Master 节点应该使用 RAID，而 Segment 节点应该使用 JBOD。这些硬件级别的系统为单一磁盘损坏提供高性能冗余，而不必进入到数据库级别的容错。RAID 和 JBOD 在磁盘级别提供了低层次的冗余。

2. Master 镜像

高可用集群中的 Master 节点有一主一从两个。和 Master 节点与 Segment 节点分开部署类似，Master 的主和从也应该部署到不同的主机，以容忍单一主机失效。客户端连接到主 Master 节点并且查询只能在主 Master 节点上执行。从 Master 节点保持与主 Master 节点的实时同步，这是通过将预写日志从主复制到从实现的。

3. 双集群

可以通过部署两套 HAWQ 集群存储相同的数据，从而增加另一级别的冗余。有两个主要方法用于保持双集群的数据同步，分别是双 ETL 和备份/还原。

双 ETL 提供一个与主集群数据完全相同的备用集群。ETL（抽取、装换与装载）指的是一个数据清洗、转换、验证和装载进数据仓库的过程。通过双 ETL，将此过程并行执行两次，每次在一个集群中执行。应该在两个集群上都进行验证，以确保双 ETL 执行成功。这种做法是最彻底的冗余，需要部署两套 HAWQ 集群与 ETL 程序。该方法带来的一个附加好处是应用利用双集群，能够同时在两个集群上查询数据，将查询吞吐量增加一倍。

用备份/还原方法维护一个双集群，需要创建一个主集群的备份，并在备用集群上还原。这种方法与双 ETL 策略相比，备用节点数据同步的时间要长得多，但优点是只需要开发更少的应用逻辑。如果数据修改和 ETL 执行的频率是每天或更低的频率，同时备份/还原时间又在可接受的范围内，那么用备份生成数据是比较理想的方式。备份/还原方法不适用于要求数据实时同步的情况。

11.2.2　Master 节点镜像

在 HAWQ 中配置一主一从两个 Master 节点，客户端连接主 Master 节点，并只能在主 Master 节点上执行查询。从 Master 是一个纯粹的容错节点，只作为主 Master 出现问题时的备用。如果主 Master 节点不可用，从 Master 节点作为热备。可以在主 Master 节点联机时，从它创建一个从 Master 节点。当主 Master 节点持续为用户提供服务时，HAWQ 可以生成主 Master 节点实例的事务快照。除了生成事务快照并部署到从 Master 节点外，HAWQ 还记录主 Master 节点的变化。HAWQ 在从 Master 节点部署了快照后，会应用更新以将从 Master 节点与主 Master 节点数据同步。

主从 Master 节点初始同步后，HAWQ 分别在主、从节点上启动 WAL Send 和 WAL Redo 服务器进程，保持主从实时同步。它们是基于预写日志（Write-Ahead Logging，WAL）的复制进程。WAL Redo 是一个从 Master 节点进程，WAL Send 是主 Master 节点进程。这两个进程使用基于 WAL 的流复制保持主从同步。因为 Master 节点不保存用户数据，只有系统目录表在主从 Master 节点间被同步。当这些系统表被更新时（如 DDL 所引起），改变自动复制到从 Master 节点保持它与当前的主 Master 节点数据一致。

HAWQ 中的 Master 节点镜像架构如图 11-3 所示。

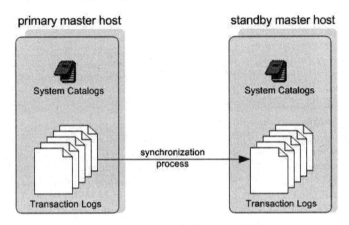

图 11-3　HAWQ 主从 Master 架构

如果主 Master 节点失效，复制进程停止。此时管理员需要使用命令行工具或者 Ambari，手工执行 Master 切换，指示从 Master 节点成为新的主 Master。在激活从 Master 节点后，复制的日志重构主 Master 节点在最后成功提交事务时的状态。从 Master 节点初始化后，被激活的从 Master 作为 HAWQ 的主节点，在指定端口接收连接请求。

可以为主、从配置同一个虚 IP 地址，这样在主从切换时，客户端程序就不需要连接到两个不同的网络地址。如果主机失效，虚 IP 可以自动交换到实际活动的主节点。虚 IP 可能需要额外的软件支持，如 Keepalived 等。下面演示如何配置从 Master 及其失败切换过程。

1. 配置从 Master

hdp2 为现有的 Master 节点，下面过程将 hdp3 配置为 hdp2 的从 Master 节点。

（1）前提配置

确保 hdp3 上已经安装了 HAWQ 并进行了以下配置：

- 创建了 gpadmin 系统用户。
- 已安装了 HAWQ 二进制包。
- 已设置了 HAWQ 相关的环境变量。
- 已配置主、从 Master 的 SSH 免密码登录。
- 已创建了 Master 数据目录。

（2）初始化从 Master 节点

登录 hdp3，清空 Master 数据目录。

```
[root@hdp3 ~]# rm -rf /data/hawq/master/*
```

登录 hdp2，初始化从 Master 节点

```
[gpadmin@hdp2 ~]$ . /usr/local/hawq/greenplum_path.sh
[gpadmin@hdp2 ~]$ hawq init standby -s hdp3
```

（3）检查 HAWQ 集群状态

- 在 hdp2 上执行 hawq state 命令检查 HAWQ 集群的状态，主 Master 状态应该是 Active，从 Master 状态是 Passive。

```
[gpadmin@hdp2 ~]$ hawq state
...
--HAWQ instance status summary
-----------------------------------------------------
--   Master instance                                = Active
--   Master standby                                 = hdp3
--   Standby master state                           = Standby host passive
--   Total segment instance count from config file  = 4
-----------------------------------------------------
--   Segment Status
-----------------------------------------------------
--   Total segments count from catalog              = 4
--   Total segment valid (at master)                = 4
--   Total segment failures (at master)             = 0
--   Total number of postmaster.pid files missing   = 0
--   Total number of postmaster.pid files found     = 4
```

- 查询 gp_segment_configuration 表验证 Segment 已经注册到主 Master 节点，查询结果如下，可以看到 hdp2 与 hdp3 的角色分别是 'm' 和 's'。

```
[gpadmin@hdp2 ~]$ psql -c 'SELECT * FROM gp_segment_configuration;'
```

registration_order	role	status	port	hostname	address	description
-1	s	u	5432	hdp3	172.16.1.126	
0	m	u	5432	hdp2	hdp2	
1	p	u	40000	localhost	127.0.0.1	
2	p	u	40000	hdp4	172.16.1.127	
3	p	u	40000	hdp3	172.16.1.126	
4	p	u	40000	hdp1	172.16.1.124	

(6 rows)

- 查询 gp_master_mirroring 系统视图检查主节点镜像的状态。该视图提供了关于 HAWQ Master 节点的 WAL Send 进程的使用信息。查询结果如下，可以看到主、从 Master 数据已经同步。

```
[gpadmin@hdp2 ~]$ psql -c 'SELECT * FROM gp_master_mirroring;'
 summary_state | detail_state |          log_time          | error_message
---------------+--------------+----------------------------+---------------
 Synchronized  |              | 2017-02-28 01:16:59+08     |
(1 row)
```

2. 手工执行失败切换

当主 Master 不可用时，需要手工执行切换，将从 Master 激活为主 Master。

（1）保证系统中已经配置从 Master 节点主机。

（2）激活从 Master 节点。登录到 HAWQ 从 Master 节点并激活它，之后从 Master 成为 HAWQ 的主 Master。

```
[gpadmin@hdp3 ~]$ hawq activate standby
```

3. 配置一个新的从 Master 节点（可选但推荐）

手工切换 Master 后，最好配置一个新的从 Master 节点，继续保持 Master 的高可用性，配置过程参考上面"配置从 Master"的内容。

4. 重新同步主、从节点

如果主、从之间的日志同步进程停止或者落后，从 Master 可能变成过时状态。这种情况下查询 gp_master_mirroring 视图，会看到 summary_state 字段输出中显示 Not Synchronized。为了将从与主重新进行同步，在主 Master 节点上执行下面的命令。该命令停止并重启主 Master 节点，然后同步从 Master 节点。

```
[gpadmin@hdp3 ~]$ hawq init standby -n
```

11.2.3 HAWQ 文件空间与 HDFS 高可用

如果在初始化 HAWQ 时没有启用 HDFS 的高可用性，可以使用下面的过程启用它。

（1）配置 HDFS 集群高可用性。
（2）收集目标文件空间的信息。
（3）停止 HAWQ 集群，并且备份系统目录。
（4）使用命令行工具迁移文件空间。
（5）重新配置${GPHOME}/etc/hdfs-client.xml 和${GPHOME}/etc/hawq-site.xml，然后同步更新所有 HAWQ 节点的配置文件。
（6）启动 HAWQ 集群，并在迁移文件空间后重新同步从 Master 节点。

1. 配置 HDFS 集群高可用性

（1）HDFS HA 概述

HDFS 中的 NameNode 非常重要，其中保存了 DataNode 上数据块存储位置的相关信息。它主要维护两个映射，一个是文件到块的对应关系，一个是块到节点的对应关系。如果

NameNode 停止工作，就无法知道数据所在的位置，整个 HDFS 将陷入瘫痪，因此保证 NameNode 的高可用性非常重要。

在 Hadoop 1 时代，只有一个 NameNode。如果该 NameNode 数据丢失或者不能工作，那么整个集群就不能恢复了。这是 Hadoop 1 中的单点故障问题，也是 Hadoop 1 不可靠的体现。图 11-4 是 Hadoop 1 的架构图。

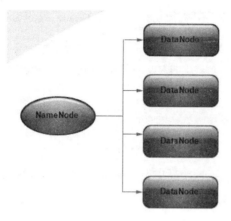

图 11-4　Hadoop1 结构

为了解决 Hadoop 1 中的单点问题，Hadoop 2 中支持两个 NameNode，每一个都有相同的职能。一个是 Active 状态，一个是 Standby 状态。当集群运行时，只有 Active NameNode 正常工作，Standby 状态的 NameNode 处于待命状态，时刻同步 Active NameNode 的数据。一旦 Active NameNode 不能工作，通过手工或者自动切换，将 Standby NameNode 转变为 Active 状态，就可以继续工作了。

Hadoop 2 中，两个 NameNode 的数据是实时共享的。HDFS 采用了一种共享机制，通过 Quorum Journal Node（JournalNode）集群或者 Network File System（NFS）进行共享。NFS 是操作系统层面的，JournalNode 是 Hadoop 层面的（主流做法）。图 11-5 为 JournalNode 架构图。

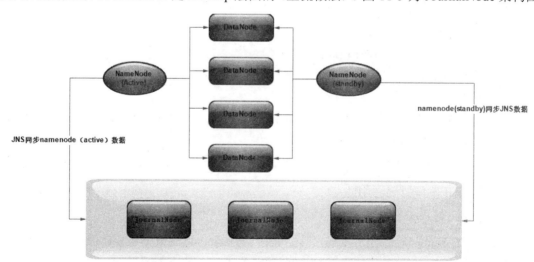

图 11-5　Hadoop2 JournalNode 架构

两个 NameNode 为了数据同步，会通过一组称作 JournalNodes 的独立进程进行相互通信。当 Active NameNode 的命名空间有任何修改时，会告知 JournalNodes 进程。Standby NameNode 读取 JNs 中的变更信息，并且一直监控 edit log 的变化，把变化应用于自己的命名空间。Standby 可以确保在集群出错时，命名空间状态已经完全同步。

对于 HA 集群而言，确保同一时刻只有一个 NameNode 处于 Active 状态非常重要。否则，两个 NameNode 的数据状态就可能产生分歧，或造成数据丢失，或产生错误的结果。为了保证这点，需要利用 ZooKeeper。首先 HDFS 集群中的两个 NameNode 都在 ZooKeeper 中注册，当 Active NameNode 出故障时，ZooKeeper 能检测到这种情况，然后它就会自动把 Standby NameNode 切换为 Active 状态。

（2）使用 Ambari 启用 HDP 的高可用性（参考 https://docs.hortonworks.com/HDPDocuments/Ambari-2.4.1.0/bk_ambari-user-guide/content/how_to_configure_namenode_high_availability.html）。

- 检查 Hadoop 集群，确保集群中至少有三台主机，并且至少运行三个 ZooKeeper 服务器。
- 检查 Hadoop 集群，确保 HDFS 和 ZooKeeper 服务不是在维护模式中。启用 NameNode HA 时，这些服务需要重启，而维护模式阻止启动和停止。如果 HDFS 或者 ZooKeeper 服务处于维护模式，NameNode HA 向导将不能完全成功。
- 在 Ambari Web 里，选择 Services→HDFS→Summary。
- 在 Service Actions 中选择 Enable NameNode HA，如图 11-6 所示。

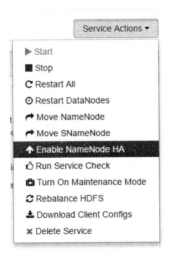

图 11-6　启用 NameNode HA

- 出现 Enable HA 向导。此向导描述了配置 NameNode HA 必须执行的自动和手工步骤。
- Get Started。此步骤给出一个处理预览，并允许用户选择一个 Nameservice ID，本例的 Nameservice ID 为 mycluster。HA 一旦配置好，就需要使用 Nameservice ID 代替 NameNode FQDN。单击 Next 继续处理，如图 11-7 所示。

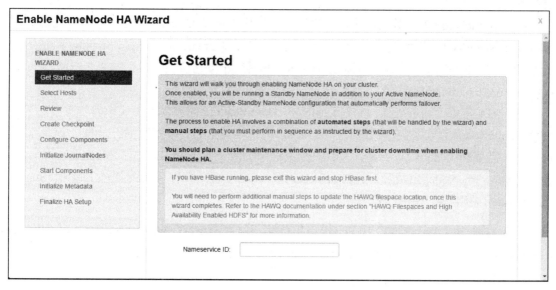

图 11-7 开始配置 NameNode HA

- Select Hosts：为 Standby NameNode 和 JournalNodes 选择主机。可以使用下拉列表调整向导建议的选项。单击 Next 继续处理。
- Review：确认主机的选择，并单击 Next，如图 11-8 所示。

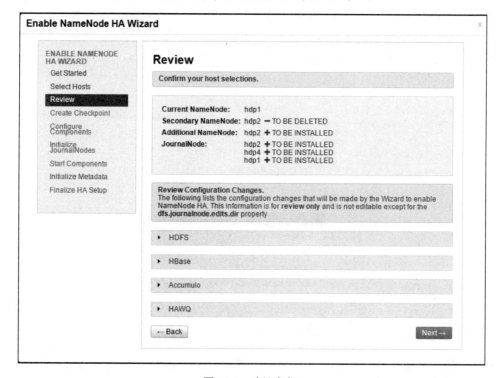

图 11-8 确认主机

- 创建检查点：此步骤中提示执行两条命令，第一条命令把 NameNode 置于安全模式，第二条命令创建一个检查点，如图 11-9 所示。需要登录当前的 NameNode 主机终端执行这两条命令。当 Ambari 检测到命令执行成功后，窗口下端的提示消息将改变。单击 Next。

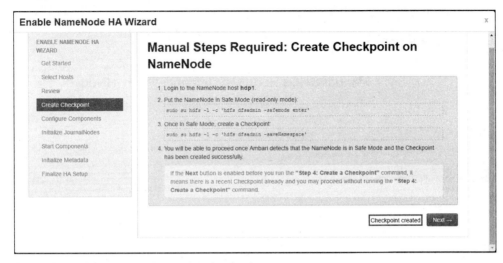

图 11-9　创建检查点

按照页面提示，在终端控制台执行两条命令：

```
[root@hdp1 ~]# sudo su hdfs -l -c 'hdfs dfsadmin -safemode enter'
Safe mode is ON
[root@hdp1 ~]# sudo su hdfs -l -c 'hdfs dfsadmin -saveNamespace'
Save namespace successful
[root@hdp1 ~]#
```

- 确认组件。向导开始配置相关组件，显示进度跟踪步骤。配置成功如图 11-10 所示。单击 Next 继续。

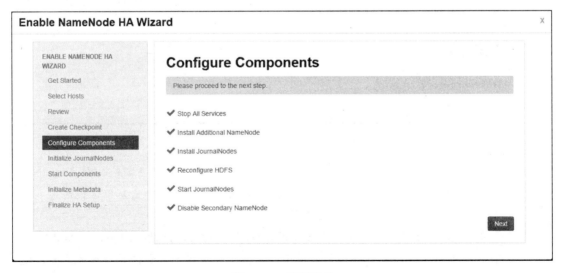

图 11-10　配置组件

- 初始化 JournalNodes。此步骤中提示执行一条指令，如图 11-11 所示。需要登录到当前的 NameNode 主机运行命令初始化 JournalNodes。当 Ambari 检测成功，窗口下端的提示消息将改变。单击 Next。

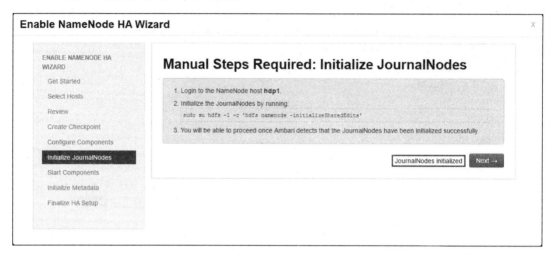

图 11-11　初始化 JournalNodes

- 启动组件：向导启动 ZooKeeper 服务器和 NameNode，显示进度跟踪步骤。执行成功后如图 11-12 所示。单击 Next 继续。

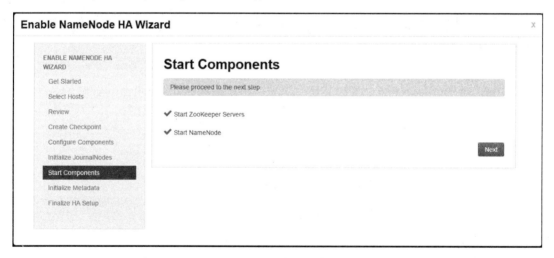

图 11-12　启动组件

- 初始化元数据。此步骤中根据提示执行命令，确保登录正确的主机（主、从 NameNode）执行每个命令。当完成了两个命令，单击 Next。显示一个弹出窗口，提醒用户确认已经执行了两个命令。单击 OK 确认。
- 最终设置。此步骤中，向导显示进度跟踪步骤。单击 Done 结束向导。在 Ambari Web GUI 重载后，可以看到一些警告提示。等待几分钟直到服务恢复。如果需要，使用 Ambari Web 重启相关组件。

- 调整 ZooKeeper 失败切换控制器的重试次数设置。浏览 Services→HDFS→Configs→Advanced core-site，设置 ha.failover-controller.active-standby-elector.zk.op.retries=120。

至此已经配置好 HDP HDFS 的高可用性。从 NameNode UI 可以看到，hdp1 和 hdp2 分别显示为 active 和 standby。如图 11-13 和图 11-14 所示。

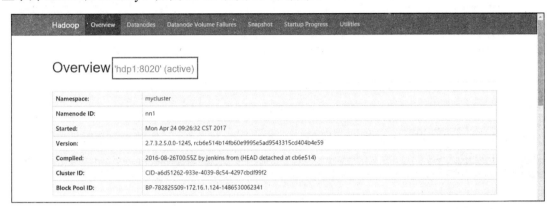

图 11-13　Active NameNode

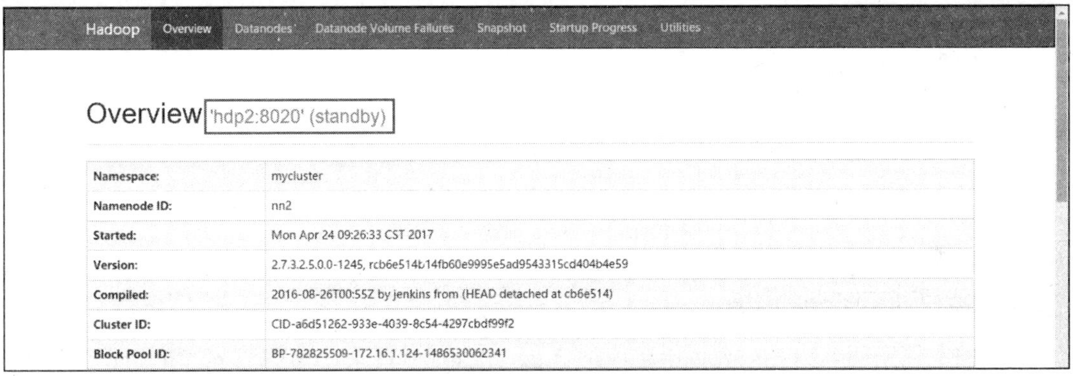

图 11-14　Standby NameNode

此时在 hdp1 上执行如下命令关闭 hdp1 上的 NameNode：

```
[hdfs@hdp1 ~]$ /usr/hdp/2.5.0.0-1245/hadoop/sbin/hadoop-daemon.sh stop namenode
stopping namenode
[hdfs@hdp1 ~]$
```

再次查看 hdp2 上的 NameNode，发现已自动切换为 Active，如图 11-15 所示。

图 11-15　Standby 切换为 Active

再次查看 hdp1 上的 NameNode，发现已自动切换为 Standby，如图 11-16 所示。

图 11-16　Active 切换为 Standby

2. 收集目标文件空间信息

默认的文件空间名为 dfs_system，存在于 pg_filespace 目录，其参数 pg_filespace_entry 包含每个文件空间的细节信息。为了将文件空间位置迁移到 HDFS HA 的位置，必须将数据迁移到集群中新的 HDFS HA 路径。使用下面的 SQL 查询收集关于 HDFS 上的文件空间位置信息。

```
select fsname, fsedbid, fselocation
  from pg_filespace as sp, pg_filespace_entry as entry, pg_filesystem as fs
  where sp.fsfsys = fs.oid and fs.fsysname = 'hdfs' and sp.oid = entry.fsefsoid
order by entry.fsedbid;
```

输出信息中包含 HDFS 路径共享相同的前缀，以及当前文件空间的位置：

```
   fsname    | fsedbid |          fselocation
-------------+---------+-------------------------------
 dfs_system  |    0    | hdfs://hdp1:8020/hawq_data
(1 row)
```

为了在 HAWQ 中使用 HDFS HA，需要文件空间名和 HDFS 路径的通用前缀信息。文件空间位置的格式类似一个 URL。如果无 HA 的文件空间位置是

hdfs://test5:9000/hawq/hawq-1459499690，并且 HDFS HA 的通用前缀是 hdfs://hdfs-cluster，那么新的文件空间位置应该是 hdfs://hdfs-cluster/hawq/hawq-1459499690。

3. 停止 HAWQ 集群并备份系统目录

用户必须手工执行这个步骤。在 HAWQ 中启用 HDFS HA 时会修改 HAWQ 的目录和永久表。因此迁移文件空间位置前，先要备份目录，以确保不会因为硬件失效或在一个操作期间（如杀掉 HAWQ 进程）丢失数据。

（1）如果 HAWQ 主节点使用了一个定制端口，输出 PGPORT 环境变量。例如：

```
export PGPORT=8020
```

（2）保存 HAWQ Master 节点目录，在 hawq-site.xml 文件中找到 hawq_master_directory 属性，赋给一个环境变量。

```
export MDATA_DIR=/data/hawq/master
```

（3）断掉所有连接。使用以下查询检查活动连接：

```
[gpadmin@hdp3 ~]$ psql -c "SELECT * FROM pg_catalog.pg_stat_activity"
```

（4）执行检查点。

```
[gpadmin@hdp3 ~]$ psql -c "CHECKPOINT"
```

（5）停止 HAWQ 集群。

```
[gpadmin@hdp3 ~]$ hawq stop cluster -a -M fast
```

（6）复制主节点数据目录到备份的位置：

```
$ cp -r ${MDATA_DIR} /catalog/backup/location
```

主节点数据目录包含子目录，一定要备份此目录。

4. 迁移文件空间位置

用户必须手工执行这个步骤。HAWQ 提供了命令行工具 hawq filespace，迁移文件空间的位置。

（1）如果 HAWQ 主节点使用了一个定制端口，输出 PGPORT 环境变量。例如：

```
export PGPORT=9000
```

（2）运行下面的命令迁移文件空间的位置：

```
[gpadmin@hdp3 master]$ hawq filespace --movefilespace default --location=hdfs://mycluster/hawq_data
```

迁移文件空间时可能出现的以下潜在错误：

- 如果提供了无效的输入，或者在修改文件空间位置时没有停止 HAWQ，可能发生非崩溃错误。检查是否已经从头正确执行了所有步骤，或者在再次执行 hawq filespace 前修正输入错误。

- 崩溃错误可能发生在硬件失效或者修改文件空间位置时杀死 HAWQ 进程失败的情况下。当发生崩溃错误时，在输出中可以看到 "PLEASE RESTORE MASTER DATA DIRECTORY" 消息。此时应该停止数据库，并且还原在 "停止 HAWQ 集群并备份系统目录"步骤中备份的${MDATA_DIR}目录。

5. 重新配置 hdfs-client.xml 和 hawq-site.xml，更新 HAWQ 使用 NameNode HA

如果使用 Ambari 管理 HDFS 和 HAWQ，不需要执行此步骤，因为 Ambari 在启用了 NameNode HA 后会自动执行这些修改。

如果使用命令行安装和管理 HAWQ 集群，参考 http://hawq.incubator.apache.org/docs/userguide/2.1.0.0-incubating/admin/HAWQFilespacesandHighAvailabilityEnabledHDFS.html#configuregphomeetchdfsclientxml，修改 HAWQ 配置以使用 NameNode HA 服务。

6. 重启 HAWQ 集群并重新同步主从 Master 节点

（1）重启 HAWQ 集群：

```
[gpadmin@hdp3 master]$ hawq start cluster -a
```

（2）迁移文件空间到新位置会使从 Master 节点无效，因此需要重新同步从 Master 节点。在主 Master 节点上，运行下面的命令保证从 Master 的目录与主 Master 重新同步。

```
[gpadmin@hdp3 master]$ hawq init standby -n -M fast
```

至此已经在 HAWQ 的文件空间中使用了 HDFS HA，再次查询文件空间信息，结果如下：

```
gpadmin=# select fsname, fsedbid, fselocation
gpadmin-#   from pg_filespace as sp,
gpadmin-#        pg_filespace_entry as entry,
gpadmin-#        pg_filesystem as fs
gpadmin-#  where sp.fsfsys = fs.oid and fs.fsysname = 'hdfs'
gpadmin-#    and sp.oid = entry.fsefsoid
gpadmin-#  order by entry.fsedbid;
   fsname   | fsedbid |        fselocation
------------+---------+-----------------------------
 dfs_system |       0 | hdfs://mycluster/hawq_data
(1 row)
```

11.2.4　HAWQ 容错服务

HAWQ 的容错服务（fault tolerance service，FTS）使得 HAWQ 可以在 Segment 节点失效时持续操作。容错服务自动运行，并且不需要额外的配置。

每个 Segment 运行一个资源管理进程，定期（默认每 30 秒）向 Master 节点的资源管理进程发送 Segment 状态。这个时间间隔由 hawq_rm_segment_heartbeat_interval 服务器配置参数所控制。当一个 Segment 碰到严重错误，例如，由于硬件问题，Segment 上的一个临时目录损坏，Segment 通过心跳报告向 Master 节点报告有一个临时目录损坏。Master 节点接收到报告后，在 gp_segment_configuration 表中将该 Segment 标记为 DOWN。所有 Segment 状态的变化都被记录到 gp_configuration_history 目录表，包括 Segment 被标记为 DOWN 的原因。当这个 Segment 被置为 DOWN，Master 节点不会在该 Segment 上运行查询执行器。失效的 Segment 与集群剩下的节点相隔离。

包括磁盘故障的其他原因也会导致一个 Segment 被标记为 DOWN。例如，HAWQ 运行在 YARN 模式中，每个 Segment 应该有一个运行的 NodeManager（Hadoop 的 YARN 服务），因此 Segment 可以被看作 HAWQ 的一个资源。但如果 Segment 上的 NodeManager 不能正常操作，那么该 Segment 会在 gp_segment_configuration 表中被标记为 DOWN。失效对应的原因被记录进 gp_configuration_history。

为查看当前 Segment 的状态，查询 gp_segment_configuration 表。如果 Segment 的状态是 DOWN，"description" 列显示原因。下面列出了几个常见的 Segment 宕机原因。

1. heartbeat timeout

主节点没有接收到来自 Segment 的心跳。如果看到这个原因，确认该 Segment 上的 HAWQ 实例是否运行。如果 Segment 在以后的时间报告心跳，Segment 被自动标记为 UP。

2. failed probing segment

Master 节点探测 Segment 以验证它是否能被正常操作，段的响应为 NO。在一个 HAWQ 实例运行时，查询分发器发现某些 Segment 上的查询执行器不能正常工作。Master 节点上的资源管理器进程向这个 Segment 发送一个消息。当 Segment 的资源管理器接收到来自 Master 节点的消息，它检查其 postmaster 进程是否工作正常，并且向 Master 节点发送一个响应消息。如果 Master 节点收到的响应消息表示该 Segment 的 postmaster 进程没有正常工作，那么 Master 节点标记 Segment 为 DOWN，原因记为 "failed probing segment."。检查失败 Segment 的日志并且尝试重启 HAWQ 实例。

3. communication error

Master 节点不能连接到 Segment。检查 Master 节点和 Segment 之间的网络连接。

4. resource manager process was reset

如果 Segment 资源管理器进程的时间戳与先前的时间戳不匹配，意味着 Segment 上的资源管理器进程被重启过。在这种情况下，HAWQ Master 节点需要回收该 Segment 上的资源并将其标记为 DOWN。如果 Master 节点从该 Segment 接收到一个新的心跳，它会自动标记为 UP。

5. no global node report

HAWQ 使用 YARN 管理资源，但没有为该 Segment 接收到集群报告。检查该 Segment 上的 NodeManager 是否可以正常操作。如果不能，尝试启动该 Segment 上的 NodeManager。在

NodeManager 启动后，运行 yarn node --list 查看该节点是否在列表中。如果存在，该 Segment 被自动置为 UP。

11.3 小结

HAWQ 可以通过 gpfdist、PXF、pg_dump、pg_restore 等工具备份/还原数据库。这些备份方法方便表的迁移和恢复人为误操作，是对 HDFS 副本集功能的有效补充。HAWQ 高可用的实现方式包括硬件冗余、Master 镜像、利用 HDFS HA 几个层面。HAWQ 建议在部署时，Master 节点应该使用 RAID，而 Segment 节点应该使用 JBOD。主从 Master 能防止 Master 节点的单点故障。当主 Master 出现问题，需要手工将切换主从 Master。HAWQ 文件空间支持 HDFS HA。容错服务（fault tolerance service，FTS）使得 HAWQ 可以在 Segment 节点失效时持续操作，该服务自动运行，不需要额外配置。

第二部分

HAWQ 实战演练

第 12 章

建立数据仓库示例模型

在本书第二部分,我们将利用第一部分讲解的 HAWQ 技术特性,使用一个完整的示例,展示如何在 HAWQ 上一步步构建自己的数据仓库,目的在于演示以 HAWQ 代替传统数据仓库的具体实现全过程。本章说明示例的业务场景、数据仓库架构、实验环境、源和目标库的建立过程、测试数据和日期维度的生成等内容。后面章节陆续介绍实现初始数据装载、定期数据装载、调度 ETL 工作流自动执行、维度表技术、事实表技术、OLAP 和数据可视化的相关方法与工具。限于篇幅,我们不再对实体关系、数据库设计范式、ETL、OLAP 等数据仓库相关的基本概念做出详细解释。

12.1 业务场景

1. 操作型数据源

示例的操作型数据源来自一个销售订单系统,初始时只有产品、客户、销售订单三个表,实体关系图如图 12-1 所示。

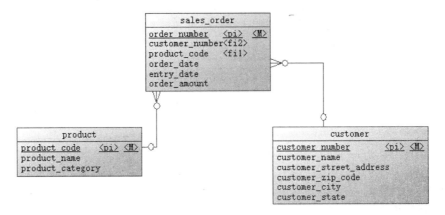

图 12-1 数据源实体关系图

这个场景中的表及其属性都很简单。产品表和客户表属于基本信息表,分别存储产品和客户信息。产品只有产品编号、产品名称、产品分类三个属性，产品编号是主键，唯一标识一个产品。客户有六个属性，除客户编号和客户名称外，还包含省、市、街道、邮编四个客户所在地区属性。客户编号是主键，唯一标识一个客户。在实际应用中，基本信息表通常由后台应用系统维护。销售订单表有六个属性，订单号是主键，唯一标识一条销售订单记录。产品编号和客户编号是两个外键，分别引用产品表和客户表的主键。另外三个属性是订单时间、登记时间和订单金额。订单时间指的是客户下订单的时间，订单金额属性指的是该笔订单需要花费的金额，这些属性的含义很清楚。订单登记时间表示订单录入的时间，大多数情况下它应该等同于订单时间。如果由于某种情况需要重新录入订单，同时还要记录原始订单的时间和重新录入的时间，或者出现某种问题，订单登记时间滞后于下订单的时间，这两个属性值就会不同。

源系统采用关系模型设计，为了减少表的数量，这个系统只做到了 2NF。地区信息依赖于邮编，所以该模型中存在传递依赖。

2. 销售订单数据仓库模型

使用以下步骤设计数据仓库模型：

（1）选择业务流程。在本示例中只涉及一个销售订单业务流程。

（2）声明粒度。ETL 处理时间周期为每天一次，这是数据仓库的最小粒度。事实表中存储按天汇总的订单数据。

（3）确认维度。显然产品和客户是销售订单的维度。日期维度用于业务集成，并为数据仓库提供重要的历史视角，每个数据仓库中都应该有一个日期维度。订单维度是特意设计的，用于说明退化维度技术。

（4）确认事实。销售订单是当前场景中唯一的事实。

示例数据仓库的实体关系图如图 12-2 所示。

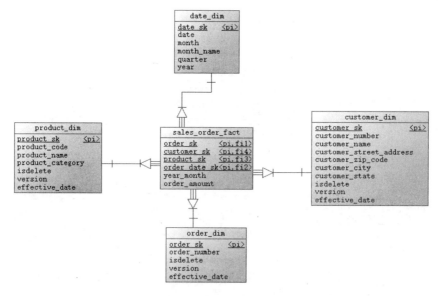

图 12-2　数据仓库实体关系图

12.2 数据仓库架构

"架构"是什么?这个问题从来就没有一个准确的答案。在软件行业,一种被普遍接受的架构定义是指系统的一个或多个结构。结构中包括软件的构建(构建是指软件的设计与实现),构建外部可以看到属性以及它们之间的相互关系。参考此定义,这里把数据仓库架构理解成构成数据仓库的组件及其之间的关系,那么就有了如图12-3所示的数据仓库架构图。

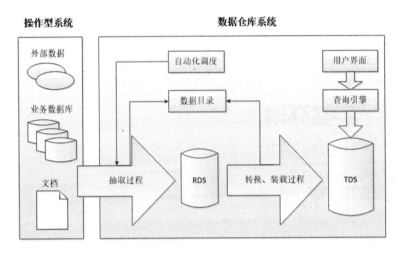

图 12-3　数据仓库架构图

图中显示的整个数据仓库环境包括操作型系统和数据仓库系统两大部分。操作型系统的数据经过抽取、转换和装载(ETL)过程进入数据仓库系统。这里把 ETL 过程分成了抽取和转换装载两个部分。抽取过程负责从操作型系统获取数据,该过程一般不做数据聚合和汇总,物理上是将操作型系统的数据全量或增量复制到数据仓库系统的 RDS 中。转换装载过程将数据进行清洗、过滤、汇总、统一格式化等一系列转换操作,使数据转为适合查询的格式,然后装载进数据仓库系统的 TDS 中。传统数据仓库的基本模式是用一些过程将操作型系统的数据抽取到文件,然后另一些过程将这些文件转化成 MySQL 这样的关系数据库的记录。最后,第三部分过程负责把数据导入进数据仓库。本例中的业务数据使用 MySQL 数据库存储。

- RDS(RAW DATA STORES)是原始数据存储的意思。它的作用主要有三个:作为数据缓冲区;提供细节数据可供查询;保留原始数据,便于跟踪和修正 ETL 错误。本例中的 RDS 使用 HAWQ 的 HDFS 外部表。
- TDS(TRANSFORMED DATA STORES)意为转换后的数据存储。这里存储真正的数据仓库中的数据。本例中的 TDS 使用 HAWQ 内部表。
- 自动化调度组件的作用是自动定期重复执行 ETL 过程。作为通用的需求,所有数据仓库系统都应该能够建立周期性自动执行的工作流作业。传统数据仓库一般利用操作系统自带的调度功能(如 Linux 的 cron 或 Windows 的计划任务)实现作业自动执行。本示例使用 Oozie 和 Falcon 完成自动调度任务。

- 数据目录有时也被称为元数据存储，它可以提供一份数据仓库中数据的清单。一个好的数据目录是让用户体验到系统易用性的关键。HAWQ 是数据库系统，自带一系列元数据表和视图。
- 查询引擎组件负责实际执行用户查询。传统数据仓库中，它可能是存储转换后数据的 MySQL 等关系数据库系统内置的查询引擎，还可能是以固定时间间隔向其导入数据的 OLAP 立方体，如 Essbase cube。HAWQ 本身就是以一个强大的查询引擎而存在，本示例使用 HAWQ 作为查询引擎正是物尽其用。
- 用户界面指的是最终用户所使用的接口程序。可能是一个 GUI 软件，如 BI 套件的中的客户端软件，也可能就是一个浏览器。本示例的用户界面使用 Zeppelin。

12.3 实验环境

1. 硬件环境

四台 VMware 虚机组成的 HDP 集群，每台机器配置如下：

- 15K RPM SAS 100GB
- Intel(R) Xeon(R) E5-2620 v2 @ 2.10GHz，双核双 CPU
- 8G 内存，8GSwap
- 10000Mb/s 虚拟网卡

2. 软件环境

- Linux: CentOS release 6.4，核心 2.6.32-358.el6.x86_64
- Ambari: 2.4.1
- Hadoop: HDP 2.5.0
- HAWQ: 2.1.1.0
- HAWQ PXF: 3.1.1
- MySQL 5.6.14

HDP 与 HAWQ 的安装部署过程参考本书第 2 章。表 12-1 汇总了各主机的角色。

表 12-1 主机角色

主机名	IP 地址	角色
hdp1	172.16.1.124	HAWQ Segment
hdp2	172.16.1.125	HAWQ Standby Master、HAWQ Segment
hdp3	172.16.1.126	HAWQ Primary Master、HAWQ Segment
hdp4	172.16.1.127	HAWQ Segment、MySQL 源库

12.4 HAWQ 相关配置

1. 创建客户端认证

编辑 Master 上的/data/hawq/master/pg_hba.conf 文件，添加 dwtest 用户：

```
[gpadmin@hdp3 ~]$ more /data/hawq/master/pg_hba.conf
local   all         gpadmin              ident
host    all         gpadmin              127.0.0.1/28            trust
host    all         gpadmin              ::1/128                 trust
host    all         gpadmin              172.16.1.126/32         trust
host    all         gpadmin              fe80::250:56ff:fea5:526f/128    trust
host    all         gpadmin              172.16.1.125/32         trust
host    all         gpadmin              172.16.1.127/32         trust
host    all         gpadmin              172.16.1.0/24           trust
host    all         wxy                  172.16.1.0/24           md5
host    all         dwtest               172.16.1.0/24           md5
[gpadmin@hdp3 ~]$
```

2. 设置并发连接数

```
-- 设置参数
hawq config -c max_connections -v 100
hawq config -c seg_max_connections -v 1000
hawq config -c max_prepared_transactions -v 200

-- 重启 HAWQ
hawq restart cluster
```

这些参数的含义与作用参见第 3 章。查看修改后的配置如下：

```
[gpadmin@hdp3 ~]$ hawq config -s max_connections
GUC     : max_connections
Value   : 100
[gpadmin@hdp3 ~]$ hawq config -s seg_max_connections
GUC     : seg_max_connections
Value   : 1000
[gpadmin@hdp3 ~]$ hawq config -s max_prepared_transactions
GUC     : max_prepared_transactions
Value   : 200
[gpadmin@hdp3 ~]$
```

3. 创建数据库用户

（1）用 gpadmin 连接 HAWQ，建立用户 dwtest，授予建库权限。

```
-- 创建用户
```

```
create role dwtest with password '123456' login createdb;
```

查看用户如下：

```
gpadmin=# \dg
                         List of roles
  Role name  |              Attributes              | Member of
-------------+--------------------------------------+-----------
 dwtest      | Create DB                            |
 gpadmin     | Superuser, Create role, Create DB    |
 wxy         | Superuser                            |

gpadmin=#
```

（2）测试登录。

连接成功后，查看数据库如下所示：

```
[gpadmin@hdp3 ~]$ psql -U dwtest -d gpadmin -h hdp3
Password for user dwtest:
psql (8.2.15)
Type "help" for help.

gpadmin=> \l
              List of databases
   Name    |  Owner  | Encoding | Access privileges
-----------+---------+----------+-------------------
 gpadmin   | gpadmin | UTF8     |
 postgres  | gpadmin | UTF8     |
 template0 | gpadmin | UTF8     |
 template1 | gpadmin | UTF8     |
(4 rows)
```

4. 创建资源队列

（1）将默认的 **pg_default** 的资源限制由 50% 改为 20%，同时将过度使用因子设置为 5。

```
alter resource queue pg_default with
  (memory_limit_cluster=20%,core_limit_cluster=20%,resource_overcommit_factor=5);
```

（2）建立一个 dwtest 用户使用的专用队列，资源限制为 80%，同时将过度使用因子设置为 2。

```
create resource queue dwtest_queue with
  (parent='pg_root', memory_limit_cluster=80%,
core_limit_cluster=80%,resource_overcommit_factor=2);
```

（3）查看资源队列配置。

```
select rsqname qname, parentoid poid, activestats stats,
      memorylimit mlimit, corelimit climit, resovercommit resoc,
      allocpolicy policy, vsegresourcequota vsresquota
  from pg_resqueue;
```

结果如下：

```
    qname     | poid | stats | mlimit | climit | resoc | policy | vsresquota
--------------+------+-------+--------+--------+-------+--------+------------
 pg_root      |    0 |    -1 | 100%   | 100%   |     2 | even   |
 pg_default   | 9800 |    20 | 20%    | 20%    |     5 | even   | mem:256mb
 dwtest_queue | 9800 |    20 | 80%    | 80%    |     2 | even   | mem:256mb
(3 rows)
```

（4）用 gpadmin 将 dwtest 用户的资源队列设置为新建的 dwtest_queue。

```
-- 修改用户资源队列
alter role dwtest resource queue dwtest_queue;
-- 查看用户资源队列
select rolname, rsqname from pg_roles, pg_resqueue
 where pg_roles.rolresqueue=pg_resqueue.oid;
```

结果如下：

```
 rolname |   rsqname
---------+--------------
 dwtest  | dwtest_queue
 wxy     | pg_default
 gpadmin | pg_default
(3 rows)
```

假设其他用户都使用默认的 pg_default 队列。采用以上定义，工作负载通过资源队列划分如下：

- 如果没有活跃用户使用 pg_default 队列，dwtest 用户可以使用 100%的资源。
- 如果有同时使用 pg_default 队列的其他用户，则 dwtest 用户使用 80%，其他用户使用 20%资源。
- 如果没有 dwtest 用户的活跃任务，其他使用 pg_default 队列的用户可以使用 100%的资源。

5. 配置资源

（1）查看集群所有节点都已启动

```
select * from gp_segment_configuration;
```

结果如下：

```
registration_order |role| status | port  | hostname |    address    |description
-------------------+----+--------+-------+----------+---------------+-----------
                -1 | s  |  u     |  5432 |  hdp2    |  172.16.1.125 |
                 0 | m  |  u     |  5432 |  hdp3    |  hdp3         |
                 1 | p  |  u     | 40000 |  hdp3    |  172.16.1.126 |
                 2 | p  |  u     | 40000 |  hdp2    |  172.16.1.125 |
                 3 | p  |  u     | 40000 |  hdp1    |  172.16.1.124 |
                 4 | p  |  u     | 40000 |  hdp4    |  172.16.1.127 |
(6 rows)
```

所有节点的状态都应该是启动状态（status='u'）。hawq_rm_rejectrequest_nseg_limit 参数保持默认值 0.25，就是说现有集群的全部 4 个 Segment 中如果有两个宕机，则 HAWQ 资源管理器将直接拒绝查询的资源请求。

（2）资源管理使用默认的独立模式

独立模式下，HAWQ 使用集群节点资源时，不考虑其他共存的应用，HAWQ 假设它能使用所有 Segment 的资源。对于专用 HAWQ 集群，独立模式是可选的方案。

```
[gpadmin@hdp3 ~]$ hawq config -s hawq_global_rm_type
GUC              : hawq_global_rm_type
Value            : none
[gpadmin@hdp3 ~]$
```

（3）设置每个 Segment 资源配额

硬件配置内存 8GB，双核双 CPU，共 4 个虚拟 CPU 核。所以每个 Segment 使用的内存与 CPU 核数配额分别配置为 8GB 和 4 核，最大限度利用资源。

```
[gpadmin@hdp3 ~]$ hawq config -s hawq_rm_memory_limit_perseg
GUC              : hawq_rm_memory_limit_perseg
Value            : 8GB
[gpadmin@hdp3 ~]$ hawq config -s hawq_rm_nvcore_limit_perseg
GUC              : hawq_rm_nvcore_limit_perseg
Value            : 4
[gpadmin@hdp3 ~]$
```

所有资源队列中虚拟段的资源限额均为默认的 256MB，每个 Segment 可以分配 32 个虚拟段。并且 8GB 是 256MB 的 32 倍，每核 2GB 内存，这种配置防止形成资源碎片。

6. 在 HDFS 上创建 HAWQ 外部表对应的目录

```
su - hdfs -c 'hdfs dfs -mkdir -p /data/ext'
su - hdfs -c 'hdfs dfs -chown -R gpadmin:gpadmin /data/ext'
su - hdfs -c 'hdfs dfs -chmod -R 777 /data/ext'
su - hdfs -c 'hdfs dfs -chmod -R 777 /user'
```

查看外部表的 HDFS 目录，结果如下：

```
[root@hdp3 ~]# su - hdfs -c 'hdfs dfs -ls /data'
Found 1 items
drwxrwxrwx   - gpadmin gpadmin          0 2017-05-05 10:48 /data/ext
[root@hdp3 ~]#
```

12.5 创建示例数据库

12.5.1 在 hdp4 上的 MySQL 中创建源库对象并生成测试数据

（1）执行下面的 SQL 语句建立源数据库表。

```sql
-- 建立源数据库
drop database if exists source;
create database source;

use source;

-- 建立客户表
create table customer (
    customer_number int not null auto_increment primary key comment '客户编号',
    customer_name varchar(50) comment '客户名称',
    customer_street_address varchar(50) comment '客户住址',
    customer_zip_code int comment '邮编',
    customer_city varchar(30) comment '所在城市',
    customer_state varchar(2) comment '所在省份'
);

-- 建立产品表
create table product (
    product_code int not null auto_increment primary key comment '产品编码',
    product_name varchar(30) comment '产品名称',
    product_category varchar(30) comment '产品类型'
);

-- 建立销售订单表
create table sales_order (
    order_number int not null auto_increment primary key comment '订单号',
    customer_number int comment '客户编号',
```

```sql
    product_code int comment '产品编码',
    order_date datetime comment '订单日期',
    entry_date datetime comment '登记日期',
    order_amount decimal(10 , 2 ) comment '销售金额',
    foreign key (customer_number)
        references customer (customer_number)
        on delete cascade on update cascade,
    foreign key (product_code)
        references product (product_code)
        on delete cascade on update cascade
);
```

（2）执行下面的 SQL 语句生成源库测试数据。

```sql
use source;

-- 生成客户表测试数据
insert into customer
(customer_name,customer_street_address,customer_zip_code,
customer_city,customer_state)
values
('really large customers', '7500 louise dr.',17050, 'mechanicsburg','pa'),
('small stores', '2500 woodland st.',17055, 'pittsburgh','pa'),
('medium retailers','1111 ritter rd.',17055,'pittsburgh','pa'),
('good companies','9500 scott st.',17050,'mechanicsburg','pa'),
('wonderful shops','3333 rossmoyne rd.',17050,'mechanicsburg','pa'),
('loyal clients','7070 ritter rd.',17055,'pittsburgh','pa'),
('distinguished partners','9999 scott st.',17050,'mechanicsburg','pa');

-- 生成产品表测试数据
insert into product (product_name,product_category)
values
('hard disk drive', 'storage'),
('floppy drive', 'storage'),
('lcd panel', 'monitor');

-- 生成100条销售订单表测试数据
drop procedure if exists generate_sales_order_data;
delimiter //
create procedure generate_sales_order_data()
begin
    drop table if exists temp_sales_order_data;
    create table temp_sales_order_data as select * from sales_order where 1=0;

    set @start_date := unix_timestamp('2016-03-01');
    set @end_date := unix_timestamp('2016-07-01');
    set @i := 1;
```

```
    while @i<=100 do
        set @customer_number := floor(1 + rand() * 6);
        set @product_code := floor(1 + rand() * 2);
        set @order_date := from_unixtime(@start_date + rand()
                    * (@end_date - @start_date));
        set @amount := floor(1000 + rand() * 9000);

        insert into temp_sales_order_data values
      (@i,@customer_number,@product_code,@order_date,@order_date,@amount);
        set @i:=@i+1;
    end while;

    truncate table sales_order;
    insert into sales_order
    select
null,customer_number,product_code,order_date,entry_date,order_amount
    from temp_sales_order_data order by order_date;
    commit;

end
//
delimiter ;

call generate_sales_order_data();
```

上面代码中创建了一个 MySQL 存储过程，生成 100 条销售订单测试数据。为了模拟实际订单的情况，订单表中的客户编号、产品编号、订单时间和订单金额都取一个范围内的随机值，订单时间与登记时间相同。因为订单表的主键是自增的，为了使主键值和订单时间字段的值顺序保持一致，引入了一个名为 temp_sales_order_data 的表，存储中间临时数据。在后面章节中都是使用此方案生成订单测试数据。

（3）创建操作源数据库的用户并授权。

```
create user 'dwtest'@'%' identified by '123456';
grant select on source.* to 'dwtest'@'%';
```

12.5.2 创建目标库对象

1. 创建目标数据库及其模式

（1）用 dwtest 用户连接 HAWQ

```
psql -U dwtest -d gpadmin -h hdp3
```

（2）建立 dw 数据库

```
create database dw;
```

（3）在 dw 库中建立三个模式

```
-- 连接 dw 数据库
\c dw
-- 创建 ext 模式
create schema ext;
-- 创建 rds 模式
create schema rds;
-- 创建 tds 模式
create schema tds;
-- 查看模式
\dn
```

（4）设置模式查找路径

```
-- 修改 dw 数据库的模式查找路径
alter database dw set search_path to ext, rds, tds;
-- 重新连接 dw 数据库
\c dw
-- 显示模式查找路径
show search_path;
```

此时 dw 库的模式查找路径如下：

```
  search_path
----------------
 ext, rds, tds
(1 row)
```

每个 HAWQ 会话在任一时刻只能连接一个数据库。ETL 处理期间，需要将 RDS 与 TDS 中的表关联查询，因此将 RDS 和 TDS 对象存放在单独的数据库中显然是不合适的。这里在 dw 库中创建了 ext、rds、tds 三个模式。前面描述数据仓库架构时只提到了 RDS 和 TDS，并指出本示例的 RDS 使用 HAWQ 的 HDFS 外部表，为什么这里创建了三个模式呢？究其原因如下：

- Sqoop 可以将 MySQL 数据导入到 HDFS 或 Hive，但目前还没有命令行工具可以将 MySQL 数据直接导入到 HAWQ 表中。所以不得不将缓冲数据存储到 HDFS，再利用 HAWQ 的外部表进行访问。
- 如果只创建两个模式分别用作 RDS 和 TDS，则会带来性能问题。变化数据捕获（CDC）时需要关联 RDS 和 TDS 的表，而 HAWQ 的外部表和内部表关联查询的速度很慢，数据量非常大的情况下，查询延时将不可忍受。
- 通常维度数据量比事实数据量小得多。在这个前提下，用 EXT 模式存储直接从 MySQL 导出的数据，包括全部维度数据和增量的事实数据，然后将这些数据装载进 RDS 模式内部表中。这一步装载的数据量并不是很大，而且没有关联逻辑，都是简单的单表查询与数据插入。在装载 TDS 内部表时，仍然关联 RDS 与 TDS 的表，但这两个模式中的表都是内部表，查询速度是可接受的。

这里使用三个模式来划分直接外部数据、源数据存储和多维数据仓库的对象，不但逻辑上非常清晰，而且兼顾了 ETL 的处理速度。

2. 创建 EXT 模式中的数据库对象

（1）用 HAWQ 管理员用户授予 dwtest 用户在 dw 库中创建外部表的权限：

```
psql -d dw -h hdp3 -c "grant all on protocol pxf to dwtest"
```

如果不授予相应权限，创建外部表时会报以下错误：

```
ERROR: permission denied for external protocol pxf
```

（2）创建 HAWQ 外部表

```
-- 设置模式查找路径
set search_path to ext;

-- 建立客户外部表
create external table customer
( customer_number int,
  customer_name varchar(30),
  customer_street_address varchar(30),
  customer_zip_code int,
  customer_city varchar(30),
  customer_state varchar(2) )
location ('pxf://mycluster/data/ext/customer?profile=hdfstextsimple')
  format 'text' (delimiter=e',');

comment on table customer is '客户外部表';
comment on column customer.customer_number is '客户编号';
comment on column customer.customer_name is '客户姓名';
comment on column customer.customer_street_address is '客户地址';
comment on column customer.customer_zip_code is '客户邮编';
comment on column customer.customer_city is '客户所在城市';
comment on column customer.customer_state is '客户所在省份';

-- 建立产品外部表
create external table product
( product_code int,
  product_name varchar(30),
  product_category varchar(30) )
location ('pxf://mycluster/data/ext/product?profile=hdfstextsimple')
  format 'text' (delimiter=e',');
```

```sql
comment on table product is '产品外部表';
comment on column product.product_code is '产品编码';
comment on column product.product_name is '产品名称';
comment on column product.product_category is '产品类型';

-- 建立销售订单外部表
create external table sales_order
( order_number int,
  customer_number int,
  product_code int,
  order_date timestamp,
  entry_date timestamp,
  order_amount decimal(10,2) )
location ('pxf://mycluster/data/ext/sales_order?profile=hdfstextsimple')
  format 'text' (delimiter=e',', null='null');

comment on table sales_order is '销售订单外部表';
comment on column sales_order.order_number is '订单号';
comment on column sales_order.customer_number is '客户编号';
comment on column sales_order.product_code is '产品编码';
comment on column sales_order.order_date is '订单日期';
comment on column sales_order.entry_date is '登记日期';
comment on column sales_order.order_amount is '销售金额';
```

说明:

- 外部表结构与 MySQL 里的源表完全对应，其字段与源表相同。
- PXF 外部数据位置指向前面 12.4 节第 6 步创建的 HDFS 目录。
- 文件格式使用逗号分隔的简单文本格式，文件中的'null'字符串代表数据库中的 NULL 值。下一章说明数据初始装载时会看到，为了让 EXT 的数据文件尽可能地小，Sqoop 使用了压缩选项，而 hdfstextsimples 属性的 PXF 外部表能自动正确读取 Sqoop 默认的 gzip 压缩文件。

3. 创建 RDS 模式中的数据库对象

```sql
-- 设置模式查找路径
set search_path to rds;

-- 建立客户原始数据表
create table customer
( customer_number int,
  customer_name varchar(30),
  customer_street_address varchar(30),
```

```sql
    customer_zip_code int,
    customer_city varchar(30),
    customer_state varchar(2) );

comment on table customer is '客户原始数据表';

-- 建立产品原始数据表
create table product
( product_code int,
  product_name varchar(30),
  product_category varchar(30) );

comment on table product is '产品原始数据表';

-- 建立销售订单原始数据表
create table sales_order
( order_number int,
  customer_number int,
  product_code int,
  order_date timestamp,
  entry_date timestamp,
  order_amount decimal(10,2) )
partition by range (entry_date)
( start (date '2016-01-01') inclusive
  end (date '2018-01-01') exclusive
  every (interval '1 month') );

comment on table sales_order is '销售订单原始数据表';
```

说明：

- RDS 模式中的表是 HAWQ 内部表。
- RDS 模式中表数据来自 EXT 表，并且是原样装载，不需要任何转换，因此其表结构与 EXT 中的外部表一致。
- HAWQ 支持 row 和 parquet 两种数据存储格式。如果单纯从性能方面考虑，似乎 parquet 列格式更适合数据仓库应用。这里使用默认的 row 格式，是因为注意到文档中这样一句话："HAWQ does not support using ALTER TABLE to ADD or DROP a column in an existing Parquet table."。在任何项目中，数据库表创建完后就再也不用增删字段的情况几乎是不可能发生的。关于行列存储的选择，http://storage.chinabyte.com/491/12390991.shtml 这篇文章进行了比较客观的论述。
- 为了保持查询弹性使用资源和更好地数据本地化，使用默认的随机数据分布策略，而没有使用哈希分布。关于 HAWQ 表数据的存储与分布，参见第 6 章 "存储管理"。

- RDS 是实际上是原始业务数据的副本，维度数据量小，可以覆盖装载全部数据，而事实数据量大，需要追加装载每天的新增数据。因此事实表采取分区表设计，每月数据一分区，以登记日期作为分区键，预创建 2017 年一年的分区。

4. 创建 TDS 模式中的数据库对象

```sql
-- 设置模式查找路径
set search_path to tds;

-- 建立客户维度表
create table customer_dim (
    customer_sk bigserial,
    customer_number int,
    customer_name varchar(50),
    customer_street_address varchar(50),
    customer_zip_code int,
    customer_city varchar(30),
    customer_state varchar(2),
    isdelete boolean default false,
    version int,
    effective_date date );

comment on table customer_dim is '客户维度表';
comment on column customer_dim.customer_sk is '客户维度代理键';
comment on column customer_dim.customer_number is '客户编号';
comment on column customer_dim.isdelete is '是否删除';
comment on column customer_dim.version is '版本';
comment on column customer_dim.effective_date is '生效日期';

-- 建立产品维度表
create table product_dim (
    product_sk bigserial,
    product_code int,
    product_name varchar(30),
    product_category varchar(30),
    isdelete boolean default false,
    version int,
    effective_date date );

comment on table product_dim is '产品维度表';
comment on column product_dim.product_sk is '产品维度代理键';
comment on column product_dim.product_code is '产品编码';
comment on column product_dim.isdelete is '是否删除';
comment on column product_dim.version is '版本';
```

```sql
comment on column product_dim.effective_date is '生效日期';

-- 建立订单维度表
create table order_dim (
    order_sk bigserial,
    order_number int,
    isdelete boolean default false,
    version int,
    effective_date date );

comment on table order_dim is '订单维度表';
comment on column order_dim.order_sk is '订单维度代理键';
comment on column order_dim.order_number is '订单号';
comment on column order_dim.isdelete is '是否删除';
comment on column order_dim.version is '版本';
comment on column order_dim.effective_date is '生效日期';

-- 建立日期维度表
create table date_dim (
    date_sk bigserial,
    date date,
    month smallint,
    month_name varchar(9),
    quarter smallint,
    year smallint );

comment on table date_dim is '日期维度表';
comment on column date_dim.date_sk is '日期维度代理键';
comment on column date_dim.date is '日期';
comment on column date_dim.month is '月份';
comment on column date_dim.month_name is '月份名称';
comment on column date_dim.quarter is '季度';
comment on column date_dim.year is '年份';

-- 建立销售订单事实表
create table sales_order_fact (
    order_sk bigint,
    customer_sk bigint,
    product_sk bigint,
    order_date_sk bigint,
    year_month int,
    order_amount decimal(10,2) )
partition by range (year_month)
```

```
( partition p201601 start (201601) inclusive ,
  partition p201602 start (201602) inclusive ,
  partition p201603 start (201603) inclusive ,
  partition p201604 start (201604) inclusive ,
  partition p201605 start (201605) inclusive ,
  partition p201606 start (201606) inclusive ,
  partition p201607 start (201607) inclusive ,
  partition p201608 start (201608) inclusive ,
  partition p201609 start (201609) inclusive ,
  partition p201610 start (201610) inclusive ,
  partition p201611 start (201611) inclusive ,
  partition p201612 start (201612) inclusive ,
partition p201701 start (201701) inclusive ,
  partition p201702 start (201702) inclusive ,
  partition p201703 start (201703) inclusive ,
  partition p201704 start (201704) inclusive ,
  partition p201705 start (201705) inclusive ,
  partition p201706 start (201706) inclusive ,
  partition p201707 start (201707) inclusive ,
  partition p201708 start (201708) inclusive ,
  partition p201709 start (201709) inclusive ,
  partition p201710 start (201710) inclusive ,
  partition p201711 start (201711) inclusive ,
  partition p201712 start (201712) inclusive
                end (201801) exclusive );

comment on table sales_order_fact is '销售订单事实表';
comment on column sales_order_fact.order_sk is '订单维度代理键';
comment on column sales_order_fact.customer_sk is '客户维度代理键';
comment on column sales_order_fact.product_sk is '产品维度代理键';
comment on column sales_order_fact.order_date_sk is '日期维度代理键';
comment on column sales_order_fact.year_month is '年月分区键';
comment on column sales_order_fact.order_amount is '销售金额';
```

说明：

- TDS 模式中的表是 HAWQ 内部表。
- 比源库多了一个日期维度表。数据仓库可以追踪历史数据，因此每个数据仓库都有日期时间相关的维度表。
- 为了捕获和表示数据变化，除日期维度表外，其他维度表比源表多了代理键、是否删除标志、版本号和版本生效日期四个字段。日期维度一次性生成数据后就不会改变，因此除了日期本身相关属性，只增加了一列代理键。
- 事实表由维度表的代理键和度量属性构成。目前只有一个销售订单金额的度量值。

- 由于事实表数据量大，事实表采取分区表设计。事实表中冗余了一列年月，作为分区键。之所以用年月做范围分区，是考虑到数据分析时经常使用年月分组进行查询和统计，这样可以有效利用分区消除提高查询性能。与 rds.sales_order 不同，这里显式定义了 2016、2017 两年的分区。
- 出于同样的考虑，与 RDS 一样，TDS 表也使用 row 存储格式和随机数据分布策略。

12.5.3 装载日期维度数据

日期维度是数据仓库中的一个特殊角色。日期维度包含时间概念，而时间是最重要的。因为数据仓库的主要功能之一就是存储和追溯历史数据，所以每个数据仓库里的数据都有一个时间特征。本例中创建一个 HAWQ 的函数，一次性预装载日期数据。

```sql
-- 生成日期维度表数据的函数
create or replace function fn_populate_date (start_dt date, end_dt date)
returns void as $$
declare
    v_date date:= start_dt;
    v_datediff int:= end_dt - start_dt;
begin
    for i in 0 .. v_datediff loop
        insert into date_dim(date, month, month_name, quarter, year)
        values(v_date, extract(month from v_date), to_char(v_date,'mon'),
extract(quarter from v_date), extract(year from v_date));
        v_date := v_date + 1;
    end loop;
    analyze date_dim;
end; $$
language plpgsql;
```

执行函数生成日期维度数据：

```sql
select fn_populate_date(date '2000-01-01', date '2020-12-31');
```

查询生成的日期：

```sql
select min(date_sk) min_sk, min(date) min_date, max(date_sk) max_sk,
max(date) max_date, count(*) c
from date_dim;
```

查询结果如下：

```
min_sk | min_date   | max_sk | max_date   |  c
-------+------------+--------+------------+------
     1 | 2000-01-01 |   7671 | 2020-12-31 | 7671
(1 row)
```

至此，我们的示例数据仓库模型搭建完成，后面章节将实现 ETL。

12.6 小结

本章我们使用一个简单而典型的销售订单示例，建立数据仓库模型。操作型源数据存储在 MySQL 数据库中，数据仓库使用 HAWQ 构建。为了满足 ETL 的性能需求，我们在 HAWQ 库中创建了三个模式，分别用作业务数据临时存储（增量）、原始数据存储（全量）和转换后的数据存储。事实表采用分区设计，并且使用冗余的年月列作为分区键。HAWQ 内部表均采用默认的 row 存储格式和随机数据分布策略。除了建立相关数据库对象，还在 MySQL 和 HAWQ 中分别生成了原始业务数据和日期维度数据。

第 13 章

◀ 初始ETL ▶

在数据仓库可以使用前,需要装载历史数据。这些历史数据是导入进数据仓库的第一个数据集合。首次装载被称为初始装载,一般是一次性工作。由最终用户来决定有多少历史数据进入数据仓库。例如,数据仓库使用的开始时间是 2017 年 3 月 1 日,而用户希望装载两年的历史数据,那么应该初始装载 2015 年 3 月 1 日 ~ 2017 年 2 月 28 日之间的源数据。在 2017 年 3 月 2 日装载 2017 年 3 月 1 日的数据(假设执行频率是每天一次),之后周期性地每天装载前一天的数据。在装载事实表前,必须先装载所有的维度表。因为事实表需要引用维度的代理键。这不仅针对初始装载,也针对定期装载。本章说明执行初始装载的步骤,包括标识源数据、维度历史的处理、用 Sqoop 抽取数据、用 HAWQ 开发和验证初始装载过程等。

13.1 用 Sqoop 初始数据抽取

Sqoop 是一个在 Hadoop 与结构化数据存储(如关系数据库)之间高效传输大批量数据的工具。它在 2012 年 3 月被成功孵化,现在已是 Apache 的顶级项目。Sqoop 有 Sqoop1 和 Sqoop2 两代,Sqoop1 最后的稳定版本是 1.4.6,Sqoop2 最后版本是 1.99.6。需要注意,1.99.6 与 1.4.6 并不兼容,而且截至目前为止,1.99.6 并不完善,不推荐在生产环境中部署。Sqoop1 架构如图 13-1 所示。HDP 2.5 中自带 Sqoop 1.4.6 版本。

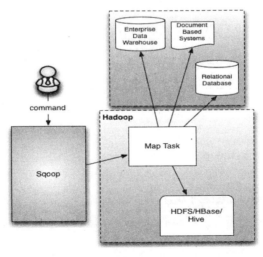

图 13-1 Sqoop1 架构

本例使用 Sqoop 将 MySQL 的数据抽取到 HDFS 上的指定目录，然后利用 HAWQ 外部表功能将 HDFS 数据文件装载到内部表中。表 13-1 汇总了示例中维度表和事实表用到的源数据表及其抽取模式。

表 13-1 数据源抽取模式

源数据表	HDFS 目录	对应 EXT 模式中的表	抽取模式
customer	/data/ext/customer	customer	整体、拉取
product	/data/ext/product	product	整体、拉取
sales_order	/data/ext/sales_order	sales_order	基于时间戳的 CDC、拉取

13.1.1 覆盖导入

对于 customer、product 这两个表采用整体拉取的方式抽数据。ETL 通常是按一个固定的时间间隔，周期性定时执行的，因此对于整体拉取的方式而言，每次导入的数据需要覆盖上次导入的数据。Sqoop 提供了 delete-target-dir 参数实现覆盖导入。该参数指示在每次抽取数据前先将目标目录删除，作用是提供了一个幂等操作的选择。所谓幂等操作指的是其执行任意多次所产生的影响均与一次执行的影响相同。这样就能在导入失败或修复 bug 后可以再次执行该操作，而不用担心重复执行会对系统造成数据混乱。

13.1.2 增量导入

Sqoop 提供了增量导入模式，用于只导入比已经导入行新的数据行。表 13-2 所示参数用来控制增量导入。

表 13-2 Sqoop 增量导入参数

参数	描述
--check-column	在确定应该导入哪些行时，指定被检查的列。列不能是 CHAR/NCHAR/VARCHAR/VARNCHAR/LONGVARCHAR/LONGNVARCHAR 数据类型
--incremental	指定 Sqoop 怎样确定哪些行是新行。有效值是 append 和 lastmodified
--last-value	指定已经导入数据的被检查列的最大值

Sqoop 支持两种类型的增量导入：append 和 lastmodified。可以使用--incremental 参数指定增量导入的类型。当被导入表的新行具有持续递增的行 id 值时，应该使用 append 模式。指定行 id 为--check-column 的列。Sqoop 导入那些被检查列的值比--last-value 给出的值大的数据行。Sqoop 支持的另一种表修改策略叫做 lastmodified 模式。当源表的数据行可能被修改，并且每次修改都会更新一个 last-modified 列为当前时间戳时，应该使用 lastmodified 模式。那些被检查列的时间戳比 last-value 给出的时间戳新的数据行被导入。

增量导入命令执行后，在控制台输出的最后部分，会打印出后续导入需要使用的 last-value。当周期性执行导入时，应该用这种方式指定--last-value 参数的值，以确保只导入新的或修改过的数据。可以通过一个增量导入的保存作业自动执行这个过程，这是适合重复执行增量导入的方式。

有了对 Sqoop 增量导入的基本了解，下面看一下如何在本示例中使用它抽取数据。对于 sales_order 这个表采用基于时间戳的 CDC 拉取方式抽数据。这里假设源系统中销售订单记录一旦入库就不再改变，或者可以忽略改变。也就是说销售订单是一个随时间变化单向追加数据的表。sales_order 表中有两个关于时间的字段，order_date 表示订单时间，entry_date 表示订单数据实际插入表里的时间，两个时间可能不同。那么用哪个字段作为 CDC 的时间戳呢？设想这样的情况，一个销售订单的订单时间是 2017 年 1 月 1 日，实际插入表里的时间是 2017 年 1 月 2 日，ETL 每天 0 点执行，抽取前一天的数据。如果按 order_date 抽取数据，条件为 where order_date >= '2017-01-02' AND order_date < '2017-01-03'，则 2017 年 1 月 3 日 0 点执行的 ETL 不会捕获到这个新增的订单数据。所以应该以 entry_date 作为 CDC 的时间戳。

13.1.3 建立初始抽取脚本

用 sqoop 操作系统用户建立初始数据抽取脚本文件~/init_extract.sh，内容如下：

```bash
#!/bin/bash

# 建立 Sqoop 增量导入作业，以 order_number 作为检查列，初始的 last-value 是 0
sqoop job --delete myjob_incremental_import
sqoop job --create myjob_incremental_import \
-- import \
--connect "jdbc:mysql://172.16.1.127:3306/source?usessl=false&user=dwtest&password=123456" \
--table sales_order \
--target-dir /data/ext/sales_order \
--compress \
--where "entry_date < current_date()" \
--incremental append \
--check-column order_number \
--last-value 0

# 全量抽取客户表
sqoop import --connect jdbc:mysql://172.16.1.127:3306/source --username dwtest --password 123456 --table customer --target-dir /data/ext/customer --delete-target-dir --compress

# 全量抽取产品表
sqoop import --connect jdbc:mysql://172.16.1.127:3306/source --username dwtest --password 123456 --table product --target-dir /data/ext/product --delete-target-dir --compress
```

```
# 首次全量抽取销售订单表
sqoop job --exec myjob_incremental_import
```

说明：

- 为了保证外部表数据量尽可能小，使用 compress 选项进行压缩，Sqoop 默认的压缩算法是 gzip，hdfstextsimples 属性的 HAWQ PXF 外部表能自动正确读取这种格式的压缩文件。
- 执行时先重建 Sqoop 增量抽取作业，指定 last-value 为 0。由于 order_number 都是大于 0 的，因此初始时会装载所有订单数据。

将文件修改为可执行模式：

```
chmod 755 ~/init_extract.sh
```

13.2 向 HAWQ 初始装载数据

13.2.1 数据源映射

表 13-3 显示了本示例需要的源数据关键信息，包括源数据表、对应的数据仓库目标表等属性。客户和产品的源数据直接与其数据仓库里的目标表、customer_dim 和 product_dim 表相对应，而销售订单事务表是多个数据仓库表的数据源。

表 13-3　数据源映射

源数据	源数据类型	文件名/表名	数据仓库中的目标表
客户	MySQL 表	customer	customer_dim
产品	MySQL 表	product	product_dim
销售订单	MySQL 表	sales_order	order_dim、sales_order_fact

13.2.2 确定 SCD 处理方法

标识出了数据源，现在要考虑维度历史的处理。渐变维（SCD）是一种在多维数据仓库中实现维度历史的技术。有三种不同的 SCD 类型：SCD 类型 1（SCD1），SCD 类型 2（SCD2），SCD 类型 3（SCD3）：

- SCD1：通过更新维度记录直接覆盖已存在的值，它不维护记录的历史。SCD1 一般用于修改错误的数据。
- SCD2：在源数据发生变化时，给维度记录建立一个新的"版本"，从而维护维度历史。SCD2 不删除、修改已存在的数据。

- SCD3：通常用作保持维度记录的几个版本。它通过给某个数据单元增加多个列来维护历史。例如，为了记录客户地址的变化，customer_dim 维度表有一个 customer_address 列和一个 previous_customer_address 列，分别记录当前和上一个版本的地址。SCD3 可以有效维护有限的历史，而不像 SCD2 那样保存全部历史。SCD3 很少使用，它只适用于数据的存储空间不足并且用户接受有限维度历史的情况。

同一维度表中的不同字段可以有不同的变化处理方式。在传统数据仓库中，对于 SCD1 一般就直接 UPDATE 更新属性，而 SCD2 则要新增记录。但 HAWQ 没有提供 UPDATE、DELETE 等 DML 操作，因此对于所有属性变化均增加一条记录，即所有维度属性都按 SCD2 方式处理。

13.2.3 实现代理键

多维数据仓库中的维度表和事实表一般都需要有一个代理键，作为这些表的主键，代理键一般由单列的自增数字序列构成。HAWQ 中的 bigserial 数据类型在功能上与 MySQL 的 auto_increment 类似，常用于定义自增列。但它的实现方法却与 Oracle 的 sequence 类似，当创建 bigserial 字段的表时，HAWQ 会自动创建一个自增的 sequence 对象，bigserial 字段自动引用 sequence 实现自增。

13.2.4 建立初始装载脚本

所有技术实现的细节都清楚后，现在编写初始数据装载脚本。需要执行两步主要操作，一是将外部表的数据装载到 RDS 模式的表中，二是向 TDS 模式中的表装载数据。用 gpadmin 操作系统用户建立初始数据装载脚本文件~/init_load.sql，内容如下：

```sql
-- 分析外部表
analyze ext.customer;
analyze ext.product;
analyze ext.sales_order;

-- 将外部数据装载到原始数据表
set search_path to rds;

truncate table customer;
truncate table product;
truncate table sales_order;

insert into customer select * from ext.customer;
insert into product select * from ext.product;
insert into sales_order select * from ext.sales_order;

-- 分析rds模式的表
```

```sql
analyze rds.customer;
analyze rds.product;
analyze rds.sales_order;

-- 装载数据仓库数据
set search_path to tds;

truncate table customer_dim;
truncate table product_dim;
truncate table order_dim;
truncate table sales_order_fact;

-- 序列初始化
alter sequence customer_dim_customer_sk_seq restart with 1;
alter sequence product_dim_product_sk_seq restart with 1;
alter sequence order_dim_order_sk_seq restart with 1;

-- 装载客户维度表
insert into customer_dim
(customer_number, customer_name, customer_street_address, customer_zip_code,
 customer_city, customer_state, version, effective_date)
select t1.customer_number, t1.customer_name, t1.customer_street_address,
       t1.customer_zip_code, t1.customer_city, t1.customer_state,
       1, '2016-03-01'
  from rds.customer t1
 order by t1.customer_number;

-- 装载产品维度表
insert into product_dim
(product_code, product_name, product_category, version, effective_date)
select product_code, product_name, product_category, 1, '2016-03-01'
  from rds.product t1
 order by t1.product_code;

-- 装载订单维度表
insert into order_dim (order_number,version,effective_date)
select order_number, 1, order_date
  from rds.sales_order t1
 order by t1.order_number;

-- 装载销售订单事实表
insert into sales_order_fact
select order_sk,customer_sk,product_sk,date_sk,e.year*100 +
```

```
e.month,order_amount
   from rds.sales_order a,
       order_dim b,
       customer_dim c,
       product_dim d,
       date_dim e
 where a.order_number = b.order_number
   and a.customer_number = c.customer_number
   and a.product_code = d.product_code
   and date(a.order_date) = e.date;

-- 分析 tds 模式的表
analyze customer_dim;
analyze product_dim;
analyze order_dim;
analyze sales_order_fact;
```

说明：

- 装载前清空表，以及重新初始化序列的目的是为了可重复执行初始装载脚本。
- 依据 HAWQ 的建议，装载数据后，执行查询前，先分析表以提高查询性能。

13.3 建立初始 ETL 脚本

前面的数据抽取脚本文件的属主是 sqoop 用户，而数据装载脚本文件的属主是 gpadmin 用户。除了这两个用户以外，还需要使用 hdfs 用户执行文件操作。为了简化多用户调用执行，用 root 用户将所有需要的操作封装到一个文件中，提供统一的初始数据装载执行入口。用 root 操作系统用户建立初始 ETL 脚本文件~/init_etl.sh，内容如下：

```
#!/bin/bash

# 为了可以重复执行初始装载过程，先使用 hdfs 用户删除销售订单外部表目录
su - hdfs -c 'hdfs dfs -rm -r /data/ext/sales_order/*'

# 使用 sqoop 用户执行初始抽取脚本
su - sqoop -c '~/init_extract.sh'

# 使用 gpadmin 用户执行初始装载脚本
su - gpadmin -c 'export PGPASSWORD=123456;psql -U dwtest -d dw -h hdp3 -f ~/init_load.sql'
```

说明：

- 使用 su 命令，以不同用户执行相应的脚本文件。
- Sqoop 中 incremental append 与 delete-target-dir 参数不能同时使用。因此为了可重复执行 Sqoop 增量抽取作业，先要用 hdfs 用户删除相应目录下的所有文件。
- 由于 mysql-connector 的版本问题，可能导致 Sqoop 执行时出现下面的异常，用最新版本的 MySQL JDBC 包替换老版本通常能解决此问题。

```
ERROR manager.SqlManager: Error reading from database: java.sql.SQLException: Streaming result set com.mysql.jdbc.RowDataDynamic@xxxxxxx is still active.
```

将文件修改为可执行模式：

```
chmod 755 ~/init_etl.sh
```

用 root 用户执行初始 ETL 脚本：

```
~/init_etl.sh
```

执行以下查询验证初始 ETL 结果：

```sql
select order_number,
       customer_name,
       product_name,
       date,
       order_amount amount
  from sales_order_fact a,
       customer_dim b,
       product_dim c,
       order_dim d,
       date_dim e
 where a.customer_sk = b.customer_sk
   and a.product_sk = c.product_sk
   and a.order_sk = d.order_sk
   and a.order_date_sk = e.date_sk
 order by order_number;
```

共装载 100 条销售订单数据，最后 5 条如下所示。

```
 ...
  96 | really large customers | hard disk drive | 2016-06-23 | 7024.00
  97 | good companies         | hard disk drive | 2016-06-23 | 6046.00
  98 | small stores           | hard disk drive | 2016-06-23 | 8018.00
  99 | loyal clients          | floppy drive    | 2016-06-25 | 8313.00
 100 | good companies         | floppy drive    | 2016-06-28 | 1161.00
(100 rows)
```

13.4 小结

数据仓库正式投入使用前，需要向其中装载历史数据，这通常是一次性操作。初始 ETL 的一般过程包括：识别数据源；数据映射；确定数据导入方式；确定 SCD 处理方法等。Sqoop 用于在关系数据库和 HDFS 间传递数据，能实现覆盖导入和增量导入。HAWQ 的 SQL 脚本用以实现数据的转换与装载。为了简化多用户调用执行，以 root 操作系统用户将所有需要的操作封装到一个文件中，提供统一的初始数据装载执行入口。

第 14 章 定期ETL

初始装载只在开始数据仓库使用前执行一次,然而,必须按时调度定期执行 ETL。与初始装载不同,定期装载一般都是增量的,而且需要捕获并记录数据的变化历史。本章说明执行定期 ETL 的步骤,包括识别源数据与装载类型、确定变化数据捕获方式、使用 HAWQ 开发和测试定期装载等过程。在某些实时性要求较高的场景中,普通的以天作为执行周期将不再适用,本章最后讨论一种准实时数据抽取方案。

14.1 变化数据捕获

数据获取处理需要重点考虑增量抽取,也被称为变化数据捕获,简称 CDC。假设一个数据仓库系统,在每天夜里的业务低峰时间从操作型源系统抽取数据,那么增量抽取只需要过去 24 小时内发生变化的数据。变化数据捕获也是建立准实时数据仓库的关键技术。

当能够识别并获得最近发生变化的数据时,抽取及其后面的转换、装载操作显然都会变得更高效,因为要处理的数据量会小很多。遗憾的是,很多源系统很难识别出最近变化的数据,或者必须侵入源系统才能做到。变化数据捕获是数据抽取中典型的技术挑战。

1. 变化数据捕获方法

常用的变化数据捕获方法有时间戳、快照、触发器和日志 4 种。相信熟悉数据库的读者对这些方法都不会陌生。时间戳方法需要源系统有相应的数据列表示最后的数据变化。快照方法可以使用数据库系统自带的机制实现,如 Oracle 的物化视图技术,也可以自己实现相关逻辑,但会比较复杂。触发器是关系数据库系统具有的特性,源表上建立的触发器会在对该表执行 insert、update、delete 等语句时被触发,触发器中的逻辑用于捕获数据的变化。日志可以使用应用日志或系统日志,这种方式对源系统不具有侵入性,但需要额外的日志解析工作。表 14-1 总结了这 4 种方案的特点。

表 14-1　四种 CDC 方案比较

比较项目	时间戳	触发器	快照	日志
能区分插入/更新	否	是	是	是
周期内，检测到多次更新	否	是	否	是
能检测到删除	否	是	是	是
不具有侵入性	否	否	否	是
支持实时	否	是	否	是
不依赖数据库	是	否	是	否

2. 识别数据源与装载类型

定期装载首先要识别数据仓库的每个事实表和每个维度表用到的并且是可用的源数据。然后决定适合装载的抽取模式和维度历史装载类型。表 14-2 汇总了本示例的这些信息。

表 14-2　定期装载类型

数据源	EXT 模式	RDS 模式	TDS 模式	抽取模式	维度历史装载类型
customer	customer	customer	customer_dim	整体、拉取	所有属性均为 SCD2
product	product	product	product_dim	整体、拉取	所有属性均为 SCD2
sales_order	sales_order	sales_order	order_dim	CDC（每天）、拉取	唯一订单号
			sales_order_fact	CDC（每天）、拉取	N/A
N/A	N/A	N/A	date_dim	N/A	预装载

3. 处理渐变维（SCD）

上一章提到，HAWQ 只有 INSERT，没有 UPDATE、DELETE 操作，因此所有维度属性都使用 SDC2 记录全部历史变化。在捕获数据变化时，需要使用维度表的当前版本数据与从业务数据库最新抽取来的数据做比较。实现方式是在维度表上建立一个当前维度版本的视图，用于比较数据变化。这种设计既可以保留所有数据变化的历史，又屏蔽了查询当前版本的复杂性。

事实表需要引用维度表的代理键，而且不一定是引用当前版本的代理键。比如有些迟到的事实，就必须找到事实发生时的维度版本。因此一个维度的所有版本区间应该构成一个连续且互斥时间范围，每个事实数据都能对应维度的唯一版本。实现方式是在维度表上建立一个维度历史版本的视图，在这个视图中增加版本过期日期导出列。任何一个版本的有效期是一个"左闭右开"的区间，也就是说该版本包含生效日期，但不包含过期日期，而是到过期日期的前一天为止。ETL 粒度为每天执行一次，也即一天内的数据变化将被忽略。

4. 设置数据处理时间窗口

对于事实表，我们采用基于时间戳的 CDC 增量装载模式，时间粒度为天。因此需要两个时间点，分别表示本次装载的起始时间点和终止时间点，这两个时间点定义了本次处理的时间窗口，即装载这个时间区间内的数据。还要说明一点，这个区间是左包含的，就是处理的数据包括起始时间点的，但不包括终止时间点的。这样设计的原因是，我们既要处理完整的数据，不能有遗漏，又不能重复装载数据，这就要求时间处理窗口既要连续，又不能存在重叠的部分。

14.2 创建维度表版本视图

1. 创建当前版本视图

```
-- 切换到 tds 模式
set search_path=tds;

-- 建立客户维度当前视图
create or replace view v_customer_dim_latest as
select customer_sk, customer_number, customer_name, customer_street_address,
       customer_zip_code, customer_city, customer_state, version,
effective_date
  from (select distinct on (customer_number) customer_number, customer_sk,
               customer_name, customer_street_address, customer_zip_code,
               customer_city, customer_state, isdelete, version, effective_date
          from customer_dim
          order by customer_number, customer_sk desc) as latest
  where isdelete is false;

-- 建立产品维度当前视图
create or replace view v_product_dim_latest as
select product_sk, product_code, product_name, product_category, version,
       effective_date
  from (select distinct on (product_code) product_code, product_sk,
               product_name, product_category, isdelete, version, effective_date
          from product_dim
          order by product_code, product_sk desc) as latest
  where isdelete is false;
```

说明：

- 如前所述，创建维度表的当前视图。这里只为客户和产品维度创建视图，而订单维度不需要当前版本视图，因为假设业务上订单数据只能增加，不能修改，所以没有版本变化。
- 使用 HAWQ 的 DISTINCT ON 语法去重。DISTINCT ON (expression [, …])把记录根据[, …]的值进行分组，分组之后仅返回每一组的第一行。需要注意的是，如果不指定 ORDER BY 子句，返回的第一条的不确定的。如果使用了 ORDER BY 子句，那么[, …]里面的值必须靠近 ORDER BY 子句的最左边。本例中我们按业务主键（分别是 customer_number 和 product_code）分组，每组按代理键（分别是 customer_sk 和 product_sk）倒排序，每组第一行即为维度的当前版本。

2. 创建历史版本视图

```sql
-- 切换到 tds 模式
set search_path=tds;

-- 建立客户维度历史视图，增加版本过期日期导出列
create or replace view v_customer_dim_his as
select *, date(lead(effective_date,1,date '2200-01-01') over (partition by customer_number order by effective_date)) expiry_date
  from customer_dim;

-- 建立产品维度历史视图，增加版本过期日期导出列
create or replace view v_product_dim_his as
select *, date(lead(effective_date,1,date '2200-01-01') over (partition by product_code order by effective_date)) expiry_date
  from product_dim;
```

说明：

- 维度历史视图增加了版本的过期日期列。
- 使用 LEAD 窗口函数实现。以业务主键（分别是 customer_number 和 product_code）分区，每个分区内按生效日期排序。LEAD 函数在一个分区内取到当前生效日期的下一个日期，该日期即为对应版本的过期日期。如果是当前版本，下一日期为空，则返回一个很大的时间值，大到足以满足数据仓库整个生命周期的需要，本示例设置的是 2200 年 1 月 1 日。

14.3 创建时间戳表

```sql
create table rds.cdc_time (last_load date, current_load date);
insert into rds.cdc_time select current_date - 1, current_date - 1;
```

说明：

- 本示例中 order_dim 维度表和 sales_order_fact 事实表使用基于时间戳的 CDC 装载模式。为此在 rds 模式中建立一个名为 cdc_time 的时间戳表，这个表里有 last_load 和 current_load 两个字段。之所以需要两个字段，是因为抽取到的数据可能会多于本次需要处理的数据。比如，两点执行 ETL 过程，则零点到两点这两个小时的数据不会在本次处理。为了确定这个截止时间点，需要给时间戳设定一个上限条件，即这里的 current_load 字段值。

- 本示例的时间粒度为每天,所以时间戳只要保留日期部分即可,因此数据类型选为 date。这两个字段的初始值是"初始加载"执行日期的前一天。当开始装载时,current_load 设置为当前日期。

14.4 用 Sqoop 定期数据抽取

用 sqoop 操作系统用户建立定期数据抽取脚本文件~/regular_extract.sh,内容如下:

```
#!/bin/bash

# 全量抽取客户表
sqoop import --connect jdbc:mysql://172.16.1.127:3306/source --username dwtest
--password 123456 --table customer --target-dir /data/ext/customer
--delete-target-dir --compress

# 全量抽取产品表
sqoop import --connect jdbc:mysql://172.16.1.127:3306/source --username dwtest
--password 123456 --table product --target-dir /data/ext/product
--delete-target-dir --compress

# 增量抽取销售订单表
sqoop job --exec myjob_incremental_import
```

这个文件与上一章介绍的初始抽取的 shell 脚本基本相同,只是去掉了创建 Sqoop 作业的命令。每次装载后,都会将已经导入的最大订单号赋予增量抽取作业的 last-value。

将文件修改为可执行模式:

```
chmod 755 ~/regular_extract.sh
```

14.5 建立定期装载 HAWQ 函数

```
create or replace function fn_regular_load ()
returns void as $$
declare
    -- 设置 scd 的生效时间
    v_cur_date date := current_date;
    v_pre_date date := current_date - 1;
    v_last_load date;
begin
```

```sql
-- 分析外部表
analyze ext.customer;
analyze ext.product;
analyze ext.sales_order;

-- 将外部表数据装载到原始数据表
truncate table rds.customer;
truncate table rds.product;

insert into rds.customer select * from ext.customer;
insert into rds.product select * from ext.product;
insert into rds.sales_order select * from ext.sales_order;

-- 分析 rds 模式的表
analyze rds.customer;
analyze rds.product;
analyze rds.sales_order;

-- 设置 cdc 的上限时间
select last_load into v_last_load from rds.cdc_time;
truncate table rds.cdc_time;
insert into rds.cdc_time select v_last_load, v_cur_date;

-- 装载客户维度
insert into tds.customer_dim
(customer_number, customer_name, customer_street_address,
 customer_zip_code, customer_city, customer_state, isdelete,
 version, effective_date)
select case flag when 'D' then a_customer_number else b_customer_number
       end customer_number,
       case flag when 'D' then a_customer_name else b_customer_name
         end customer_name,
       case flag when 'D' then a_customer_street_address
           else b_customer_street_address
         end customer_street_address,
       case flag when 'D' then a_customer_zip_code else b_customer_zip_code
         end customer_zip_code,
       case flag when 'D' then a_customer_city else b_customer_city
         end customer_city,
       case flag when 'D' then a_customer_state else b_customer_state
         end customer_state,
       case flag when 'D' then true else false
         end isdelete,
```

```sql
                  case flag when 'D' then a_version when 'I' then 1 else a_version + 1
                   end v,
                  v_pre_date
            from (select a.customer_number a_customer_number,
                         a.customer_name a_customer_name,
                         a.customer_street_address a_customer_street_address,
                         a.customer_zip_code a_customer_zip_code,
                         a.customer_city a_customer_city,
                         a.customer_state a_customer_state,
                         a.version a_version,
                         b.customer_number b_customer_number,
                         b.customer_name b_customer_name,
                         b.customer_street_address b_customer_street_address,
                         b.customer_zip_code b_customer_zip_code,
                         b.customer_city b_customer_city,
                         b.customer_state b_customer_state,
                         case when a.customer_number is null then 'I'
                              when b.customer_number is null then 'D'
                              else 'U'
                         end flag
                    from v_customer_dim_latest a
                    full join rds.customer b on a.customer_number = b.customer_number
                   where a.customer_number is null -- 新增
                      or b.customer_number is null -- 删除
                      or (a.customer_number = b.customer_number
                     and not
 (a.customer_name = b.customer_name
                      and a.customer_street_address = b.customer_street_address
                      and a.customer_zip_code = b.customer_zip_code
                      and a.customer_city = b.customer_city
                      and a.customer_state = b.customer_state))) t
             order by coalesce(a_customer_number, 999999999999),
b_customer_number limit 999999999999;

    -- 装载产品维度
    insert into tds.product_dim
    (product_code, product_name, product_category, isdelete,version,
effective_date)
    select case flag when 'D' then a_product_code else b_product_code
           end product_code,
           case flag when 'D' then a_product_name else b_product_name
           end product_name,
           case flag when 'D' then a_product_category else b_product_category
```

```sql
           end product_category,
       case flag when 'D' then true else false
        end isdelete,
       case flag when 'D' then a_version when 'I' then 1 else a_version + 1
        end v,
       v_pre_date
  from (select a.product_code a_product_code,
               a.product_name a_product_name,
               a.product_category a_product_category,
               a.version a_version,
               b.product_code b_product_code,
               b.product_name b_product_name,
               b.product_category b_product_category,
               case when a.product_code is null then 'I'
                    when b.product_code is null then 'D'
                    else 'U'
                end flag
          from v_product_dim_latest a
          full join rds.product b on a.product_code = b.product_code
         where a.product_code is null  -- 新增
            or b.product_code is null  -- 删除
            or (a.product_code = b.product_code
                and not
                   (a.product_name = b.product_name
                    and a.product_category = b.product_category))) t
    order by coalesce(a_product_code, 999999999999), b_product_code
limit 999999999999;

-- 装载 order 维度
insert into order_dim (order_number, version, effective_date)
select t.order_number, t.v, t.effective_date
  from (select order_number, 1 v, order_date effective_date
          from rds.sales_order, rds.cdc_time
         where entry_date >= last_load and entry_date < current_load) t;

-- 装载销售订单事实表
insert into sales_order_fact
select order_sk, customer_sk, product_sk, date_sk, year * 100 + month,
       order_amount
  from rds.sales_order a,
       order_dim b,
       v_customer_dim_his c,
       v_product_dim_his d,
```

```
                date_dim e,
                rds.cdc_time f
         where a.order_number = b.order_number
           and a.customer_number = c.customer_number
           and a.order_date >= c.effective_date
           and a.order_date < c.expiry_date
           and a.product_code = d.product_code
           and a.order_date >= d.effective_date
           and a.order_date < d.expiry_date
           and date(a.order_date) = e.date
           and a.entry_date >= f.last_load and a.entry_date < f.current_load;

    -- 分析 tds 模式的表
    analyze customer_dim;
    analyze product_dim;
    analyze order_dim;
    analyze sales_order_fact;

    -- 更新时间戳表的 last_load 字段
    truncate table rds.cdc_time;
    insert into rds.cdc_time select v_cur_date, v_cur_date;

end;
$$
language plpgsql;
```

说明：

- 该函数分成两大部分，一是装载 RDS 模式的表，二是处理 TDS 的表。
- 同初始装载一样，RDS 模式表的数据来自从 EXT 模式的外部表，rds.customer 和 rds.product 全量装载，rds.sales_order 增量装载。
- 脚本中设置三个变量，v_last_load 和 v_cur_date 分别赋予起始日期、终止日期，并且将时间戳表 rds.cdc_time 的 last_load 和 current_load 字段分别设置为起始日期和终止日期。v_pre_date 表示版本过期日期。
- 维度表数据可能是新增、修改或删除。这里用 FULL JOIN 连接原始数据表与维度当前版本视图，统一处理这三种情况。外查询中使用 CASE 语句判断属于哪种情况，分别取得不同的字段值。
- 为了保证数据插入维度表时，代理键与业务主键保持相同的顺序，必须使用 "order by coalesce(a_product_code, 999999999999), b_product_code limit 999999999999;" 类似的语句。
- 订单维度增量装载，没有历史版本问题。
- 装载事实表时连接维度历史视图，引用事实数据所对应的维度代理键。该代理键可以

通过维度版本的生效日期、过期日期区间唯一确定。
- 装载数据后，执行查询前，分析表以提高查询性能。
- 数据装载完成后，更新数据处理时间窗口。

14.6 建立定期 ETL 脚本

用 root 操作系统用户建立定期 ETL 脚本文件~/regular_etl.sh，内容如下：

```
#!/bin/bash

# 外部表只保存销售订单增量数据
su - hdfs -c 'hdfs dfs -rm -r /data/ext/sales_order/*'

# 使用 sqoop 用户执行定期抽取脚本
su - sqoop -c '~/regular_extract.sh'

# 使用 gpadmin 用户执行定期装载函数
su - gpadmin -c 'export PGPASSWORD=123456;psql -U dwtest -d dw -h hdp3 -c "set search_path=tds;select fn_regular_load ();"'
```

该文件的作用与初始 ETL 的 shell 脚本基本相同，为定期 ETL 提供统一的执行入口。

将文件修改为可执行模式：

```
chmod 755 ~/regular_etl.sh
```

14.7 测试

14.7.1 准备测试数据

在 hdp4 上的 MySQL 数据库中执行下面的 SQL 脚本，准备源数据库中的客户、产品和销售订单测试数据。

```
use source;

/***
客户数据的改变如下：
客户 6 的街道号改为 7777 ritter rd。（原来是 7070 ritter rd）
客户 7 的姓名改为 distinguished agencies。（原来是 distinguished partners）
新增第八个客户。
```

```
***/
update customer set customer_street_address = '7777 ritter rd.' 
where customer_number = 6 ;
update customer set customer_name = 'distinguished agencies' 
where customer_number = 7 ;
insert into customer (customer_name, customer_street_address, 
customer_zip_code, customer_city, customer_state) 
values ('subsidiaries', '10000 wetline blvd.', 17055, 'pittsburgh', 'pa') ;

/***
产品数据的改变如下：
产品 3 的名称改为 flat panel。（原来是 lcd panel）
新增第四个产品。
***/
update product set product_name = 'flat panel' where product_code = 3 ;
insert into product (product_name, product_category) 
values ('keyboard', 'peripheral');

/***
新增订单日期为 2017 年 5 月 4 日的 16 条订单。
***/
set @start_date := unix_timestamp('2017-05-04');
set @end_date := unix_timestamp('2017-05-05');
drop table if exists temp_sales_order_data;
create table temp_sales_order_data as select * from sales_order where 1=0;

set @order_date := from_unixtime(@start_date + rand() * (@end_date - @start_date));
set @amount := floor(1000 + rand() * 9000);
insert into temp_sales_order_data values (101, 1, 1, @order_date, @order_date, @amount);

... 共插入 16 条数据 ...

insert into sales_order
select null,customer_number,product_code,order_date,entry_date,order_amount
from temp_sales_order_data order by order_date;

commit ;
```

14.7.2 执行定期 ETL 脚本

用 root 用户执行定期 ETL 脚本。

~/regular_etl.sh

14.7.3 确认 ETL 过程正确执行

（1）查询客户维度当前视图

```
select customer_sk c_sk, customer_number c_num, customer_name c_name,
       customer_street_address c_str_add, version v, effective_date eff_date
  from v_customer_dim_latest
 order by customer_number;
```

查询结果如下，可以看到视图包含所有客户的最新信息。客户 6、7、8 这三条记录的生效日期为 2017-05-04。

c_sk	c_num	c_name	c_str_add	v	eff_date
1	1	really large customers	7500 louise dr.	1	2016-03-01
2	2	small stores	2500 woodland st.	1	2016-03-01
3	3	medium retailers	1111 ritter rd.	1	2016-03-01
4	4	good companies	9500 scott st.	1	2016-03-01
5	5	wonderful shops	3333 rossmoyne rd.	1	2016-03-01
8	6	loyal clients	7777 ritter rd.	2	2017-05-04
9	7	distinguished agencies	9999 scott st.	2	2017-05-04
10	8	subsidiaries	10000 wetline blvd.	1	2017-05-04

(8 rows)

（2）查询客户维度历史视图

```
select customer_sk c_sk, customer_number c_num, customer_name c_name,
       customer_street_address c_str_add, version v, effective_date eff_date,
       expiry_date exp_date, isdelete isdel
  from v_customer_dim_his
 order by customer_number, version;
```

查询结果如下，可以看到视图包含所有客户的历史版本，以及版本对应的时间段，当前版本的过期日期为 2200-01-01 。客户 6、7 因为修改了信息，具有两个版本，老版本的过期日期和新版本的生效日期均为 2017-05-04。

c_sk	c_num	c_name	c_str_add	v	eff_date	exp_date	isdel
1	1	really large customers	7500 louise dr.	1	2016-03-01	2200-01-01	f
2	2	small stores	2500 woodland st.	1	2016-03-01	2200-01-01	f
3	3	medium retailers	1111 ritter rd.	1	2016-03-01	2200-01-01	f
4	4	good companies	9500 scott st.	1	2016-03-01	2200-01-01	f
5	5	wonderful shops	3333 rossmoyne rd.	1	2016-03-01	2200-01-01	f
6	6	loyal clients	7070 ritter rd.	1	2016-03-01	2017-05-04	f

```
         8 |      6 | loyal clients          | 7777 ritter rd.    | 2 | 2017-05-04 | 2200-01-01 | f
         7 |      7 | distinguished partners | 9999 scott st.     | 1 | 2016-03-01 | 2017-05-04 | f
         9 |      7 | distinguished agencies | 9999 scott st.     | 2 | 2017-05-04 | 2200-01-01 | f
        10 |      8 | subsidiaries           | 10000 wetline blvd.| 1 | 2017-05-04 | 2200-01-01 | f
(10 rows)
```

（3）查询产品维度当前视图

```
select product_sk, product_code, product_name, version, effective_date
  from v_product_dim_latest
 order by product_code;
```

查询结果如下：

```
 product_sk | product_code | product_name    | version | effective_date
------------+--------------+-----------------+---------+----------------
          1 |            1 | hard disk drive |       1 | 2016-03-01
          2 |            2 | floppy drive    |       1 | 2016-03-01
          4 |            3 | flat panel      |       2 | 2017-05-04
          5 |            4 | keyboard        |       1 | 2017-05-04
(4 rows)
```

（4）查询产品维度历史视图

```
select product_sk p_sk, product_code p_code, product_name p_name, version v,
       effective_date eff_date, expiry_date exp_date
  from v_product_dim_his
 order by product_code, version;
```

查询结果如下：

```
 p_sk | p_code | p_name          | v | eff_date   | exp_date
------+--------+-----------------+---+------------+------------
    1 |      1 | hard disk drive | 1 | 2016-03-01 | 2200-01-01
    2 |      2 | floppy drive    | 1 | 2016-03-01 | 2200-01-01
    3 |      3 | lcd panel       | 1 | 2016-03-01 | 2017-05-04
    4 |      3 | flat panel      | 2 | 2017-05-04 | 2200-01-01
    5 |      4 | keyboard        | 1 | 2017-05-04 | 2200-01-01
(5 rows)
```

（5）查询订单维度表和事实表记录数，结果都是16，说明新装载了16条订单记录

```
select count(*) from order_dim;
select count(*) from sales_order_fact;
```

（6）查询事实表数据

```
select * from sales_order_fact
 where order_sk > 100
 order by order_sk;
```

查询结果如下。可以看到，customer_sk 没有 6、7，而是 8、9，10 为新增；product_sk 用 4 代替 3，5 为新增。

```
 order_sk | customer_sk | product_sk | order_date_sk | year_month | order_amount
----------+-------------+------------+---------------+------------+--------------
      101 |           4 |          5 |          6334 |     201705 |      1683.00
      102 |           4 |          5 |          6334 |     201705 |      4917.00
      103 |           8 |          2 |          6334 |     201705 |      3516.00
      104 |           5 |          1 |          6334 |     201705 |      4666.00
      105 |           8 |          2 |          6334 |     201705 |      8396.00
      106 |           2 |          2 |          6334 |     201705 |      8306.00
      107 |           9 |          4 |          6334 |     201705 |      9539.00
      108 |           2 |          2 |          6334 |     201705 |      2942.00
      109 |          10 |          5 |          6334 |     201705 |      6682.00
      110 |           1 |          1 |          6334 |     201705 |      5456.00
      111 |           9 |          4 |          6334 |     201705 |      7289.00
      112 |           3 |          4 |          6334 |     201705 |      6863.00
      113 |           5 |          2 |          6334 |     201705 |      7613.00
      114 |           3 |          4 |          6334 |     201705 |      9707.00
      115 |           1 |          1 |          6334 |     201705 |      2449.00
      116 |          10 |          5 |          6334 |     201705 |      7844.00
(16 rows)
```

（7）查询时间窗口表

```
select * from rds.cdc_time;
```

查询结果如下，可以看到时间窗口已经更新。

```
 last_load  | current_load
------------+--------------
 2017-05-05 | 2017-05-05
(1 row)
```

14.8 动态分区滚动

rds.sales_order 和 tds.sales_order_fact 都是按月做的范围分区，需要进一步设计滚动分区维护策略。通过维护一个数据滚动窗口，删除老分区，添加新分区，将老分区的数据迁移到数据仓库以外的次级存储，以节省系统开销。下面的 HAWQ 函数按照转储最老分区数据、删除最老分区数据、建立新分区的步骤动态滚动分区。

```
-- 创建动态滚动分区的函数
create or replace function tds.fn_rolling_partition(p_year_month_start date)
```

```
returns int as $body$
   declare
     v_min_partitiontablename name;
     v_year_month_end date := p_year_month_start + interval '1 month';
   v_year_month_start_int int := extract(year from p_year_month_start) * 100
   + extract(month from p_year_month_start);
   v_year_month_end_int int := extract(year from v_year_month_end) * 100
   + extract(month from v_year_month_end);
     sqlstring varchar(1000);
   begin
     -- 处理 rds.sales_order
     -- 转储最早一个月的数据，
     select partitiontablename into v_min_partitiontablename
       from pg_partitions
      where tablename='sales_order' and partitionrank = 1;

     sqlstring = 'copy (select * from ' || v_min_partitiontablename || ') to
''/home/gpadmin/sales_order_' || cast(v_year_month_start_int as varchar) ||
'.txt'' with delimiter ''|'';';
     execute sqlstring;
     -- raise notice '%', sqlstring;

     -- 删除最早月份对应的分区
     sqlstring := 'alter table sales_order drop partition for (rank(1));';
     execute sqlstring;

     -- 增加下一个月份的新分区
     sqlstring := 'alter table sales_order add partition start (date '''||
p_year_month_start ||''') inclusive end (date '''||v_year_month_end ||''')
exclusive;';
     execute sqlstring;
     -- raise notice '%', sqlstring;


     -- 处理 tds.sales_order_fact
     -- 转储最早一个月的数据，
     select partitiontablename into v_min_partitiontablename
       from pg_partitions
      where tablename='sales_order_fact' and partitionrank = 1;

     sqlstring = 'copy (select * from ' || v_min_partitiontablename || ') to
''/home/gpadmin/sales_order_fact_' || cast(v_year_month_start_int as varchar) ||
'.txt'' with delimiter ''|'';';
```

```
    execute sqlstring;
    -- raise notice '%', sqlstring;

    -- 删除最早月份对应的分区
    sqlstring := 'alter table sales_order_fact drop partition for (rank(1));';
    execute sqlstring;

    -- 增加下一个月份的新分区
    sqlstring := 'alter table sales_order_fact add partition start
('||cast(v_year_month_start_int as varchar)||') inclusive end
('||cast(v_year_month_end_int as varchar)||') exclusive;';
    execute sqlstring;
    -- raise notice '%', sqlstring;

    -- 正常返回 1
    return 1;

-- 异常返回 0
exception when others then
    raise exception '%: %', sqlstate, sqlerrm;
    return 0;
end
$body$ language plpgsql;
```

将执行该函数的 psql 命令行放到 cron 中自动执行。下面的例子表示每月 1 号 2 点执行分区滚动操作。假设数据仓库中只保留最近一年的销售数据。

```
0 2 1 * * psql -d dw -c "set search_path=rds,tds; select
fn_rolling_partition(date(date_trunc('month',current_date) + interval '1
month'));" > rolling_partition.log 2>&1
```

14.9 准实时数据抽取

前面的 ETL 过程中都使用 Sqoop 从 MySQL 数据库增量抽取数据到 HDFS，然后用 HAWQ 的外部表进行访问。这种方式只需要很少的配置即可完成数据抽取任务，但缺点同样明显，那就是实时性。Sqoop 使用 MapReduce 读写数据，而 MapReduce 是为了批处理场景设计的，目标是大吞吐量，并不太关心低延时问题。就像示例中所做的，每天定时增量抽取数据一次。

Flume 是一个海量日志采集、聚合和传输的系统，支持在日志系统中定制各类数据发送方，用于收集数据。同时，Flume 提供对数据进行简单处理，并写到各种数据接收方的能力。Flume 以流方式处理数据，可作为代理持续运行。当新的数据可用时，Flume 能够立即获取数据并输出至目标，这样就可以在很大程度上解决实时性问题。Flume 最初只是一个日志收集器，但随着 flume-ng-sql-source 插件的出现，使得 Flume 从关系数据库采集数据成为可能。下面简单介绍 Flume，并详细说明如何配置 Flume 将 MySQL 表数据准实时抽取到 HDFS。

1. Flume 简介

Flume 是分布式的日志收集系统，它将各个服务器中的数据收集起来并发送到指定的目标，比如说送到 HDFS。简单来说 Flume 就是一个收集日志的工具。

（1）Event 的概念

Flume 的核心功能是把数据从源（source）收集起来，再将收集到的数据送到指定的目的地（sink）。为了保证输送的过程一定成功，在送到目的地（sink）之前，会先缓存数据（channel），待数据真正到达目的地（sink）后，Flume 再删除自己缓存的数据。

在整个数据的传输过程中，流动的是 event，即事务保证是在 event 级别进行的。那么什么是 event 呢？Event 将传输的数据进行封装，是 Flume 传输数据的基本单位，如果是文本文件，通常一个 event 就是一行记录。Event 也是事务的基本单位。Event 从 source，流向 channel，再到 sink，本身为一个字节数组，并可携带 headers（头信息）信息。Event 代表着一个数据的最小完整单元，从外部数据源来，向外部目的地去。

（2）Flume 架构

Flume 架构如图 14-1 所示。

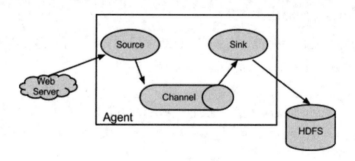

图 14-1　Flume 架构

图中的 Agent 本身是一个 Java 进程，运行在日志收集节点，也就是 Flume 服务器节点。Agent 里面包含三个核心的组件：source、channel 和 sink，类似生产者、仓库、消费者的架构。

- source：source 组件是专门用来收集数据的，可以处理各种类型、各种格式的日志数据，包括 avro、thrift、exec、jms、spooling directory、netcat、sequence generator、syslog、http、legacy、自定义。
- channel：source 组件把数据收集以后，临时存放在 channel 中，即 channel 组件在 agent 中是专门用来存放临时数据的，对采集到的数据进行简单的缓存，可以存放在 memory、jdbc、file 等等。
- sink：sink 组件是用于把数据发送到目的地的组件，目的地包括 HDFS、logger、Avro、

Thrift、ipc、file、null、HBase、Solr、自定义。

(3) Flume 的运行机制

Flume 的核心就是 agent，这个 agent 对外有两个进行交互的地方，一个是接受数据输入的 source，一个是数据输出的 sink，sink 负责将数据发送到外部指定的目的地。source 接收到数据之后，将数据发送给 channel，chanel 作为一个数据缓冲区会临时存放这些数据，随后 sink 会将 channel 中的数据发送到指定的地方，例如 HDFS 等。注意，只有在 sink 将 channel 中的数据成功发送出去之后，channel 才会将临时数据进行删除，这种机制保证了数据传输的可靠性与安全性。

2. 配置与测试

我们的 HDP 安装中包含 Flume，只要配置 Flume 服务即可。

（1）建立 MySQL 测试表并添加数据

```
use test;

create table wlslog
(id          int not null,
 time_stamp  varchar(40),
 category    varchar(40),
 type        varchar(40),
 servername  varchar(40),
 code        varchar(40),
 msg         varchar(40),
 primary key ( id ) );

insert into wlslog(id,time_stamp,category,type,servername,code,msg)
values(1,'apr-8-2014-7:06:16-pm-pdt','notice','weblogicserver','adminserver','bea-000365','server state changed to standby');
 insert into wlslog(id,time_stamp,category,type,servername,code,msg)
values(2,'apr-8-2014-7:06:17-pm-pdt','notice','weblogicserver','adminserver','bea-000365','server state changed to starting');
 insert into wlslog(id,time_stamp,category,type,servername,code,msg)
values(3,'apr-8-2014-7:06:18-pm-pdt','notice','weblogicserver','adminserver','bea-000365','server state changed to admin');
 insert into wlslog(id,time_stamp,category,type,servername,code,msg)
values(4,'apr-8-2014-7:06:19-pm-pdt','notice','weblogicserver','adminserver','bea-000365','server state changed to resuming');
 insert into wlslog(id,time_stamp,category,type,servername,code,msg)
values(5,'apr-8-2014-7:06:20-pm-pdt','notice','weblogicserver','adminserver','bea-000361','started weblogic adminserver');
 insert into wlslog(id,time_stamp,category,type,servername,code,msg)
```

```
values(6,'apr-8-2014-7:06:21-pm-pdt','notice','weblogicserver','adminserver','bea-000365','server state changed to running');
    insert into wlslog(id,time_stamp,category,type,servername,code,msg)
values(7,'apr-8-2014-7:06:22-pm-pdt','notice','weblogicserver','adminserver','bea-000360','server started in running mode');
    commit;
```

（2）建立相关目录与文件

创建本地状态文件：

```
mkdir -p /var/lib/flume
cd /var/lib/flume
touch sql-source.status
chmod -R 777 /var/lib/flume
```

建立 HDFS 目标目录：

```
hdfs dfs -mkdir -p /flume/mysql
hdfs dfs -chmod -R 777 /flume/mysql
```

（3）准备 JAR 包

从 http://book2s.com/java/jar/f/flume-ng-sql-source/download-flume-ng-sql-source-1.3.7.html 下载 flume-ng-sql-source-1.3.7.jar 文件，并复制到 Flume 库目录。

```
cp flume-ng-sql-source-1.3.7.jar /usr/hdp/current/flume-server/lib/
```

将 MySQL JDBC 驱动 JAR 包也复制到 Flume 库目录。

```
cp mysql-connector-java-5.1.17.jar /usr/hdp/current/flume-server/lib/mysql-connector-java.jar
```

（4）建立 HAWQ 外部表

```
create external table ext_wlslog
  (id          int,
   time_stamp  varchar(40),
   category    varchar(40),
   type        varchar(40),
   servername  varchar(40),
   code        varchar(40),
   msg         varchar(40)
  ) location ('pxf://mycluster/flume/mysql?profile=hdfstextmulti') format 'csv' (quote=e'"');
```

（5）配置 Flume

在 Ambari → Flume → Configs → flume.conf 中配置如下属性：

```
agent.channels.ch1.type = memory
agent.sources.sql-source.channels = ch1
agent.channels = ch1
agent.sinks = HDFS

agent.sources = sql-source
agent.sources.sql-source.type = org.keedio.flume.source.SQLSource
agent.sources.sql-source.connection.url = jdbc:mysql://172.16.1.127:3306/test
agent.sources.sql-source.user = root
agent.sources.sql-source.password = 123456
agent.sources.sql-source.table = wlslog
agent.sources.sql-source.columns.to.select = *

agent.sources.sql-source.incremental.column.name = id
agent.sources.sql-source.incremental.value = 0
agent.sources.sql-source.run.query.delay=5000
agent.sources.sql-source.status.file.path = /var/lib/flume
agent.sources.sql-source.status.file.name = sql-source.status

agent.sinks.HDFS.channel = ch1
agent.sinks.HDFS.type = hdfs
agent.sinks.HDFS.hdfs.path = hdfs://mycluster/flume/mysql
agent.sinks.HDFS.hdfs.fileType = DataStream
agent.sinks.HDFS.hdfs.writeFormat = Text
agent.sinks.HDFS.hdfs.rollSize = 268435456
agent.sinks.HDFS.hdfs.rollInterval = 0
agent.sinks.HDFS.hdfs.rollCount = 0
```

Flume 在 flume.conf 文件中指定 source、channel 和 sink 相关的配置，各属性描述如表 14-3 所示。

表 14-3 Flume 配置文件属性

属性	描述
agent.channels.ch1.type	Agent 的 channel 类型
agent.sources.sql-source.channels	Source 对应的 channel 名称
agent.channels	Channel 名称
agent.sinks	Sink 名称
agent.sources	Source 名称
agent.sources.sql-source.type	Source 类型
agent.sources.sql-source.connection.url	数据库 URL
agent.sources.sql-source.user	数据库用户名
agent.sources.sql-source.password	数据库密码

属性	描述
agent.sources.sql-source.table	数据库表名
agent.sources.sql-source.columns.to.select	查询的列
agent.sources.sql-source.incremental.column.name	增量列名
agent.sources.sql-source.incremental.value	增量初始值
agent.sources.sql-source.run.query.delay	发起查询的时间间隔，单位是毫秒
agent.sources.sql-source.status.file.path	状态文件路径
agent.sources.sql-source.status.file.name	状态文件名称
agent.sinks.HDFS.channel	Sink 对应的 channel 名称
agent.sinks.HDFS.type	Sink 类型
agent.sinks.HDFS.hdfs.path	Sink 路径
agent.sinks.HDFS.hdfs.fileType	流数据的文件类型
agent.sinks.HDFS.hdfs.writeFormat	数据写入格式
agent.sinks.HDFS.hdfs.rollSize	目标文件轮转大小，单位是字节
agent.sinks.HDFS.hdfs.rollInterval	hdfs sink 间隔多长将临时文件滚动成最终目标文件，单位是秒；如果设置成 0，则表示不根据时间来滚动文件
agent.sinks.HDFS.hdfs.rollCount	当 events 数据达到该数量时候，将临时文件滚动成目标文件；如果设置成 0，则表示不根据 events 数据来滚动文件

（6）运行 Flume 代理

保存上一步的设置，然后重启 Flume 服务，如图 14-2 所示。

图 14-2 运行 Flume 代理

重启后，状态文件已经记录了最新的 id 值 7：

```
[root@hdp4 flume]# ll
总用量 4
-rw-r--r-- 1 flume hadoop 17 6月  23 16:20 sql-source.status
-rwxrwxrwx 1 root  root    0 6月  23 15:35 sql-source.status.bak.1498206038870
[root@hdp4 flume]# more sql-source.status
{"LastIndex":"7"}
[root@hdp4 flume]#
```

查看目标路径，生成了一个临时文件，其中有 7 条记录：

```
[root@hdp4 flume]# hdfs dfs -ls /flume/mysql/*
-rw-r--r--   3 flume hdfs        829 2017-06-23 16:20 /flume/mysql/FlumeData.1498206041064.tmp
[root@hdp4 flume]# hdfs dfs -cat /flume/mysql/*
"1","apr-8-2014-7:06:16-pm-pdt","notice","weblogicserver","adminserver","bea-000365","server state changed to standby"
"2","apr-8-2014-7:06:17-pm-pdt","notice","weblogicserver","adminserver","bea-000365","server state changed to starting"
"3","apr-8-2014-7:06:18-pm-pdt","notice","weblogicserver","adminserver","bea-000365","server state changed to admin"
"4","apr-8-2014-7:06:19-pm-pdt","notice","weblogicserver","adminserver","bea-000365","server state changed to resuming"
"5","apr-8-2014-7:06:20-pm-pdt","notice","weblogicserver","adminserver","bea-000361","started weblogic adminserver"
"6","apr-8-2014-7:06:21-pm-pdt","notice","weblogicserver","adminserver","bea-000365","server state changed to running"
"7","apr-8-2014-7:06:22-pm-pdt","notice","weblogicserver","adminserver","bea-000360","server started in running mode"
```

查询 HAWQ 外部表，结果也有全部 7 条数据：

```
test=# select id, time_stamp, code, msg from ext_wlslog;
 id |        time_stamp         |    code    |              msg
----+---------------------------+------------+--------------------------------
  1 | apr-8-2014-7:06:16-pm-pdt | bea-000365 | server state changed to standby
  2 | apr-8-2014-7:06:17-pm-pdt | bea-000365 | server state changed to starting
  3 | apr-8-2014-7:06:18-pm-pdt | bea-000365 | server state changed to admin
  4 | apr-8-2014-7:06:19-pm-pdt | bea-000365 | server state changed to resuming
  5 | apr-8-2014-7:06:20-pm-pdt | bea-000361 | started weblogic adminserver
  6 | apr-8-2014-7:06:21-pm-pdt | bea-000365 | server state changed to running
  7 | apr-8-2014-7:06:22-pm-pdt | bea-000360 | server started in running mode
(7 rows)
```

至此，初始数据抽取已经完成。

（7）测试准实时增量数据抽取

在源表中新增 id 为 8、9、10 的三条记录。

```
use test;
insert into wlslog(id,time_stamp,category,type,servername,code,msg) 
values(8,'apr-8-2014-7:06:22-pm-pdt','notice','weblogicserver','adminserver','bea-000360','server started in running mode');
insert into wlslog(id,time_stamp,category,type,servername,code,msg) 
values(9,'apr-8-2014-7:06:22-pm-pdt','notice','weblogicserver','adminserver','bea-000360','server started in running mode');
insert into wlslog(id,time_stamp,category,type,servername,code,msg) 
values(10,'apr-8-2014-7:06:22-pm-pdt','notice','weblogicserver','adminserver','bea-000360','server started in running mode');
commit;
```

5 秒之后查询 HAWQ 外部表，可以看到已经查询出全部 10 条数据，准实时增量抽取成功。

```
test=# select id, time_stamp, code, msg from ext_wlslog;
 id |       time_stamp         |    code    |              msg
----+--------------------------+------------+-------------------------------
 ...
  6 | apr-8-2014-7:06:21-pm-pdt | bea-000365 | server state changed to running
  7 | apr-8-2014-7:06:22-pm-pdt | bea-000360 | server started in running mode
  8 | apr-8-2014-7:06:22-pm-pdt | bea-000360 | server started in running mode
  9 | apr-8-2014-7:06:22-pm-pdt | bea-000360 | server started in running mode
 10 | apr-8-2014-7:06:22-pm-pdt | bea-000360 | server started in running mode
(10 rows)
```

3. 方案优缺点

利用 Flume 采集关系数据库表数据最大的优点是配置简单，不用编程。相比 tungsten-replicator 的复杂性，Flume 只要在 flume.conf 文件中配置 source、channel 及 sink 的相关属性，已经没什么难度了。而与现在很火的 Canal 比较，虽然不够灵活，但毕竟一行代码也不用写。再有该方案采用普通 SQL 轮询方式实现，具有通用性，适用于所有关系库数据源。

这种方案的缺点与其优点一样突出，主要体现在以下几方面。

- 在源库上执行了查询，具有入侵性。
- 通过轮询方式实现增量，只能做到准实时，而且轮询间隔越短，对源库的影响越大。
- 只能识别新增数据，检测不到删除与更新。
- 要求源库必须有用于表示增量的字段。

即便有诸多局限，Flume 抽取关系库数据的方案还是有一定的价值，特别是在要求快速部署、简化编程，又能满足需求的应用场景，对传统的 Sqoop 方式也不失为一种有效的补充。

14.10 小结

数据仓库开始使用后，一般都要定期执行 ETL 过程，为数据仓库提供新的数据。时间戳、触发器、快照和日志是常见的变化数据捕获方法，它们各有优缺点，销售订单示例采用时间戳 CDC 方式。本章详述了如何使用 Sqoop 和 HAWQ 函数完成定期 ETL，给出了完整的实现和测试代码。有些应用场景需要较高的实时性，我们讨论了一种利用 Flume 将关系库数据准实时抽取到 HDFS 的方案。虽然不够完善，但该方案作为 Sqoop 的补充，还是具有一定的价值。

第 15 章
自动调度执行ETL作业

一旦数据仓库开始使用,就需要不断从源系统给数据仓库提供新数据。为了确保数据流的稳定,需要使用所在平台上可用的任务调度器来调度 ETL 定期自动执行。调度模块是 ETL 系统必不可少的组成部分,它不但是数据仓库的基本需求,也对项目的成功起着举足轻重的作用。本章说明如何使用 HDP 中的 Oozie 和 Falcon 服务实现 ETL 执行自动化。

15.1 Oozie 简介

Oozie 是一个管理 Hadoop 作业、可伸缩、可扩展、可靠的工作流调度系统,它内部定义了三种作业:工作流作业、协调器作业和 Bundle 作业。工作流作业是由一系列动作构成的有向无环图(DAGs),协调器作业是按时间频率周期性触发 Oozie 工作流的作业,Bundle 管理协调器作业。Oozie 支持的用户作业类型有 Java map-reduce、Streaming map-reduce、Pig、Hive、Sqoop 和 Distcp,及其 Java 程序和 shell 脚本或命令等特定的系统作业。

1. 为什么使用 Oozie

使用 Oozie 主要基于以下两点原因:

- Hadoop 中执行的任务有时候需要把多个 MapReduce 作业连接到一起执行,或者需要多个作业并行处理。Oozie 可以把多个 MapReduce 作业组合到一个逻辑工作单元中,从而完成更大型的任务。
- 从调度的角度看,如果使用 crontab 的方式调用多个工作流作业,可能需要编写大量的脚本,还要通过脚本来控制好各个工作流作业的执行时序问题,不但不好维护,而且监控也不方便。基于这样的背景,Oozie 提出了 Coordinator 的概念,它能够将每个工作流作业作为一个动作来运行,相当于工作流定义中的一个执行节点,这样就能够将多个工作流作业组成一个称为 Coordinator Job 的作业,并指定触发时间和频率,还可以配置数据集、并发数等。

2. Oozie 架构

Oozie 架构如图 15-1 所示。

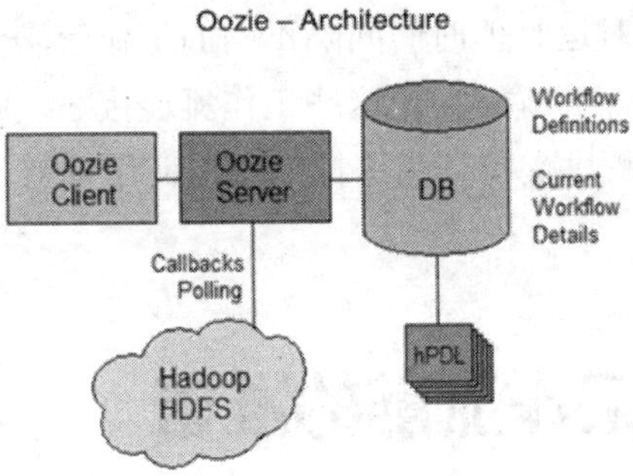

图 15-1　Oozie 架构

Oozie 是一种 Java Web 应用程序，它运行在 Java Servlet 容器，即 Tomcat 中，并使用数据库来存储以下内容：（1）工作流定义；（2）当前运行的工作流实例，包括实例的状态和变量。Oozie 工作流是放置在 DAG（有向无环图 Direct Acyclic Graph）中的一组动作，例如，Hadoop 的 Map/Reduce 作业、Pig 作业等。DAG 控制动作的依赖关系，指定了动作执行的顺序。Oozie 使用 hPDL 这种 XML 流程定义语言来描述这个图。

hPDL 是一种很简洁的语言，它只会使用少数流程控制节点和动作节点。控制节点会定义执行的流程，并包含工作流的起点和终点（start、end 和 fail 节点）以及控制工作流执行路径的机制（decision、fork 和 join 节点）。动作节点是实际执行操作的部分，通过它们工作流会触发执行计算或者处理任务。

所有由动作节点触发的计算和处理任务都不在 Oozie 中运行。它们是由 Hadoop 的 MapReduce 框架执行的。这种低耦合的设计方法让 Oozie 可以有效利用 Hadoop 的负载平衡、灾难恢复等机制。这些任务主要是串行执行的，只有文件系统动作例外，它是并行处理的。这意味着对于大多数工作流动作触发的计算或处理任务类型来说，在工作流操作转换到工作流的下一个节点之前都需要等待，直到前面节点的计算或处理任务结束了之后才能够继续。Oozie 可以通过两种不同的方式来检测计算或处理任务是否完成：回调与轮询。当 Oozie 启动了计算或处理任务时，它会为任务提供唯一的回调 URL，然后任务会在完成的时候给这个特定的 URL 发送通知。在任务无法触发回调 URL 的情况下（可能是因为任何原因，比方说网络闪断等），或者当任务的类型无法在完成时触发回调 URL 的时候，Oozie 有一种机制，可以对计算或处理任务进行轮询，从而能够判断任务是否完成。

Oozie 工作流可以参数化，例如在工作流定义中使用像${inputDir}之类的变量。提交工作流操作时，我们必须提供参数值。如果经过合适的参数化，比如使用不同的输出目录，那么多个同样的工作流操作可以并发执行。

一些工作流是根据需要触发的，但是大多数情况下，我们有必要基于一定的时间段、数据可用性或外部事件来运行它们。Oozie 协调系统（Coordinator system）让用户可以基于这些参数来定义工作流执行计划。Oozie 协调程序让我们可以用谓词的方式对工作流执行触发器进行建模，谓词可以是时间条件、数据条件、内部事件或外部事件。工作流作业会在谓词得到满足的时候启动。不难看出，这里的谓词，其作用和 SQL 语句的 WHERE 子句中的谓词类似，本质上都是在满足某些条件时触发某种事件。

有时，我们还需要连接定时运行但时间间隔不同的工作流操作，多个以不同频率运行的工作流的输出会成为下一个工作流的输入。把这些工作流连接在一起，会让系统把它作为数据应用的管道来引用。Oozie 协调程序支持创建这样的数据应用管道。

15.2 建立工作流前的准备

我们的定期 ETL 需要使用 Oozie 中的 FS、Sqoop 和 SSH 三种动作，其中增量数据抽取要用到 Sqoop job。由于 Oozie 在执行这些动作时存在一些特殊要求，因此在定义工作流前先要进行适当的配置。

1. 启动 Oozie 服务

示例实验环境使用的 HDP2.5.0，在安装之时就已经配置并启动了 Oozie 服务。可以在 Ambari Web 控制台中交互式配置、启动、停止、重启 Oozie 服务。

2. 配置 Sqoop 的 metastore

默认时，Sqoop metastore 自动连接存储在~/.sqoop/.目录下的本地嵌入式数据库。然而要在 Oozie 中执行 Sqoop job，需要 Sqoop 使用共享的元数据存储，否则会报类似如下的错误：

```
ERROR org.apache.sqoop.metastore.hsqldb.HsqldbJobStorage  - Cannot restore job
```

本例中我们使用 hdp2 上的 MySQL 数据库存储 Sqoop 元数据，下面是配置步骤。

（1）记录当前 Sqoop 作业的 last.value 值，该值在后面重建 Sqoop 作业时会用到。

```
last_value=`sqoop job --show myjob_incremental_import | grep incremental.last.value | awk '{print $3}'`
```

（2）在 MySQL 中创建 Sqoop 的元数据存储数据库。

```
create database sqoop;
create user 'sqoop'@'hdp2' identified by 'sqoop';
grant all privileges on sqoop.* to 'sqoop'@'hdp2';
flush privileges;
```

（3）配置 Sqoop 的元数据存储参数。

在 Ambari 的 Sqoop → Configs → Custom sqoop-site 中添加如图 15-2 所示的参数。

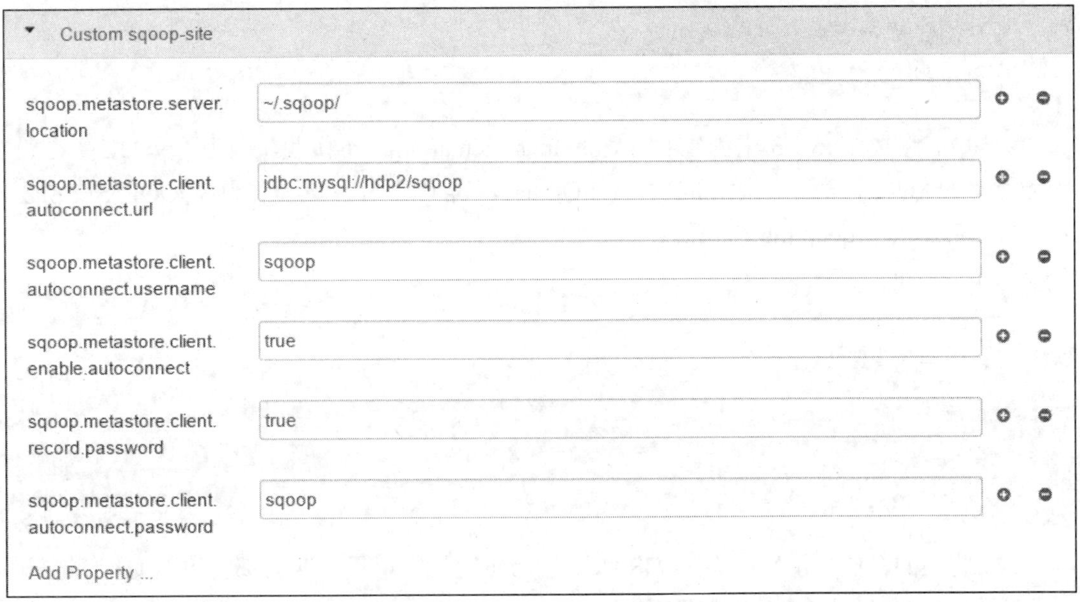

Custom sqoop-site	
sqoop.metastore.server.location	~/.sqoop/
sqoop.metastore.client.autoconnect.url	jdbc:mysql://hdp2/sqoop
sqoop.metastore.client.autoconnect.username	sqoop
sqoop.metastore.client.enable.autoconnect	true
sqoop.metastore.client.record.password	true
sqoop.metastore.client.autoconnect.password	sqoop

图 15-2 Sqoop 元数据存储参数

各参数含义如下：

- sqoop.metastore.server.location：指定元数据服务器位置，初始化建表时需要。
- sqoop.metastore.client.autoconnect.url：客户端自动连接的数据库的 URL。
- sqoop.metastore.client.autoconnect.username：连接数据库的用户名。
- sqoop.metastore.client.enable.autoconnect：启用客户端自动连接数据库。
- sqoop.metastore.client.record.password：在数据库中保存密码，不需要密码即可执行 sqoop job 脚本。
- sqoop.metastore.client.autoconnect.password：连接数据库的密码。

（4）重启 Sqoop 服务。

在 Ambari 中重启 Sqoop 服务，重启完成后，MySQL 的 sqoop 库中有了一个名为 SQOOP_ROOT 的空表。

```
mysql> show tables;
+-----------------+
| Tables_in_sqoop |
+-----------------+
| SQOOP_ROOT      |
+-----------------+
1 row in set (0.00 sec)
```

（5）预装载 SQOOP 表。

```
use sqoop;
insert into SQOOP_ROOT values (NULL, 'sqoop.hsqldb.job.storage.version', '0');
```

（6）创建初始表。

```
sqoop job --list
```

此时并不会返回 13.1.3 中创建的 myjob_incremental_import 作业，因为 MySQL 中还没有 Sqoop 元数据信息。该命令执行完成后，MySQL 的 sqoop 库中有了一个名为 SQOOP_SESSIONS 的空表，该表存储 sqoop job 相关信息。

```
mysql> show tables;
+------------------+
| Tables_in_sqoop  |
+------------------+
| SQOOP_ROOT       |
| SQOOP_SESSIONS   |
+------------------+
2 rows in set (0.00 sec)
```

（7）将表的存储引擎修改为 MYISAM，因为每次执行增量抽取后都会更新 last_value 值，如果使用 Innodb 可能引起事务锁超时错误。

```
alter table SQOOP_ROOT engine=myisam;
alter table SQOOP_SESSIONS engine=myisam;
```

3. 创建 myjob_incremental_import 作业

```
sqoop job --create myjob_incremental_import \
-- import \
--connect "jdbc:mysql://172.16.1.127:3306/source?usessl=false&user=dwtest&password=123456" \
--table sales_order \
--target-dir /data/ext/sales_order \
--compress \
--where "entry_date < current_date()" \
--incremental append \
--check-column order_number \
--last-value $last_value
```

执行上面的命令不会报作业已存在的错误，因为 MySQL 中还没有 Sqoop 元数据，已经存在的作业信息存储在本地嵌入式数据库中。上面的命令执行后，SQOOP_SESSIONS 表中存储了 Sqoop job 的信息。

```
select * from SQOOP_SESSIONS\G
...
*************************** 53. row ***************************
 job_name: myjob_incremental_import
 propname: sqoop.property.set.id
  propval: 0
propclass: schema
*************************** 54. row ***************************
 job_name: myjob_incremental_import
```

```
   propname: sqoop.tool
    propval: import
 propclass: schema
*************************** 55. row ***************************
  job_name: myjob_incremental_import
  propname: temporary.dirRoot
   propval: _sqoop
 propclass: SqoopOptions
*************************** 56. row ***************************
  job_name: myjob_incremental_import
  propname: verbose
   propval: false
 propclass: SqoopOptions
56 rows in set (0.00 sec)
```

此时再次执行 sqoop job –list 命令，可以看到刚刚创建的 Sqoop 作业。

```
sqoop job --list
...
Available jobs:
  myjob_incremental_import
```

关于使用 MySQL 作为 Sqoop 元数据存储的配置，可以参考 https://community.hortonworks.com/articles/55937/using-sqoop-with-mysql-as-metastore.html。

4．准备 java-json.jar 文件

Oozie 中执行 Sqoop 时如果缺少 java-json.jar 文件，会报类似如下的错误：

```
Failing Oozie Launcher, Main class [org.apache.oozie.action.hadoop.SqoopMain], main() threw exception, org/json/JSONObject
```

我们实验环境的 HDP2.5.0 安装中没有该文件，需要自行下载，然后复制到相关目录。

```
cp java-json.jar /usr/hdp/current/sqoop-client/lib/
su - hdfs -c 'hdfs dfs -put /usr/hdp/current/sqoop-client/lib/java-json.jar /user/oozie/share/lib/lib_20170208131207/sqoop/'
```

5．配置 SSH 免密码登录

实际的数据装载过程是通过 HAWQ 的用户自定义函数实现的，自然工作流中要执行包含 psql 命令行的本地 shell 脚本文件。这需要明确调用的 shell 使用的是本地的 shell，可以通过 Oozie 中的 SSH 动作指定本地文件。在使用 SSH 这个动作的时候，可能会遇到 AUTH_FAILED:Not able to perform operation 的问题，解决该问题要对 Oozie 的服务器做免密码登录处理。

（1）修改/etc/passwd 文件

HDP 默认运行 Oozie Server 的用户是 oozie，因此在/etc/passwd 中更改 oozie 用户，使得其可登录。我们的环境配置为：

```
oozie:x:506:504:Oozie user:/home/oozie:/bin/bash
```

（2）从 oozie 用户到 root 用户做免密码登录

本示例是用 root 操作系统用户提交 Oozie 任务的，因此这里要对从 oozie 用户到 root 用户做免密码登录。

```
su - oozie
ssh-keygen
… 一路回车生成密钥文件 …
su -
# 将 oozie 的公钥复制到 root 的 authorized_keys 文件中
cat /home/oozie/.ssh/id_rsa.pub >> authorized_keys
```

完成以上配置后，在 oozie 用户下可以免密码 ssh root@hdp2。关于 Oozie 调用本地 shell 脚本可以参考 http://www.cognoschina.net/Article/121421。

15.3 用 Oozie 建立定期 ETL 工作流

1. 建立 workflow.xml 文件

建立内容如下的 workflow.xml 文件：

```xml
<?xml version="1.0" encoding="UTF-8"?>
<workflow-app xmlns="uri:oozie:workflow:0.4" name="RegularETL">
    <start to="hdfsCommands"/>
    <action name="hdfsCommands">
        <fs>
            <delete path='${nameNode}/data/ext/sales_order/*'/>
        </fs>
        <ok to="fork-node"/>
        <error to="fail"/>
    </action>
    <fork name="fork-node">
        <path start="sqoop-customer" />
        <path start="sqoop-product" />
        <path start="sqoop-sales_order" />
    </fork>
    <action name="sqoop-customer">
        <sqoop xmlns="uri:oozie:sqoop-action:0.2">
            <job-tracker>${jobTracker}</job-tracker>
            <name-node>${nameNode}</name-node>
            <arg>import</arg>
            <arg>--connect</arg>
            <arg>jdbc:mysql://172.16.1.127:3306/source?useSSL=false</arg>
            <arg>--username</arg>
            <arg>dwtest</arg>
            <arg>--password</arg>
            <arg>123456</arg>
            <arg>--table</arg>
```

```xml
                <arg>customer</arg>
                <arg>--target-dir</arg>
                <arg>/data/ext/customer</arg>
                <arg>--delete-target-dir</arg>
                <arg>--compress</arg>
            </sqoop>
            <ok to="joining"/>
            <error to="fail"/>
        </action>
        <action name="sqoop-product">
            <sqoop xmlns="uri:oozie:sqoop-action:0.2">
                <job-tracker>${jobTracker}</job-tracker>
                <name-node>${nameNode}</name-node>
                <arg>import</arg>
                <arg>--connect</arg>
                <arg>jdbc:mysql://172.16.1.127:3306/source?useSSL=false</arg>
                <arg>--username</arg>
                <arg>dwtest</arg>
                <arg>--password</arg>
                <arg>123456</arg>
                <arg>--table</arg>
                <arg>product</arg>
                <arg>--target-dir</arg>
                <arg>/data/ext/product</arg>
                <arg>--delete-target-dir</arg>
                <arg>--compress</arg>
            </sqoop>
            <ok to="joining"/>
            <error to="fail"/>
        </action>
        <action name="sqoop-sales_order">
            <sqoop xmlns="uri:oozie:sqoop-action:0.2">
                <job-tracker>${jobTracker}</job-tracker>
                <name-node>${nameNode}</name-node>
                <command>job --meta-connect jdbc:mysql://hdp2/sqoop?user=sqoop&password=sqoop --exec myjob_incremental_import</command>
                <archive>/user/oozie/share/lib/lib_20170208131207/sqoop/java-json.jar#java-json.jar</archive>
            </sqoop>
            <ok to="joining"/>
            <error to="fail"/>
        </action>
        <join name="joining" to="psql-node"/>
        <action name="psql-node">
            <ssh xmlns="uri:oozie:ssh-action:0.1">
                <host>${focusNodeLogin}</host>
                <command>${myScript}</command>
                <capture-output/>
            </ssh>
```

```
            <ok to="end"/>
            <error to="fail"/>
        </action>
        <kill name="fail">
            <message>Sqoop failed, error message[${wf:errorMessage(wf:lastErrorNode())}]</message>
        </kill>
        <end name="end"/>
    </workflow-app>
```

这个工作流的 DAG 如图 15-3 所示。

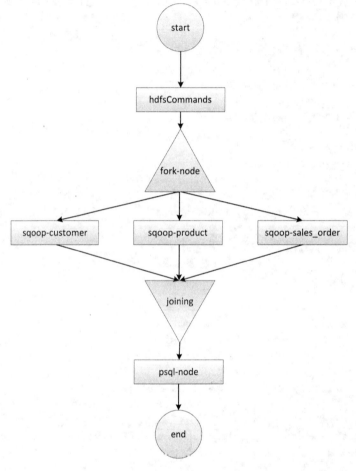

图 15-3　Oozie 工作流 DAG

上面的 XML 文件使用 hPDL 的语法定义了一个名为 RegularETL 的工作流。该工作流包括 10 个节点，其中有 5 个控制节点和 5 个动作节点。控制节点有：工作流的起点 start、终点 end、失败处理节点 fail（DAG 图中未显示），两个执行路径控制节点 fork-node 和 joining。动作节点有：一个 FS 动作节点 hdfsCommands 用于删除增量抽取的 HDFS 数据目录；三个并行处理的 Sqoop 动作节点 sqoop-customer、sqoop-product、sqoop-sales_order 用作数据抽取；一个 SSH 动作节点 psql-node 调用本地 shell 脚本，执行 HAWQ 数据装载。

Oozie 的工作流节点分为控制节点和动作节点两类。控制节点控制着工作流的开始、结束和作业的执行路径。动作节点触发计算或处理任务的执行。节点的名字必须符合[a-zA-Z][\-_a-zA-Z0-0]*这种正则表达式模式，并且不能超过 20 个字符。为了能让 Falcon 调用 Oozie 工作流，工作流名称不要带下划线等字符。

工作流定义中可以使用形式参数。当工作流被 Oozie 执行时，所有形参都必须提供具体的值。参数定义使用 JSP 2.0 的语法，参数不仅可以是单个变量，还支持函数和复合表达式。参数可用于指定动作节点和 decision 节点的配置值、XML 属性值和 XML 元素值，但是不能在节点名称、XML 属性名称、XML 元素名称和节点的转向元素中使用参数。上面工作流中的 ${jobTracker}和${nameNode}两个参数，分别指定 YARN 资源管理器的主机/端口和 HDFS NameNode 的主机/端口（如果配置了 HDFS HA，NameNode 使用 Nameservice ID）。${focusNodeLogin}指定本地 shell 脚本所在主机，${myScript}指定本地 shell 脚本文件全路径。

Oozie 的工作流作业本身还提供了丰富的内建函数，Oozie 将它们统称为表达式语言函数（Expression Language Functions，简称 EL 函数）。通过这些函数可以对动作节点和 decision 节点的谓词进行更复杂的参数化。上面定义的工作流中使用了 wf:errorMessage 和 wf:lastErrorNode 两个内建函数。wf:errorMessage 函数返回特定节点的错误消息，如果没有错误则返回空字符串。错误消息常被用于排错和通知的目的。wf:lastErrorNode 函数返回最后出错的节点名称，如果没有错误则返回空字符串。

2. 部署工作流

这里所说的部署就是把相关文件上传到 HDFS 的对应目录中。我们需要上传工作流定义文件，还要上传 file、archive、script 元素中指定的文件。可以使用 hdfs dfs -put 命令将本地文件上传到 HDFS，-f 参数的作用是，如果目标位置已经存在同名的文件，则用上传的文件覆盖已存在的文件。

```
# 上传工作流文件
hdfs dfs -put -f workflow.xml /user/oozie/
# 上传 MySQL JDBC 驱动文件到 Oozie 的共享库目录中
hdfs dfs -put /var/lib/ambari-agent/tmp/mysql-connector-java-5.1.38-bin.jar /user/oozie/share/lib/lib_20170208131207/sqoop/
```

3. 建立本地 shell 脚本文件

建立内容如下的/root/regular_etl.sh 文件：

```
#!/bin/bash
# 使用 gpadmin 用户执行定期装载函数
su - gpadmin -c 'export PGPASSWORD=123456;psql -U dwtest -d dw -h hdp3 -c "set search_path=tds;select fn_regular_load();"'
```

该 shell 文件内容很简单，可执行的就一行，调用 psql 执行 HAWQ 定期数据装载函数。

15.4 Falcon 简介

Apache Falcon 是一个面向 Hadoop 的、新的数据处理和管理平台，设计用于数据移动、数据管道协调、生命周期管理和数据发现。它使终端用户可以快速地将他们的数据及其相关的处理和管理任务"上载（onboard）"到 Hadoop 集群。Falcon 解决了大数据领域中一个非常重要和关键的问题，已经升级为 Apache 顶级项目。Falcon 有一个完善的路线图，可以减少应用程序开发人员编写复杂数据管理和处理应用程序的痛苦。

用户会发现，在 Apache Falcon 中，"基础设施端点（infrastructure endpoint）"、数据集（也称 Feed）、处理规则均是声明式的。这种声明式配置显式定义了实体之间的依赖关系。这也是该平台的一个特点，它本身只维护依赖关系，而并不做任何繁重的工作。所有的功能和工作流状态管理需求都委托给工作流调度程序来完成。

1. Falcon 架构

图 15-4 是 Falcon 的架构图。

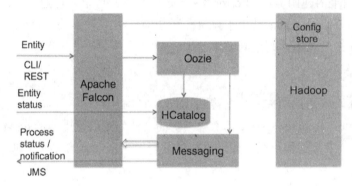

图 15-4　Falcon 架构

从上图可以看出，Apache Falcon：

- 在 Hadoop 环境中各种数据和"处理元素（processing element）"之间建立了联系。
- 可以与 Hive/HCatalog 集成。
- 根据可用的 Feed 组向最终用户发送通知。

2. 调度器

Falcon 选择 Oozie 作为默认的调度器。Hadoop 上的许多数据处理需要基于数据可用性或时间进行调度，当前 Oozie 本身就支持这些功能。同时 Falcon 系统又是开放的，可以整合其他调度器。Falcon process 调度流程如图 15-5 所示。

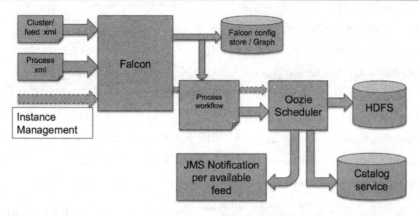

图 15-5　Falcon process 调度流程

15.5 用 Falcon process 调度 Oozie 工作流

本示例中，只使用 Falcon 的 process 功能，调用前面定义的 Oozie 工作流定期自动执行。以下为配置步骤。

1. 启动 Falcon 服务

示例实验环境使用的 HDP2.5.0，在安装之时就已经配置并启动了 Falcon 服务。可以在 Ambari Web 控制台中交互式配置、启动、停止、重启 Falcon 服务。

2. 建立 Falcon Cluster 使用的 HDFS 目录

```
# 建立目录
hdfs dfs -mkdir /apps/falcon/primaryCluster
hdfs dfs -mkdir /apps/falcon/primaryCluster/staging
hdfs dfs -mkdir /apps/falcon/primaryCluster/working

# 修改属主
hdfs dfs -chown -R falcon:users /apps/falcon/*

# 修改权限
hdfs dfs -chmod -R 777 /apps/falcon/primaryCluster/staging
hdfs dfs -chmod -R 755 /apps/falcon/primaryCluster/working
```

3. 建立 Falcon Cluster

Falcon 里的 Cluster 定义集群上各种资源的默认访问点，还定义 Falcon 作业使用的默认工作目录。在 Falcon Web UI 中，单击 Create → Cluster，在界面中填写 Cluster 相关信息，这里定义如下：

- Cluster Name：集群的唯一标识，填写 primaryCluster。

- Data Center or Colo Name：数据中心名称，填写 primaryColo。
- Tags：标签用于对实体进行分组和定位，填写 EntityType 和 Cluster。
- File System Read Endpoint Address：NameNode 地址，填写 http://hdp1:50070。
- File System Default Address：文件系统地址，由于配置了 HDFS HA，此处填写 hdfs://mycluster。
- Yarn Resource Manager Address：YARN 资源管理器地址，填写 hdp2:8050。
- Workflow Address：工作流地址，填写 http://hdp2:11000/oozie/。
- Message Broker Address：消息代理地址，填写 tcp://hdp2:61616?daemon=true。

其他属性使用默认值，所有信息确认后保存 Cluster 定义。创建 Falcon Cluster 可以参考 https://hortonworks.com/hadoop-tutorial/create-falcon-cluster/。

4. 建立 Falcon process

在 Falcon Web UI 中，单击 Create → Process，在界面中填写 Process 相关信息，这里定义如下：

- Process Name：处理名称，填写 RegularETL。
- Engine：执行引擎，选择 Oozie。
- Workflow Name：工作流名称，填写 RegularETL。此名称是在 Oozie 的 workflow.xml 中定义的名称。
- Workflow Path：工作流目录，填写 /user/oozie。该路径是 workflow.xml 文件所在的 HDFS 目录。
- Cluster：集群名称，选择 primaryCluster。该集群是上一步建立的 Cluster。
- Startd：执行开始时间，选 2017/5/18 01:00 PM，下午 1 点开始执行。
- End：执行结束时间，使用默认的 2099/12/31 11:59 AM。
- Repeat Every：重复执行周期，使用默认的 30 minutes，每半小时执行一次。本示例实际应该选 1 Days，半小时执行一次主要方便看 Process 执行结果。
- Timezone：选择（GMT +8:00）。

在 Oozie 的 workflow.xml 工作流文件中使用了 ${jobTracker}、${nameNode}、${focusNodeLogin}、${myScript} 等形式参数。工作流被 Oozie 执行时，所有形参都必须提供具体的值。这些值在创建 process 时的 ADVANCED OPTIONS → Properties 指定。这里的配置如图 15-6 所示。所有信息确认后保存 process 定义。

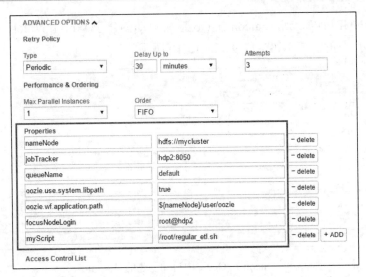

图 15-6　定义 Falcon process 属性

定义 Falcon Process 可以参考 https://hortonworks.com/hadoop-tutorial/defining-processing-data-end-end-data-pipeline-apache-falcon/。

5. 执行 Falcon process

首次执行 process 前，先将 Sqoop 的目标数据目录改为完全读写模式，否则可能报权限错误。这是初始化性质的一次性操作，之后不再需要这步。

```
su - hdfs -c 'hdfs dfs -chmod -R 777 /data/ext'
```

等到下午一点开始第一次执行 RegularETL Process，之后每半小时执行一次。Falcon 的执行结果如图 15-7 所示。

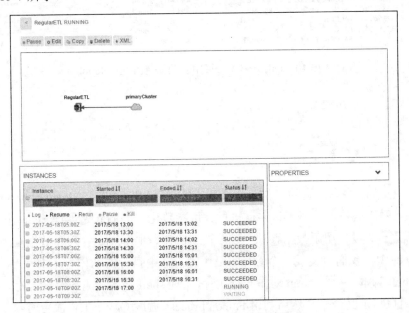

图 15-7　Falcon process 执行结果

在 Oozie Web UI 可以看到，Falcon 在 Oozie 中自动创建了 Workflow Job、Coordinator Job 和 Bundle Job，分别如图 15-8、图 15-9、图 15-10 所示。

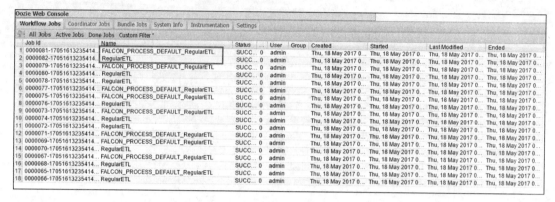

图 15-8　Falcon 在 Oozie 中自动创建的 Workflow Job

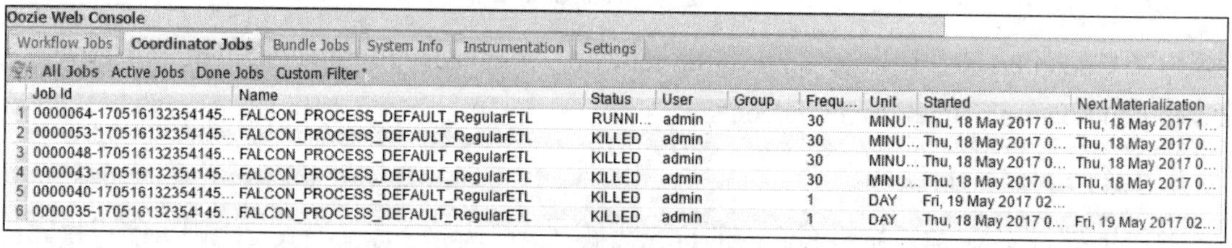

图 15-9　Falcon 在 Oozie 中自动创建的 Coordinator Job

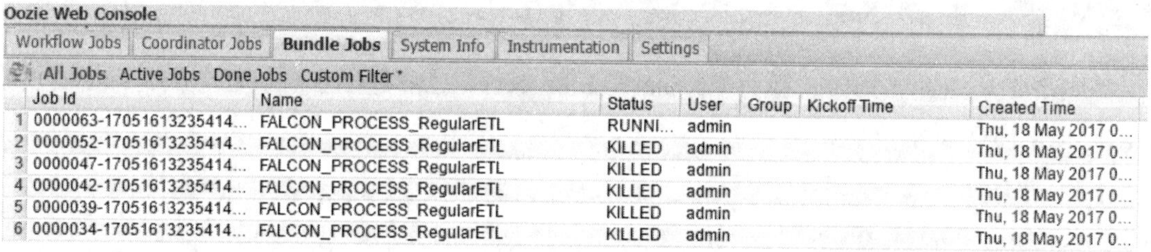

图 15-10　Falcon 在 Oozie 中自动创建的 Bundle Job

15.6　小结

Oozie 是 Hadoop 生态圈里的工作流系统，支持 Workflow、Coordinator 和 Bundle 三种作业。Oozie 的主要特点是自动调度、并行执行、可参数化。Falcon 是面向 Hadoop 的新一代数据处理和管理平台，设计用于数据移动、数据管道协调、生命周期管理和数据发现。在销售订单示例数据仓库中，我们定义了一个 Oozie 工作流，其中执行的动作包括 HDFS 命令、Sqoop 作业和本地 shell 脚本。三个 Sqoop 动作并行执行，本地 shell 脚本调用 psql 命令行，执行 HAWQ 的定期 ETL 用户自定义函数。然后利用 Falcon process 调度 Oozie 工作流定时自动执行。

第 16 章
维度表技术

前面章节中,我们实现了多维数据仓库的基本功能,如使用 Sqoop 和 HAWQ 实现 ETL 过程,使用 Oozie 和 Falcon 定期执行 ETL 任务等。本章将继续讨论常见的维度表技术。我们以最简单的"增加列"开始,继而讨论维度子集、角色扮演维度、层次维度、退化维度、杂项维度、维度合并、分段维度等基本的维度表技术。这些技术都是在实际应用中经常用到的。在说明这些技术的相关概念和使用场景后,我们以销售订单数据仓库为例,给出实现代码和测试过程,必要时会对 ETL 脚本做出适当修改。

16.1 增加列

业务的扩展或变化是不可避免的,尤其像互联网行业,需求变更已经成为常态,唯一不变的就是变化本身,其中最常碰到的扩展是给一个已经存在的表增加列。

以销售订单为例,假设因为业务需要,在操作型源系统的客户表中增加了送货地址的四个字段,并在销售订单表中增加了销售数量字段。由于数据源表增加了字段,数据仓库中的表也要随之修改。本节说明如何在客户维度表和销售订单事实表上添加列,并在新列上应用 SCD2,以及对定时装载脚本所做的修改。图 16-1 显示了增加列后的数据仓库模式。

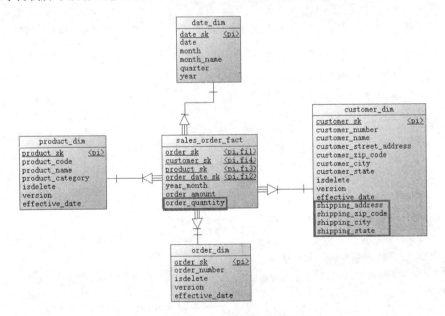

图 16-1 增加列后的数据仓库模式

1. 修改数据库表结构

（1）修改源数据库表结构

执行下面的 SQL 语句修改 MySQL 中源数据库模式。

```
use source;

-- 在客户表最后增加四列
alter table customer
  add shipping_address varchar(30) after customer_state,
  add shipping_zip_code int after shipping_address,
  add shipping_city varchar(30) after shipping_zip_code,
  add shipping_state varchar(2) after shipping_city ;

-- 在销售订单表最后增加一列
alter table sales_order add order_quantity int after order_amount ;
```

以上语句给客户表增加了四列，表示客户的送货地址。销售订单表在销售金额列后面增加了销售数量列。注意 after 关键字，这是 MySQL 对标准 SQL 的扩展，HAWQ 目前还不支持这种扩展，只能把新增列加到已有列的后面。在关系理论中，列是没有顺序的。

（2）修改 ext 模式中的表结构

HAWQ 外部表目前不支持 ALTER TABLE 语句，报错如下：

```
dw=> alter table ext.customer add column shipping_address varchar(30);
ERROR:  "customer" is an external table
HINT:  Use ALTER EXTERNAL TABLE instead
dw=> alter external table ext.customer add column shipping_address varchar(30);ERROR:  Cannot support alter external table statement yet
```

因此要增加列只能重建 HAWQ 外部表。我们在数据抽取时都是覆盖外部表，其中的数据只是临时性的，重建表不涉及数据问题，并不会造成很大影响。

```
-- 设置模式查找路径
set search_path to ext;

-- 删除客户外部表
drop external table customer;
-- 建立客户外部表
create external table customer
( customer_number int,
  customer_name varchar(30),
  customer_street_address varchar(30),
  customer_zip_code int,
  customer_city varchar(30),
  customer_state varchar(2),
```

```
    shipping_address varchar(30),
    shipping_zip_code int,
    shipping_city varchar(30),
    shipping_state varchar(2) )
location ('pxf://mycluster/data/ext/customer?profile=hdfstextsimple')
    format 'text' (delimiter=e',');

-- 删除销售订单外部表
drop external table sales_order;
-- 建立销售订单外部表
create external table sales_order
( order_number int,
  customer_number int,
  product_code int,
  order_date timestamp,
  entry_date timestamp,
  order_amount decimal(10 , 2 ),
  order_quantity int )
location ('pxf://mycluster/data/ext/sales_order?profile=hdfstextsimple')
    format 'text' (delimiter=e',', null='null');
```

需要注意的是 ext 表中列的顺序要和源数据库严格保持一致。因为客户表和产品表是全量覆盖抽取数据，所以如果源和目标顺序不一样，将产生错误的结果。

（3）修改 rds 模式中的表结构

HAWQ 允许使用 ALTER TABLE 语句为内部表增加列。与 MySQL 不同，HAWQ 每条 ALTER TABLE 语句只能增加一列，因此增加四列需要执行四次 ALTER TABLE 语句，并且在增加列时需要指定新增列的默认值，否则会报类似如下的错误：

```
ERROR: ADD COLUMN with no default value in append-only tables is not yet supported.
```

使用下面的 SQL 语句修改 rds 模式中的表结构。

```
alter table rds.customer add column shipping_address varchar(30) default null;
alter table rds.customer add column shipping_zip_code int default null;
alter table rds.customer add column shipping_city varchar(30) default null;
alter table rds.customer add column shipping_state varchar(2) default null;

comment on column rds.customer.shipping_address is '送货地址';
comment on column rds.customer.shipping_zip_code is '送货邮编';
comment on column rds.customer.shipping_city is '送货城市';
comment on column rds.customer.shipping_state is '送货省份';

alter table rds.sales_order add column order_quantity int default null;
```

```sql
comment on column rds.sales_order.order_quantity is '销售数量';
```

（4）修改 tds 模式中的表结构

使用下面的 SQL 语句修改 tds 模式中的表结构。

```sql
alter table tds.customer_dim add column shipping_address varchar(30) default null;
alter table tds.customer_dim add column shipping_zip_code int default null;
alter table tds.customer_dim add column shipping_city varchar(30) default null;
alter table tds.customer_dim add column shipping_state varchar(2) default null;
comment on column tds.customer_dim.shipping_address is '送货地址';
comment on column tds.customer_dim.shipping_zip_code is '送货邮编';
comment on column tds.customer_dim.shipping_city is '送货城市';
comment on column tds.customer_dim.shipping_state is '送货省份';

alter table tds.sales_order_fact add column order_quantity int default null;
comment on column tds.sales_order_fact.order_quantity is '销售数量';
```

2. 重建相关视图

HAWQ 不允许修改视图的列数，错误信息如下：

```
ERROR: cannot change number of columns in view
```

因此需要使用下面的 SQL 语句重建客户维度表的当前视图和历史视图，增加四列。

（1）重建客户维度当前视图

```sql
-- 切换到 tds 模式
set search_path=tds;

-- 删除视图
drop view v_customer_dim_latest;

-- 建立视图
create or replace view v_customer_dim_latest as
select customer_sk, customer_number, customer_name, customer_street_address,
       customer_zip_code, customer_city, customer_state, shipping_address,
       shipping_zip_code, shipping_city, shipping_state, version, effective_date
  from (select distinct on (customer_number) customer_number, customer_sk,
               customer_name, customer_street_address, customer_zip_code,
               customer_city, customer_state, shipping_address,
               shipping_zip_code, shipping_city, shipping_state,
               isdelete, version, effective_date
          from customer_dim
```

```
         order by customer_number, customer_sk desc) as latest
   where isdelete is false;
```

（2）重建客户维度历史视图

```
-- 切换到 tds 模式
set search_path=tds;

-- 删除视图
drop view v_customer_dim_his;

-- 建立视图，增加版本过期日期导出列
create or replace view v_customer_dim_his as
select *, date(lead(effective_date,1,date '2200-01-01') over (partition by customer_number order by effective_date)) expiry_date
  from customer_dim;
```

3. 修改定期装载函数 fn_regular_load

增加列后，对定期装载函数 fn_regular_load 也要做相应的修改，增加对新增数据列的处理。这里只需要对客户维度表和销售订单事实表的部分进行修改，修改后的函数如下（只列出修改的部分）。

```
create or replace function fn_regular_load ()
returns void as $$
declare
    -- 设置 scd 的生效时间
    ...
begin
    -- 分析外部表
    ...
    -- 将外部表数据装载到原始数据表
    ...
    -- 分析 rds 模式的表
    ...
    -- 设置 cdc 的上限时间
    ...
    -- 装载客户维度
    insert into tds.customer_dim
    (customer_number, customer_name, customer_street_address,
     customer_zip_code, customer_city, customer_state,
     shipping_address, shipping_zip_code, shipping_city, shipping_state,
     isdelete, version, effective_date)
    select case flag when 'D' then a_customer_number else b_customer_number
           end customer_number,
```

```sql
                case flag when 'D' then a_customer_name else b_customer_name
                 end customer_name,
                case flag when 'D' then a_customer_street_address
                    else b_customer_street_address
                 end customer_street_address,
                case flag when 'D' then a_customer_zip_code else b_customer_zip_code
 end customer_zip_code,
                case flag when 'D' then a_customer_city else b_customer_city
                 end customer_city,
                case flag when 'D' then a_customer_state else b_customer_state
                 end customer_state,
                case flag when 'D' then a_shipping_address else b_shipping_address
 end shipping_address,
                case flag when 'D' then a_shipping_zip_code else b_shipping_zip_code
 end shipping_zip_code,
                case flag when 'D' then a_shipping_city else b_shipping_city
                 end shipping_city,
                case flag when 'D' then a_shipping_state else b_shipping_state
                 end shipping_state,
                case flag when 'D' then true else false
                 end isdelete,
                case flag when 'D' then a_version when 'I' then 1 else a_version + 1
 end v,
                v_pre_date
         from (select a.customer_number a_customer_number,
                      a.customer_name a_customer_name,
                      a.customer_street_address a_customer_street_address,
                      a.customer_zip_code a_customer_zip_code,
                      a.customer_city a_customer_city,
                      a.customer_state a_customer_state,
                      a.shipping_address a_shipping_address,
                      a.shipping_zip_code a_shipping_zip_code,
                      a.shipping_city a_shipping_city,
                      a.shipping_state a_shipping_state,
                      a.version a_version,
                      b.customer_number b_customer_number,
                      b.customer_name b_customer_name,
                      b.customer_street_address b_customer_street_address,
                      b.customer_zip_code b_customer_zip_code,
                      b.customer_city b_customer_city,
                      b.customer_state b_customer_state,
                      b.shipping_address b_shipping_address,
                      b.shipping_zip_code b_shipping_zip_code,
```

```sql
                    b.shipping_city b_shipping_city,
                    b.shipping_state b_shipping_state,
                    case when a.customer_number is null then 'I'
                         when b.customer_number is null then 'D'
                         else 'U'
                    end flag
               from v_customer_dim_latest a
               full join rds.customer b on a.customer_number = b.customer_number
where a.customer_number is null -- 新增
               or b.customer_number is null -- 删除
               or (a.customer_number = b.customer_number
                   and not (coalesce(a.customer_name,'') =
coalesce(b.customer_name,'')
                   and coalesce(a.customer_street_address,'') =
coalesce(b.customer_street_address,'')
                   and coalesce(a.customer_zip_code,0) =
coalesce(b.customer_zip_code,0)
                   and coalesce(a.customer_city,'') =
coalesce(b.customer_city,'')
                   and coalesce(a.customer_state,'') =
coalesce(b.customer_state,'')
                   and coalesce(a.shipping_address,'') =
coalesce(b.shipping_address,'')
                   and coalesce(a.shipping_zip_code,0) =
coalesce(b.shipping_zip_code,0)
                   and coalesce(a.shipping_city,'') =
coalesce(b.shipping_city,'')
                   and coalesce(a.shipping_state,'') =
coalesce(b.shipping_state,'') ))) t
          order by coalesce(a_customer_number, 999999999999), b_customer_number
   limit 999999999999;

    -- 装载产品维度
    ...
    -- 装载 order 维度
    ...
    -- 装载销售订单事实表
    insert into sales_order_fact
    select order_sk, customer_sk, product_sk, date_sk, year * 100 + month,
order_amount, order_quantity
       from rds.sales_order a,
            order_dim b,
            v_customer_dim_his c,
```

```
                v_product_dim_his d,
                date_dim e,
                rds.cdc_time f
          where a.order_number = b.order_number
            and a.customer_number = c.customer_number
            and a.order_date >= c.effective_date
            and a.order_date < c.expiry_date
            and a.product_code = d.product_code
            and a.order_date >= d.effective_date
            and a.order_date < d.expiry_date
            and date(a.order_date) = e.date
            and a.entry_date >= f.last_load and a.entry_date < f.current_load;
       -- 分析 tds 模式的表
       ...
       -- 更新时间戳表的 last_load 字段
       ...
end; $$
language plpgsql;
```

同客户地址一样，新增的送货地址列也是用 SCD2 新增历史版本。与 14.5 节建立的定期装载函数中相同部分比较，会发现在比较客户属性时使用了 coalesce 函数。

源系统库中，客户地址和送货地址列都是允许为空的，这样的设计是出于灵活性和容错性的考虑。我们以送货地址为例进行讨论。使用"a.shipping_address = b.shipping_address"条件判断送货地址是否更改，根据等号两边的值是否为空，会出现以下三种情况：

（1）shipping_address 和 b.shipping_address 都不为空。这种情况下如果两者相等则返回 true，说明地址没有变化，否则返回 false，说明地址改变了，逻辑正确。

（2）shipping_address 和 b.shipping_address 都为空。两者的比较会演变成 null=null，根据 HAWQ 对"="操作符的定义，会返回 NULL。此时如果其他属性没变，则比较演变为 NOT (NULL AND TRUE)，否则演变为 NOT (NULL AND FALSE)，前者返回 NULL，后者返回 TRUE。这符合我们的逻辑。

（3）shipping_address 和 b.shipping_address 只有一个为空。就是说地址列从 NULL 变成非 NULL，或者从非 NULL 变成 NULL，这种情况明显应该新增一个版本，但根据"="的定义，此时 a.shipping_address=b.shipping_address 返回值是 NULL，查询不会返回行，不符合我们的需求。

基于以上分析，这里使用 HAWQ 的 coalesce 函数处理 NULL 值（类似于 Oracle 的 NVL 或 SQL Server 的 ISNULL）将 NULL 值比较转化为标量值比较。空值的逻辑判断有其特殊性，为了避免不必要的麻烦，数据库设计时应该尽量将字段设计成非空，必要时用默认值代替 NULL，并将此作为一个基本的设计原则。

4. 测试

（1）在源库中增加测试数据

执行下面的 SQL 语句，在 MySQL 的源数据库中增加客户和销售订单测试数据。

```sql
use source;

-- 默认的送货地址与客户地址相同
update customer
   set shipping_address = customer_street_address,
       shipping_zip_code = customer_zip_code,
       shipping_city = customer_city,
       shipping_state = customer_state ;

-- 新增一个客户
insert into customer
(customer_name, customer_street_address, customer_zip_code, customer_city,
customer_state,shipping_address,shipping_zip_code,shipping_city,shipping_state)
values
  ('online distributors', '2323 louise dr.', 17055, 'pittsburgh', 'pa',
  '2323 louise dr.', 17055, 'pittsburgh', 'pa') ;

-- 新增订单日期为昨天的 9 条订单。
set @start_date := unix_timestamp(date_add(current_date, interval -1 day));
set @end_date := unix_timestamp(current_date);
drop table if exists temp_sales_order_data;
create table temp_sales_order_data as select * from sales_order where 1=0;
set @order_date := from_unixtime(@start_date + rand() * (@end_date -
@start_date));
set @amount := floor(1000 + rand() * 9000);
set @quantity := floor(10 + rand() * 90);
insert into temp_sales_order_data
values (117, 1, 1, @order_date, @order_date, @amount, @quantity);

… 新增 9 条订单 …

insert into sales_order
select null,customer_number,product_code,order_date,entry_date,
order_amount,order_quantity
  from temp_sales_order_data
 order by order_date;

commit ;
```

（2）执行定期 ETL 脚本

```
su - hdfs -c 'hdfs dfs -chmod -R 777 /data/ext'
~/regular_etl.sh
```

regular_etl.sh 是在 14.6 节建立的定期 ETL shell 脚本文件，在此不需要做修改。

（3）查询数据，确认 ETL 过程正确执行

regular_etl.sh 成功执行后，查询 v_customer_dim_latest 和 v_customer_dim_his 视图，应该看到由于源库中为送货地址增加了默认值，每条客户记录都新增了一个版本，老的过期记录的送货地址为空。9 号客户是新加的，具有送货地址。查询 sales_order_fact 表，应该只有 9 个订单有销售数量，老的销售数据数量字段为空。

16.2 维度子集

有些需求不需要最细节的数据。例如更想要某个月的销售汇总，而不是某天的数据。再比如相对于全部的销售数据，可能对某些特定状态的数据更感兴趣等。此时事实数据需要关联到特定的维度，这些特定维度包含在从细节维度选择的行中，所以叫维度子集。维度子集比细节维度的数据少，因此更易使用，查询也更快。

有时称细节维度为基本维度，维度子集为子维度，基本维度表与子维度表具有相同的属性或内容，称这样的维度表具有一致性。一致的维度具有一致的维度关键字、一致的属性列名字、一致的属性定义以及一致的属性值。如果属性的含义不同或者包含不同的值，维度表就不是一致的。

子维度是一种一致性维度，由基本维度的列与行的子集构成。当构建聚合事实表，或者需要获取粒度级别较高的数据时，通常用到子维度。对基本维度和子维度表来说，属性是公共的，其标识和定义相同，两个表中的值相同，然而，基本维度和子维度表的主键是不同的。还有另外一种情况，就是当两个维度具有同样粒度级别的细节数据，但其中一个仅表示行的部分子集时，也需要一致性维度子集。

ETL 数据流应当根据基本维度建立一致性子维度，而不是独立于基本维度，以确保一致性。本节中将准备两个特定子维度，月份维度与 Pennsylvania 州客户维度。它们均取自现有的维度，月份维度是日期维度的子集，Pennsylvania 州客户维度是客户维度的子集。

1. 建立包含属性子集的子维度

（1）建立月份维度表

```
-- 设置模式查找路径
set search_path to tds;

-- 建立月份维度表
```

```sql
create table month_dim (
    month_sk bigserial,
    month smallint,
    month_name varchar(9),
    quarter smallint,
    year smallint );

comment on table month_dim is '月份维度表';
```

(2)初始装载月份维度数据

本示例中，以下语句将生成252条月份数据。

```sql
insert into month_dim (month, month_name, quarter, year)
select distinct month, month_name, quarter, year
 from date_dim
 order by year, month
 limit 99999999999999;

analyze month_dim;
```

(3)建立追加日期数据的函数

该函数用于向日期维度表和月份维度表追加数据。如果日期所在的月份没在月份维度中，那么该月份会被装载到月份维度中。

```sql
create or replace function fn_append_date (end_dt date)
returns void as $$
declare
    v_date date;
    v_datediff int;
begin
    select max(date) + 1 into v_date from date_dim;
    v_datediff := end_dt - v_date;

    for i in 0 .. v_datediff
    loop
        insert into date_dim(date, month, month_name, quarter, year)
        values(v_date, extract(month from v_date), to_char(v_date,'mon'),
extract(quarter from v_date), extract(year from v_date));
        v_date := v_date + 1;
    end loop;
    analyze date_dim;

    insert into month_dim (month, month_name, quarter, year)
    select * from
    (select distinct month, month_name, quarter, year
```

```
    from date_dim
  except all
  select month, month_name, quarter, year
    from month_dim) t
  order by year, month
  limit 99999999999999;
  analyze month_dim;

end; $$
language plpgsql;
```

(4)测试追加日期数据的函数

执行以下语句追加生成一年的日期数据。

```
select fn_append_date(date '2021-12-31');
```

执行下面的查询,应该看到日期维度表新增2021年的365条记录。

```
select * from date_dim where date > date '2020-12-31' order by date;
```

执行下面的查询,应该看到月份维度表新增2021年的12条记录。

```
select * from month_dim where year > 2020 order by year,month;
```

2. 建立包含行子集的子维度

当两个维度处于同一细节粒度,但是其中一个仅仅是行的子集时,会产生另外一种一致性维度构造子集。例如,销售订单示例中,客户维度表包含多个州的客户信息。对于不同州的销售分析可能需要浏览客户维度的子集,需要分析的维度仅包含部分客户数据。通过使用行的子集,不会破坏整个客户集合。当然,与该子集连接的事实表必须被限制在同样的客户子集中。月份维度是一个上卷维度,包含基本维度的上层数据,而特定维度子集是基本维度的行子集。执行下面的脚本建立特定维度表,并导入Pennsylvania(PA)客户维度子集数据。

(1)建立PA客户维度表

```
create table pa_customer_dim
(customer_sk bigserial,
 customer_number int,
 customer_name varchar(50),
 customer_street_address varchar(50),
 customer_zip_code int,
 customer_city varchar(30),
 customer_state varchar(2),
 isdelete boolean default false,
 version int,
 effective_date date,
 shipping_address varchar(50),
```

```
 shipping_zip_code int,
 shipping_city varchar(30),
 shipping_state varchar(2));

comment on table pa_customer_dim is 'PA 客户维度表';
```

PA 客户维度子集与月份维度子集有两点区别：

- pa_customer_dim 表和 customer_dim 表有完全相同的列，而 month_dim 不包含 date_dim 表的日期列。
- pa_customer_dim 表的代理键就是客户维度的代理键，而 month_dim 表里的月份维度代理键并不来自日期维度，而是独立生成的。

（2）修改定期装载函数

通常在基本维度表装载数据后，进行包含其行子集的子维度表的数据装载。因此修改定期装载函数 fn_regular_load，增加对 PA 客户维度的处理，修改后的 fn_regular_load 函数如下（只列出修改的部分）。

```
create or replace function fn_regular_load ()
returns void as $$
declare
    -- 设置 scd 的生效时间
    ...
begin
    -- 分析外部表
    ...
    -- 将外部表数据装载到原始数据表
    ...
    -- 分析 rds 模式的表
    ...
    -- 设置 cdc 的上限时间
    ...
    -- 装载客户维度
    ...
    -- 重载 PA 客户维度
    truncate table pa_customer_dim;
    insert into pa_customer_dim
    select customer_sk, customer_number, customer_name, customer_street_address,           customer_zip_code, customer_city, customer_state, isdelete, version,           effective_date, shipping_address, shipping_zip_code, shipping_city,           shipping_state
      from customer_dim
     where customer_state = 'pa';
```

```
        -- 装载产品维度
        ...
        -- 装载order维度
        ...
        -- 装载销售订单事实表
        ...
        -- 分析tds模式的表
        ...
        -- 更新时间戳表的last_load字段
        ...
    end; $$
    language plpgsql;
```

上面的函数在处理完客户维度表后，装载PA客户维度。每次重新覆盖pa_customer_dim表中的所有数据。先用truncate table语句清空表，然后用insert into ... select语句，从客户维度表中选取Pennsylvania州的数据，并插入到pa_customer_dim表中。

（3）测试定期数据装载函数

用以下步骤测试PA客户子维度的数据装载。

步骤01 执行下面的SQL语句向客户源数据里添加一个PA的客户和四个OH的客户。

```
use source;
insert into customer
(customer_name, customer_street_address, customer_zip_code,
 customer_city, customer_state, shipping_address,
 shipping_zip_code, shipping_city, shipping_state)
values
('pa customer', '1111 louise dr.', '17050', 'mechanicsburg', 'pa', '1111 louise dr.', '17050', 'mechanicsburg', 'pa'),
('bigger customers', '7777 ridge rd.', '44102', 'cleveland', 'oh', '7777 ridge rd.', '44102', 'cleveland', 'oh'),
('smaller stores', '8888 jennings fwy.', '44102', 'cleveland', 'oh', '8888 jennings fwy.', '44102', 'cleveland', 'oh'),
('small-medium retailers', '9999 memphis ave.', '44102', 'cleveland', 'oh', '9999 memphis ave.', '44102', 'cleveland', 'oh'),
('oh customer', '6666 ridge rd.', '44102', 'cleveland', 'oh', '6666 ridge rd.', '44102','cleveland', 'oh') ;
commit;
```

步骤02 使用下面的命令执行定期装载。

```
~/regular_etl.sh
```

步骤03 使用下面的查询验证结果，pa_customer_dim表此时应该有20条记录。

```
select customer_name, customer_state, version, effective_date
  from tds.pa_customer_dim;
```

3. 使用视图实现维度子集

为了实现维度子集，我们创建了新的子维度表。这种实现方式有两个主要问题，一是需要额外的存储空间，因为新创建的子维度是物理表；二是存在数据不一致的潜在风险。本质上，只要相同的数据存储多份，就会有数据不一致的可能。这也就是为什么在数据库设计时要强调规范化以最小化数据冗余的原因之一。为了解决这些问题，还有一种常用的做法是在基本维度上建立视图生成子维度。下面是创建子维度视图的 SQL 语句。

```
-- 建立月份维度视图
create view v_month_dim as
select row_number() over (order by t1.year,t1.month) month_sk, t1.*
  from (select distinct month, month_name, quarter, year from date_dim) t1;

-- 建立 PA 维度视图
create view v_pa_customer_dim as
select * from customer_dim where customer_state = 'pa';

-- 建立 PA 维度当前视图
create view v_pa_customer_dim_latest as
select * from v_customer_dim_latest where customer_state = 'pa';

-- 建立 PA 维度历史视图
create view v_pa_customer_dim_his as
select * from v_customer_dim_his where customer_state = 'pa';
```

这种方法的主要优点是：实现简单，只要创建视图，不需要修改原来脚本中的逻辑；不占用存储空间，因为视图不真正存储数据；消除了数据不一致的可能，因为数据只有一份。虽然优点很多，但此方法的缺点也十分明显：当基本维度表和子维度表的数据量相差悬殊时，性能会比物理表差得多；如果定义视图的查询很复杂，并且视图很多的话，可能会对元数据存储系统造成压力，严重影响查询性能。

注意，视图是与存储无关的纯粹的逻辑对象，HAWQ 不支持物化视图。当查询引用了一个视图，视图的定义被评估后产生一个行集，用作查询后续的处理。这只是一个概念性的描述，实际上，作为查询优化的一部分，HAWQ 可能把视图的定义和查询结合起来考虑，而不一定是先生成视图所定义的行集。例如，优化器可能将查询的过滤条件下推到视图中。

一旦视图建立，它的结构就是固定的，之后底层表的结构改变，如添加字段等，不会反映到视图的结构中。如果底层表被删除了，或者表结构改变成一种与视图定义不兼容的形式，视图将变为无效状态，其上的查询将失败。

视图是只读的，不能对视图使用 LOAD 或 INSERT 语句装载数据，但可以使用 alter view 语句修改视图的某些元数据。视图定义中可以包含 order by 和 limit 子句，例如，如果一个视图定义中指定了 limit 5，而查询语句为 select * from v limit 10，那么至多会返回 5 行记录。

16.3 角色扮演维度

单个物理维度可以被事实表多次引用，每个引用连接逻辑上存在差异的角色维度。例如，事实表可以有多个日期，每个日期通过外键引用不同的日期维度，原则上每个外键表示不同的日期维度视图，这样引用具有不同的含义。这些不同的维度视图具有唯一的代理键列名，被称为角色，相关维度被称为角色扮演维度。

当一个事实表多次引用一个维度表时会用到角色扮演维度。例如，一个销售订单有一个是订单日期，还有一个请求交付日期，这时就需要引用日期维度表两次。我们期望在每个事实表中设置日期维度，因为总是希望按照时间来分析业务情况。在事务型事实表中，主要的日期列是事务日期，例如，订单日期。有时会发现其他日期也可能与每个事实关联，例如，订单事务的请求交付日期。每个日期应该成为事实表的外键。

本节说明两类角色扮演维度的实现，分别是表别名和数据库视图。表别名是在 SQL 语句里引用维度表多次，每次引用都赋予维度表一个别名。而数据库视图，则是按照事实表需要引用维度表的次数，建立相同数量的视图。我们先修改销售订单数据库模式，添加一个请求交付日期字段，并对数据抽取和装载脚本做相应的修改。这些表结构修改好后，插入测试数据，演示别名和视图在角色扮演维度中的用法。

1. 修改数据库模式

（1）修改源库表结构

执行下面的语句，给源库中销售订单表 sales_order 增加 request_delivery_date 字段。

```
use source;
alter table sales_order add request_delivery_date datetime after order_date ;
```

（2）修改数据仓库表结构

```
-- 修改外部表
drop external table ext.sales_order;
create external table ext.sales_order
( order_number int,
  customer_number int,
  product_code int,
  order_date timestamp,
  request_delivery_date timestamp,
  entry_date timestamp,
  order_amount decimal(10 , 2 ),
  order_quantity int )
location ('pxf://mycluster/data/ext/sales_order?profile=hdfstextsimple')
format 'text' (delimiter=e',', null='null');
```

```
    comment on table ext.sales_order is '销售订单外部表';

    -- 修改 rds.sales_order
    alter table rds.sales_order add column request_delivery_date timestamp default
null;
    comment on column rds.sales_order.request_delivery_date is '请求交付日期';

    -- 修改 tds.sales_order_fact
    alter table tds.sales_order_fact add column request_delivery_date_sk bigint
default null;
    comment on column tds.sales_order_fact.request_delivery_date_sk is '请求交付
日期维度代理键';
    comment on column tds.sales_order_fact.order_date_sk is '订单日期维度代理键';
```

增加列的过程已经在16.1节详细讨论过。HAWQ不支持给外部表增加列，因此需要重建表。在销售订单外部表上增加请求交付日期字段，数据类型是 timestamp，对应源库表上的 datetime 类型。注意外部表中列的顺序要和源表中列定义的顺序保持一致。RDS 和 TDS 中的内部表直接使用 ALTER TABLE 语句增加请求交付日期列。因为 HAWQ 的 ADD COLUMN 不支持 after 语法，新增字段会加到所有已存在字段的后面。修改后数据仓库模式如图 16-2 所示。

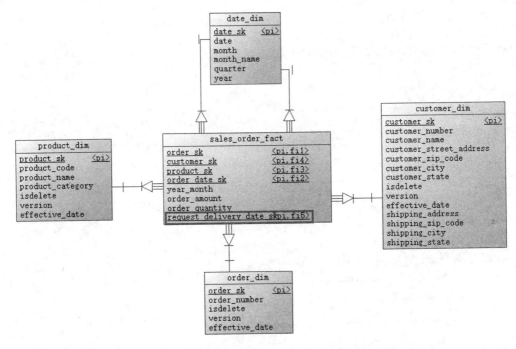

图 16-2 数据仓库中增加请求交付日期属性

从图中可以看到，销售订单事实表和日期维度表之间有两条连线，表示订单日期和请求交付日期都是引用日期维度表的外键。注意，虽然图中显示了表之间的关联关系，但 HAWQ 中并不支持主外键数据库约束。

2. 修改定期数据装载函数

定期装载函数需要增加对请求交付日期列的处理，修改后的函数如下所示（只显示修改的部分）。

```sql
create or replace function fn_regular_load ()
returns void as $$
declare
    -- 设置 scd 的生效时间
    ...
begin
    -- 分析外部表
    ...
    -- 将外部表数据装载到原始数据表
    ...
insert into rds.sales_order
select order_number, customer_number, product_code, order_date,
       entry_date, order_amount, order_quantity, request_delivery_date
  from ext.sales_order;

    -- 分析 rds 模式的表
    ...
    -- 设置 cdc 的上限时间
    ...
    -- 装载客户维度
    ...
    -- 重载 PA 客户维度
    ...
    -- 装载产品维度
    ...
    -- 装载 order 维度
    ...
    -- 装载销售订单事实表
    insert into sales_order_fact
    select order_sk, customer_sk, product_sk, e.date_sk, e.year * 100 + e.month,
           order_amount, order_quantity, f.date_sk
      from rds.sales_order a,
           order_dim b,
           v_customer_dim_his c,
           v_product_dim_his d,
           date_dim e,
           date_dim f,
           rds.cdc_time g
     where a.order_number = b.order_number
```

```
            and a.customer_number = c.customer_number
            and a.order_date >= c.effective_date
            and a.order_date < c.expiry_date
            and a.product_code = d.product_code
            and a.order_date >= d.effective_date
            and a.order_date < d.expiry_date
            and date(a.order_date) = e.date
            and date(a.request_delivery_date) = f.date
            and a.entry_date >= g.last_load and a.entry_date < g.current_load;

    -- 分析 tds 模式的表
    ...
    -- 更新时间戳表的 last_load 字段
    ...
end; $$
language plpgsql;
```

函数做了以下两点修改：

- 装载 rds.sales_order 时显式指定列的顺序，因为外部表与内部表列的顺序不一致。
- 在装载销售订单事实表时，关联了日期维度表两次，分别赋予别名 e 和 f。事实表和两个日期维度表关联，取得日期代理键。e.date_sk 表示订单日期代理键，f.date_sk 表示请求交付日期的代理键。

3. 测试

（1）在源库中生成测试数据

执行下面的 SQL 语句在源库中增加三个带有交货日期的销售订单。

```
use source;
/*** 新增订单日期为昨天的 3 条订单。***/
  set @start_date := unix_timestamp(date_add(current_date, interval -1 day));
set @end_date := unix_timestamp(current_date);

  drop table if exists temp_sales_order_data;
  create table temp_sales_order_data as select * from sales_order where 1=0;
  set @order_date := from_unixtime(@start_date + rand() * (@end_date - @start_date));
  set @request_delivery_date := from_unixtime(unix_timestamp(date_add(current_date, interval 5 day)) + rand() * 86400);
  set @amount := floor(1000 + rand() * 9000);
  set @quantity := floor(10 + rand() * 90);
  insert into temp_sales_order_data
  values (126, 1, 1, @order_date,
```

```
@request_delivery_date, @order_date, @amount, @quantity);

… 插入 3 条订单记录 …

insert into sales_order
select null,customer_number,product_code,order_date,
request_delivery_date,entry_date,order_amount,order_quantity
from temp_sales_order_data order by order_date;

commit ;
```

(2) 执行定期装载函数并查看结果

```
~/regular_etl.sh
```

使用下面的查询验证结果。

```
select a.order_sk, request_delivery_date_sk, c.date
  from sales_order_fact a, date_dim b, date_dim c
 where a.order_date_sk = b.date_sk
   and a.request_delivery_date_sk = c.date_sk ;
```

查询结果如下：

```
 order_sk | request_delivery_date_sk |    date
----------+--------------------------+------------
      126 |                     6360 | 2017-05-30
      127 |                     6360 | 2017-05-30
      128 |                     6360 | 2017-05-30
(3 rows)
```

可以看到只有三个新的销售订单具有 request_delivery_date_sk 值，6360 对应的日期是 2017 年 5 月 30 日。

4. 使用角色扮演维度查询

(1) 使用表别名查询

```
select order_date_dim.date order_date,
       request_delivery_date_dim.date request_delivery_date,
       sum(order_amount),count(*)
  from sales_order_fact a,
       date_dim order_date_dim,
       date_dim request_delivery_date_dim
 where a.order_date_sk = order_date_dim.date_sk
   and a.request_delivery_date_sk = request_delivery_date_dim.date_sk
 group by order_date_dim.date , request_delivery_date_dim.date
 order by order_date_dim.date , request_delivery_date_dim.date;
```

（2）使用视图查询

```
-- 创建订单日期视图
create view v_order_date_dim
(order_date_sk,
 order_date,
 month,
 month_name,
 quarter,
 year)
as select * from date_dim;
-- 创建请求交付日期视图
create view v_request_delivery_date_dim
(request_delivery_date_sk,
 request_delivery_date,
 month,
 month_name,
 quarter,
 year)
as select * from date_dim;
-- 查询
select order_date,request_delivery_date,sum(order_amount),count(*)
  from sales_order_fact a,v_order_date_dim b,v_request_delivery_date_dim c
 where a.order_date_sk = b.order_date_sk
   and a.request_delivery_date_sk = c.request_delivery_date_sk
 group by order_date , request_delivery_date
 order by order_date , request_delivery_date;
```

上面两种实现方式是等价的，查询结果如下：

```
order_date | request_delivery_date |   sum    | count
-----------+-----------------------+----------+-------
2017-05-24 | 2017-05-30            | 12104.00 |     3
(1 row1)
```

尽管不能连接到单一的日期维度表，但可以建立并管理单独的物理日期维度表，然后使用视图或别名建立两个不同日期维度的描述。注意在每个视图或别名列中需要唯一的标识。例如，订单日期属性应该具有唯一标识 order_date，以便与请求交付日期 request_delivery_date 区别。别名与视图在查询中的作用并没有本质的区别，都是为了从逻辑上区分同一个物理维度表。许多 BI 工具也支持在语义层使用别名。但是，如果有多个 BI 工具，连同直接基于 SQL 的访问，都同时在组织中使用的话，不建议采用语义层别名的方法。当某个维度在单一事实表中同时出现多次时，则会存在维度模型的角色扮演。基本维度可能作为单一物理表存在，但是每种角色应该被当成标识不同的视图展现到 BI 工具中。

5. 一种有问题的设计

为处理多日期问题，一些设计者试图建立单一日期维度表，该表使用一个键表示每个订单日期和请求交付日期的组合，例如：

```
create table date_dim (date_sk int, order_date date, delivery_date date);
create table sales_order_fact (date_sk int, order_amount int);
```

这种方法存在两个的问题。首先，如果需要处理所有日期维度的组合情况，则包含大约每年 365 行的清晰、简单的日期维度表将会极度膨胀。例如，订单日期和请求交付日期存在如下多对多关系：

订单日期	请求交付日期
2017-05-26	2017-05-29
2017-05-27	2017-05-29
2017-05-28	2017-05-29
2017-05-26	2017-05-30
2017-05-27	2017-05-30
2017-05-28	2017-05-30
2017-05-26	2017-05-31
2017-05-27	2017-05-31
2017-05-28	2017-05-31

如果使用角色扮演维度，日期维度表中只需要 2017-05-26 ~ 2017-05-31 6 条记录。而采用单一日期表设计方案，每一个组合都要唯一标识，明显需要九条记录。当两种日期及其组合很多时，这两种方案的日期维度表记录数会相去甚远。

其次，合并的日期维度表不再适合其他经常使用的日、周、月等日期维度。日期维度表每行记录的含义不再指唯一一天，因此无法在同一张表中标识出周、月等一致性维度，进而无法简单地处理按时间维度的上卷、聚合等需求。

16.4 层次维度

大多数维度都具有一个或多个层次。示例数据仓库中的日期维度就有一个四级层次：年、季度、月和日。这些级别用 date_dim 表里的列表示。日期维度是一个单路径层次，因为除了年-季度-月-日这条路径外，它没有任何其他层次。为了识别数据仓库里一个维度的层次，首先要理解维度中列的含义，然后识别两个或多个列是否具有相同的主题。年、季度、月和日具有相同的主题，因为它们都是关于日期的。具有相同主题的列形成一个组，组中的一列必须包含至少一个组内的其他成员（除了最低级别的列），前面提到的组中，月包含日。这些列的链条形成了一个层次，例如，年-季度-月-日这个链条是一个日期维度的层次。除了日期维度，邮编维度中的地理位置信息，产品维度的产品与产品分类，也都构成层次关系。表 16-1 显示了三个维度的层次。

表 16-1 销售订单数据仓库中的层次维度

customer_dim		product_dim	date_dim
customer_street_address	shipping_address	product_name	date
customer_zip_code	shipping_zip_code	product_category	month
customer_city	shipping_city		quarter
customer_state	shipping_state		year

本节描述处理层次关系的方法，包括在固定深度的层次上进行分组和钻取查询，多路径层次和参差不齐层次的处理等，从最基本的情况开始讨论。

16.4.1 固定深度的层次

固定深度层次是一种一对多关系，如一年中有四个季度，一个季度包含三个月等等。当固定深度层次定义完成后，层次就具有了固定的名称，层次级别作为维度表中的不同属性出现。只要满足上述条件，固定深度层次就是最容易理解和查询的层次关系，固定层次也能够提供可预测的、快速的查询性能。可以在固定深度层次上进行分组和钻取查询。

分组查询是把度量按照一个维度的一个或多个级别进行分组聚合。下面的脚本是一个分组查询的例子。该查询按产品（product_category 列）和日期维度的三个层次级别（year、quarter 和 month 列）分组返回销售金额。

```
select product_category,year,quarter,month,sum(order_amount) s_amount
  from v_sales_order_fact a,product_dim b,date_dim c
 where a.product_sk = b.product_sk
   and a.year_month = c.year * 100 + c.month
 group by product_category, year, quarter, month
 order by product_category, year, quarter, month;
```

这是一个非常简单的分组查询，结果输出的每一行度量（销售订单金额）都沿着年-季度-月的层次分组。

与分组查询类似，钻取查询也把度量按照一个维度的一个或多个级别进行分组。但与分组查询不同的是，分组查询只显示分组后最低级别，即本例中月级别上的度量，而钻取查询显示分组后维度每一个级别的度量。下面使用 UNION ALL 和 GROUPING SETS 两种方法进行钻取查询，结果显示了每个日期维度级别，即年、季度和月各级别的订单汇总金额。

```
-- 使用union all
select product_category, time, order_amount
  from (select product_category,
               case when sequence = 1 then 'year: '||time
                    when sequence = 2 then 'quarter: '||time
                    else 'month: '||time
               end time,
```

```sql
                order_amount,
                sequence,
                date
         from (select product_category, min(date) date, year time, 1 sequence, sum(order_amount) order_amount
                 from v_sales_order_fact a, product_dim b, date_dim c
                 where a.product_sk = b.product_sk
                   and a.year_month = c.year * 100 + c.month
                 group by product_category , year
                union all
                select product_category, min(date) date, quarter time, 2 sequence, sum(order_amount) order_amount
                 from v_sales_order_fact a, product_dim b, date_dim c
                 where a.product_sk = b.product_sk
                   and a.year_month = c.year * 100 + c.month
                 group by product_category , year , quarter
                union all
                select product_category, min(date) date, month time, 3 sequence, sum(order_amount) order_amount
                 from v_sales_order_fact a, product_dim b, date_dim c
                 where a.product_sk = b.product_sk
                   and a.year_month = c.year * 100 + c.month
                 group by product_category , year , quarter , month) x) y
  order by product_category , date , sequence , time;
-- 使用grouping sets
select product_category,
       case when gid = 3 then 'year: '||year
            when gid = 1 then 'quarter: '||quarter
            else 'month: '||month
        end time,
       order_amount
  from (select product_category, year, quarter, month, min(date) date, sum(order_amount) order_amount,
        grouping(product_category,year,quarter,month) gid
         from v_sales_order_fact a, product_dim b, date_dim c
         where a.product_sk = b.product_sk
           and a.year_month = c.year * 100 + c.month
         group by grouping sets
((product_category,year,quarter,month),(product_category,year,quarter),(product_category,year))) x
  order by product_category , date , gid desc, time;
```

以上两种不同写法的查询语句结果相同：

```
 product_category |     time      |  order_amount
------------------+---------------+---------------
 monitor          | year: 2017    |    1635343.00
 monitor          | quarter: 2    |    1635343.00
 monitor          | month: 5      |    1635343.00
 peripheral       | year: 2017    |    2079433.00
 peripheral       | quarter: 2    |    2079433.00
 peripheral       | month: 5      |    1792823.00
 peripheral       | month: 6      |     286620.00
 storage          | year: 2016    |   17467030.00
 storage          | quarter: 1    |    5925898.00
 storage          | month: 3      |    5925898.00
 storage          | quarter: 2    |   11541132.00
 storage          | month: 4      |    4633260.00
 storage          | month: 5      |    3415332.00
 storage          | month: 6      |    3492540.00
 storage          | year: 2017    |    4701194.00
 storage          | quarter: 2    |    4701194.00
 storage          | month: 5      |    4427234.00
 storage          | month: 6      |     273960.00
(18 rows)
```

第一条语句的子查询中使用 union all 集合操作，将年、季度、月三个级别的汇总数据联合成一个结果集。注意 union all 的每个查询必须包含相同个数和类型的字段。附加的 min(date) 和 sequence 导出列用于对输出结果排序显示。这种写法使用标准的 SQL 语法，具有通用性。

第二条语句使用 HAWQ 提供的 grouping 函数和 group by grouping sets 子句。grouping set 对列出的每一个字段组进行 group by 操作，如果字段组为空，则不进行分组处理。因此该语句会生成按产品类型、年、季度、月；类型、年、季度；类型、年分组的聚合数据行。grouping(<column> [, …])函数用于区分查询结果中的 null 值是属于列本身的还是聚合的结果行。该函数为每个参数产生一位 0 或 1，1 代表结果行是聚合行，0 表示结果行是正常分组数据行。函数值使用了位图策略（bitvector，位向量），即它的二进制形式中的每一位表示对应列是否参与分组，如果某一列参与了分组，对应位就被置为 1，否则为 0。最后将二进制数转换为十进制数返回。通过这种方式可以区分出数据本身中的 null 值。

16.4.2 多路径层次

多路径层次是对单路径层次的扩展。现在数据仓库的月维度只有一条层次路径，即年-季度-月这条路径。现在增加一个新的"促销期"级别，并且加一个新的年-促销期-月的层次路径。这时月维度将有两条层次路径，因此是多路径层次维度。下面的脚本给 month_dim 表添加一个叫做 campaign_session 的新列，并建立 rds.campaign_session 过渡表。

```
alter table tds.month_dim add column campaign_session varchar(30) default null;
```

```
comment on column tds.month_dim.campaign_session is '促销期';

create table rds.campaign_session
(campaign_session varchar(30),month smallint,year smallint);
```

假设所有促销期都不跨年,并且一个促销期可以包含一个或多个月份,但一个月份只能属于一个促销期。为了理解促销期如何工作,表 16-2 给出了一个促销期定义的例子。

表 16-2 2017 年促销期

促销期	月份
2017 年第一促销期	1 月—4 月
2017 年第二促销期	5 月—7 月
2017 年第三促销期	8 月
2017 年第四促销期	9 月—12 月

每个促销期有一个或多个月。一个促销期也许并不是正好一个季度,也就是说,促销期级别不能上卷到季度,但是促销期可以上卷至年级别。假设 2017 年促销期的数据如下,并保存在/home/gpadmin/campaign_session.csv 文件中。

```
2017 First Campaign,1,2017
2017 First Campaign,2,2017
2017 First Campaign,3,2017
2017 First Campaign,4,2017
2017 Second Campaign,5,2017
2017 Second Campaign,6,2017
2017 Second Campaign,7,2017
2017 Third Campaign,8,2017
2017 Last Campaign,9,2017
2017 Last Campaign,10,2017
2017 Last Campaign,11,2017
2017 Last Campaign,12,2017
```

现在可以执行下面的语句把 2017 年的促销期数据装载进月维度。本地文件必须在 HAWQ master 主机上的本地目录中,并且 copy 命令需要使用 gpadmin 用户执行。

```
copy rds.campaign_session from '/home/gpadmin/campaign_session.csv' with delimiter ',';

set search_path = tds;

create table tmp as
select t1.month_sk month_sk, t1.month month1, t1.month_name month_name,
       t1.quarter quarter, t1.year year1, t2.campaign_session campaign_session
```

```
from month_dim t1
    left join rds.campaign_session t2 on t1.year = t2.year
    and t1.month = t2.month;

truncate table month_dim;
insert into month_dim select * from tmp;
drop table tmp;
```

此时查询月份维度表，应该看到 2017 年的促销期已经有数据了，其他年份的 campaign_session 字段值为 null。

16.4.3 参差不齐的层次

在一个或多个级别上没有数据的层次称为不完全层次。例如，在特定月份没有促销期，那么月维度就具有不完全促销期层次。下面是一个不完全促销期的例子，数据存储在 ragged_campaign.csv 文件中。2017 年 1 月、4 月、6 月、9 月、10 月、11 月和 12 月没有促销期。

```
,1,2017
2017 Early Spring Campaign,2,2017
2017 Early Spring Campaign,3,2017
,4,2017
2017 Spring Campaign,5,2017
,6,2017
2017 Last Campaign,7,2017
2017 Last Campaign,8,2017
,9,2017
,10,2017
,11,2017
,12,2017
```

重新向 month_dim 表装载促销期数据。

```
truncate table rds.campaign_session;
copy rds.campaign_session from '/home/gpadmin/ragged_campaign.csv' with delimiter ',';

set search_path = tds;

create table tmp as
select t1.month_sk month_sk, t1.month month1, t1.month_name month_name,
       t1.quarter quarter, t1.year year1, null campaign_session
  from month_dim t1;

truncate table month_dim;
```

```
insert into month_dim
select t1.month_sk, t1.month1, t1.month_name, t1.quarter, t1.year1,
       case when t2.campaign_session != '' then t2.campaign_session
            else t1.month_name
        end campaign_session
  from tmp t1 left join rds.campaign_session t2
    on t1.year1 = t2.year and t1.month1 = t2.month;

drop table tmp;
```

在有促销期的月份，campaign_session 列填写促销期名称，而对于没有促销期的月份，该列填写月份名称。轻微参差不齐层次没有固定的层次深度，但层次深度有限。如地理层次深度通常包含 3～6 层。与其使用复杂的机制构建难以预测的可变深度层次，不如将其变换为固定深度位置设计，针对不同的维度属性确立最大深度，然后基于业务规则放置属性值。

下面的语句查询年-促销期-月层次。

```
select product_category,
       case when gid = 3 then cast(year as varchar(10))
            when gid = 1 then campaign_session
            else month_name
        end time,
        order_amount
  from (select product_category, year, campaign_session, month, month_name,
              sum(month_order_amount) order_amount,
              sum(month_order_quantity) order_quantity,
              grouping(product_category,year,campaign_session,month) gid,
              min(month) min_month
         from v_month_end_sales_order_fact a, product_dim b, month_dim c
        where a.product_sk = b.product_sk
          and a.year_month = c.year * 100 + c.month
          and c.year = 2017
        group by grouping sets
((product_category,year,campaign_session,month,month_name),(product_category,year,campaign_session),(product_category,year))) x
 order by product_category, min_month, gid desc, month;
```

查询结果如下：

```
product_category  |          time         | order_amount
------------------+-----------------------+---------------
monitor           | 2017                  |     52753.00
monitor           | 2017 Spring Campaign  |     52753.00
monitor           | may                   |     52753.00
peripheral        | 2017                  |     67387.00
```

```
     peripheral            | 2017 Spring Campaign |     57833.00
     peripheral            | may                  |     57833.00
     peripheral            | jun                  |      9554.00
     peripheral            | jun                  |      9554.00
     storage               | 2017                 |    151946.00
     storage               | 2017 Spring Campaign |    142814.00
     storage               | may                  |    142814.00
     storage               | jun                  |      9132.00
     storage               | jun                  |      9132.00
(13 rows)
```

min_month 列用于排序。在有促销期月份的路径，月级别的行的汇总与促销期级别的行相同。而对于没有促销期的月份，其促销期级别的行与月级别的行相同。也就是说，在没有促销期级别的月份，月上卷了它们自己。例如，2017 年 6 月没有促销期，所以在输出看到，每种产品分类有两个相同的 6 月的行，其中后一行是月份级别的行，前一行表示是没有促销期的行。对于没有促销期的月份，促销期行的销售订单金额（输出里的 order_amount 列）与月分行的相同。

16.5 退化维度

退化维度技术减少维度的数量，简化多维数据仓库模式。简单的模式比复杂的更容易理解，也有更好的查询性能。有时，维度表中除了业务主键外没有其他内容。在本销售订单示例中，订单维度表除了订单号，没有任何其他属性，而订单号是事务表的主键，这种维度就是退化维度。业务系统中的主键通常是不允许修改的。销售订单只能新增，不能修改已经存在的订单号，也不会删除订单记录。因此订单维度表也不会有历史数据版本问题。

销售订单事实表中的每行记录都包括作为退化维度的订单号代理键。在操作型系统中，销售订单表是最细节事务表，订单号是订单表的主键，每条订单都可以通过订单号定位，订单中的其他属性，如客户、产品等，都依赖于订单号。也就是说，订单号把与订单属性有关的表联系起来。但是，在维度模型中，事实表中的订单号代理键通常与订单属性的其他表没有直接关联，而是将订单事实表所有关心的属性分类到不同的维度中。例如，订单日期关联到日期维度，客户关联到客户维度等。在事实表中保留订单号最主要的原因是用于连接数据仓库与操作型系统，它也可以起到事实表主键的作用。某些情况下，可能会有一个或两个属性仍然属于订单而不属于其他维度。当然，此时订单维度就不再是退化维度了。

退化维度常被保留作为操作型事务的标识符。实际上可以将订单号作为一个属性加入到事实表中。这样订单维度就没有数据仓库需要的任何数据，此时就可以退化订单维度。需要把退化维度的相关数据迁移到事实表中，然后删除退化的维度。操作型事务中的控制号码，如订单号码、发票号码、提货单号码等通常产生空的维度并表示为事务事实表中的退化维度。

1. 退化订单维度

使用维度退化技术时先要识别数据，分析从来不用的数据列。订单维度的 order_number 列就是这样的一列。但如果用户想看事务的细节，还需要订单号。因此，在退化订单维度前，要把订单号迁移到 sales_order_fact 事实表。图 16-3 显示了修改后的模式。

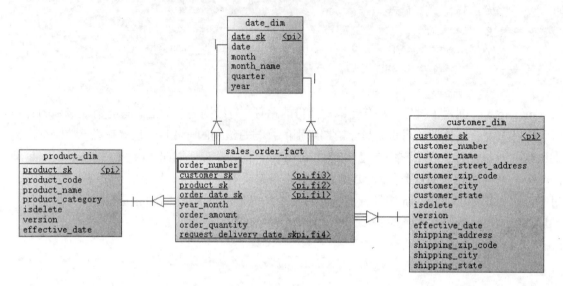

图 16-3　退化订单维度

按顺序执行下面的四步退化 order_dim 维度表：

（1）给 sales_order_fact 表添加 order_number 列。

（2）把 order_dim 表里的订单号迁移到 sales_order_fact 表。

（3）删除 sales_order_fact 表里的 order_sk 列。

（4）删除 order_dim 表。

下面的语句完成所有退化订单维度所需的步骤。

```
set search_path=tds;

alter table sales_order_fact rename to sales_order_fact_old;
create table sales_order_fact as
select t2.order_number, t1.customer_sk, t1.product_sk, t1.order_date_sk,
       t1.year_month, t1.order_amount, t1.order_quantity,
       t1.request_delivery_date_sk
  from sales_order_fact_old t1 inner join order_dim t2 on t1.order_sk = t2.order_sk;

comment on table sales_order_fact is '销售订单事实表';
comment on column sales_order_fact.order_number is '订单号';
comment on column sales_order_fact.customer_sk is '客户维度代理键';
```

```
comment on column sales_order_fact.product_sk is '产品维度代理键';
comment on column sales_order_fact.order_date_sk is '日期维度代理键';
comment on column sales_order_fact.year_month is '年月分区键';
comment on column sales_order_fact.order_amount is '销售金额';
comment on column sales_order_fact.order_quantity is '数量';
comment on column sales_order_fact.request_delivery_date_sk is '请求交付日期代理键';

drop table sales_order_fact_old;
drop table order_dim;
```

HAWQ 没有提供 UPDATE 功能,因此要更新已有数据的订单号,只能重新装载所有数据。本例中,订单号维度表中代理键和订单号业务主键的值相同,其实可以简单地将事实表的 order_sk 字段改名为 order_number。但这只是一种特殊情况,通常代理键和业务主键的值是不同的,因此这里依然使用标准的方式重新生成数据。

2. 修改定期数据装载函数

退化一个维度后需要做的另一件事就是修改定期数据装载函数。需要把订单号加入到销售订单事实表,而不再需要导入订单维度。修改后的函数如下(只列出修改的部分)。

```
create or replace function fn_regular_load ()
returns void as $$
declare
    -- 设置 scd 的生效时间
    ...
begin
    -- 分析外部表
    ...
    -- 将外部表数据装载到原始数据表
    ...
    -- 分析 rds 模式的表
    ...
    -- 设置 cdc 的上限时间
    ...
    -- 装载客户维度
    ...
    -- 重载 PA 客户维度
    ...
    -- 装载产品维度
    ...

    去掉装载 order_dim 维度表的语句
```

```
    -- 装载销售订单事实表
    insert into sales_order_fact
    select a.order_number,
           customer_sk,
           product_sk,
           e.date_sk,
           e.year * 100 + e.month,
           order_amount,
           order_quantity,
           f.date_sk
      from rds.sales_order a,
           v_customer_dim_his c,
           v_product_dim_his d,
           date_dim e,
           date_dim f,
           rds.cdc_time g
     where a.customer_number = c.customer_number
       and a.order_date >= c.effective_date
       and a.order_date < c.expiry_date
       and a.product_code = d.product_code
       and a.order_date >= d.effective_date
       and a.order_date < d.expiry_date
       and date(a.order_date) = e.date
       and date(a.request_delivery_date) = f.date
       and a.entry_date >= g.last_load and a.entry_date < g.current_load;
    -- 分析 tds 模式的表
    ...
    -- 更新时间戳表的 last_load 字段
    ...
end; $$
language plpgsql;
```

函数做了以下两点修改：

- 去掉装载和分析 order_dim 维度表的语句。
- 事实表中的 order_number 字段字节从 rds.sales_order 表获得。

3. 测试

（1）准备测试数据

执行下面的 SQL 语句在源库中增加两条销售订单记录。

```
use source;

set @start_date := unix_timestamp('2017-05-25');
```

```
    set @end_date := unix_timestamp('2017-05-25 12:00:00');
    set @order_date := from_unixtime(@start_date + rand() * (@end_date -
@start_date));
    set @request_delivery_date :=
from_unixtime(unix_timestamp(date_add(current_date, interval 5 day)) + rand() *
86400);
    set @amount := floor(1000 + rand() * 9000);
    set @quantity := floor(10 + rand() * 90);

    insert into sales_order values
(null,1,1,@order_date,@request_delivery_date,@order_date,@amount,@quantity);

    set @start_date := unix_timestamp('2017-05-25 12:00:01');
    set @end_date := unix_timestamp('2017-05-26');
    set @order_date := from_unixtime(@start_date + rand() * (@end_date -
@start_date));
    set @request_delivery_date :=
from_unixtime(unix_timestamp(date_add(current_date, interval 5 day)) + rand() *
86400);
    set @amount := floor(1000 + rand() * 9000);
    set @quantity := floor(10 + rand() * 90);

    insert into sales_order values
(null,1,1,@order_date,@request_delivery_date,@order_date,@amount,@quantity);

    commit ;
```

以上语句在源库上生成 2017 年 5 月 25 日的两条销售订单。为了保证自增订单号与订单时间顺序相同，注意一下@order_date 变量的赋值。

（2）执行定期装载函数并查看结果

```
~/regular_etl.sh
```

脚本执行成功后，查询 sales_order_fact 表，验证新增的两条订单是否正确装载。

```
select a.order_number onum, customer_name cname, product_name pname,
  e.date odate, f.date rdate,
     order_amount amount
  from sales_order_fact a,
     customer_dim b,
     product_dim c,
     date_dim e,
     date_dim f
 where a.customer_sk = b.customer_sk
```

```
        and a.product_sk = c.product_sk
        and a.order_date_sk = e.date_sk
        and a.request_delivery_date_sk = f.date_sk
 order by order_number desc
 limit 5;
```

查询结果如下，可以看到新增两条记录的订单号被正确装载。

```
 onum |         cname          |      pname      |   odate    |   rdate    | amount
------+------------------------+-----------------+------------+------------+--------
  130 | really large customers | hard disk drive | 2017-05-25 | 2017-05-31 | 3531.00
  129 | really large customers | hard disk drive | 2017-05-25 | 2017-05-31 | 4358.00
  128 | really large customers | hard disk drive | 2017-05-24 | 2017-05-30 | 8003.00
  127 | medium retailers       | flat panel      | 2017-05-24 | 2017-05-30 | 2153.00
  126 | small stores           | floppy drive    | 2017-05-24 | 2017-05-30 | 1948.00
(5 rows)
```

16.6 杂项维度

1. 什么是杂项维度

简单地说，杂项维度就是一种包含的数据具有很少可能值的维度。事务型商业过程通常产生一系列混杂的、低基数的标志位或状态信息。与其为每个标志或属性定义不同的维度，不如建立单独的、将不同维度合并到一起的杂项维度。这些维度，通常在一个模式中标记为事务型概要维度，一般不需要所有属性可能值的笛卡儿积，但应该至少包含实际发生在源数据中的组合值。例如，在销售订单中，可能存在有很多离散数据（yes-no 这种开关类型的值）：

- verification_ind（如果订单已经被审核，值为 yes）
- credit_check_flag（表示此订单的客户信用状态是否已经被检查）
- new_customer_ind（如果这是新客户的首个订单，值为 yes）
- web_order_flag（表示一个订单是在线上订单还是线下订单）

这类数据常被用于增强销售分析，其特点是属性可能很多但每种属性的可能值却很少。

2. 处理杂项维度的常用方法

在建模复杂的操作型源系统时，经常会遭遇大量五花八门的标志或状态信息，它们包含小范围的离散值。处理这些较低基数的标志或状态位可以采用以下几种方法。

（1）忽略这些标志和指标

姑且将这种回避问题的处理方式也算作方法之一吧。在开发 ETL 系统时，ETL 开发小组可以向业务用户询问有关忽略这些标志的必要问题，如果它们是微不足道的。但是这样的方案通常立即就被否决了，因为有人偶尔还需要它们。

（2）保持事实表行中的标志位不变

以销售订单为例，和源数据库一样，我们可以在事实表中也建立这四个标志位字段。在装载事实表时，除了订单号以外，同时装载这四个字段的数据，这些字段没有对应的维度表，而是作为订单的属性保留在事实表中。

这种处理方法简单直接，装载过程不需要做大量修改，也不需要建立相关的维度表。但是一般我们不希望在事实表中存储难以识别的标志位，尤其是当每个标志位还配有一个文字描述字段时。不要在事实表行中存储包含大量字符的描述符，因为每一行都会有文字描述，它们可能会使表快速膨胀。在行中保留一些文本标志是令人反感的，比较好的做法是分离出单独的维度表保存这些标志位字段的数据，它们的数据量很小，并且极少改变。事实表通过维度表的代理键引用这些标志。

（3）将每个标志位放入其自己的维度中

例如，为销售订单的四个标志位分别建立四个对应的维度表。装载事实表数据前先处理这四个维度表，必要时生成新的代理键，然后在事实表中引用这些代理键。这种方法是将杂项维度当作普通维度来处理，多数情况下这也是不合适的。

首先，当类似的标志或状态位字段比较多时，需要建立很多的维度表，其次事实表的外键数也会大量增加。处理这些新增的维度表和外键需要大量修改数据装载脚本，还会增加出错的机会，同时会给 ETL 的开发、维护、测试过程带来很大的工作量。最后，杂项维度的数据有自己明显的特点，即属性多但每个属性的值少，并且极少修改，这种特点决定了它应该与普通维度的处理区分开。

作为一个经验值，如果外键的数量处于合理的范围中，即不超过 20 个，则在事实表中增加不同的外键是可以接受的。若外键列表已经很长，则应该避免将更多外键加入事实表中。

（4）将标志位字段存储到订单维度中

可以将标志位字段添加到订单维度表中。上一节我们将订单维度表作为退化维度删除了，因为它除了订单号，没有其他任何属性。与其将订单号当成是退化维度，不如视其为将低基数标志或状态作为属性的普通维度。事实表通过引用订单维度表的代理键，关联到所有的标志位信息。

尽管该方法精确地表示了数据关系，但依然存在前面讨论的问题。在订单维度表中，每条业务订单都会存在对应的一条销售订单记录，该维度表的记录数会膨胀到跟事实表一样多，而在如此多的数据中，每个标志位字段都存在大量的冗余。通常维度表应该比事实表小得多。

（5）使用杂项维度

处理这些标志位的适当替换方法是将它们包装为一个杂项维度，其中放置各种离散的标志或状态数据。对杂项维度数据量的估算会影响其建模策略。如果某个简单的杂项维度包含 10 个二值标识，则最多将包含 1024（2^10）行。杂项维度可提供所有标识的组合，并用于基于这些标识的约束和报表。事实表与杂项维度之间存在一个单一的、小型的代理键。

如果具有高度非关联的属性，包含更多的数量值，则将它们合并为单一的杂项维度是不合适的。假设存在 5 个标识，每个仅包含 3 个值，则单一杂项维度是这些属性的最佳选择，因为维度最多仅有 243（3^5）行。但如果 5 个没有关联的标识，每个具有 100 个可能值，建议建立不同维度，因为单一杂项维度表最大可能存在 1 亿（100^5）行。

关于杂项维度的一个微妙的问题是，在杂项维度中行的组合确定并已知的前提下，是应该事先为所有组合的完全笛卡儿积建立行，还是建立杂项维度行，只用于保存那些在源系统中出现的组合情况的数据。答案要看大概有多少可能的组合，最大行数是多少。一般来说，理论上组合的数量较小，比如只有几百行时，可以预装载所有组合的数据。而组合的数量大，那么在数据获取时，当遇到新标志或指标时再建立杂项维度行。当然，如果源数据中用到了全体组合时，那别无选择，只能预先装载好全部杂项维度数据。

3. 新增销售订单属性杂项维度

图 16-4 显示了增加杂项维度表后的数据仓库模式。

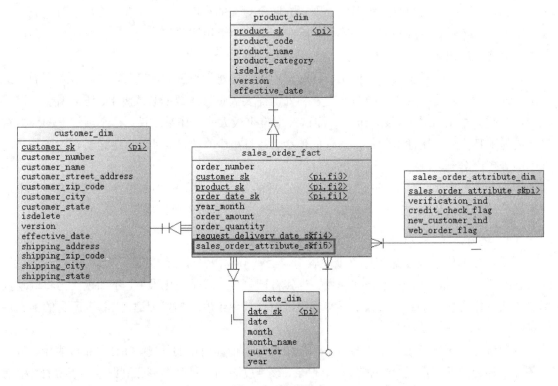

图 16-4 杂项维度

给现有数据仓库新增一个销售订单属性杂项维度。需要新增一个名为 sales_order_attribute_dim 的杂项维度表，该表包括四个 yes-no 列：verification_ind、credit_check_flag、new_customer_ind 和 web_order_flag，各列的含义已经在本节开头说明。每个列可以有两个可能值中的一个，Y 或 N，因此 sales_order_attribute_dim 表最多有 16（2^4）行。假设这 16 行已经包含了所有可能的组合，并且不考虑杂项维度修改的情况，则可以预装载这个维度，并且只需装载一次。

执行下面的语句修改数据库模式。

```sql
-- MySQL
-- 给源库的销售订单表增加对应的属性
use source;
alter table sales_order
  add verification_ind char (1) after product_code,
  add credit_check_flag char (1) after verification_ind,
  add new_customer_ind char (1) after credit_check_flag,
  add web_order_flag char (1) after new_customer_ind ;

-- HAWQ
-- 重建外部表，增加杂项属性，列的顺序必须和源表一致
set search_path=ext;
drop external table sales_order;
create external table sales_order
( order_number int,
  customer_number int,
  product_code int,
  verification_ind char(1),
  credit_check_flag char(1),
  new_customer_ind char(1),
  web_order_flag char(1),
  order_date timestamp,
  request_delivery_date timestamp,
  entry_date timestamp,
  order_amount decimal(10 , 2 ),
  order_quantity int )
location ('pxf://mycluster/data/ext/sales_order?profile=hdfstextsimple')
  format 'text' (delimiter=e',', null='null');

comment on table sales_order is '销售订单外部表';

-- 给销售订单过渡表增加对应的属性
set search_path=rds;
alter table sales_order add column verification_ind char(1) default null;
alter table sales_order add column credit_check_flag char(1) default null;
alter table sales_order add column new_customer_ind char(1) default null;
alter table sales_order add column web_order_flag char(1) default null;

comment on column sales_order.verification_ind is '审核标志';
comment on column sales_order.credit_check_flag is '信用检查标志';
```

```sql
comment on column sales_order.new_customer_ind is '客户首个订单标志';
comment on column sales_order.web_order_flag is '线上订单标志';

set search_path=tds;
-- 建立杂项维度表
create table sales_order_attribute_dim (
    sales_order_attribute_sk int,
    verification_ind char(1),
    credit_check_flag char(1),
    new_customer_ind char(1),
    web_order_flag char(1) );

comment on table sales_order_attribute_dim is '杂项维度表';

-- 生成杂项维度数据，共插入16条记录
insert into sales_order_attribute_dim values (1, 'n', 'n', 'n', 'n');
insert into sales_order_attribute_dim values (2, 'n', 'n', 'n', 'y');
insert into sales_order_attribute_dim values (3, 'n', 'n', 'y', 'n');
insert into sales_order_attribute_dim values (4, 'n', 'n', 'y', 'y');
insert into sales_order_attribute_dim values (5, 'n', 'y', 'n', 'n');
insert into sales_order_attribute_dim values (6, 'n', 'y', 'n', 'y');
insert into sales_order_attribute_dim values (7, 'n', 'y', 'y', 'n');
insert into sales_order_attribute_dim values (8, 'n', 'y', 'y', 'y');
insert into sales_order_attribute_dim values (9, 'y', 'n', 'n', 'n');
insert into sales_order_attribute_dim values (10, 'y', 'n', 'n', 'y');
insert into sales_order_attribute_dim values (11, 'y', 'n', 'y', 'n');
insert into sales_order_attribute_dim values (12, 'y', 'n', 'y', 'y');
insert into sales_order_attribute_dim values (13, 'y', 'y', 'n', 'n');
insert into sales_order_attribute_dim values (14, 'y', 'y', 'n', 'y');
insert into sales_order_attribute_dim values (15, 'y', 'y', 'y', 'n');
insert into sales_order_attribute_dim values (16, 'y', 'y', 'y', 'y');

-- 建立杂项维度外键
alter table sales_order_fact add column sales_order_attribute_sk int default null;
comment on column sales_order_fact.sales_order_attribute_sk is '杂项维度代理键';
```

4. 修改定期数据装载函数

由于有了一个新的维度表，必须修改定期数据装载函数。下面显示了修改后的 fn_regular_load 函数（只列出修改的部分）。

```sql
create or replace function fn_regular_load ()
returns void as $$
```

```sql
declare
    -- 设置 scd 的生效时间
    ...
begin
    -- 分析外部表
    ...
    -- 将外部表数据装载到原始数据表
    ...
    insert into rds.sales_order
    select order_number,
           customer_number,
           product_code,
           order_date,
           entry_date,
           order_amount,
           order_quantity,
           request_delivery_date,
           verification_ind,
           credit_check_flag,
           new_customer_ind,
           web_order_flag
      from ext.sales_order;

    -- 分析 rds 模式的表
    ...
    -- 设置 cdc 的上限时间
    ...
    -- 装载客户维度
    ...
    -- 重载 PA 客户维度
    ...
    -- 装载产品维度
    ...
    -- 装载销售订单事实表
    insert into sales_order_fact
    select a.order_number,
           customer_sk,
           product_sk,
           e.date_sk,
           e.year * 100 + e.month,
           order_amount,
           order_quantity,
           f.date_sk,
```

```
                g.sales_order_attribute_sk
    from rds.sales_order a,
        v_customer_dim_his c,
        v_product_dim_his d,
        date_dim e,
        date_dim f,
        sales_order_attribute_dim g,
        rds.cdc_time h
   where a.customer_number = c.customer_number
     and a.order_date >= c.effective_date
     and a.order_date < c.expiry_date
     and a.product_code = d.product_code
     and a.order_date >= d.effective_date
     and a.order_date < d.expiry_date
     and date(a.order_date) = e.date
     and date(a.request_delivery_date) = f.date
     and a.verification_ind = g.verification_ind
     and a.credit_check_flag = g.credit_check_flag
     and a.new_customer_ind = g.new_customer_ind
     and a.web_order_flag = g.web_order_flag
     and a.entry_date >= h.last_load and a.entry_date < h.current_load;
 -- 分析 tds 模式的表
 ...
 -- 更新时间戳表的 last_load 字段
 ...
end; $$
language plpgsql;
```

函数做了以下两点修改：

- 装载 rds.sales_order 时增加了四个杂项属性。
- 装载事实表时，关联了 sales_order_attribute_dim 维度表，为事实表中装载杂项维度代理键。

杂项属性维度数据已经预装载，所以在定期装载脚本中只需要修改处理事实表的部分。源数据中有四个属性列，而事实表中只对应一列，因此需要使用四列关联条件的组合确定杂项维度表的代理键值，并装载到事实表中。

5. 测试

（1）准备测试数据

执行下面的语句添加 8 个销售订单。

```
use source;
drop table if exists temp_sales_order_data;
```

```sql
    create table temp_sales_order_data as select * from sales_order where 1=0;

    set @start_date := unix_timestamp(date_add(current_date, interval -1 day));
set @end_date := unix_timestamp(current_date);

    set @order_date := from_unixtime(@start_date + rand() * (@end_date - @start_date));
    set @request_delivery_date := 
from_unixtime(unix_timestamp(date_add(current_date, interval 5 day)) + rand() * 86400);
    set @amount := floor(1000 + rand() * 9000);
    set @quantity := floor(10 + rand() * 90);

    insert into temp_sales_order_data
    values (1, 1, 1, 'y', 'y', 'n', 'y', @order_date, @request_delivery_date, @order_date, @amount, @quantity);

    … 一共添加各种属性组合的八条记录 …

    insert into sales_order
    select null, customer_number, product_code, verification_ind, credit_check_flag,     new_customer_ind, web_order_flag, order_date, request_delivery_date,         entry_date, order_amount, order_quantity
      from temp_sales_order_data t1
     order by t1.order_date;
    commit;
```

（2）执行定期装载函数并查看结果

```
    ~/regular_etl.sh
```

可以使用下面的分析性查询确认装载是否正确。该查询分析出检查了信用状态的新用户所占的比例。

```sql
    select round(cast(checked as float) / (checked + not_checked) * 100)||' % '
from 
    (select sum(case when credit_check_flag='y' then 1 else 0 end) checked,
sum(case when credit_check_flag='n' then 1 else 0 end) not_checked       from
sales_order_fact a, sales_order_attribute_dim b
      where new_customer_ind = 'y'
        and a.sales_order_attribute_sk = b.sales_order_attribute_sk) t;
```

查询结果如下。

```
    ?column?
    ----------
```

```
75 %
(1 row)
```

16.7 维度合并

有一种合并维度的情况，就是本来属性相同的维度，因为某种原因被设计成重复的维度属性。随着数据仓库中维度的增加，我们会发现有些通用的数据存在于多个维度中。例如，在销售订单示例中，客户维度的客户地址相关信息、送货地址相关信息里都有邮编、城市和省份。下面说明如何把客户维度里的两个邮编相关信息合并到一个新的维度中。

1. 修改数据仓库表结构

为了合并维度，需要改变数据仓库表结构。图 16-5 显示了修改后的结构。新增了一个 zip_code_dim 邮编信息维度表，sales_order_fact 事实表的结构也做了相应的修改。

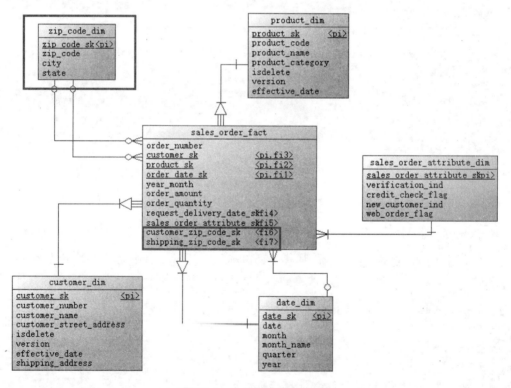

图 16-5　合并邮编信息维度

zip_code_dim 维度表与销售订单事实表相关联。这个关系替换了事实表与客户维度的关系。sales_order_fact 表需要两个关系，一个关联到客户地址邮编，另一个关联到送货地址邮编，相应地增加了两个外键字段。假设邮编相关信息不会修改，因此 zip_code_dim 表中没有是否删除、版本号、生效日期等 SCD 属性。

下面的语句用于修改数据仓库模式,所做的修改如下:

- 创建邮编维度表 zip_code_dim。
- 初始装载邮编相关数据。
- 创建 v_customer_zip_code_dim 和 v_shipping_zip_code_dim 视图。
- sales_order_fact 表上增加 customer_zip_code_sk 和 shipping_zip_code_sk 列。
- 基于已有的客户邮编和送货邮编初始装载两个邮编代理键。
- 在 customer_dim 表上删除客户和送货邮编以及他们的城市和州列。
- 在 pa_customer_dim 上删除客户的城市、州和邮编列。

```sql
set search_path=tds;

-- 建立邮编维度表
create table zip_code_dim
(zip_code_sk serial,
 zip_code int,
 city varchar(30),
 state varchar(2) );

comment on table zip_code_dim is '邮编维度表';

-- 初始装载邮编相关数据
insert into zip_code_dim (zip_code, city, state)
select distinct *
  from (select customer_zip_code, customer_city, customer_state
          from customer_dim
         where customer_zip_code is not null
         union all
        select shipping_zip_code, shipping_city, shipping_state
          from customer_dim
         where shipping_zip_code is not null) t1;

-- 创建视图
create view v_customer_zip_code_dim
(customer_zip_code_sk, customer_zip_code, customer_city, customer_state) as
select * from zip_code_dim;

create view v_shipping_zip_code_dim
(shipping_zip_code_sk, shipping_zip_code, shipping_city, shipping_state) as
select * from zip_code_dim;

-- 添加邮编代理键
alter table sales_order_fact add column customer_zip_code_sk int default null;
```

```sql
alter table sales_order_fact add column shipping_zip_code_sk int default null;

-- 初始装载两个邮编代理键
create table sales_order_fact_bak as select * from sales_order_fact;
truncate table sales_order_fact;

insert into sales_order_fact
select t1.order_number, t1.customer_sk, t1.product_sk, t1.order_date_sk,
       t1.year_month, t1.order_amount, t1.order_quantity,
       t1.request_delivery_date_sk, t1.sales_order_attribute_sk,
       t2.customer_zip_code_sk, t3.shipping_zip_code_sk
  from sales_order_fact_bak t1
  left join
  (select a.order_number order_number,
c.customer_zip_code_sk customer_zip_code_sk
    from sales_order_fact_bak a,
         customer_dim b,
         v_customer_zip_code_dim c
    where a.customer_sk = b.customer_sk
      and b.customer_zip_code = c.customer_zip_code) t2
on t1.order_number = t2.order_number
  left join
  (select a.order_number order_number,
c.shipping_zip_code_sk shipping_zip_code_sk
    from sales_order_fact_bak a,
         customer_dim b,
         v_shipping_zip_code_dim c
    where a.customer_sk = b.customer_sk
      and b.shipping_zip_code = c.shipping_zip_code) t3
on t1.order_number = t3.order_number;

drop table sales_order_fact_bak;

-- 在customer_dim和pa_customer_dim表上删除客户和送货邮编以及他们的城市和州列。alter table customer_dim drop column customer_zip_code cascade;
  alter table customer_dim drop column customer_city;
  alter table customer_dim drop column customer_state;
  alter table customer_dim drop column shipping_zip_code;
  alter table customer_dim drop column shipping_city;
  alter table customer_dim drop column shipping_state;

  alter table pa_customer_dim drop column customer_zip_code;
  alter table pa_customer_dim drop column customer_city;
```

```sql
alter table pa_customer_dim drop column customer_state;
alter table pa_customer_dim drop column shipping_zip_code;
alter table pa_customer_dim drop column shipping_city;
alter table pa_customer_dim drop column shipping_state;

-- 重建相关视图
create or replace view v_customer_dim_latest as
select customer_sk, customer_number, customer_name, customer_street_address, version, effective_date, shipping_address
   from (select distinct on (customer_number) customer_number, customer_sk, customer_name, customer_street_address, isdelete, version,
            effective_date, shipping_address
         from customer_dim
       order by customer_number, customer_sk desc) as latest
   where isdelete is false;

create or replace view v_customer_dim_his as
  select *, date(lead(effective_date,1,date '2200-01-01') over (partition by customer_number order by effective_date)) expiry_date
    from customer_dim;

create or replace view v_pa_customer_dim_latest as
select customer_sk, customer_number, customer_name, customer_street_address, version, effective_date, shipping_address
   from (select distinct on (customer_number) customer_number, customer_sk, customer_name, customer_street_address, isdelete, version,
            effective_date, shipping_address
         from pa_customer_dim
       order by customer_number, customer_sk desc) as latest
   where isdelete is false;

create or replace view v_pa_customer_dim_his as
  select *, date(lead(effective_date,1,date '2200-01-01') over (partition by customer_number order by effective_date)) expiry_date
    from pa_customer_dim;
```

说明：

- 邮编维度的初始数据是从客户维度表中来，这只是为了演示数据装载的过程。客户的邮编信息很可能覆盖不到所有邮编，所以更好的方法是装载一个完整的邮编信息表。由于客户地址和送货地址可能存在交叉的情况，因此使用 distinct 去重。送货地址的三个字段是后加的，在此之前数据的送货地址为空，邮编维度表中不能含有 NULL 值，所以要加上 where shipping_zip_code is not null 过滤条件去除邮编信息为 NULL

的数据行。
- 基于邮编维度表创建客户邮编和送货邮编视图，分别用作两个地理信息的角色扮演维度。
- 把数据备份表 sales_order_fact_bak 中的数据装载回销售订单事实表，同时需要关联两个邮编角色维度视图，查询出两个代理键，装载到事实表中。注意老的事实表与新的邮编维度表是通过客户维度表关联起来的，所以在子查询中需要三表连接，然后用两个左外连接查询出所有原事实表数据，装载到新的增加了邮编维度代理键的事实表中。
- 在 customer_dim 和 pa_customer_dim 表上删除列时，需要使用 cascade 子句同时删除依赖它的视图，之后重建相关视图。

2. 修改定期数据装载函数

定期装载函数有三个地方的修改：

- 删除客户维度装载里所有邮编信息相关的列，因为客户维度里不再有客户邮编和送货邮编相关信息。
- 在事实表中引用客户邮编视图和送货邮编视图中的代理键。
- 修改 pa_customer_dim 装载，因为需要从销售订单事实表的 customer_zip_code_sk 获取客户邮编。

修改后的 fn_regular_load 函数如下（只列出修改的部分）。

```
create or replace function fn_regular_load ()
returns void as $$
declare
    -- 设置 scd 的生效时间
    ...
begin
    -- 分析外部表
    ...
    -- 将外部表数据装载到原始数据表
    ...
    -- 分析 rds 模式的表
    ...
    -- 设置 cdc 的上限时间
    ...
-- 装载客户维度

-- 只需要注意去掉邮编信息的六个字段，别的逻辑没变

    -- 装载产品维度
    ...
    -- 装载销售订单事实表
    insert into sales_order_fact
    select a.order_number,
           customer_sk,
           product_sk,
           e.date_sk,
```

```sql
                e.year * 100 + e.month,
                order_amount,
                order_quantity,
                f.date_sk,
                g.sales_order_attribute_sk,
                h.customer_zip_code_sk,
                i.shipping_zip_code_sk
           from rds.sales_order a,
                v_customer_dim_his c,
                v_product_dim_his d,
                date_dim e,
                date_dim f,
                sales_order_attribute_dim g,
                v_customer_zip_code_dim h,
                v_shipping_zip_code_dim i,
                rds.customer j,
                rds.cdc_time k
          where a.customer_number = c.customer_number
            and a.order_date >= c.effective_date
            and a.order_date < c.expiry_date
            and a.product_code = d.product_code
            and a.order_date >= d.effective_date
            and a.order_date < d.expiry_date
            and date(a.order_date) = e.date
            and date(a.request_delivery_date) = f.date
            and a.verification_ind = g.verification_ind
            and a.credit_check_flag = g.credit_check_flag
            and a.new_customer_ind = g.new_customer_ind
            and a.web_order_flag = g.web_order_flag
            and a.customer_number = j.customer_number
            and j.customer_zip_code = h.customer_zip_code
            and j.shipping_zip_code = i.shipping_zip_code
            and a.entry_date >= k.last_load and a.entry_date < k.current_load;
    -- 重载 PA 客户维度
    truncate table pa_customer_dim;
    insert into pa_customer_dim
    select distinct a.*
      from customer_dim a,
           sales_order_fact b,
           v_customer_zip_code_dim c
     where c.customer_state = 'pa'
       and b.customer_zip_code_sk = c.customer_zip_code_sk
       and a.customer_sk = b.customer_sk;

    -- 分析 tds 模式的表
    ...
    -- 更新时间戳表的 last_load 字段
    ...
end; $$
language plpgsql;
```

上面的函数需要注意两个地方。装载事实表数据时，除了关联两个邮编维度视图外，还要关联过渡区的 rds.customer 表。这是因为要取得邮编维度代理键，必须连接邮编代码字段，而邮编代码已经从客户维度表中删除，只有在源数据的客户表中保留。第二个是 PA 子维度的装载。州代码已经从客户维度表删除，被放到了新的邮编维度表中，而客户维度和邮编维度并没有直接关系，它们是通过事实表的客户代理键和邮编代理键产生联系，因此必须关联事实表、客户维度表、邮编维度表三个表才能取出 PA 子维度数据。这也就是把 PA 子维度的装载放到了事实表装载之后的原因。

3. 测试

按照以下步骤测试修改后的定期装载函数（代码从略）。

（1）对源数据的客户邮编相关信息做一些修改。
（2）装载新的客户数据前，查询最后的客户和送货邮编，后面可以用改变后的信息和此查询的输出作对比。
（3）新增销售订单源数据。
（4）执行定期装载。
（5）查询客户维度表、销售订单事实表和 PA 子维度表，确认数据已经正确装载。

16.8 分段维度

1. 分段维度简介

在客户维度中，最具有分析价值的属性就是各种分类，这些属性的变化范围比较大。对某个个体客户来说，可能的分类属性包括：性别、年龄、民族、职业、收入和状态，例如，新客户、活跃客户、不活跃客户、已流失客户等。在这些分类属性中，有些能够定义成包含连续值的分段，如年龄和收入这种数值型的属性，天然就可以分成连续的数值区间，而像状态这种描述性的属性，可能需要用户根据自己的实际业务仔细定义，通常定义的根据是某种可度量的数值。

组织还可能使用为其客户打分的方法刻画客户行为。分段维度模型通常以不同方式按照积分将客户分类，如基于他们的购买行为、支付行为、流失走向等。每个客户用所得的分数标记。一个常用的客户评分及分析系统是考察客户行为的相关度（R）、频繁度（F）和强度（I），该方法被称为 RFI 方法。有时将强度替换为消费度（M），因此也被称为 RFM 度量。相关度是指客户上次购买或访问网站距今的天数。频繁度是指一段时间内客户购买或访问网站的次数，通常是过去一年的情况。强度是指客户在某一固定时间周期中消费的总金额。在处理大型客户数据时，某个客户的行为可以按照如图 16-6 所示的 RFI 多维数据仓库建模。在此图中，每个维度形成一条数轴，某个轴的积分度量值从 1～5，代表某个分组的实际值，三条数轴组合构成客户积分立方体，每个客户的积分都在这个立方体之中。

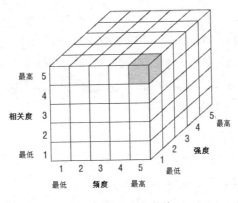

图 16-6　RFI 立方体

定义有意义的分组非常重要。应该由业务人员和数据仓库开发团队共同定义可能会利用的行为标识，更复杂的场景可能包含信用行为和回报情况，例如定义如下 8 个客户标识：

- A：活跃客户，信誉良好，产品回报多
- B：活跃客户，信誉良好，产品回报一般
- C：最近的新客户，尚未建立信誉等级
- D：偶尔出现的客户，信誉良好
- E：偶尔出现的客户，信誉不好
- F：以前的优秀客户，最近不常见
- G：只逛不买的客户，几乎没有效益
- H：其他客户

至此可以考察客户时间序列数据，并将某个客户关联到报表期间的最近分类中。例如，某个客户在最近 10 个考察期间的情况可以表示为：CCCDDAAABB。这一行为时间序列标记来自于固定周期度量过程，观察值是文本类型的，不能计算或求平均值，但是它们可以被查询。如可以发现在以前的第 5 个、第 4 个或第 3 个周期中获得 A，且在第 2 个或第 1 个周期中获得 B 的所有客户。通过这样的进展分析还可以发现那些可能失去的有价值的客户，进而用于提高产品回报率。

行为标记可能不会被当成普通事实存储，因为它虽然由事实表的度量所定义，但其本身不是度量值。行为标记的主要作用在于为前面描述的例子制定复杂的查询模式。推荐的处理行为标记的方法是为客户维度建立分段属性的时间序列。这样 BI 接口比较简单，因为列都在同一个表中，性能也较好，因为可以对它们建立时间戳索引。除了为每个行为标记时间周期建立不同的列，建立单一的包含多个连续行为标记的连接字符串，也是较好的一种方法，例如，CCCDDAAABB。该列支持通配符模糊搜索模式，"D 后紧跟着 B"可以简单实现为"where flag like '%DB%'"。

2. 销售订单分段维度

下面以销售订单为例，说明分段维度的实现技术。分段维度包含连续的分段度量值。年度

销售订单分段维度可能包含有叫做"低""中""高"的三个档次,各档定义分别为消费额在0.01 到 3000、3000.01 到 6000.00、6000.01 到 99999999.99 区间。如果一个客户的年度销售订单金额累计为 1000,则被归为"低"档。分段维度可以存储多个分段集合,可能有一个用于促销分析的分段集合,另一个用于市场细分,可能还有一个用于销售区域计划。分段一般由用户定义,而且很少能从源事务数据直接获得。

（1）年度销售订单多维模型

为了实现年度订单分段维度,我们需要两个新的多维模型,如图 16-7 所示。第一个星型模式由 annual_sales_order_fact 事实表、customer_dim 维度表构成。年度销售额事实表存储客户一年的消费总额,数据从现有的销售订单事实表汇总而来。第二个星型模式由 annual_customer_segment_fact 事实表、annual_order_segement_dim 维度表、customer_dim 维度表构成。客户年度分段事实表中没有度量,只有来自两个相关维度表的代理键,因此它是一个无事实的事实表,存储的数据实际上就是前面所说的行为标记时间序列。17.3 节将详细讨论无事实事实表技术。年度订单分段维度表用于存储分段的定义,在本例中,它只与年度分段事实表有关系。

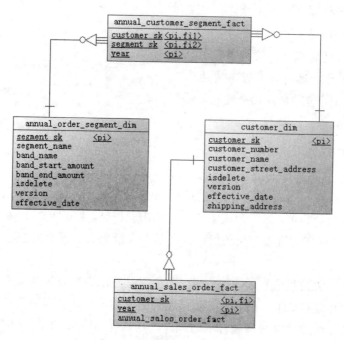

图 16-7　年度销售额分段维度

如果多个分段的属性相同,可以将它们存储到单一维度表中,因为分段通常只有很小的基数。本例中 annual_order_segment_dim 表存储了"project"和"grid"两种分段集合,它们都是按照客户的年度销售订单金额将其分类。分段维度按消费金额的定义如表 16-3 所示,project 分六段,grid 分三段。

表 16-3 客户年度消费分段维度定义

分段类别	分段名称	开始值	结束值
Project	Bottom	0.01	2500.00
Project	Low	2500.01	3000.00
Project	mid-low	3000.01	4000.00
Project	Mid	4000.00	5500.00
Project	mid-high	5500.01	6500.00
Project	Top	6500.01	99999999.99
Grid	Low	0.01	3000.00
Grid	Mid	3000.01	6000.00
Grid	High	6000.01	99999999.99

每一分段有一个开始值和一个结束值。分段的粒度就是本段和下段之间的间隙。粒度必须是度量的最小可能值，在销售订单示例中，金额的最小值是 0.01。最后一个分段的结束值是销售订单金额可能的最大值。下面的语句用于建立分段维度。新建了三个表，分别是分段维度表、年度销售事实表和年度客户消费分段事实表，并向分段维度表插入 9 条分段定义数据。假设分段维度表需要 SCD 处理，于是该表有删除标志、版本号、生效日期等附加属性，并建立了该表的当前视图和历史视图。

```
set search_path=tds;

-- 建立分段维度表
create table annual_order_segment_dim (
    segment_sk serial,
    segment_name varchar(30),
    band_name varchar(50),
    band_start_amount numeric(10,2),
    band_end_amount numeric(10,2),
    isdelete boolean default false,
    version int default 1,
    effective_date date default current_date );

-- 添加分段定义数据
insert into annual_order_segment_dim
(segment_name, band_name, band_start_amount, band_end_amount)
values ('project', 'bottom', 0.01, 2500.00),
('project', 'low', 2500.01, 3000.00),
('project', 'mid-low', 3000.01, 4000.00),
('project', 'mid', 4000.01, 5500.00),
('project', 'mid_high', 5500.01, 6500.00),
```

```sql
('project', 'top', 6500.01, 99999999.99),
('grid', 'low', 0.01, 3000),
('grid', 'med', 3000.01, 6000.00),
('grid', 'high', 6000.01, 99999999.99);

-- 建立分段维度当前视图
create or replace view v_annual_order_segment_dim_latest as
select segment_sk, segment_name, band_name, band_start_amount,
       band_end_amount, version, effective_date
  from (select distinct on (segment_name, band_name) segment_sk, segment_name,
band_name, band_start_amount, band_end_amount, isdelete,
             version, effective_date
          from annual_order_segment_dim
         order by segment_name, band_name, segment_sk desc) as latest
 where isdelete is false;

-- 建立分段维度历史视图
create or replace view v_annual_order_segment_dim_his as
select *, date(lead(effective_date,1,date '2200-01-01') over (partition by
segment_name, band_name order by effective_date)) expiry_date
  from annual_order_segment_dim;

-- 建立年度销售订单事实表
create table annual_sales_order_fact (
    customer_sk int,
    year int,
    annual_order_amount numeric(10,2) );

-- 建立年度销售订单分段事实表
create table annual_customer_segment_fact (
    segment_sk int,
    customer_sk int,
    year int );
```

（2）初始装载

执行下面的语句初始装载分段相关数据。

```sql
insert into annual_sales_order_fact
select customer_sk,
       year_month/100,
       sum(order_amount)
  from sales_order_fact
 where year_month/100 < 2017
 group by customer_sk, year_month/100;
```

```
insert into annual_customer_segment_fact
select d.segment_sk,
       a.customer_sk,
       a.year
  from annual_sales_order_fact a,
       v_annual_order_segment_dim_latest d
 where annual_order_amount >= band_start_amount
   and annual_order_amount <= band_end_amount;
```

因为装载过程不能导入当年的数据，所以使用 year < 2017 过滤条件。这里是按客户代理键 customer_sk 分组求和来判断分段，实际情况可能是以 customer_number 进行分组的，因为无论客户的 SCD 属性如何变化，一般还是认为是一个客户。将年度销售事实表里与分段维度表关联，把客户、分段维度的代理键插入年度客户消费分段事实表。注意，数据装载过程中并没有引用客户维度表，因为客户代理键可以直接从销售订单事实表得到。分段定义中，每个分段结束值与下一分段的开始值是连续的，并且分段之间不存在数据重叠，所以装载分段事实表时，订单金额判断条件两端都使用闭区间。

执行初始装载脚本后，使用下面的语句查询客户分段事实表，确认装载的数据是正确的。

```
select csk, y, amt, string_agg(sn||':'||bn,' / ')
  from (select a.customer_sk csk, a.year y, annual_order_amount amt,
               segment_name sn, band_name bn
          from annual_customer_segment_fact a,
               v_annual_order_segment_dim_latest b,
               annual_sales_order_fact c
         where a.segment_sk = b.segment_sk
           and a.customer_sk = c.customer_sk
           and a.year = c.year) t
 group by csk, y, amt
 order by y, amt desc;
```

（3）定期装载

定期装载与初始装载类似。年度销售事实表里的数据被导入分段事实表。每年调度执行下面的定期装载脚本，此脚本装载前一年的销售数据。

```
insert into annual_sales_order_fact
select customer_sk, year_month/100, sum(order_amount)
  from sales_order_fact
 where year_month/100 = extract(year from current_date) - 1
 group by customer_sk, year_month/100;

insert into annual_customer_segment_fact
select b.segment_sk, a.customer_sk, a.year
  from annual_sales_order_fact a,
       v_annual_order_segment_dim_latest b
 where a.year = extract(year from current_date) - 1
   and annual_order_amount >= band_start_amount
   and annual_order_amount <= band_end_amount;
```

16.9 小结

修改数据仓库模式时，要注意空值的处理，必要时使用 coalesce 函数将 null 转换为常量。子维度通常有包含属性子集的子维度和包含行子集的子维度两种，常用视图实现。单个物理维度可以被事实表多次引用，每个引用连接逻辑上存在差异的角色扮演维度。视图和表别名是实现角色扮演维度的两种常用方法。处理层次维度时，经常使用 grouping、grouping sets 等函数或语句。除了业务主键外没有其他内容的维度表通常是退化维度。将业务主键作为一个属性加入到事实表中是处理退化维度的适当方式。杂项维度是一种包含的数据具有很少可能值的维度。有时与其为每个标志或属性定义不同的维度，不如建立单独的、将不同维度合并到一起的杂项维度表。如果几个相关维度的基数都很小，或者具有多个公共属性时，可以考虑将它们进行维度合并。分段维度的定义中包含连续的分段度量值，通常用作客户维度的行为标记时间序列，分析客户行为。

第 17 章 事实表技术

发生在业务系统中的操作型事务,其所产生的可度量数值,存储在事实表中,从最细节粒度级别看,事实表和操作型事务表的数据有一一对应的关系。因此,数据仓库中事实表的设计应该依赖于业务系统,而不受可能产生的最终报表影响。除数字类型的度量外,事实表总是包含所引用维度表的外键,也能包含可选的退化维度键或时间戳。数据分析的实质就是基于事实表开展计算和聚合操作。

事实表中的数字度量值可划分为可加、半可加、不可加三类。可加性度量可以按照与事实表关联的任意维度汇总,就是说按任何维度汇总得到的度量和是相同的,事实表中的大部分度量属于此类。半可加度量可以对某些维度汇总,但不能对所有维度汇总。余额是常见的半可加度量,除时间维度外,它们可以跨所有维度进行加法操作。另外还有些度量是完全不可加的,例如比例。对非可加度量,较好的处理方法是尽可能存储构成非可加度量的可加分量,如构成比例的分子和分母,并将这些分量汇总到最终的结果集合中,而对不可加度量的计算通常发生在 BI 层或 OLAP 层。

事实表中可以存在空值度量。所有聚合函数,如 sum、count、min、max、avg 等均可针对空值度量计算,其中 sum、count(字段名)、min、max、avg 会忽略空值,而 count(1)或 count(*)在计数时会将空值包含在内。然而,事实表中的外键不能存在空值,否则会导致违反参照完整性的情况发生。关联的维度表必须用默认代理键而不是空值表示未知的条件。

很多情况下数据仓库需要装载如下三种不同类型的事实表。

- 事务事实表:以每个事务或事件为单位,例如一个销售订单记录、一笔转账记录等,作为事实表里的一行数据。这类事实表可能包含精确的时间戳和退化维度键,其度量值必须与事务粒度保持一致。销售订单数据仓库中的 sales_order_fact 表就是事务事实表。
- 周期快照事实表:这种事实表里并不保存全部数据,只保存固定时间间隔的数据,例如每天或每月的销售额,或每月的账户余额等。
- 累积快照事实表:累积快照用于跟踪事实表的变化。例如,数据仓库可能需要累积或存储销售订单,从下订单的时间开始,到订单中的商品被打包、运输和到达的各阶段的时间点数据来跟踪订单生命周期的进展情况。当这个过程进行时,随着以上各种时间的出现,事实表里的记录也要不断更新。

17.1 周期快照

1. 周期快照简介

周期快照事实表中的每行汇总了发生在某一标准周期，如一天、一周或一月的多个度量。其粒度是周期性的时间段，而不是单个事务。周期快照事实表通常包含许多数据的总计，因为任何与事实表时间范围一致的记录都会被包含在内。在这些事实表中，外键的密度是均匀的，因为即使周期内没有活动发生，通常也会在事实表中为每个维度插入包含 0 或空值的行。

周期快照在一个给定的时间对事实表进行一段时期的总计。有些数据仓库用户，尤其是业务管理者或者运营部门，经常要看某个特定时间点的汇总数据。下面在示例数据仓库中创建一个月销售订单周期快照，用于按产品统计每个月总的销售订单金额和产品销售数量。

2. 建立周期快照表

假设需求是要按产品统计每个月的销售金额和销售数量。单从功能上看，此数据能够从事务事实表中直接查询得到。例如，要取得 2017 年 5 月的销售数据，可使用下面的查询：

```sql
select b.month_sk, a.product_sk, sum(order_amount), sum(order_quantity)
  from sales_order_fact a,
       month_dim b,
       v_order_date_dim d
 where a.order_date_sk = d.order_date_sk
   and b.month = d.month
   and b.year = d.year
   and b.month = 5
   and b.year = 2017
group by b.month_sk, a.product_sk ;
```

只要将年、月参数传递给这条查询语句，就可以获得任何年月的统计数据。但即便是在如此简单的场景下，我们仍然需要建立独立的周期快照事实表。事务事实表的数据量总会很大，如果每当需要月销售统计数据时，都从最细粒度的事实表查询，那么性能将会差到不堪忍受的程度。再者，月统计数据往往只是下一步数据分析的输入信息，有时把更复杂的逻辑放到一个单一的查询语句中效率会更差。因此，好的做法是将事务型事实表作为一个基石事实数据，以此为基础，向上逐层建立需要的快照事实表。

新的周期快照事实表中有两个度量值，month_order_amount 和 month_order_quantity。这两个值是不能加到 sales_order_fact 表中的，因为 sales_order_fact 表和新的度量值有不同的时间属性，也即数据的粒度不同。sales_order_fact 表包含的是单一事务记录。新的度量值是每月的汇总数据，它们是可加的。使用下面的语句建立 month_end_sales_order_fact 表。

```
set search_path=tds;
```

```sql
create table month_end_sales_order_fact (
    year_month int,
    product_sk bigint,
    month_order_amount numeric(10,2),
    month_order_quantity bigint )
partition by range (year_month)
( partition p201601 start (201601) inclusive ,
  partition p201602 start (201602) inclusive ,
  partition p201603 start (201603) inclusive ,
  partition p201604 start (201604) inclusive ,
  partition p201605 start (201605) inclusive ,
  partition p201606 start (201606) inclusive ,
  partition p201607 start (201607) inclusive ,
  partition p201608 start (201608) inclusive ,
  partition p201609 start (201609) inclusive ,
  partition p201610 start (201610) inclusive ,
  partition p201611 start (201611) inclusive ,
  partition p201612 start (201612) inclusive ,
  partition p201701 start (201701) inclusive ,
  partition p201702 start (201702) inclusive ,
  partition p201703 start (201703) inclusive ,
  partition p201704 start (201704) inclusive ,
  partition p201705 start (201705) inclusive ,
  partition p201706 start (201706) inclusive ,
  partition p201707 start (201707) inclusive ,
  partition p201708 start (201708) inclusive ,
  partition p201709 start (201709) inclusive ,
  partition p201710 start (201710) inclusive ,
  partition p201711 start (201711) inclusive ,
  partition p201712 start (201712) inclusive
              end (201801) exclusive );

comment on table month_end_sales_order_fact is '月销售周期快照表';
comment on column month_end_sales_order_fact.year_month is '年月';
comment on column month_end_sales_order_fact.product_sk is '产品代理键';
comment on column month_end_sales_order_fact.month_order_amount is '销售金额';
comment on column month_end_sales_order_fact.month_order_quantity is '销售数量';
```

和销售订单事实表一样，月销售周期快照表也以年月做分区。这样做主要有两点好处：

- 按年月查询周期快照表时，可以利用分区消除提高性能。
- 便于实现重复执行定期装载过程。HAWQ 没有 DELETE 语句，但是可以单独清空分区对应的子表。

3. 装载周期快照表

建立了 month_end_sales_order_fact 表后，现在需要向表中装载数据。实际装载时，月销售周期快照事实表的数据源是已有的销售订单事务事实表，而不需要关联产品维度表。之所以可以这样做，是因为总是先处理事务事实表，再处理周期快照事实表，并且事务事实表中的产品代理键就是当时有效的产品描述。这样做还有一个好处是，不必要非在1号装载上月的数据，这点在后面修改工作流时详细说明。执行下面的语句初始装载月销售数据。

```
insert into month_end_sales_order_fact
select year_month,product_sk,sum(order_amount),sum(order_quantity)
  from sales_order_fact
 group by year_month,product_sk;
```

fn_month_sum 函数用于定期装载月销售订单周期快照事实表，函数定义如下。

```
create or replace function tds.fn_month_sum(p_year_month int)
returns void as $$
declare
    sqlstring varchar(1000);
begin
    -- 幂等操作，先删除上月数据
sqlstring := 'truncate table month_end_sales_order_fact_1_prt_p'
|| cast(p_year_month as varchar);
    execute sqlstring;

    -- 插入上月销售汇总数据
    insert into month_end_sales_order_fact
    select t1.year_month, t2.product_sk, coalesce(t2.month_order_amount,0),
        coalesce(t2.month_order_quantity,0)
      from (select year * 100 + month year_month
             from month_dim
            where year * 100 + month = p_year_month) t1
      left join (select year_month, product_sk,
sum(order_amount) month_order_amount,
sum(order_quantity) month_order_quantity
               from sales_order_fact
              where year_month = p_year_month
              group by year_month,product_sk) t2
        on t1.year_month = t2.year_month;

end; $$
language plpgsql;
```

执行以下语句装载上个月的销售汇总数据。该语句可重复执行，汇总数据不会重复累加。

```
select tds.fn_month_sum(cast(extract(year from current_date - interval '1 
month') * 100 + extract(month from current_date - interval '1 month') as int));
```

周期快照表的外键密度是均匀的，因此这里使用外连接关联月份维度和事务事实表。即使上个月没有任何销售记录，周期快照中仍然会有一行记录。在这种情况下，周期快照记录中只有年月，而产品代理键的值为空，度量为 0。查询销售订单事实表时可以利用分区消除提高性能。每个月给定的任何一天，在每天销售订单定期装载执行完后，执行 fn_month_sum 函数，装载上个月的销售订单汇总数据。为此需要修改 Oozie 的工作流定义。

4. 修改工作流

（1）修改 Oozie 工作流作业配置文件

需要在 15.3 节中创建的 workflow.xml 工作流定义文件中增加月底销售周期快照的数据装载部分，修改后的文件内容如下（只列出增加的部分）。

```xml
<?xml version="1.0" encoding="UTF-8"?>
<workflow-app xmlns="uri:oozie:workflow:0.4" name="RegularETL">
… start 节点 …
… fs 节点 …
… fork 节点 …
… 三个并行执行的 Sqoop 节点 …
… join 节点 …
… 执行定期装载的 ssh 节点 …

    <decision name="decision-node">
      <switch>
        <case to="month-sum">
            ${date eq '02'}
        </case>
        <default to="end"/>
      </switch>
    </decision>

    <action name="month-sum">
      <ssh xmlns="uri:oozie:ssh-action:0.1">
          <host>${focusNodeLogin}</host>
          <command>${myScript1}</command>
          <capture-output/>
      </ssh>
      <ok to="end"/>
      <error to="fail"/>
    </action>

… kill 节点 …
```

```
… end 节点 …
</workflow-app>
```

在该配置文件中增加了一个名为 decision-node 的 decision 控制节点，用来判断 date 参数的值。当 date 等于'02'时，转到 month-sum 动作节点，否则转到 end 节点结束工作流。month-sum 是一个 SSH 动作节点，执行 fn_month_sum 函数装载周期快照事实表，成功执行后转到 end 节点结束。很明显，本例中 decision 节点的作用就是控制在并且只在一个月当中的某一天执行周期快照表的数据装载，其他日期不做这步操作。之所以这里是'02'是为了方便测试。fn_month_sum 函数接收年月作为参数，因此不必要非得 1 号执行，任何一天都可以。这个工作流定义保证了每月汇总只有在每天汇总执行完后才执行，并且每月只执行一次。工作流的 DAG 如图 17-1 所示。

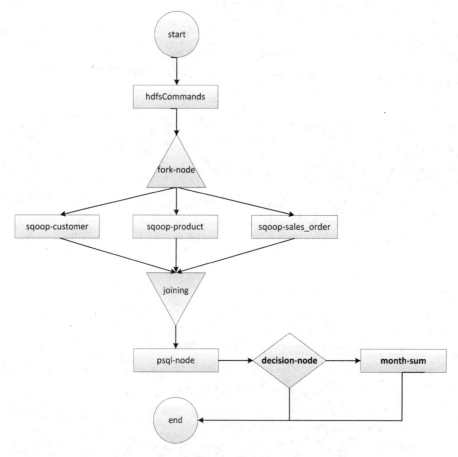

图 17-1 增加了周期快照装载的 Oozie 工作流

（2）部署工作流

```
hdfs dfs -put -f workflow.xml /user/oozie/
```

（3）在 Falcon process 的 ADVANCED OPTIONS 中增加属性

需要在调度作业配置中增加 myScript1 和 date 两个属性的定义，如图 17-2 所示。

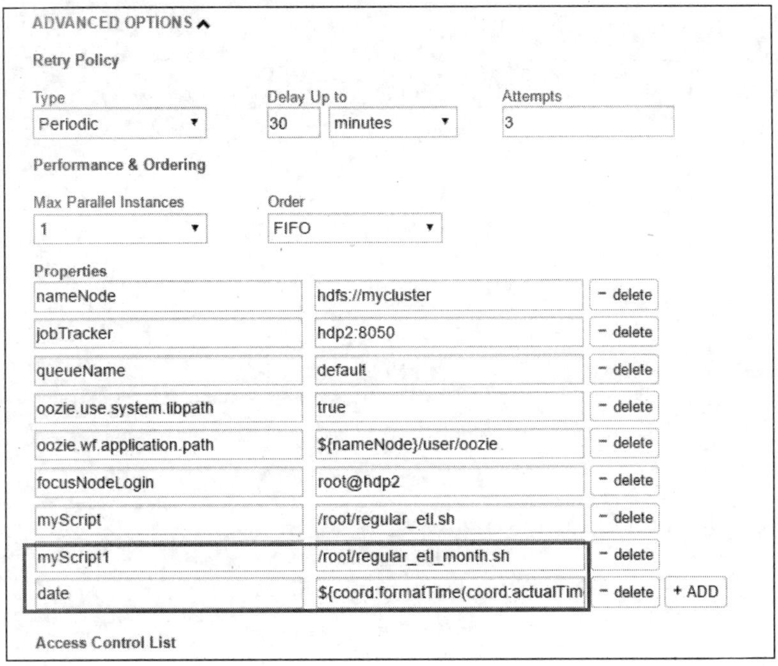

图 17-2 在 Falcon process 的 ADVANCED OPTIONS 中增加属性

myScript1 属性的值为/root/regular_etl_month.sh，是调用 psql 的 shell 脚本文件。date 属性的值为${coord:formatTime(coord:actualTime(), "dd")}，用 Oozie 的系统函数取得工作流执行时的月中日期。Falcon 调度执行工作流时，这些属性的值会作为实参传入 workflow.xml 工作流定义文件中。

（4）编写快照表数据装载脚本

/root/regular_etl_month.sh 文件的内容如下。

```
#!/bin/bash

# 使用 gpadmin 用户执行月周期快照装载函数
su - gpadmin -c 'export PGPASSWORD=123456;psql -U dwtest -d dw -h hdp3 -c "set search_path=tds;select fn_month_sum(cast(extract(year from current_date - interval '\''1 month'\'') * 100 + extract(month from current_date - interval '\''1 month'\'') as int))"'
```

该文件以 root 用户执行，需要注意 shell 中引号嵌套的用法。

5. 测试

首先清空上个月的周期快照数据。

```
truncate table month_end_sales_order_fact_1_prt_p201705;
```

然后在 Falcon Web UI 中执行 process。执行成功后查询 month_end_sales_order_fact 表，结

果如下。可以看到，已经生成了上个月的销售汇总周期快照数据。

```
 year_month | product_sk | month_order_amount | month_order_quantity 
------------+------------+--------------------+----------------------
     201603 |          1 |           89488.00 |                     
     201603 |          2 |          101670.00 |                     
     201604 |          2 |           75204.00 |                     
     201604 |          1 |           79238.00 |                     
     201605 |          2 |           65415.00 |                     
     201605 |          1 |           44757.00 |                     
     201606 |          2 |           41790.00 |                     
     201606 |          1 |           74628.00 |                     
     201705 |          5 |           57833.00 |                  391
     201705 |          1 |           85107.00 |                  582
     201705 |          2 |           52083.00 |                  339
     201705 |          4 |           49666.00 |                  393
(12 rows)
```

17.2 累积快照

1. 累积快照简介

累积快照事实表用于定义业务过程开始、结束以及期间的可区分的里程碑事件。通常在此类事实表中针对过程中的关键步骤都包含日期外键，并包含每个步骤的度量，这些度量的产生一般都会滞后于数据行的创建时间。累积快照事实表中的一行，对应某一具体业务的多个状态。例如，当订单产生时会插入一行。当该订单的状态改变时，累积事实表行被访问并修改。这种对累积快照事实表行的一致性修改在三种类型的事实表（事务、周期快照、累积快照）中具有独特性，对于前面两类事实表只追加数据，不会对已经存在的行进行更新操作。除了日期外键与每个关键过程步骤关联外，累积快照事实表中还可以包含其他维度和可选退化维度的外键。

累积快照事实表在库存、采购、销售、电商等业务领域都有广泛应用。比如在电商订单里面，下单的时候只有下单时间，但是在支付的时候，又会有支付时间，同理，还有发货时间，完成时间等。下面以销售订单数据仓库为例，讨论累积快照事实表的实现。

假设希望跟踪以下五个销售订单的里程碑：下订单、分配库房、打包、配送和收货，分别用状态 N、A、P、S、R 表示。这五个里程碑的日期及其各自的数量来自源数据库的销售订单表。一个订单完整的生命周期由五行数据描述：下订单时生成一条销售订单记录；订单商品被分配到相应库房时，新增一条记录，存储分配时间和分配数量；产品打包时新增一条记录，存储打包时间和数量；类似的，订单配送和订单客户收货时也都分别新增一条记录，保存各自的时间戳与数量。为简化示例，不考虑每种状态出现多条记录的情况（例如，一条订单中的产品可能是在不同时间点分多次出库），并且假设这五个里程碑是以严格的时间顺序正向进行的。

对订单的每种状态新增记录只是处理这种场景的多种设计方案之一。如果里程碑的定义良好并且不会轻易改变,也可以考虑在源订单事务表中新增每种状态对应的数据列,例如,新增 8 列,保存每个状态的时间戳和数量。新增列的好处是仍然能够保证订单号的唯一性,并保持相对较少的记录数。但是,这种方案还需要额外增加一个 last_modified 字段记录订单的最后修改时间,用于 Sqoop 增量数据抽取。因为每条订单在状态变更时都会被更新,所以订单号字段已经不能作为变化数据捕获的比较依据。

2. 建立累积快照表

(1) 修改源库表结构

执行下面的语句将源数据库中销售订单事务表结构做相应改变,以处理 5 种不同的状态。

```
use source;
-- 修改销售订单事务表
alter table sales_order
    change order_date status_date datetime,
    add order_status varchar(1) after status_date,
    change order_quantity quantity int;

-- 删除 sales_order 表的主键
alter table sales_order change order_number order_number int not null;
alter table sales_order drop primary key;

-- 建立新的主键
alter table sales_order add id int unsigned not null auto_increment primary key comment '主键' first;
```

说明:

- 将 order_date 字段改名为 status_date,因为日期不再单纯指订单日期,而是指变为某种状态日期。
- 将 order_quantity 字段改名为 quantity,因为数量变为某种状态对应的数量。
- 在 status_date 字段后增加 order_status 字段,存储 N、A、P、S、R 等订单状态之一。它描述了 status_date 列对应的状态值,例如,如果一条记录的状态为 N,则 status_date 列是下订单的日期。如果状态是 R,status_date 列是收货日期。
- 每种状态都会有一条订单记录,这些记录具有相同的订单号,因此订单号不能再作为事务表的主键,需要删除 order_number 字段上的自增属性与主键约束。
- 添加自增 id 字段作为销售订单表的主键,它是表中的第一个字段。

(2) 重建销售订单外部表

执行下面的语句重建销售订单外部表,使其与源表结构一致。

```
set search_path=ext;
drop external table sales_order;
```

```sql
create external table sales_order
( id bigint,
  order_number int,
  customer_number int,
  product_code int,
  verification_ind char(1),
  credit_check_flag char(1),
  new_customer_ind char(1),
  web_order_flag char(1),
  status_date timestamp,
  order_status char(1),
  request_delivery_date timestamp,
  entry_date timestamp,
  order_amount decimal(10 , 2 ),
  quantity int )
location ('pxf://mycluster/data/ext/sales_order?profile=hdfstextsimple')
format 'text' (delimiter=e',', null='null');

comment on table sales_order is '销售订单外部表';
```

（3）修改销售订单原始数据存储表

```sql
set search_path=rds;
alter table sales_order rename order_date to status_date;
alter table sales_order rename order_quantity to quantity;
alter table sales_order add column order_status char(1) default null;

comment on column sales_order.status_date is '状态日期';
comment on column sales_order.quantity is '数量';
comment on column sales_order.order_status is '订单状态';
```

说明：

- 将销售订单事实表中 order_date 和 order_quantity 字段的名称修改为与源表一致。
- 增加订单状态字段。
- rds.sales_order 并没有增加 id 列，原因有两个：一是该列只作为增量检查列，不用在原始数据表中存储；二是不需要再重新导入已有数据。

（4）修改销售订单事实表

```sql
set search_path=tds;
alter table sales_order_fact rename order_date_sk to status_date_sk;
alter table sales_order_fact rename order_quantity to quantity;
alter table sales_order_fact add column order_status char(1) default null;

comment on column sales_order_fact.status_date_sk is '状态日期外键';
```

```sql
comment on column sales_order_fact.quantity is '数量';
comment on column sales_order_fact.order_status is '订单状态';

create view v_sales_order_fact as
select order_number, customer_sk, product_sk, year_month, order_amount,
       request_delivery_date_sk, sales_order_attribute_sk,
       customer_zip_code_sk, shipping_zip_code_sk,
       max(case order_status when 'N' then status_date_sk else null end) nd,
       max(case order_status when 'N' then quantity else null end) nq,
       max(case order_status when 'A' then status_date_sk else null end) ad,
       max(case order_status when 'A' then quantity else null end) aq,
       max(case order_status when 'P' then status_date_sk else null end) pd,
       max(case order_status when 'P' then quantity else null end) pq,
       max(case order_status when 'S' then status_date_sk else null end) sd,
       max(case order_status when 'S' then quantity else null end) sq,
       max(case order_status when 'R' then status_date_sk else null end) rd,
       max(case order_status when 'R' then quantity else null end) rq
  from sales_order_fact
 group by order_number, customer_sk, product_sk, year_month, order_amount,
          request_delivery_date_sk, sales_order_attribute_sk,
          customer_zip_code_sk, shipping_zip_code_sk;

-- 建立四个日期维度视图
create view v_allocate_date_dim
(allocate_date_sk, allocate_date, month, month_name, quarter, year)
as select * from date_dim ;

create view v_packing_date_dim
(packing_date_sk, packing_date, month, month_name, quarter, year)
as select * from date_dim ;

create view v_ship_date_dim
(ship_date_sk, ship_date, month, month_name, quarter, year)
as select * from date_dim ;

create view v_receive_date_dim
(receive_date_sk, receive_date, month, month_name, quarter, year)
as select * from date_dim ;
```

说明：

- 对销售订单事实表结构的修改与 rds.sales_order 类似。
- 新建了一个视图 v_sales_order_fact，将 5 个状态及其数量做行转列。
- 建立 4 个日期角色扮演维度视图，用来获取相应状态的日期代理键。

3. 重建增量抽取 Sqoop 作业

使用下面的脚本重建 Sqoop 作业,因为源表会有多个相同的 order_number,所以不能再用它作为检查字段,将检查字段改为 id。

```
last_value=`sqoop job --show myjob_incremental_import | grep incremental.last.value | awk '{print $3}'`
  sqoop job --delete myjob_incremental_import
  sqoop job --create myjob_incremental_import -- import --connect "jdbc:mysql://172.16.1.127:3306/source?usessl=false&user=dwtest&password=123456" --table sales_order --target-dir /data/ext/sales_order --compress --where "entry_date < current_date()" --incremental append --check-column id --last-value $last_value
```

4. 修改定期数据装载函数

需要依据数据库模式修改定期装载函数,修改后的函数如下(只列出修改的部分)。

```
create or replace function fn_regular_load ()
returns void as $$
declare
    -- 设置 scd 的生效时间
    ...
begin
    -- 分析外部表
    ...
    -- 将外部表数据装载到原始数据表
    ...

insert into rds.sales_order
    select order_number, customer_number, product_code, status_date, entry_date,
           order_amount, quantity, request_delivery_date, verification_ind,
           credit_check_flag, new_customer_ind, web_order_flag, order_status
      from ext.sales_order;

    -- 分析 rds 模式的表
    ...
    -- 设置 cdc 的上限时间
    ...
    -- 装载客户维度
    ...
    -- 装载产品维度
    ...
    -- 装载销售订单事实表
```

```sql
    insert into sales_order_fact
    select a.order_number, customer_sk, product_sk, e.date_sk,
           e.year * 100 + e.month, order_amount, quantity, f.date_sk,
           g.sales_order_attribute_sk, h.customer_zip_code_sk,
           i.shipping_zip_code_sk, a.order_status
      from rds.sales_order a,
           v_customer_dim_his c,
           v_product_dim_his d,
           date_dim e,
           date_dim f,
           sales_order_attribute_dim g,
           v_customer_zip_code_dim h,
           v_shipping_zip_code_dim i,
           rds.customer j,
           rds.cdc_time k
     where a.customer_number = c.customer_number
       and a.status_date >= c.effective_date
       and a.status_date < c.expiry_date
       and a.product_code = d.product_code
       and a.status_date >= d.effective_date
       and a.status_date < d.expiry_date
       and date(a.status_date) = e.date
       and date(a.request_delivery_date) = f.date
       and a.verification_ind = g.verification_ind
       and a.credit_check_flag = g.credit_check_flag
       and a.new_customer_ind = g.new_customer_ind
       and a.web_order_flag = g.web_order_flag
       and a.customer_number = j.customer_number
       and j.customer_zip_code = h.customer_zip_code
       and j.shipping_zip_code = i.shipping_zip_code
       and a.entry_date >= k.last_load and a.entry_date < k.current_load;

    -- 重载 PA 客户维度
    ...
    -- 分析 tds 模式的表
    ...
    -- 更新时间戳表的 last_load 字段
    ...
end; $$
language plpgsql;
```

需要修改定期数据装载中的相应列名。在装载事务事实表时，只用 entry_date >= last_load and entry_date < current_load 条件就可以过滤出所有新录入的、包括五种状态的订单，因为每

种状态的订单都有自己对应的录入时间。HAWQ 不能更新已有的表数据，因此在装载时只新增数据，然后通过视图转化为固定状态列的格式。注意，本示例中的累积周期快照视图仍然是以订单号字段作为逻辑上的主键。

5. 测试

在源数据库的销售订单事务表中新增两个销售订单记录。

```
use source;

set @order_date := from_unixtime(unix_timestamp('2017-06-02 00:00:01') + rand() * (unix_timestamp('2017-06-02 12:00:00') - unix_timestamp('2017-06-02 00:00:01')));
set @request_delivery_date := from_unixtime(unix_timestamp(date_add(current_date, interval 5 day)) + rand() * 86400);
set @amount := floor(1000 + rand() * 9000);
set @quantity := floor(10 + rand() * 90);

insert into source.sales_order values
   (null, 141, 1, 1, 'y', 'y', 'y', 'y', @order_date, 'N', @request_delivery_date, @order_date, @amount, @quantity);

set @order_date := from_unixtime(unix_timestamp('2017-06-02 12:00:00') + rand() * (unix_timestamp('2017-06-03 00:00:00') - unix_timestamp('2017-06-02 12:00:00')));
set @request_delivery_date := from_unixtime(unix_timestamp(date_add(current_date, interval 5 day)) + rand() * 86400);
set @amount := floor(1000 + rand() * 9000);
set @quantity := floor(10 + rand() * 90);

insert into source.sales_order values
   (null, 142, 2, 2, 'y', 'y', 'y', 'y', @order_date, 'N', @request_delivery_date, @order_date, @amount, @quantity);

commit;
```

设置时间窗口。

```
truncate table rds.cdc_time;
insert into rds.cdc_time select date '2017-06-02', date '2017-06-02';
```

执行定期装载脚本。

```
~/regular_etl.sh
```

查询 v_sales_order_fact 里的两个销售订单，确认定期装载成功。

```
select a.order_number, c.order_date, d.allocate_date, e.packing_date,
       f.ship_date, g.receive_date
  from v_sales_order_fact a
  left join v_order_date_dim c on a.nd = c.order_date_sk
  left join v_allocate_date_dim d on a.ad = d.allocate_date_sk
  left join v_packing_date_dim e on a.pd = e.packing_date_sk
  left join v_ship_date_dim f on a.sd = f.ship_date_sk
  left join v_receive_date_dim g on a.rd = g.receive_date_sk
 where a.order_number > 140
 order by order_number;
```

查询结果如下，只有 order_date 列有值，其他日期都是空，因为这两个订单是新增的，并且还没有分配库房、打包、配送或收货。

```
 order_number | order_date |allocate_date|packing_date |ship_date| receive_date
--------------+------------+-------------+-------------+---------+-------------
          141 | 2017-06-02 |             |             |         |
          142 | 2017-06-02 |             |             |         |
(2 rows)
```

添加销售订单作为这两个订单的分配库房和/或打包的里程碑。

```
use source;

set @order_date := from_unixtime(unix_timestamp('2017-06-03 00:00:00') + rand() 
* (unix_timestamp('2017-06-03 12:00:00') - unix_timestamp('2017-06-03 
00:00:00')));
  insert into sales_order
  select null, order_number, customer_number, product_code, verification_ind,
         credit_check_flag, new_customer_ind, web_order_flag, @order_date, 'A',
         request_delivery_date, @order_date, order_amount, quantity
    from sales_order
   where order_number = 141;

set @order_date := from_unixtime(unix_timestamp('2017-06-03 12:00:00') + rand() 
* (unix_timestamp('2017-06-04 00:00:00') - unix_timestamp('2017-06-03 
12:00:00')));
  insert into sales_order
  select null, order_number, customer_number, product_code, verification_ind,
         credit_check_flag, new_customer_ind, web_order_flag, @order_date, 'P',
         request_delivery_date, @order_date, order_amount, quantity
    from sales_order
   where id = 143;
```

```
    set @order_date := from_unixtime(unix_timestamp('2017-06-03 12:00:00') + rand()
* (unix_timestamp('2017-06-04 00:00:00') - unix_timestamp('2017-06-03
12:00:00')));
    insert into sales_order
    select null, order_number, customer_number, product_code, verification_ind,
        credit_check_flag, new_customer_ind, web_order_flag, @order_date, 'A',
        request_delivery_date, @order_date, order_amount, quantity
    from sales_order
    where order_number = 142;

    commit;
```

设置时间窗口。

```
    truncate table rds.cdc_time;
    insert into rds.cdc_time select date '2017-06-03', date '2017-06-03';
```

执行定期装载脚本。

```
    ~/regular_etl.sh
```

查询 v_sales_order_fact 表里的两个销售订单，确认定期装载成功。查询结果如下。第一个订单具有了 allocate_date 和 packing_date，第二个只具有 allocate_date。

```
 order_number | order_date |allocate_date|packing_date| ship_date  | receive_date
--------------+------------+-------------+------------+------------+-------------
         141  | 2017-06-02 | 2017-06-03  | 2017-06-03 |            |
         142  | 2017-06-02 | 2017-06-03  |            |            |
(2 rows)
```

添加销售订单作为这两个订单后面的里程碑：打包、配送和/或收货。注意 4 个日期可能相同。

```
    use source;

    set @order_date := from_unixtime(unix_timestamp('2017-06-04 00:00:00') + rand()
* (unix_timestamp('2017-06-04 12:00:00') - unix_timestamp('2017-06-04
00:00:00')));
    insert into sales_order
    select null, order_number, customer_number, product_code, verification_ind,
        credit_check_flag, new_customer_ind, web_order_flag, @order_date, 'S',
        request_delivery_date, @order_date, order_amount, quantity
    from sales_order
    where order_number = 141
    order by id desc
    limit 1;
```

```
    set @order_date := from_unixtime(unix_timestamp('2017-06-04 12:00:00') + rand()
* (unix_timestamp('2017-06-05 00:00:00') - unix_timestamp('2017-06-04
12:00:00')));
    insert into sales_order
    select null, order_number, customer_number, product_code, verification_ind,
           credit_check_flag, new_customer_ind, web_order_flag, @order_date, 'R',
           request_delivery_date, @order_date, order_amount, quantity
      from sales_order
     where order_number = 141
     order by id desc
     limit 1;

    set @order_date := from_unixtime(unix_timestamp('2017-06-04 12:00:00') + rand()
* (unix_timestamp('2017-06-05 00:00:00') - unix_timestamp('2017-06-04
12:00:00')));
    insert into sales_order
    select null, order_number, customer_number, product_code, verification_ind,
           credit_check_flag, new_customer_ind, web_order_flag, @order_date, 'P',
           request_delivery_date, @order_date, order_amount, quantity
      from sales_order
     where order_number = 142
     order by id desc
     limit 1;

    commit;
```

设置时间窗口。

```
    truncate table rds.cdc_time;
    insert into rds.cdc_time select date '2017-06-04', date '2017-06-04';
```

执行定期装载脚本。

```
~/regular_etl.sh
```

查询 v_sales_order_fact 表里的两个销售订单，确认定期装载成功。查询结果如下。第一个订单号为 141 的订单，具有了全部日期，这意味着订单已完成（客户已经收货）。第二个订单已经打包，但是还没有配送。

```
 order_number | order_date | allocate_date | packing_date | ship_date  | receive_date
--------------+------------+---------------+--------------+------------+-------------
          141 | 2017-06-02 | 2017-06-03    | 2017-06-03   | 2017-06-04 | 2017-06-04
          142 | 2017-06-02 | 2017-06-03    | 2017-06-04   |            |
(2 rows)
```

6. 修改周期快照表装载函数

累积快照将原来的一个数量 order_quantity 变为了每种状态对应一个数量，因此需要修改周期快照表装载函数 fn_month_sum。该函数汇总月底订单金额和数量，我们必须重新定义数量。假设需要统计的是新增订单中的数量，修改后的函数如下。

```
create or replace function tds.fn_month_sum(p_year_month int)
returns void as $$
declare
    sqlstring varchar(1000);
begin
    -- 幂等操作，先删除上月数据
sqlstring := 'truncate table month_end_sales_order_fact_1_prt_p'
|| cast(p_year_month as varchar);
    execute sqlstring;

    -- 插入上月销售汇总数据
    insert into month_end_sales_order_fact
    select t1.year_month, t2.product_sk, coalesce(t2.month_order_amount,0),
         coalesce(t2.month_order_quantity,0)
      from (select p_year_month year_month) t1
      left join (select year_month, product_sk,
sum(order_amount) month_order_amount,
sum(quantity) month_order_quantity
                from sales_order_fact
               where year_month = p_year_month
and coalesce(order_status,'N') = 'N'
               group by year_month,product_sk) t2
         on t1.year_month = t2.year_month;
end; $$
language plpgsql;
```

17.3 无事实的事实表

1. 无事实的事实表简介

在多维数据仓库建模中，有一种事实表叫做"无事实的事实表"（也称无事实事实表）。普通事实表中，通常会保存若干维度外键和多个数字型度量，度量是事实表的关键所在。然而在无事实的事实表中没有这些度量值，只有多个维度外键。表面上看，无事实事实表是没有意义的，因为作为事实表，毕竟最重要的就是度量。但在数据仓库中，这类事实表有其特殊用途。无事实的事实表通常用来跟踪某种事件或者说明某些活动的范围。

无事实的事实表可以用来跟踪事件的发生。例如，在给定的某一天中发生的学生参加课程的事件，可能没有可记录的数字化事实，但该事实行带有一个包含日期、学生、教师、地点、课程等定义良好的外键。利用无事实的事实表可以按各种维度计数上课这个事件。

无事实的事实表还可以用来说明某些活动的范围，常被用于回答"什么未发生"这样的问题。例如：促销范围事实表。通常销售事实表可以回答如促销商品的销售情况，可是无法回答的一个重要问题是：处于促销状态但尚未销售的产品包括哪些？销售事实表所记录的仅仅是实际卖出的产品。事实表行中不包括由于没有销售行为而销售数量为零的行，因为如果将包含零值的产品都加到事实表中，那么事实表将变得非常巨大。这时，通过建立促销范围事实表，将商场需要促销的商品单独建立事实表保存，然后通过这个促销范围事实表和销售事实表即可得出哪些促销商品没有销售出去。

为确定当前促销的产品中哪些尚未卖出，需要两步过程：首先，查询促销无事实的事实表，确定给定时间内促销的产品。然后从销售事实表中确定哪些产品已经卖出去了。答案就是上述两个列表的差集。这样的促销范围事实表只是用来说明促销活动的范围，其中没有任何事实度量。建立一个单独的促销商品维度表能否可以达到同样的效果呢？促销无事实的事实表包含多个维度的主键，可以是日期、产品、商店、促销等，将这些键作为促销商品的属性是不合适的，因为每个维度都有自己的属性集合。

促销无事实事实表看起来与销售事实表相似。然而，它们的粒度存在显著差别。假设促销是以一周为持续期，在促销范围事实表中，将为每周每个商店中促销的产品加载一行，无论产品是否卖出。该事实表能够确保看到被促销定义的键之间的关系，而与其他事件，如产品销售无关。

下面以销售订单数据仓库为例，说明如何处理源数据中没有度量的需求。建立一个无事实的事实表，用来统计每天发布的新产品数量。产品源数据不包含产品数量信息，如果系统需要得到历史某一天新增产品的数量，很显然不能简单地从数据仓库中得到。这时就要用到无事实的事实表技术。使用此技术可以通过持续跟踪产品发布事件来计算产品的数量。可以创建一个只有产品（计什么数）和日期（什么时候计数）维度代理键的事实表。之所以叫做无事实的事实表是因为表本身并没有数字型度量值。这里定义的新增产品是指在某一给定日期，源产品表中新插入的产品记录，不包括由于 SCD2 新增的产品版本记录。注意，单从这个简单需求来看，也可以通过查询产品维度表获取结果。这里只为演示无事实事实表的实现过程。

2. 建立新产品发布的无事实事实表

在 tds 模式中新建一个产品发布的无事实事实表 product_count_fact，该表中只包含两个字段，分别是引用日期维度表和产品维度表的外键，同时这两个字段也构成了无事实事实表的逻辑主键。图 17-3 显示了跟踪产品发布数量的表。

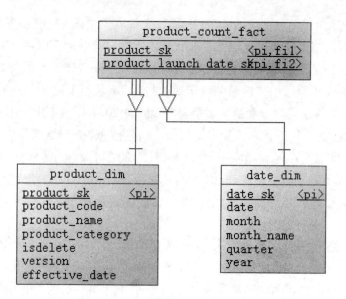

图 17-3 无事实的事实表

执行下面的语句，在数据仓库模式中创建产品发布日期视图及其无事实事实表。

```
set search_path=tds;

create view product_launch_date_dim
(product_launch_date_sk, product_launch_date, month_name, month, quarter, year) as
select distinct date_sk, date, month_name, month, quarter, year
   from product_dim a, date_dim b
  where a.effective_date = b.date
    and a.version = 1;

create table product_count_fact (
    product_sk int,
product_launch_date_sk int);
```

说明：

- 与之前创建的很多日期角色扮演维度不同，产品发布日期视图只获取产品生效日期，而不是日期维度里的所有记录。因此在定义视图的查询语句中关联了产品维度和日期维度两个表。product_launch_date_dim 维度是日期维度表的子集。
- 从字段定义上看，产品维度表中的生效日期明显就是新产品的发布日期。
- version = 1 过滤掉由于 SCD2 新增的产品版本记录。

3. 初始装载无事实事实表

下面的语句从产品维度表向无事实事实表装载已有的产品发布信息。insert 语句添加所有产品的第一个版本，即产品的首次发布日期。

```
insert into product_count_fact
select a.product_sk product_sk, b.date_sk date_sk
  from product_dim a,date_dim b
 where a.effective_date = b.date and a.version = 1;
```

使用下面的语句查询 product_count_fact 表以确认正确执行了初始装载。

```
select product_sk,product_launch_date_sk
  from tds.product_count_fact
 order by product_sk;
```

查询结果如下：

```
 product_sk | product_launch_date_sk
------------+------------------------
          1 |                   5905
          2 |                   5905
          3 |                   5905
          5 |                   6351
(4 rows)
```

4. 修改定期数据装载函数

修改了数据仓库模式后，还需要针对性的修改定期装载函数，在处理产品维度表后增加了装载 product_count_fact 表的语句。下面显示了修改后的定期装载函数（只列出修改的部分）。

```
create or replace function fn_regular_load ()
returns void as $$
declare
    -- 设置 scd 的生效时间
    ...
begin
    -- 分析外部表
    ...
    -- 将外部表数据装载到原始数据表
    ...
    -- 分析 rds 模式的表
    ...
    -- 设置 cdc 的上限时间
    ...
    -- 装载客户维度
    ...
    -- 装载产品维度
...

    -- 装载新增产品数量无事实事实表
    insert into tds.product_count_fact
```

```
    select a.product_sk, b.date_sk
      from tds.product_dim a, tds.date_dim b
     where a.version = 1
       and a.effective_date = v_pre_date
       and a.effective_date = b.date;

    -- 装载销售订单事实表
    ...
    -- 重载PA客户维度
    ...
    -- 分析tds模式的表
    ...
    -- 更新时间戳表的last_load字段
    ...
end; $$
language plpgsql;
```

5. 测试

修改源数据库的 product 表数据，把产品编码为 1 的产品名称改为 'Regular Hard Disk Drive'，并新增一个产品 'High End Hard Disk Drive'（产品编码为 5）。执行下面的语句完成此修改。

```
use source;

update product set product_name = 'Regular Hard Disk Drive' where product_code=1;
insert into product values (5, 'High End Hard Disk Drive', 'Storage');

commit;
```

执行定期装载。

```
~/regular_etl.sh
```

通过查询 product_count_fact 表确认定期装载执行正确。

```
select c.product_sk psk,
       c.product_code pc,
       b.product_launch_date_sk plsk,
       b.product_launch_date pld
  from product_count_fact a,
       product_launch_date_dim b,
       product_dim c
 where a.product_launch_date_sk = b.product_launch_date_sk
   and a.product_sk = c.product_sk
 order by pc, pld;
```

查询结果如下。可以看到只是增加了一条新产品记录，原有数据没有变化。

```
 psk | pc | plsk |    pld
-----+----+------+------------
   1 |  1 | 5905 | 2016-03-01
   2 |  2 | 5905 | 2016-03-01
   3 |  3 | 5905 | 2016-03-01
   5 |  4 | 6351 | 2017-05-21
   7 |  5 | 6366 | 2017-06-05
(5 rows)
```

无事实事实表是没有任何度量的事实表，它本质上是一组维度的交集。用这种事实表记录相关维度之间存在多对多关系，但是关系上没有数字或者文本的事实。无事实事实表为数据仓库设计提供了更多的灵活性。

17.4 迟到的事实

1. 迟到的事实简介

数据仓库通常建立在一种理想的假设情况下，这就是数据仓库的度量（事实记录）与度量的环境（维度记录）同时出现在数据仓库中。当同时拥有事实记录和正确的当前维度行时，就能够从容地首先维护维度键，然后在对应的事实表行中使用这些最新的键。然而，各种各样的原因会导致需要 ETL 系统处理迟到的事实数据。例如，某些线下的业务，数据进入操作型系统的时间会滞后于事务发生的时间。再或者出现某些极端情况，如源数据库系统出现故障，直到恢复后才能补上故障期间产生的数据。

在销售订单示例中，晚于订单日期进入源数据的销售订单可以看作是一个迟到事实的例子。销售订单数据被装载进其对应的事实表时，装载日期晚于销售订单产生的日期，因此是一个迟到的事实。本例中因为定期装载的是前一天的数据，所以这里的"晚于"指的是事务数据延迟两天及其以上才到达 ETL 系统。

必须对标准的 ETL 过程进行特殊修改以处理迟到的事实。首先，当迟到度量事件出现时，不得不反向搜索维度表历史记录，以确定事务发生时间点的有效的维度代理键，因为当前的维度内容无法匹配输入行的情况。此外，还需要调整后续事实行中的所有半可加度量，例如，由于迟到的事实导致客户当前余额的改变。迟到事实可能还会引起周期快照事实表的数据更新，如果 2017 年 5 月的销售订单金额已经计算并存储在 month_end_sales_order_fact 快照表中，这时一个迟到的 5 月订单在 6 月某天被装载，那么 2017 年 5 月的快照金额必须因迟到事实而重新计算。

下面就以销售订单数据仓库为例，说明如何处理迟到的事实。

2. 修改数据仓库表结构

在 17.1 节中建立的月销售周期快照表，其数据来自已经处理过的销售订单事务事实表。因此为了确定事实表中的一条销售订单记录是否是迟到的，需要把源数据中的登记日期列装载进销售订单事实表。为此在销售订单事实表上添加登记日期代理键列。为了获取登记日期代理键的值，还要使用维度角色扮演技术添加登记日期维度表。执行下面的语句在销售订单事实表里添加名为 entry_date_sk 的日期代理键列，并且从日期维度表创建一个叫做 v_entry_date_dim 的数据库视图。

```sql
set search_path=tds;

-- 给销售订单事实表增加登记日期代理键
alter table sales_order_fact add column entry_date_sk int default null;
comment on column sales_order_fact.entry_date_sk is '登记日期代理键';

-- 建立登记日期维度视图
create view v_entry_date_dim
(entry_date_sk, entry_date, month_name, month, quarter, year)
as
select date_sk, date, month_name, month, quarter, year
  from date_dim;
```

3. 修改定期数据装载函数

在创建了登记日期维度视图，并给销售订单事实表添加了登记日期代理键列后，需要修改数据仓库定期装载脚本来装载登记日期。修改后的装载函数如下（只列出修改的部分）。注意 sales_order 源数据表及其对应的过渡表中都已经含有登记日期，只是以前没有将其装载进数据仓库。

```sql
create or replace function fn_regular_load ()
returns void as $$
declare
    -- 设置 scd 的生效时间
    ...
begin
    -- 分析外部表
    ...
    -- 将外部表数据装载到原始数据表
    ...
    -- 分析 rds 模式的表
    ...
    -- 设置 cdc 的上限时间
    ...
    -- 装载客户维度
```

```
...
-- 装载产品维度
...
-- 装载新增产品数量无事实事实表
...
-- 装载销售订单事实表
insert into sales_order_fact
select a.order_number,
       customer_sk,
       product_sk,
       e.date_sk,
       e.year * 100 + e.month,
       order_amount,
       quantity,
       f.date_sk,
       g.sales_order_attribute_sk,
       h.customer_zip_code_sk,
       i.shipping_zip_code_sk,
       a.order_status,
       l.entry_date_sk
  from rds.sales_order a,
       v_customer_dim_his c,
       v_product_dim_his d,
       date_dim e,
       date_dim f,
       sales_order_attribute_dim g,
       v_customer_zip_code_dim h,
       v_shipping_zip_code_dim i,
       rds.customer j,
       rds.cdc_time k,
       v_entry_date_dim l
 where a.customer_number = c.customer_number
   and a.status_date >= c.effective_date
   and a.status_date < c.expiry_date
   and a.product_code = d.product_code
   and a.status_date >= d.effective_date
   and a.status_date < d.expiry_date
   and date(a.status_date) = e.date
   and date(a.request_delivery_date) = f.date
   and date(a.entry_date) = l.entry_date
   and a.verification_ind = g.verification_ind
   and a.credit_check_flag = g.credit_check_flag
   and a.new_customer_ind = g.new_customer_ind
```

```
            and a.web_order_flag = g.web_order_flag
            and a.customer_number = j.customer_number
            and j.customer_zip_code = h.customer_zip_code
            and j.shipping_zip_code = i.shipping_zip_code
            and a.entry_date >= k.last_load and a.entry_date < k.current_load;

    -- 重载 PA 客户维度
    ...
    -- 分析 tds 模式的表
    ...
    -- 更新时间戳表的 last_load 字段
    ...
end; $$
language plpgsql;
```

在装载函数中使用销售订单过渡表的状态日期字段限定当时的维度代理键。例如，为了获取事务发生时的客户代理键，筛选条件为：

```
status_date >= v_customer_dim_his.effective_date
and status_date < v_customer_dim_his.expiry_date
```

之所以可以这样做，原因在于本示例满足以下两个前提条件：在最初源数据库的销售订单表中，status_date 存储的是状态发生时的时间；维度的生效时间与过期时间构成一条连续且不重叠的时间轴，任意 status_date 日期只能落到唯一的生效时间、过期时间区间内。

4. 修改装载周期快照事实表的函数

17.1 节中创建的 fn_month_sum 函数用于装载月销售周期快照事实表。迟到的事实记录会对周期快照中已经生成的月销售汇总数据产生影响，因此必须做适当的修改。月销售周期快照表存储的是某月某产品汇总的销售数量和销售金额，表中有年月、产品代理键、销售金额、销售数量四个字段。由于迟到事实的出现，需要将事务事实表中的数据划分为两类：上月的周期快照和更早的周期快照。

fn_month_sum 函数先删除在生成上个月的汇总数据再重新生成，此时上月的迟到数据可以正确汇总。对于上上个月或更早的迟到数据，需要将迟到的数据累加到已有的周期快照上。下面修改 fn_month_sum 函数，使之能够自动处理任意时间的迟到事实数据。HAWQ 不能行级更新或删除数据，因此为了实现所谓的幂等操作，需要标识出迟到事实记录对应的事实表逻辑主键，在重复执行周期快照装载函数时过滤掉已经装载过的迟到数据。

（1）给周期快照事实表增加事务事实表的逻辑主键，正常数据（非迟到）对应的 order_number 字段值为空。

```
alter table month_end_sales_order_fact add order_number bigint default null;
```

（2）修改周期快照事实表装载函数

```sql
create or replace function tds.fn_month_sum(p_year_month int)
returns void as $$
declare
    sqlstring varchar(1000);
begin
    -- 幂等操作，先删除上月数据
sqlstring := 'truncate table month_end_sales_order_fact_1_prt_p'
|| cast(p_year_month as varchar);
    execute sqlstring;

    -- 插入上月销售汇总数据
    insert into month_end_sales_order_fact
    select t1.year_month, t2.product_sk, coalesce(t2.month_order_amount,0),
coalesce(t2.month_order_quantity,0), null
        from (select p_year_month year_month) t1
        left join (select year_month, product_sk,
sum(order_amount) month_order_amount,
sum(quantity) month_order_quantity
                    from sales_order_fact
                    where year_month = p_year_month
and coalesce(order_status,'N') = 'N'
                    group by year_month,product_sk) t2
            on t1.year_month = t2.year_month;

    -- 装载迟到的数据
    insert into month_end_sales_order_fact
    select year_month, product_sk, order_amount, quantity, order_number
      from (select t1.year_month, t1.product_sk, t1.order_amount, t1.quantity,
t1.order_number
            from sales_order_fact t1, v_entry_date_dim t2
            where coalesce(t1.entry_date_sk, t1.status_date_sk) =
t2.entry_date_sk
              and t2.year*100 + t2.month = p_year_month
              and t1.year_month < p_year_month
              and coalesce(t1.order_status,'N') = 'N'
              and not exists (select 1 from month_end_sales_order_fact t3
                              where t1.order_number =
t3.order_number)              ) t1;
  end; $$
  language plpgsql;
```

说明：

- t2.year*100 + t2.month = p_year_month and t1.year_month < p_year_month 处理上个月之前的迟到数据。
- not exists (select 1 from month_end_sales_order_fact t3 where t1.order_number = t3.order_number) 处理尚未装载的迟到数据，用于实现幂等操作。

（3）建立视图进行二次汇总

```
create view v_month_end_sales_order_fact as
select year_month, product_sk,
sum(month_order_amount) month_order_amount,
sum(month_order_quantity) month_order_quantity
  from month_end_sales_order_fact
 group by year_month, product_sk;
```

5. 测试

在执行定期装载前使用下面的语句查询 month_end_sales_order_fact 表。之后可以对比'前'（不包含迟到事实）'后'（包含了迟到事实）的数据，以确认装载的正确性。

```
select year_month,
       product_name,
       month_order_amount amt,
       month_order_quantity qty
  from month_end_sales_order_fact a,
       product_dim b
 where a.product_sk = b.product_sk
   and year_month = cast(extract(year from current_date - interval '1 month') * 100
                  + extract(month from current_date - interval '1 month') as int)
 order by year_month, product_name;
```

查询结果如下：

```
 year_month | product_name    | amt      | qty
------------+-----------------+----------+-----
     201705 | flat panel      | 49666.00 | 393
     201705 | floppy drive    | 52083.00 | 339
     201705 | hard disk drive | 85107.00 | 582
     201705 | keyboard        | 57833.00 | 391
(4 rows)
```

下一步执行下面的语句准备销售订单测试数据。将三个销售订单装载进销售订单源数据，一个是迟到的在 month_end_sales_order_fact 中已存在的产品，一个是迟到的在 month_end_sales_order_fact 中不存在的产品，另一个是非迟到的正常产品。这里需要注意，产品维度是 SCD2 处理的，所以在添加销售订单时，新增订单时间一定要在产品维度的生效与过期时间区间内。

```sql
    use source;

    -- 迟到已存在
    set @order_date := from_unixtime(unix_timestamp('2017-05-10') + rand() * 
(unix_timestamp('2017-05-11') - unix_timestamp('2017-05-10')));
    set @request_delivery_date := 
from_unixtime(unix_timestamp(date_add(@order_date, interval 5 day)) + rand() * 
86400);
    set @entry_date := from_unixtime(unix_timestamp('2017-06-07') + rand() * 
(unix_timestamp('2017-06-08') - unix_timestamp('2017-06-07')));
    set @amount := floor(1000 + rand() * 9000);
    set @quantity := floor(10 + rand() * 90);

    insert into source.sales_order values
      (null, 143, 6, 2, 'y', 'y', 'y', 'y', @order_date, 'N', 
@request_delivery_date, @entry_date, @amount, @quantity);

    -- 迟到不存在
    set @order_date := from_unixtime(unix_timestamp('2017-05-10') + rand() * 
(unix_timestamp('2017-05-11') - unix_timestamp('2017-05-10')));
    set @request_delivery_date := 
from_unixtime(unix_timestamp(date_add(@order_date, interval 5 day)) + rand() * 
86400);
    set @entry_date := from_unixtime(unix_timestamp('2017-06-07') + rand() * 
(unix_timestamp('2017-06-08') - unix_timestamp('2017-06-07')));
    set @amount := floor(1000 + rand() * 9000);
    set @quantity := floor(10 + rand() * 90);

    insert into source.sales_order values
      (null, 144, 6, 3, 'y', 'y', 'y', 'y', @order_date, 'N', 
@request_delivery_date, @entry_date, @amount, @quantity);

    -- 非迟到
    set @entry_date := from_unixtime(unix_timestamp('2017-06-07') + rand() * 
(unix_timestamp('2017-06-08') - unix_timestamp('2017-06-07')));
    set @request_delivery_date := 
from_unixtime(unix_timestamp(date_add(@entry_date, interval 5 day)) + rand() * 
86400);
    set @amount := floor(1000 + rand() * 9000);
    set @quantity := floor(10 + rand() * 90);
```

```
insert into source.sales_order values
  (null, 145, 12, 4, 'y', 'y', 'y', 'y', @entry_date, 'N',
@request_delivery_date, @entry_date, @amount, @quantity);

commit;
```

执行定期装载脚本。

```
~/regular_etl.sh
```

现在已经准备好运行修改后的月底快照装载。手工执行下面的命令执行月底销售订单事实表装载函数导入 2017 年 5 月的快照。

```
su - gpadmin -c 'export PGPASSWORD=123456;psql -U dwtest -d dw -h hdp3 -c "set search_path=tds;select fn_month_sum(cast(extract(year from current_date - interval '\''1 month'\'') * 100 + extract(month from current_date - interval '\''1 month'\'') as int))"'
```

执行与测试开始时相同的查询获取包含了迟到事实月底销售订单数据，查询结果如下。

```
year_month  | product_name     | amt       | qty
------------+------------------+-----------+-----
     201705 | flat panel       | 49666.00  | 393
     201705 | floppy drive     | 57707.00  | 361
     201705 | hard disk drive  | 85107.00  | 582
     201705 | keyboard         | 57833.00  | 391
     201705 | lcd panel        |  3087.00  |  11
(5 rows)
```

对比'前''后'查询的结果可以看到：

- 2017 年 5 月 Floppy Drive 的销售金额已经从 52083 变为 57707，这是由于迟到的产品销售订单增加了 5624 的销售金额。销售数量也相应地增加了。
- 2017 年 5 月的 LCD Panel（也是迟到的产品）被添加。

17.5 累积度量

累积度量指的是聚合从序列内第一个元素到当前元素的数据，例如统计从每年的一月到当前月份的累积销售额。本节说明如何在销售订单示例中实现累积月销售数量和金额，并对数据仓库模式、初始装载、定期装载做相应地修改。累积度量是半可加的，而且它的初始装载要复杂一些。

1. 建立累积度量事实表

执行下面的语句创建 month_end_balance_fact 事实表,用来存储销售订单金额和数量的月累积值。

```
set search_path=tds;
create table month_end_balance_fact (
    year_month int,
    product_sk int,
    month_end_amount_balance numeric(10,2),
    month_end_quantity_balance int );

comment on table month_end_balance_fact is '累积度量事实表';
comment on column month_end_balance_fact.year_month is '年月';
comment on column month_end_balance_fact.product_sk is '产品代理键';
comment on column month_end_balance_fact.month_end_amount_balance is '金额';
comment on column month_end_balance_fact.month_end_quantity_balance is '数量';
```

2. 初始装载

现在要把 month_end_sales_order_fact 表里的数据装载进 month_end_balance_fact 表,下面显示了初始装载 month_end_balance_fact 表的语句,装载累积的月销售订单汇总数据,从每年的一月累积到当月,累积数据不跨年。

```
insert into month_end_balance_fact
select a.year_month, b.product_sk,
     sum(b.month_order_amount) month_order_amount,
     sum(b.month_order_quantity) month_order_quantity
 from (select distinct year_month, year_month/100 year1,
          year_month - year_month/100*100 month1
        from v_month_end_sales_order_fact) a,
     (select *, year_month/100 year1, year_month - year_month/100*100 month1,
          max(year_month) over () max_year_month
        from v_month_end_sales_order_fact) b
 where a.year_month <= b.max_year_month
   and a.year1 = b.year1 and b.month1 <= a.month1
 group by a.year_month, b.product_sk;
```

子查询获取 month_end_sales_order_fact 表的数据,及其年月和最大月份代理键。外层查询汇总每年一月到当月的累积销售数据,a.year_month <= b.max_year_month 条件用于限定只统计到现存的最大月份为止。为了确认初始装载是否正确,在执行完初始装载脚本后,分别查询 month_end_sales_order_fact 和 month_end_balance_fact 表。

查询周期快照:

```sql
select year_month, product_sk psk, month_order_amount amt,
       month_order_quantity qty
  from v_month_end_sales_order_fact
 order by year_month, psk;
```

查询结果如下：

```
 year_month  | psk |    amt     | qty
-------------+-----+------------+-----
      201603 |   1 |   89488.00 |
      201603 |   2 |  101670.00 |
      201604 |   1 |   79238.00 |
      201604 |   2 |   75204.00 |
      201605 |   1 |   44757.00 |
      201605 |   2 |   65415.00 |
      201606 |   1 |   74628.00 |
      201606 |   2 |   41790.00 |
      201705 |   1 |   85107.00 | 582
      201705 |   2 |   57707.00 | 361
      201705 |   3 |    3087.00 |  11
      201705 |   4 |   49666.00 | 393
      201705 |   5 |   57833.00 | 391
(13 rows)
```

查询累积度量：

```sql
select year_month, product_sk psk, month_end_amount_balance amt,
       month_end_quantity_balance qty
  from month_end_balance_fact
 order by year_month, psk;
```

查询结果如下：

```
 year_month | psk |    amt     | qty
------------+-----+------------+-----
     201603 |  1  |   89488.00 |
     201603 |  2  |  101670.00 |
     201604 |  1  |  168726.00 |
     201604 |  2  |  176874.00 |
     201605 |  1  |  213483.00 |
     201605 |  2  |  242289.00 |
     201606 |  1  |  288111.00 |
     201606 |  2  |  284079.00 |
     201705 |  1  |   85107.00 | 582
     201705 |  2  |   57707.00 | 361
     201705 |  3  |    3087.00 |  11
```

```
    201705  |  4  |  49666.00  |  393
    201705  |  5  |  57833.00  |  391
(13 rows)
```

可以看到，2016 年 3 月的商品销售金额被累积到了 2016 年 4 月，2016 年 3 月和 4 月的商品销售金额被累积到了 2016 年 5 月，等等。

3. 定期装载

下面所示的 month_balance_sum.sql 脚本用于定期装载销售订单累积度量，每个月执行一次，装载上个月的数据。可以在执行完月周期快照表定期装载后执行该脚本。

```sql
insert into month_end_balance_fact
select year_month, product_sk, sum(month_order_amount),
       sum(month_order_quantity)
  from (select *
          from v_month_end_sales_order_fact
         where year_month = :v_year_month
        union all
        select :v_year_month, product_sk product_sk,
               month_end_amount_balance month_order_amount,
               month_end_quantity_balance month_order_quantity
          from month_end_balance_fact
         where year_month in (select max(case when :v_year_month
 - :v_year_month/100*100 = 1 then 0 else year_month end)
                                from month_end_balance_fact)) t
 group by year_month, product_sk;
```

子查询将累积度量表和月周期快照表做并集操作，增加上月的累积数据。最外层查询执行销售数据按月和产品的分组聚合。最内层的 case 语句用于在每年一月时重新归零再累积。v_year_month 是年月参数。

4. 测试

执行月周期快照函数，装载 2017 年 6 月的数据。

```sql
select fn_month_sum(201706);
```

执行累积度量定期装载脚本，以 shell 命令 "date +%Y%m" 的输出作为年月参数传入 month_balance_sum.sql 文件中。

```
su - gpadmin -c 'export PGPASSWORD=123456;psql -U dwtest -d dw -h hdp3 -v v_year_month='`'`date +%Y%m`'`' -f ~/month_balance_sum.sql'
```

执行和前面初始装载后相同的查询，周期快照表和累积度量表的查询结果分别如下所示。

```
 year_month | psk |    amt   | qty
------------+-----+----------+-----
```

```
 201603    | 1 |  89488.00 |
 201603    | 2 | 101670.00 |
 201604    | 1 |  79238.00 |
 201604    | 2 |  75204.00 |
 201605    | 1 |  44757.00 |
 201605    | 2 |  65415.00 |
 201606    | 1 |  74628.00 |
 201606    | 2 |  41790.00 |
 201705    | 1 |  85107.00 | 582
 201705    | 2 |  57707.00 | 361
 201705    | 3 |   3087.00 |  11
 201705    | 4 |  49666.00 | 393
 201705    | 5 |  57833.00 | 391
 201706    | 1 |   2263.00 |  80
 201706    | 2 |   6869.00 |  30
 201706    | 5 |   9554.00 |  38
(16 rows)

year_month | psk |    amt    | qty
-----------+-----+-----------+-----
 201603    | 1 |  89488.00 |
 201603    | 2 | 101670.00 |
 201604    | 1 | 168726.00 |
 201604    | 2 | 176874.00 |
 201605    | 1 | 213483.00 |
 201605    | 2 | 242289.00 |
 201606    | 1 | 288111.00 |
 201606    | 2 | 284079.00 |
 201705    | 1 |  85107.00 | 582
 201705    | 2 |  57707.00 | 361
 201705    | 3 |   3087.00 |  11
 201705    | 4 |  49666.00 | 393
 201705    | 5 |  57833.00 | 391
 201706    | 1 |  87370.00 | 662
 201706    | 2 |  64576.00 | 391
 201706    | 3 |   3087.00 |  11
 201706    | 4 |  49666.00 | 393
 201706    | 5 |  67387.00 | 429
(18 rows)
```

可以看到，2017年5月的商品销售金额和数量被累积到了2017年6月。产品1、2、5累加了5、6两个月的销售数据，产品3、4在6月没有销售，所以5月的数据顺延到6月。

5. 查询

累积度量必须小心使用，因为它是"半可加"的。一个半可加度量在某些维度（通常是时间维度）上是不可加的。例如，可以通过产品正确地累加月底累积销售金额。

```
dw=> select year_month, sum(month_end_amount_balance) s
dw->   from month_end_balance_fact
dw->   group by year_month
dw->   order by year_month;
 year_month |     s
------------+-----------
     201603 | 191158.00
     201604 | 345600.00
     201605 | 455772.00
     201606 | 572190.00
     201705 | 253400.00
     201706 | 272086.00
(6 rows)
```

而当通过月份累加月底金额时：

```
dw=> select product_name, sum(month_end_amount_balance) s
dw->   from month_end_balance_fact a,
dw->        product_dim b
dw->   where a.product_sk = b.product_sk
dw->   group by product_name
dw->   order by product_name;
  product_name    |     s
------------------+-----------
 flat panel       |  99332.00
 floppy drive     | 927195.00
 hard disk drive  | 932285.00
 keyboard        | 125220.00
 lcd panel        |   6174.00
(5 rows)
```

以上查询结果是错误的。正确的结果应该和下面的在 month_end_sales_order_fact 表上进行的查询结果相同。

```
dw=> select product_name, sum(month_order_amount) s
dw->   from month_end_sales_order_fact a,
dw->        product_dim b
dw->   where a.product_sk = b.product_sk
dw->   group by product_name
dw->   order by product_name;
  product_name    |    s
```

```
    flat panel       |  49666.00
    floppy drive     | 348655.00
    hard disk drive  | 375481.00
    keyboard         |  67387.00
    lcd panel        |   3087.00
(5 rows)
```

注意，迟到的事实对累积度量的影响非常大。例如，2016 年 1 月的数据到了 2017 年 1 月才进入数据仓库，那么 2016 年 2 月以后每个月的累积度量都要改变。如果重点考虑迟到事实数据和 HAWQ 无法行级更新的限制，也许使用查询视图方式实现累积度量是更好的选择。

```
create view v_month_end_balance_fact as
select a.year_month, b.product_sk,
       sum(b.month_order_amount) month_order_amount,
       sum(b.month_order_quantity) month_order_quantity
  from (select distinct year_month, year_month/100 year1,
               year_month - year_month/100*100 month1
          from v_month_end_sales_order_fact) a,
       (select *, year_month/100 year1, year_month - year_month/100*100 month1,
               max(year_month) over () max_year_month
          from month_end_sales_order_fact) b
 where a.year_month <= b.max_year_month
   and a.year1 = b.year1 and b.month1 <= a.month1
 group by a.year_month, b.product_sk;
```

17.6 小结

事务事实表、周期快照事实表和累积快照事实表是多维数据仓库中常见的三种事实表。定期历史数据可以通过周期快照获取，细节数据被保存到事务粒度事实表中，而对于具有多个定义良好里程碑的处理工作流，则可以使用累积快照。无事实事实表是没有任何度量的事实表，它本质上是一组维度的交集。用这种事实表记录相关维度之间存在多对多关系，但是关系上没有数字或者文本的事实。无事实事实表为数据仓库设计提供了更多的灵活性。迟到的事实指的是到达 ETL 系统的时间晚于事务发生时间的度量数据。必须对标准的 ETL 过程进行特殊修改以处理迟到的事实。需要确定事务发生时间点的有效的维度代理键，还要调整后续事实行中的所有半可加度量。此外，迟到事实可能还会引起周期快照事实表的数据更新。累积度量指的是聚合从序列内第一个元素到当前元素的数据。累积度量是半可加的，因此对累积度量执行聚合计算时要格外注意分组的维度。

第 18 章

◀ 联机分析处理 ▶

前面两章通过实例演示了常见的维度表和事实表技术,主要目的是为了说明 HAWQ 及其 Hadoop 生态圈工具,如 Sqoop、Oozie、Falcon 等,完全有能力处理传统多维数据仓库中碰到的各种情况。但是,从完整的数据仓库生命周期角度看,还有很重要的一部分没有涉及,那就是数据分析与结果展现,本章将讨论这方面的问题。在介绍了联机分析处理的相关概念后,我们会结合销售订单示例,列举典型的数据分析问题,使用 HAWQ 具体实现,并用 Zeppelin 进行交互式查询与数据图形化展示。

18.1 联机分析处理简介

18.1.1 概念

联机分析处理又被称为 OLAP,是英文 On-Line Analytical Processing 的缩写。此概念最早由关系数据库之父 E.F.Codd 于 1993 年提出。OLAP 允许以一种称为多维数据集的结构,访问业务数据源经过聚合和组织整理后的数据。以此为标准,OLAP 作为单独的一类技术同联机事务处理(On-Line Transaction Processing,OLTP)得以明显区分。

在计算领域,OLAP 是一种快速应答多维分析查询的方法,也是商业智能的一个组成部分,与之相关的概念还包括数据仓库、报表系统、数据挖掘等。数据仓库用于数据的存储和组织,OLAP 集中于数据的分析,数据挖掘则致力于知识的自动发现,报表系统侧重于数据的展现。OLAP 系统从数据仓库中的集成数据出发,构建面向分析的多维数据模型,再使用多维分析方法从多个不同的视角对多维数据集合进行分析比较,分析活动以数据驱动。通过使用 OLAP 工具,用户可以从多个视角交互式地查询多维数据。

OLAP 由三个基本的分析操作构成:合并(上卷)、下钻和切片。合并是指数据的聚合,即数据可以在一个或多个维度上进行累积和计算。例如,所有的营业部数据被上卷到销售部门以分析销售趋势。下钻是一种由汇总数据向下浏览细节数据的技术。比如用户可以从产品分类的销售数据下钻查看单个产品的销售情况。切片则是这样一种特性,通过它用户可以获取

OLAP 立方体中的特定数据集合，并从不同的视角观察这些数据。这些观察数据的视角就是我们所说的维度。例如通过经销商、日期、客户、产品或区域等等，查看同一销售事实。

OLAP 系统的核心是 OLAP 立方体，或称为多维立方体或超立方体。它由被称为度量的数值事实组成，这些度量被维度划分归类。一个 OLAP 立方体的例子如图 18-1 所示，数据单元位于立方体的交叉点上，每个数据单元跨越产品、时间、地区等多个维度。通常使用一个矩阵接口操作 OLAP 立方体，例如电子表格程序的数据透视表，可以按维度分组执行聚合或求平均值等操作。立方体的元数据一般由关系数据库中的星型模式或雪花模式生成，度量来自事实表的记录，维度来自维度表。

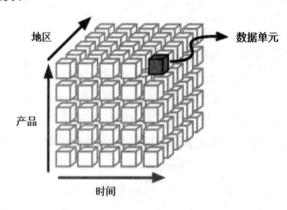

图 18-1　OLAP 立方体

18.1.2　分类

通常可以将联机分析处理系统分为 MOLAP、ROLAP、HOLAP 三种类型。

1. MOLAP

MOLAP（multi-dimensional online analytical processing）是一种典型的 OLAP 形式，甚至有时就被用来表示 OLAP。MOLAP 将数据存储在一个经过优化的多维数组中，而不是存储在关系数据库中。某些 MOLAP 工具要求预先计算并存储计算后的结果数据，这种操作方式被称为预处理。MOLAP 工具一般将预计算后的数据集合作为一个数据立方体使用。对于给定范围的问题，立方体中的数据包含所有可能的答案。预处理的好处是可以对问题做出非常快速的响应。但另一方面，依赖于预计算的聚合程度，装载新数据可能会花费很长的时间。另外还有些 MOLAP 工具，尤其是那些实现了某些数据库功能的 MOLAP 工具，并不预先计算原始数据，而是在需要时才进行计算。

MOLAP 的优点：

- 优化的数据存储、多维数据索引和缓存带来的快速查询性能。
- 相对于关系数据库，可以通过压缩技术，使数据存储需要更小的磁盘空间。
- MOLAP 工具一般能够自动进行高级别的数据聚合。
- 对于低基数维度的数据集合是紧凑的。

- 数组模型提供了原生的索引功能。

MOLAP 的缺点：

- 某些 MOLAP 解决方案中的处理步骤可能需要很长的时间，尤其是当数据量很大时。要解决这个问题，基本只能增量处理变化的数据，而不是预处理整个数据集合。
- 可能引入较多的数据冗余。

MOLAP 产品：

商业的 MOLAP 产品主要有 Cognos Powerplay、Oracle Database OLAP Option、MicroStrategy、Microsoft Analysis Services、Essbase 等。

2. ROLAP

ROLAP 直接使用关系数据库存储数据，不需要执行预计算。基础的事实数据及其维度表作为关系表被存储，而聚合信息存储在新创建的附加表中。ROLAP 以数据库规范化设计为基础，操作存储在关系数据库中的数据，实现传统的 OLAP 数据切片和分块功能。本质上讲，每种数据切片或分块行为都等同于在 SQL 语句中增加一个"WHERE"子句的过滤条件。ROLAP 不使用预计算的数据立方体，取而代之的是查询标准的关系数据库表，返回回答问题所需的数据。与预计算的 MOLAP 不同，ROLAP 工具有能力回答任意相关的数据分析问题，因为该技术不受立方体内容的限制。通过 ROLAP 还能够下钻到数据库中存储的最细节的数据。

由于 ROLAP 使用关系数据库，通常数据库模式必须经过仔细设计。为 OLTP 应用设计的数据库不能直接作为 ROLAP 数据库使用，这种投机取巧的做法并不能使 ROLAP 良好工作。因此 ROLAP 仍然需要创建额外的数据复制。但不管怎样，ROLAP 毕竟用的是数据库，各种各样的数据库设计与优化技术都可以被有效利用。

ROLAP 的优点：

- 在处理大量数据时，ROLAP 更具可伸缩性，尤其当模型中包含的维度具有很高基数，如维度表中有上百万的成员时。
- 有很多可选用的数据装载工具，并且能够针对特定的数据模型精细调整 ETL 代码，数据装载所需时间通常比自动化的 MOLAP 装载少得多。
- 因为数据存储于标准关系数据库中，可以使用 SQL 报表工具访问数据，而不必是专有的 OLAP 工具。
- ROLAP 更适合处理非聚合的事实，例如文本型描述。在 MOLAP 工具中查询文本型元素时性能会相对较差。
- 通过将数据存储从多维模型中解耦出来，相对于用使用严格的维度模型，这种更普通的关系模型增加了成功建模的可能性。
- ROLAP 方法可以利用数据库的权限控制，例如通过行级安全性设置，可以用事先设定的条件过滤查询结果。如 Oracle 的 VPD 技术，能够根据连接的用户自动在查询的 SQL 语句中拼接 WHERE 谓词条件。

ROLAP 的缺点：

- 业界普遍认为 ROLAP 工具比 MOLAP 查询速度慢。
- 聚合表的数据装载必须由用户自己定制的 ETL 代码控制。ROLAP 工具不能自动完成这个任务，这意味着额外的开发工作量。
- 如果跳过创建聚合表的步骤，查询性能会大打折扣，因为不得不查询大量的细节数据表。虽然可以通过适当建立聚合表缓解性能问题，但对所有维度表及其属性的组合创建聚合表是不切实际的。
- ROLAP 依赖于针对通用查询或缓存目标的数据库，因此并没有提供某些 MOLAP 工具所具有的特殊技术，如透视表等。但是现代 ROLAP 工具可以利用 SQL 语言中的 CUBE、ROLLUP 操作或其他 SQL OLAP 扩展。随着这些 SQL 扩展的逐步完善，MOLAP 工具的优势也不那么明显了。
- 因为 ROLAP 工具的所有计算都依赖于 SQL，对于某些不易转化为 SQL 的计算密集型模型，ROLAP 不再适用。如地理位置计算的场景。

ROLAP 产品：

使用 ROLAP 的商业产品包括 Microsoft Analysis Services、MicroStrategy、SAP Business Objects、Oracle Business Intelligence Suite Enterprise Edition、Tableau Software 等等。也有开源的 ROLAP 服务器，如 Mondrian。

3. HOLAP

在额外的 ETL 开发成本与缓慢的查询性能之间难以选择，正是因为这种情况，现在大部分商业 OLAP 工具都使用一种混合型（Hybrid）方法，它允许模型设计者决定哪些数据存储在 MOLAP 中，哪些数据存储在 ROLAP 中。除了把数据划分成传统关系型存储和专有存储，业界对混合型 OLAP 并没有清晰的定义。例如，某些厂商的 HOLAP 数据库使用关系表存储大量的细节数据，而是用专用表保存少量的聚合数据。HOLAP 结合了 MOLAP 和 ROLAP 两种方法的优点，可以同时利用预计算的多维立方体和关系数据源。HOLAP 有以下两种划分数据的策略。

- 垂直分区。这种模式的 HOLAP 将聚合数据存储在 MOLAP 中，以支持良好的查询性能，而把细节数据存储在 ROLAP 中以减少立方体处理所需时间。
- 水平分区。这种模式的 HOLAP 按数据热度划分，将某些最近使用的数据分片存储在 MOLAP 中，而将老的数据存储在 ROLAP。

18.1.3 性能

OLAP 分析所需的原始数据量是非常庞大的。一个分析模型，往往会涉及数千万或数亿条甚至更多的数据，而且分析模型中包含多个维度数据，这些维度又可以由用户作任意的组合。这样的结果就是大量实时运算导致过长的响应时间。想象一个 1000 万条记录的分析模型，如果一次提取 4 个维度进行组合分析，每个维度有 10 个不同的取值，理论上的运算次数将达到

10 的 12 次方。这样的运算量将导致数十分钟乃至更长的等待时间。如果用户对维组合次序进行调整，或增加、或减少某些维度的话，又将是一个重新计算过程。

从上面的分析中可以得出结论，如果不能解决 OLAP 运算效率问题的话，OLAP 将只会是一个没有实用价值的概念。在 OLAP 的发展历史中，常见的解决方案是用多维数据库代替关系数据库设计，将数据根据维度进行最大限度的聚合运算，运算中会考虑到各种维度组合情况，运算结果将生成一个数据立方体，并保存在磁盘上，用这种预运算方式提高 OLAP 的速度。例如 Kylin 就是使用这种以空间换时间的方式来提高查询速度。而 HAWQ 在性能上的优势，也使它较为适合 OLAP 应用。HAWQ 与 Impala 和 Hive 的性能对比，参见 1.4.2 节。

18.2 联机分析处理实例

18.2.1 销售订单

要做好 OLAP 类的应用，需要对业务数据有深入的理解。只有了解了业务，才能知道需要分析哪些指标，从而有的放矢地剖析相关数据，得出可信的结论来辅助决策。下面就以销售订单数据仓库为例，提出若干问题，然后使用 HAWQ 查询数据以回答这些问题：

（1）每种产品类型以及单个产品的累积销售量和销售额是多少？
（2）每种产品类型以及单个产品在每个省、每个城市的月销售量和销售额趋势是什么？
（3）每种产品类型销售量和销售额的同比如何？
（4）每个省以及每个城市的客户数量及其消费金额汇总是多少？
（5）迟到订单的比例是多少？
（6）客户年消费金额的平均数和中位数是多少？
（7）客户年消费金额分布处于 25%、50%、75% 位置的消费金额是多少？
（8）客户年消费金额为"高"、"中"、"低"档的人数及消费金额所占比例是多少？
（9）每个城市按销售金额排在前三位的商品是什么？
（10）所有产品的销售百分比排名？

1. 每种产品类型以及单个产品的累积销售量和销售额是多少

使用 HAWQ 的 group by rollup 求小计和总计。

```
dw=> select t2.product_category, t2.product_name, sum(nq), sum(order_amount)
dw->   from v_sales_order_fact t1, product_dim t2
dw->  where t1.product_sk = t2.product_sk
dw->  group by rollup (t2.product_category, t2.product_name)
dw->  order by t2.product_category, t2.product_name;
 product_category |  product_name   | sum  |   sum
------------------+-----------------+------+-----------
```

```
 monitor    | flat panel       |     | 49666.00
 monitor    | lcd panel        |  11 |  3087.00
 monitor    |                  |  11 | 52753.00
 peripheral | keyboard         |  38 | 67387.00
 peripheral |                  |  38 | 67387.00
 storage    | floppy drive     |  52 |348655.00
 storage    | hard disk drive  |  80 |375481.00
 storage    |                  | 132 |724136.00
            |                  | 181 |844276.00
(9 rows)
```

2. 每种产品类型以及单个产品在每个省、每个城市的月销售量和销售额是多少

查询语句与上一个问题类似，只是多关联了邮编维度表，并且在 group by rollup 中增加了省、市两列。

```
dw=> select t2.product_category, t2.product_name, t3.state, t3.city,
dw->        sum(nq), sum(order_amount)
dw->   from v_sales_order_fact t1, product_dim t2, zip_code_dim t3
dw->  where t1.product_sk = t2.product_sk
dw->    and t1.customer_zip_code_sk = t3.zip_code_sk
dw->  group by rollup (t2.product_category, t2.product_name, t3.state, t3.city)
dw->  order by t2.product_category, t2.product_name, t3.state, t3.city;
 product_category | product_name  | state |     city       |  sum |   sum
------------------+---------------+-------+----------------+------+---------
 monitor          | flat panel    | oh    | cleveland      |      |  7431.00
 monitor          | flat panel    | oh    |                |      |  7431.00
 monitor          | flat panel    | pa    | mechanicsburg  |      | 10630.00
 monitor          | flat panel    | pa    | pittsburgh     |      | 31605.00
 monitor          | flat panel    | pa    |                |      | 42235.00
 monitor          | flat panel    |       |                |      | 49666.00
 monitor          | lcd panel     | pa    | pittsburgh     |  11  |  3087.00
 monitor          | lcd panel     | pa    |                |  11  |  3087.00
 monitor          | lcd panel     |       |                |  11  |  3087.00
 monitor          |               |       |                |  11  | 52753.00
 peripheral       | keyboard      | oh    | cleveland      |  38  | 10875.00
 peripheral       | keyboard      | oh    |                |  38  | 10875.00
 peripheral       | keyboard      | pa    | mechanicsburg  |      | 29629.00
 peripheral       | keyboard      | pa    | pittsburgh     |      | 26883.00
 peripheral       | keyboard      | pa    |                |      | 56512.00
 peripheral       | keyboard      |       |                |  38  | 67387.00
 peripheral       |               |       |                |  38  | 67387.00
 storage          | floppy drive  | oh    | cleveland      |      |  8229.00
 storage          | floppy drive  | oh    |                |      |  8229.00
```

```
 storage           | floppy drive    | pa  | mechanicsburg |     | 140410.00
 storage           | floppy drive    | pa  | pittsburgh    |  52 | 200016.00
 storage           | floppy drive    | pa  |               |  52 | 340426.00
 storage           | floppy drive    |     |               |  52 | 348655.00
 storage           | hard disk drive | oh  | cleveland     |     |   8646.00
 storage           | hard disk drive | oh  |               |     |   8646.00
 storage           | hard disk drive | pa  | mechanicsburg |  80 | 194444.00
 storage           | hard disk drive | pa  | pittsburgh    |     | 172391.00
 storage           | hard disk drive | pa  |               |  80 | 366835.00
 storage           | hard disk drive |     |               |  80 | 375481.00
 storage           |                 |     |               | 132 | 724136.00
                   |                 |     |               | 181 | 844276.00
(31 rows)
```

3. 每种产品类型销售量和销售额的同比如何

需要查询周期快照 v_month_end_sales_order_fact。

```
dw=> select t2.product_category,
dw->        t1.year_month,
dw->        sum(quantity1) quantity_cur,
dw->        sum(quantity2) quantity_pre,
dw->        round((sum(quantity1) - sum(quantity2)) / sum(quantity2),2)
dw->        pct_quantity,
dw->        sum(amount1) amount_cur,
dw->        sum(amount2) amount_pre,
dw->        round((sum(amount1) - sum(amount2)) / sum(amount2),2) pct_amount
dw->   from (select t1.product_sk,
dw(>               t1.year_month,
dw(>               t1.month_order_quantity quantity1,
dw(>               t2.month_order_quantity quantity2,
dw(>               t1.month_order_amount amount1,
dw(>               t2.month_order_amount amount2
dw(>          from v_month_end_sales_order_fact t1
dw(>          join v_month_end_sales_order_fact t2
dw(>            on t1.product_sk = t2.product_sk
dw(>           and t1.year_month/100 = t2.year_month/100 + 1
dw(>           and t1.year_month - t1.year_month/100*100 =
dw(>               t2.year_month - t2.year_month/100*100) t1,
dw->        product_dim t2
dw->  where t1.product_sk = t2.product_sk
dw->  group by t2.product_category, t1.year_month
dw->  order by t2.product_category, t1.year_month;
 product_category | year_month | quantity_cur | quantity_pre | pct_quantity |
amount_cur | amount_pre | pct_amount
```

```
------------+--------+--------+--------+------------+------------+------
 storage    | 201705 |   943  |        |  142814.00 |  110172.00 |  0.30
 storage    | 201706 |   110  |        |    9132.00 |  116418.00 | -0.92
(2 rows)
```

4. 每个省以及每个城市的客户数量及其消费金额汇总是多少

```
dw=> select t2.state,
dw->        t2.city,
dw->        count(distinct customer_sk) sum_customer_num,
dw->        sum(order_amount) sum_order_amount
dw->   from v_sales_order_fact t1, zip_code_dim t2
dw->  where t1.customer_zip_code_sk = t2.zip_code_sk
dw->  group by rollup (t2.state, t2.city)
dw->  order by t2.state, t2.city;
 state |     city      | sum_customer_num | sum_order_amount
-------+---------------+------------------+------------------
 oh    | cleveland     |                4 |         35181.00
 oh    |               |                4 |         35181.00
 pa    | mechanicsburg |                8 |        375113.00
 pa    | pittsburgh    |               12 |        433982.00
 pa    |               |               20 |        809095.00
       |               |               24 |        844276.00
(6 rows)
```

5. 迟到订单的比例是多少

注意，sum_late 需要显式转化为 numeric 数据类型。

```
dw=> select sum_total, sum_late,
dw->        round(cast(sum_late as numeric)/sum_total,4) late_pct
dw->   from (select sum(case when status_date_sk < entry_date_sk then 1
dw(>                         else 0 end) sum_late,
dw(>                count(*) sum_total
dw(>           from sales_order_fact) t;
 sum_total | sum_late | late_pct
-----------+----------+----------
       151 |        2 |   0.0132
(1 row)
```

6. 客户年消费金额的平均数和中位数是多少

分别使用两种方法求得平均数和中位数。HAWQ 为分析型应用提供了丰富的聚合函数。

```
dw=> select round(avg(sum_order_amount),2) avg_amount,
dw->        round(sum(sum_order_amount)/count(customer_sk),2) avg_amount1,
dw->        percentile_cont(0.5) within group (order by sum_order_amount)
```

```
dw->        median_amount,
dw->        median(sum_order_amount) median_amount1
dw->   from (select customer_sk,sum(order_amount) sum_order_amount
dw(>          from v_sales_order_fact
dw(>         group by customer_sk) t1;
 avg_amount | avg_amount1 | median_amount | median_amount1
------------+-------------+---------------+----------------
   35178.17 |    35178.17 |         14277 |          14277
(1 row)
```

7. 客户年消费金额分布处于 25%、50%、75%位置的消费金额是多少

```
dw=> select percentile_cont(0.25) within group (order by sum_order_amount desc)
dw->        max_amount_25,
dw->        percentile_cont(0.50) within group (order by sum_order_amount desc)
dw->        max_amount_50,
dw->        percentile_cont(0.75) within group (order by sum_order_amount desc)
dw->        max_amount_75
dw->   from (select customer_sk,sum(order_amount) sum_order_amount
dw(>          from v_sales_order_fact
dw(>         group by customer_sk) t1;
 max_amount_25 | max_amount_50 | max_amount_75
---------------+---------------+---------------
       50536.5 |         14277 |       8342.25
(1 row)
```

8. 客户年消费金额为"高""中""低"档的人数及消费金额所占比例是多少

使用 16.8 节中定义的分段进行查询。

```
dw=> select year1,
dw->        bn,
dw->        c_count,
dw->        sum_band,
dw->        sum_total,
dw->        round(sum_band/sum_total,4) band_pct
dw->   from (select count(a.customer_sk) c_count,
dw(>                sum(annual_order_amount) sum_band,
dw(>                a.year year1,
dw(>                band_name bn
dw(>           from annual_customer_segment_fact a,
dw(>                annual_order_segment_dim b,
dw(>                annual_sales_order_fact d
dw(>          where a.segment_sk = b.segment_sk
dw(>            and a.customer_sk = d.customer_sk
dw(>            and a.year = d.year
```

```
dw(>              and b.segment_name = 'grid'
dw(>           group by a.year, bn) t1,
dw->        (select sum(annual_order_amount) sum_total
dw(>            from annual_sales_order_fact) t2
dw-> order by year1, bn;
 year1 |  bn  | c_count | sum_band  | sum_total | band_pct
-------+------+---------+-----------+-----------+----------
  2016 | high |       6 | 572190.00 | 572190.00 |   1.0000
(1 row)
```

9. 每个城市按销售金额排在前三位的商品是什么

使用 HAWQ 提供的窗口函数 row_number()，按城市分区，按销售额倒序，取得销售排名。

```
dw=> select case when t1.rn =1 then t1.city end city,
dw->        t2.product_name,
dw->        t1.sum_order_amount,
dw->        t1.rn
dw->   from (select city,
dw(>                product_sk,
dw(>                sum_order_amount,
dw(>                row_number()
dw(>                over (partition by city order by sum_order_amount desc) rn
dw(>           from (select t2.state||':'||t2.city city,
dw(>                        product_sk,
dw(>                        sum(order_amount) sum_order_amount
dw(>                   from v_sales_order_fact t1, zip_code_dim t2
dw(>                  where t1.customer_zip_code_sk = t2.zip_code_sk
dw(>                  group by t2.state||':'||t2.city, product_sk) t) t1
dw->   inner join product_dim t2 on t1.product_sk = t2.product_sk
dw->   where t1.rn <= 3
dw->   order by t1.city, t1.rn;
      city       |   product_name   | sum_order_amount | rn
-----------------+------------------+------------------+----
 oh:cleveland    | keyboard         |         10875.00 |  1
                 | hard disk drive  |          8646.00 |  2
                 | floppy drive     |          8229.00 |  3
 pa:mechanicsburg| hard disk drive  |        194444.00 |  1
                 | floppy drive     |        140410.00 |  2
                 | keyboard         |         29629.00 |  3
 pa:pittsburgh   | floppy drive     |        200016.00 |  1
                 | hard disk drive  |        172391.00 |  2
                 | flat panel       |         31605.00 |  3
(9 rows)
```

10. 所有产品的销售百分比排名

```
dw=> select product_name,
dw->        sum_order_amount,
dw->        percent_rank() over (order by sum_order_amount desc) rank
dw->   from (select product_sk,sum(order_amount) sum_order_amount
dw(>           from v_sales_order_fact
dw(>          group by product_sk) t1, product_dim t2
dw->  where t1.product_sk = t2.product_sk
dw->  order by rank;
 product_name    | sum_order_amount | rank
-----------------+------------------+------
 hard disk drive |        375481.00 |    0
 floppy drive    |        348655.00 | 0.25
 keyboard        |         67387.00 |  0.5
 flat panel      |         49666.00 | 0.75
 lcd panel       |          3087.00 |    1
(5 rows)
```

18.2.2 行列转置

OLAP 或报表系统中有一类常见需求就是行列转置。HAWQ 提供的内建函数和 SQL 过程语言编程功能，使行列转置操作的实现变得更为简单。

1. 行转列

（1）固定列数的行转列

原始数据如下：

```
test=# select * from score;
 name | subject | score
------+---------+-------
 张三 | 语文    |    80
 张三 | 数学    |    70
 张三 | 英语    |    60
 李四 | 语文    |    90
 李四 | 数学    |   100
 李四 | 英语    |    80
(6 rows)
```

要得到以下结果：

```
 name | 语文 | 数学 | 英语
------+------+------+------
 张三 |   80 |   70 |   60
 李四 |   90 |  100 |   80
```

① 使用标准SQL

```
test=# select name,
test-#        max(case when subject = '语文' then score else 0 end) as "语文", test-# max(case when subject = '数学' then score else 0 end) as "数学", test-#        max(case when subject = '英语' then score else 0 end) as "英语" test-#   from score
test-#  group by name order by name;
 name | 语文 | 数学 | 英语
------+------+------+------
 张三 |   80 |   70 |   60
 李四 |   90 |  100 |   80
(2 rows)
```

此方法简单并具有通用性,所有 SQL 数据库都支持。

② 使用内建聚合函数

```
test=# select name,
test-#        split_part(split_part(tmp,',',3),':',2) as "语文",
test-#        split_part(split_part(tmp,',',1),':',2) as "数学",
test-#        split_part(split_part(tmp,',',2),':',2) as "英语"
test-#   from
test-#   (select name,
test(#           string_agg(subject||':'||score,',' order by subject) as tmp
test(#      from score
test(#     group by name) as t
test-#  order by name;
 name | 语文 | 数学 | 英语
------+------+------+------
 张三 |   80 |   70 |   60
 李四 |   90 |  100 |   80
(2 rows)
```

在子查询中按 name 列分组聚合,使用 string_agg 函数将同一 name 的 subject 和 score 按 subject 顺序连接成字符串。subject 与 score 用 ':' 连接,段分隔符为 ','。子查询的结果为:

```
test=# select name,string_agg(subject||':'||score,',' order by subject) as tmp
test-#   from score
test-#  group by name;
 name |              tmp
------+-------------------------------
 张三 | 数学:70,英语:60,语文:80
 李四 | 数学:100,英语:80,语文:90
(2 rows)
```

外层查询使用两个嵌套的 split_part 函数,将字符串分隔成列。内层 split_part 取得

subject:score，外层 split_part 取得相应 subject 的 score。这种方法利用了 HAWQ 内建的聚合函数，实现简洁。

（2）不定列数的行转列

原始数据如下：

```
test=# select * from t1;
 c1 | c2 | c3
----+----+----
  1 | 我 | 1
  1 | 是 | 2
  1 | 谁 | 3
  2 | 不 | 1
  2 | 知 | 2
  3 | 道 | 1
(6 rows)
```

要得到以下的结果，其中假设列数是不定的：

```
 c1 | c2 | c3 | c4
----+----+----+----
  1 | 我 | 是 | 谁
  2 | 不 | 知 |
  3 | 道 |    |
```

使用 PostgreSQL 的 tablefunc 扩展模块很容易实现这个交叉表需求：

```
postgres=# select * from
postgres-# crosstab3('select c1::text, c3, c2::text from t1 order by c1, c3');
 row_name | category_1 | category_2 | category_3
----------+------------+------------+------------
 1        | 我         | 是         | 谁
 2        | 不         | 知         |
 3        | 道         |            |
(3 行记录)
```

遗憾的是，HAWQ 还不支持 tablefunc 扩展。由于结果集列数不固定，必须使用动态 SQL 实现，建立如下的 PLPGSQL 函数：

```
create or replace function fn_crosstab(refcursor) returns refcursor
as $body$
declare
    v_colnum int;
    v_sqlstring varchar(2000) := 'select c1 ';
begin
    -- 获得最大列数
```

```
        select max(c) into v_colnum from (select count(*) c from t1 group by c1) t;
        for i in 1 .. v_colnum loop
            v_sqlstring := v_sqlstring || ', split_part(tmp,'','',' 
|| cast(i as varchar(2)) || ') c'
            || cast(i+1 as varchar(2));
        end loop;

v_sqlstring := v_sqlstring
|| ' from (select c1,
string_agg(c2,'','' order by c3) as tmp
from t1 group by c1) t
order by c1';

        -- raise notice '%', v_sqlstring;
        open $1 for execute v_sqlstring;
        return $1;

end;
$body$ language plpgsql;
```

调用函数：

```
test=# begin;
BEGIN
test=# select fn_crosstab('cur1');
 fn_crosstab
-------------
 cur1
(1 row)

test=# fetch all in cur1;
 c1 | c2 | c3 | c4
----+----+----+----
  1 | 我 | 是 | 谁
  2 | 不 | 知 |
  3 | 道 |    |
(3 rows)

test=# commit;
COMMIT
test=#
```

服务器游标默认只能在一个事务中存在，事务结束自动销毁。如果没用 BEGIN 开启一个事务，任何一条语句都是一个事务（类似于 MySQL 的 autocommit），则 select fn_crosstab('cur1') 所建立的游标会被立即销毁。

2. 列转行

（1）单行变多行

原始数据如下：

```
test=# select * from book;
 id | name |   tag
----+------+---------
  1 | Java | aa,bb,cc
  2 | C++  | dd,ee
(2 rows)
```

要得到以下的结果：

```
name | tag | rn
-----+-----+----
Java | aa  | 1
Java | bb  | 2
Java | cc  | 3
C++  | dd  | 1
C++  | ee  | 2
```

HAWQ 2.1.1.0 基于 PostgreSQL 8.2.15，因此还不包含 generate_subscripts()、array_length()、unnest(array) with ordinality 等函数功能。为了给每个 name 的 tag 按原始位置增加序号，需要建立以下函数，返回数组值及其对应的下标：

```
create or replace function f_unnest_ord
(anyarray, out val anyelement, out ordinality integer)
returns setof record language sql immutable as
'select $1[i], i - array_lower($1,1) + 1
   from generate_series(array_lower($1,1), array_upper($1,1)) i';
```

然后执行查询：

```
test=# select name, (rec).val tag, (rec).ordinality rn
test-# from
test-#   (select *, f_unnest_ord(arr) as rec
test(#    from (select id, name, string_to_array(tag, ',') arr from book) t)
ttest-# order by id, rn;
 name | tag | rn
------+-----+----
 Java | aa  | 1
 Java | bb  | 2
 Java | cc  | 3
```

```
 C++  | dd |  1
 C++  | ee |  2
(5 rows)
```

（2）多列转多行

原始数据如下：

```
test=# select * from t1;
 c1 | c2 | c3 | c4
----+----+----+----
  1 | 我 | 是 | 谁
  2 | 不 | 知 |
  3 | 道 |    |
(3 rows)
```

要得到以下结果：

```
 c1 | c2 | c3
----+----+----
  1 | 我 |  1
  1 | 是 |  2
  1 | 谁 |  3
  2 | 不 |  1
  2 | 知 |  2
  3 | 道 |  1
```

可以看到，原数据只有三行，而结果是六行数据。要获得希望的结果，最重要的是如何从现有的行构造出新的数据行。下面用三种方法实现。

① 直接的方法——union

用 SQL 的并集操作符 union 是最容易想到的方法。

```
test=# select *
test-#   from (select c1,c2,1 c3 from t1
test(#         union all
test(#        select c1,c3,2 from t1
test(#         union all
test(#        select c1,c4,3 from t1) t
test-#  where c2 <> ''
test-#  order by c1, c3;
 c1 | c2 | c3
----+----+----
  1 | 我 |  1
  1 | 是 |  2
  1 | 谁 |  3
  2 | 不 |  1
```

```
 2 | 知 | 2
 3 | 道 | 1
(6 rows)
```

② 灵活的方法——笛卡儿积

union 虽然直接，但太过死板。如果列很多，需要叠加很多的 union all，凸显乏味。更灵活的方法是通过笛卡儿积运算构造数据行。这种方法的关键在于需要一个所需行数的辅助表。许多关系数据库都提供相应的方法，例如 Oracle 用 connect by level、MySQL 用数字辅助表、PostgreSQL 用 generate_serie 函数等。

```
test=# select *
test-#   from (select c1,
test(#            case when t2=1 then c2
test(#                 when t2=2 then c3
test(#                 else c4
test(#            end c2,
test(#            t2 c3
test(#        from (select * from t1, generate_series(1,3) t2) t) t
test-#  where c2 <> ''
test-#  order by c1, c3;
 c1 | c2 | c3
----+----+----
  1 | 我 | 1
  1 | 是 | 2
  1 | 谁 | 3
  2 | 不 | 1
  2 | 知 | 2
  3 | 道 | 1
(6 rows)
```

③ 独特的方法——unnest

前面两种是相对通用的方法，关系数据库的 SQL 都支持，而 unnest 是 PostgreSQL 独有的函数。有了前面的基础，这个实现就比较简单了，只要执行下面的查询即可。

```
test=# select *
test-#   from (select c1,split_part(unnest(c2),':',1) c2,
test(#                split_part(unnest(c2),':',2) c3
test(#          from (select c1,string_to_array(c2,',') c2
test(#                  from (select c1,
test(#                               coalesce(c2,'')||':1,'||
test(#                               coalesce(c3,'')||':2,'||
test(#                               coalesce(c4,'')||':3' c2
test(#                          from t1) t) t) t
test-#  where c2 <> ''
test-#  order by c1, c3;
 c1 | c2 | c3
----+----+----
  1 | 我 | 1
  1 | 是 | 2
```

```
 1 | 谁 | 3
 2 | 不 | 1
 2 | 知 | 2
 3 | 道 | 1
(6 rows)
```

18.3 交互查询与图形化显示

18.3.1 Zeppelin 简介

Zeppelin 是一个基于 Web 的软件,用于交互式地数据分析。它一开始是 Apache 软件基金会的孵化项目,2016 年 5 月正式成为顶级项目。Zeppelin 描述自己是一个可以进行数据摄取、数据发现、数据分析、数据可视化的笔记本,用以帮助开发者、数据科学家以及相关用户更有效地处理数据,而不必使用复杂的命令行,也不必关心集群的实现细节。Zeppelin 的架构如图 18-2 所示。

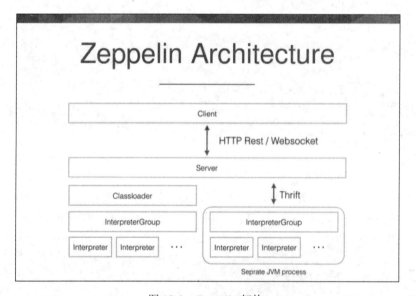

图 18-2　Zeppelin 架构

从上图中可以看到,Zeppelin 具有客户端/服务器架构,客户端一般就是指浏览器。服务器接收客户端的请求,并将请求通过 Thrift 协议发送给翻译器组。翻译器组物理表现为 JVM 进程,负责实际处理客户端的请求并与服务器进行通信。翻译器是一个插件式的体系结构,允许任何语言或后端数据处理程序以插件的形式添加到 Zeppelin 中。特别需要指出的是,Zeppelin 内建 Spark 翻译器,因此不需要构建单独的模块、插件或库。翻译器的架构如图 18-3 所示。

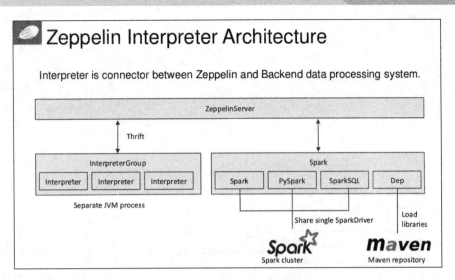

图 18-3　Zeppelin 翻译器架构

当前的 Zeppelin 已经支持很多翻译器，如 Zeppelin 0.6.0 版本自带的翻译器有 alluxio、cassandra、file、hbase、ignite、kylin、md、phoenix、sh、tajo、angular、elasticsearch、flink、hive、jdbc、lens、psql、spark 等 18 种之多。插件式架构允许用户在 Zeppelin 中使用自己熟悉的特定程序语言或数据处理方式。例如，通过使用%spark 翻译器，可以在 Zeppelin 中使用 Scala 语言代码。在数据可视化方面，Zeppelin 已经包含一些基本的图表，如柱状图、饼图、线形图、散点图等，任何支持的后端语言输出都可以被图形化表示。

在 Zeppelin 中，用户建立的每一个查询叫做一个 note，note 的 URL 在多用户间共享，Zeppelin 将向所有用户实时广播 note 的变化。Zeppelin 还提供一个只显示查询结果的 URL，该网页不包括任何菜单和按钮。用这种方式可以方便地将结果页作为一帧嵌入到自己的 Web 站点中。

18.3.2　使用 Zeppelin 执行 HAWQ 查询

1. 安装 Zeppelin

HDP 2.5.0 安装包中已经集成了 Zeppelin 0.6.0，因此不需要单独进行复杂的安装配置，只要启动 Zeppelin 服务就可以了。

2. 配置 Zeppelin 支持 HAWQ

Zeppelin 0.6.0 通过 JDBC 解析 HAWQ 查询，只需进行简单的配置即可，步骤如下。

（1）在 Ambari 控制台主页面中，单击 Services → Zeppelin Notebook → Quick Links → Zeppelin UI，打开 Zeppelin UI 主页面。

（2）在 Zeppelin UI 主页面中，单击 anonymous → interpreter，进入翻译器页面。

（3）单击 edit 编辑 jdbc 翻译器，配置 default.driver、default.password、default.url、default.user 四个属性的值，这里的配置如图 18-4 所示。

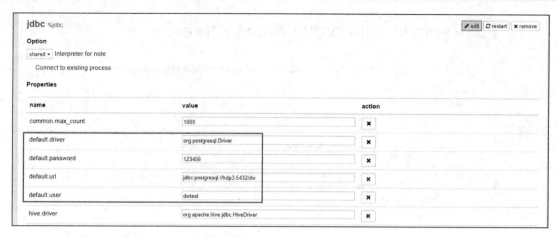

图 18-4　配置 JDBC 翻译器属性

（4）配置好后单击 Save 保存配置，然后单击 restart 重启 jdbc 翻译器，至此配置完成。

3. 在 Zeppelin 中执行 HAWQ 查询

单击 Notebook → Create new note，新建一个 note，在其中输入查询语句，如"每种产品类型以及单个产品在每个省、每个城市的月销售量和销售额是多少？"的查询。

```
%jdbc
select t2.product_category, t2.product_name, t3.state, t3.city,
sum(nq) sq, sum(order_amount) sa
  from v_sales_order_fact t1, product_dim t2, zip_code_dim t3
 where t1.product_sk = t2.product_sk
   and t1.customer_zip_code_sk = t3.zip_code_sk
 group by t2.product_category, t2.product_name, t3.state, t3.city
 order by t2.product_category, t2.product_name, t3.state, t3.city;
```

运行结果的表格、柱状图、饼图、堆叠图、线形图、散点图分别如图 18-5~图 18-10 所示。

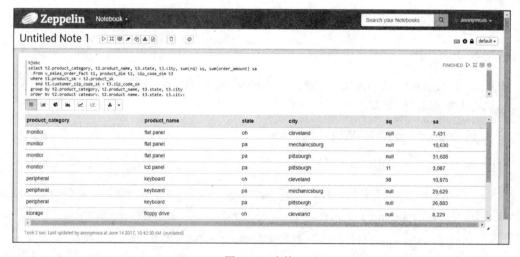

图 18-5　表格

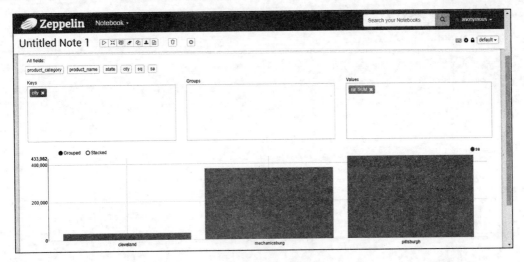

图 18-6　柱状图

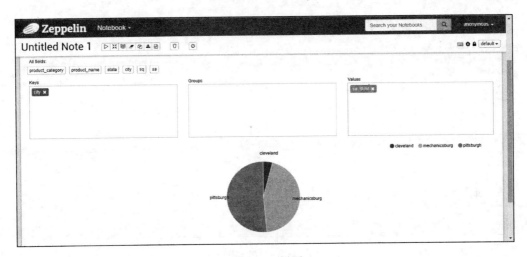

图 18-7　饼图

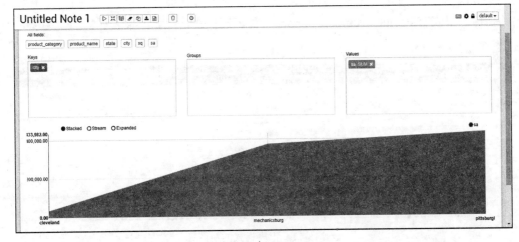

图 18-8　堆叠图

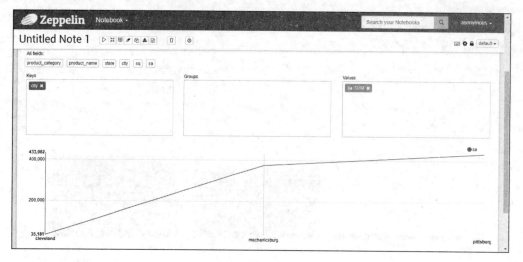

图 18-9　线形图

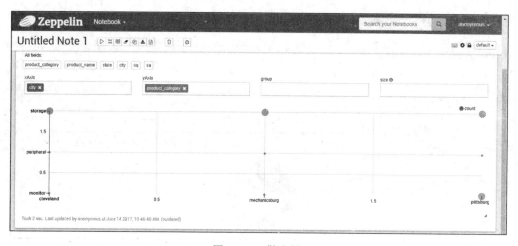

图 18-10　散点图

一个 note 中可以独立执行多个查询语句。图形显示可以根据不同的"settings"联机分析不同的指标。报表有 default、simple、report 三种可选样式。例如，报表样式的饼图表示如图 18-11 所示。

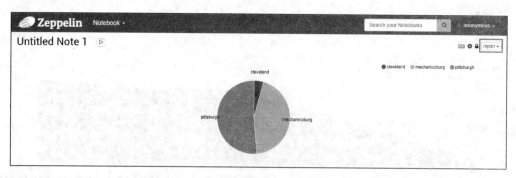

图 18-11　报表样式的饼图

可以单击如图 18-12 红框中所示的链接单独引用此报表。

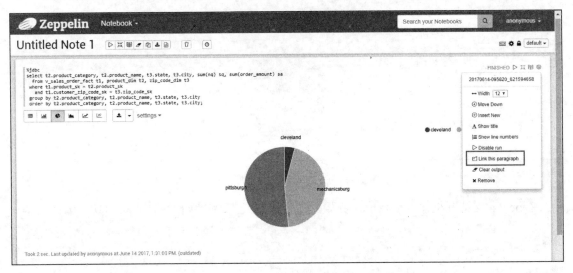

图 18-12　链接报表

单独的页面能根据查询或设置的修改而实时变化，比如将 Values 由 sa 列改为 sq 列，饼图表变为图 18-13 的样子。

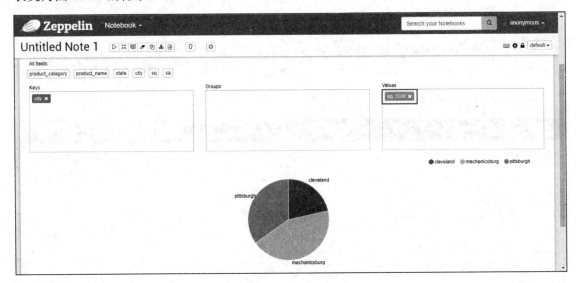

图 18-13　查询设置

单独链接的页面也随之自动发生变化，如图 18-14 所示。

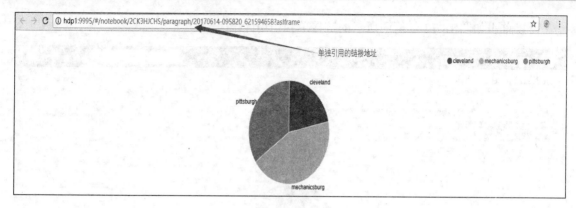

图 18-14　单独链接的页面联动改变

Zeppelin 支持联机输入变量值，例如，要查询某一年的销售情况，查询语句改为：

```
%jdbc
select t2.product_category, t2.product_name, t3.state, t3.city,
sum(nq) sq, sum(order_amount) sa
  from v_sales_order_fact t1, product_dim t2, zip_code_dim t3
 where t1.product_sk = t2.product_sk
   and t1.customer_zip_code_sk = t3.zip_code_sk
   and t1.year_month/100 = ${year}
 group by t2.product_category, t2.product_name, t3.state, t3.city
 order by t2.product_category, t2.product_name, t3.state, t3.city;
```

运行查询时会在页面中出现一个输入框，填入适当的变量值运行查询，如图 18-15 所示。

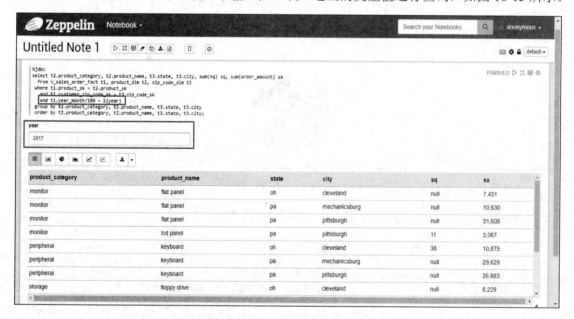

图 18-15　包含联机变量的查询

甚至可以动态定义查询的列，例如查询语句改为：

```
%jdbc
select
${checkbox:fields=t2.product_category, t2.product_category|t2.product_name},
  t3.state, t3.city, sum(nq) sq, sum(order_amount) sa
    from v_sales_order_fact t1, product_dim t2, zip_code_dim t3
  where t1.product_sk = t2.product_sk
    and t1.customer_zip_code_sk = t3.zip_code_sk
    and t1.year_month/100 = ${year}
  group by
${checkbox:fields=t2.product_category, t2.product_category|t2.product_name},
t3.state, t3.city
  order by
${checkbox:fields=t2.product_category, t2.product_category|t2.product_name},
t3.state, t3.city;
```

查询运行时出现字段复选框，如图 18-16 所示。

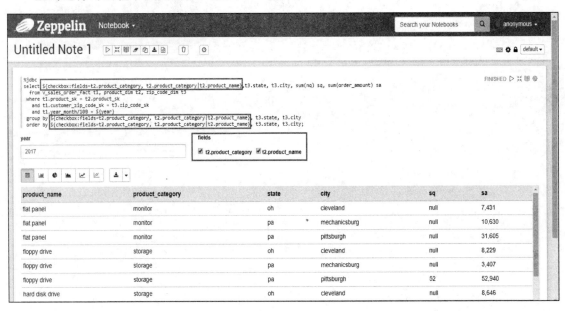

图 18-16　动态查询

18.4 小结

联机分析处理系统从数据仓库中的集成数据出发，构建面向分析的多维数据模型，再使用多维分析方法从多个不同的视角对数据集合进行分析比较，分析活动以数据驱动，通常分为 MOLAP、ROLAP、HOLAP 三种类型。HAWQ 由于性能优势和健全的函数功能，比较适合做联机分析处理。

Zeppelin 是 Hadoop 的数据可视化组件，支持的后端数据查询程序较多，原生支持 Spark。Zeppelin 采用插件式的翻译器，通过插件，可以添加任何后端语言及其数据处理程序。通过配置 JDBC，Zeppelin 可以支持 HAWQ 查询。

第三部分

HAWQ 数据挖掘

第 19 章
◀ 整合HAWQ与MADlib ▶

本书最后一部分介绍如何使用HAWQ进行数据挖掘。能用单纯的SQL解决数据挖掘问题，这也是HAWQ有别于其他SQL-on-Hadoop产品的一大亮点。我们从HAWQ与MADlib的整合开始说明，然后以实例详细介绍使用HAWQ和MADlib实现典型的数据挖掘方法。先让我们了解一下数据挖掘的基本概念。

数据挖掘（Data Mining），是从大量的、不完全的、有噪声的、模糊的、随机的实际应用数据中，提取隐含在其中的、人们事先不知道的、但又是潜在有用的信息和知识的过程。与传统数据分析（如查询、报表、OLAP等）的本质区别是，数据挖掘是在没有明确假设的前提下去挖掘信息、发现知识。数据挖掘使数据库技术进入到一个更高的阶段。概括来说，数据挖掘技术具有以下几个特点：

- 处理的数据规模庞大，达到GB、TB，甚至更大数量级。
- 查询一般是决策制定者（用户）提出的即时随机查询，往往不能形成精确的查询要求，需要靠系统本身寻找其可能感兴趣的东西。
- 在一些应用（如商业投资等）中，由于数据变化迅速，因此要求数据挖掘能快速做出反应以随时提供决策支持。
- 数据挖掘中，规则的发现基于统计规律。所发现的规则不必适用于所有数据，而是当达到某一临界值时即认为有效。因此，利用数据挖掘技术可能会发现大量的规则。
- 数据挖掘所发现的规则是动态的，它只反映了当前状态的数据库具有的规则，随着不断向数据库中加入新数据，需要随时对其进行更新。

数据挖掘过程的核心是利用算法对处理好的输入输出数据进行训练，并得到模型，然后再对模型进行验证，使得模型能够在一定程度上刻画出数据由输入到输出的关系，然后再利用该模型，对新输入的数据进行计算，从而得到新的输出，对这个输出就可以进行解释和应用了。虽然模型可能不易理解或很难直观看到，但它是基于大量数据训练并经过验证的，因此能够反映输入和输出数据之间的大致关系，这种关系（模型）就是我们需要的知识。从以上原理可以看出，数据挖掘有科学依据，挖掘的结果也是值得信任的。数据挖掘的内容总是集中在关联、回归、分类、聚类、预测、诊断六个方面，后面几章的示例也基本围绕这些方面进行讨论。

19.1 MADlib 简介

HAWQ 的数据挖掘是通过 MADlib 实现的。MADlib 是 Pivotal 公司与伯克利大学合作开发的一个开源机器学习库，提供了精确的数据并行实现、统计和机器学习方法，对结构化和非结构化数据进行分析。可以非常方便地将 MADlib 加载到数据库中，扩展数据库的分析功能。2015 年 7 月 MADlib 成为 Apache 软件基金会的孵化项目，其最新版本为 MADlib 1.11，可以用在 PostgreSQL、Greenplum 和 HAWQ 等数据库系统中。

1. 设计思想

驱动 MADlib 架构的主要思想与 Hadoop 是一致的，主要体现在以下方面：

- 操作数据库内的本地数据，不在多个运行时环境中进行不必要的数据移动。
- 充分利用数据库引擎功能，但将机器学习逻辑从特定数据库的实现细节中分离出来。
- 利用 MPP 无共享技术提供的并行性和可扩展性，如 Greenplum 数据库和 HAWQ。
- 执行的维护活动对 Apache 社区和正在进行的学术研究开放。

2. 支持的模型

（1）分类

如果所需的输出实质上是分类的，可以使用分类方法建立模型，预测新数据会属于哪一类。分类的目标是能够将输入记录标记为正确的类别。例如，假设有描述人口统计的数据，以及个人申请贷款和贷款违约历史数据，那么我们就能建立一个模型，描述新的人口统计数据集合贷款违约的可能性。此场景下输出的分类为"违约"和"正常"两类。

（2）回归

如果所需的输出具有连续性，我们使用回归方法建立模型，预测输出值。例如，如果有真实的描述房地产属性的数据，我们就可以建立一个模型，预测基于房屋已知特征的售价。因为输出反映了连续的数值而不是分类，所以该场景是一个回归问题。

（3）聚类

识别数据分组，一组中的数据项比其他组的数据项更相似。例如，在客户细分分析中，目标是识别客户行为相似特征组，以便针对不同特征的客户设计各种营销活动，以达到市场目的。如果提前了解客户细分情况，这将是一个受控的分类任务。当我们让数据识别自身分组时，这就是一个聚类任务。

（4）主题建模

主题建模与聚类相似，也是确定彼此相似的数据组。但这里的相似通常特指在文本领域中，具有相同主题的文档。

（5）关联规则挖掘

关联规则挖掘又叫做购物篮分析或频繁项集挖掘。相对于随机发生，确定哪些事项更经常一起发生,指出事项之间的潜在关系。例如,在一个网店应用中,关联规则挖掘可用于确定哪些商品倾向于被一起售出,然后将这些商品输入到客户推荐引擎中,提供促销机会,就像著名的啤酒与尿布的故事。

(6)描述性统计

描述性统计不提供模型,因此不被认为是一种机器学习方法。但描述性统计有助于向分析人员提供信息以了解基础数据,为数据提供有价值的解释,可能影响数据模型的选择。例如,计算数据集中每个变量内的数据分布,可以帮助分析理解哪些变量应被视为分类变量,哪些变量是连续性变量,以及值的分布情况。

(7)模型验证

如果不了解一个模型的准确性就开始使用它,很容易导致糟糕的结果。正因如此,理解模型存在的问题,并用测试数据评估模型的精度尤为重要。需要将训练数据和测试数据分离,频繁进行数据分析,验证统计模型的有效性,评估模型不过分拟合训练数据。N-fold 交叉验证方法经常被使用。

3. 功能

MADlib 的功能特色如图 19-1 所示。

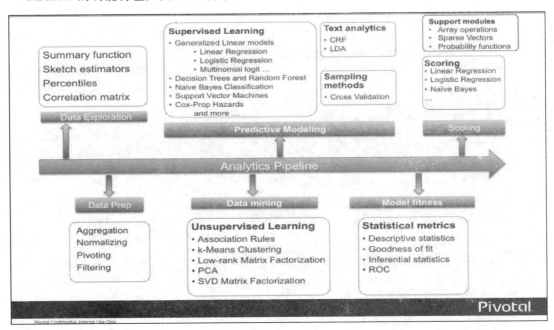

图 19-1　MADlib 功能

(1) Data Types and Transformations(数据类型转换)

```
Arrays and Matrices(数组与矩阵)
    o   Array Operations(数组运算)
    o   Matrix Operations(矩阵运算)
    o   Matrix Factorization(低矩阵分解)
```

```
            o   Low-rank Matrix Factorization（低阶矩阵分解）
            o   Singular Value Decomposition（SVD，奇异值分解）
        o   Norms and Distance functions（规范和距离函数）
        o   Sparse Vectors（稀疏向量）
    Dimensionality Reduction（降维）
        o   Principal Component Analysis（PCA 主成分分析）
        o   Principal Component Projection（PCP 主成分投影）
Encoding Categorical Variables（编码分类变量）
Stemming（切词）
```

（2）Model Evaluation（模型评估）

```
Cross Validation（交叉验证）
```

（3）Statistics（统计）

```
Descriptive Statistics（描述性统计）
    o   Pearson's Correlation（皮尔斯相关性）
    o   Summary（摘要汇总）
Inferential Statistics（推断性统计）
    o   Hypothesis Tests（假设检验）
Probability Functions（概率函数）
```

（4）Supervised Learning（监督学习算法）

```
Conditional Random Field（条件随机场）
Regression Models（回归模型）
    o   Clustered Variance（聚类方差）
    o   Cox-Proportional Hazards Regression（Cox 比率风险回归模型）
    o   Elastic Net Regularization（Elastic Net 回归）
    o   Generalized Linear Models
    o   Linear Regression（线性回归）
    o   Logistic Regression（逻辑回归）
    o   Marginal Effects（边际效应）
    o   Multinomial Regression（多项式回归）
    o   Ordinal Regression（有序回归）
    o   Robust Variance（鲁棒方差）
Support Vector Machines（SVM，支持向量机）
Tree Methods（树模型）
    o   Decision Tree（决策树）
    o   Random Forest（随机森林）
```

（5）Time Series Analysis（时间序列分析）

```
ARIMA（自回归积分滑动平均模型）
```

（6）Unsupervised Learning（无监督学习）

```
Association Rules（关联规则）
    o   Apriori Algorithm（Apriori 算法）
Clustering（聚类）
    o   k-Means Clustering（k-Means）
Topic Modelling（主题模型）
    o   Latent Dirichlet Allocation（LDA）
```

（7）Utility Functions（效用函数）

```
Developer Database Functions（开发者数据库函数）
Linear Solvers（线性求解器）
    o    Dense Linear Systems（稠密线性系统）
    o    Sparse Linear Systems（稀疏线性系统）
Path Functions（路径函数）
PMML Export（PMML 输出）
Text Analysis（文本分析）
    o    Term Frequency（词频，TF）
```

19.2 安装与卸载 MADlib

1. 确定安装平台

MADlib 最新发布版本是 1.11，可以安装在 PostgreSQL、Greenplum 和 HAWQ 中，在不同的数据库中安装过程也不尽相同。我们的实验环境是安装在 HAWQ2.1.1.0 中。

2. 下载 MADlib 二进制压缩包

下载地址为：https://network.pivotal.io/products/pivotal-hdb。2.1.1.0 版本的 HAWQ 提供了四个安装文件，如图 19-2 所示。经过测试，只有 MADlib 1.10.0 版本的文件可以正常安装。

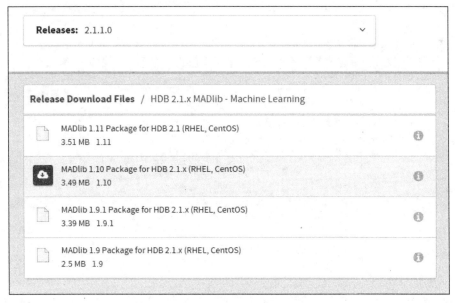

图 19-2　下载 MADlib 安装文件

3. 安装 MADlib

以下命令需要使用 gpadmin 用户，在 HAWQ 的 Master 主机上执行。

（1）解压缩

```
tar -zxvf madlib-ossv1.10.0_pv1.9.7_hawq2.1-rhel5-x86_64.tar.gz
```

（2）安装 MADlib 的 gppkg 文件

```
gppkg -i madlib-ossv1.10.0_pv1.9.7_hawq2.1-rhel5-x86_64.gppkg
```

该命令在 HAWQ 集群的所有节点（Master 和 Segment）上创建 MADlib 的安装目录和文件，默认目录为/usr/local/hawq_2_1_1_0/madlib。

（3）在指定数据库中部署 MADlib

```
$GPHOME/madlib/bin/madpack install -c /dm -s madlib -p hawq
```

该命令在 HAWQ 的 dm 数据库中建立 madlib schema，-p 参数指定平台为 HAWQ。命令执行后可以查看在 madlib schema 中创建的数据库对象。

```
dm=# set search_path=madlib;
SET
dm=# \dt
                List of relations
 Schema |       Name       | Type  |  Owner  |   Storage
--------+------------------+-------+---------+-------------
 madlib | migrationhistory | table | gpadmin | append only
(1 row)

dm=# \ds
                   List of relations
 Schema |          Name           |   Type   |  Owner  | Storage
--------+-------------------------+----------+---------+--------
 madlib | migrationhistory_id_seq | sequence | gpadmin | heap
(1 row)

dm=# select type,count(*)
dm-#   from (select p.proname as name,
dm(#            case when p.proisagg then 'agg'
dm(#                 when p.prorettype
dm(#                     = 'pg_catalog.trigger'::pg_catalog.regtype
dm(#                 then 'trigger'
dm(#                 else 'normal'
dm(#            end as type
dm(#          from pg_catalog.pg_proc p, pg_catalog.pg_namespace n
dm(#         where n.oid = p.pronamespace and n.nspname='madlib') t
dm-#  group by rollup (type);
  type  | count
--------+-------
 agg    |   135
 normal |  1324
        |  1459
(3 rows)
```

可以看到，MADlib 部署应用程序 madpack 首先创建数据库模式 madlib，然后在该模式中创建数据库对象，包括一个表，一个序列，1324 个普通函数，135 个聚合函数。所有的机器学习和数据挖掘模型、算法、操作和功能都是通过调用这些函数实际执行的。

（4）验证安装

```
$GPHOME/madlib/bin/madpack install-check -c /dm -s madlib -p hawq
```

该命令通过执行 29 个模块的 77 个案例，验证所有模块都能正常工作。命令输出如下，如果看到所有案例都已经正常执行，说明 MADlib 安装成功。

```
[gpadmin@hdp3 Madlib]$ $GPHOME/madlib/bin/madpack install-check -c /dm -s madlib -p hawq
madpack.py : INFO : Detected HAWQ version 2.1.
TEST CASE RESULT|Module: array_ops|array_ops.sql_in|PASS|Time: 1851 milliseconds
TEST CASE RESULT|Module: bayes|gaussian_naive_bayes.sql_in|PASS|Time: 24222 milliseconds
TEST CASE RESULT|Module: bayes|bayes.sql_in|PASS|Time: 70634 milliseconds
...
TEST CASE RESULT|Module: pca|pca.sql_in|PASS|Time: 523230 milliseconds
TEST CASE RESULT|Module: validation|cross_validation.sql_in|PASS|Time: 33685 milliseconds
[gpadmin@hdp3 Madlib]$
```

4. 卸载

卸载过程基本上是安装的逆过程。

（1）删除 madlib 模式

方法 1，使用 madpack 部署应用程序删除模式。

```
$GPHOME/madlib/bin/madpack uninstall -c /dm -s madlib -p hawq
```

方法 2，使用 SQL 命令手工删除模式。

```
drop schema madlib cascade;
```

（2）删除其他遗留数据库对象

- 删除模式

如果测试中途出错，数据库中可能包含测试的模式，这些模式名称的前缀都是 madlib_installcheck_，只能手工执行 SQL 命令删除这些模式，如：

```
drop schema madlib_installcheck_kmeans cascade;
```

- 删除用户

如果存在遗留的测试用户，则删除它，如：

```
drop user if exists madlib_1100_installcheck;
```

(3) 删除 MADlib rpm 包

查询包名：

```
gppkg -q --all
```

输出如下：

```
[gpadmin@hdp3 Madlib]$ gppkg -q --all
20170630:16:19:53:076493 gppkg:hdp3:gpadmin-[INFO]:-Starting gppkg with args:
-q --all
madlib-ossv1.10.0_pv1.9.7_hawq2.1
```

删除 rpm 包：

```
gppkg -r madlib-ossv1.10.0_pv1.9.7_hawq2.1
```

19.3 MADlib 基础

和其他数据挖掘语言或工具一样，MADlib 操作的基本对象也是向量与矩阵。对向量和矩阵的操作是通过一系列函数完成的。本节将介绍 MADlib 中向量与矩阵的概念，并举出一些很简单的函数调用示例。虽然很重要，但限于篇幅，这里不详细解释每个函数相关参数的具体意义及其函数的数学含义，读者可查询参考文献中列出的 MADlib 官方文档地址获得详细信息。还可以使用 psql 的联机帮助，查看函数的参数、返回值和函数体等信息，例如：\df matrix_sparsify 或\df+ matrix_sparsify。我们将侧重于应用，因为学会这些函数的基本用法是后面进行数据挖掘的基础。

19.3.1 向量

1. 定义

这里不讨论向量严格的数学定义。在 MADlib 中，可以把向量简单理解为矩阵的一种特殊形式。矩阵是 MADlib 中数据的基本格式，当矩阵只有一维时，就是向量，1 行 n 列的矩阵称为行向量，m 行 1 列的矩阵称为列向量，1 行 1 列的矩阵称为标量。

2. 函数

MADlib 的线性代数模块（linalg module）包括基本的线性代数操作的实用函数。利用线性代数函数可以很方便地实现新算法。这些函数操作向量（1 维 FLOAT8 数组）和矩阵（2 维 FLOAT8 数组）。注意，这类函数只接受 FLOAT8 数组参数，因此在调用函数时，需要将其他类型的数组转换为 FLOAT8[]。

（1）函数概览

MADlib 中的线性代数函数主要包括范数、距离、矩阵、聚合几类。表 19-1 列出了相关函数的简要说明。

表 19-1　MADlib 线性代数函数

函数名称	描述	参数	返回值
norm1()	向量的 1 范数，$\|\vec{a}\|_1$	x vector $\vec{x}=(x_1,...,x_n)$	$\|x\|_1=\sum_{i=1}^{n}\|x_i\|$
norm2()	向量的 2 范数，$\|\vec{a}\|_2$	x vector $\vec{x}=(x_1,...,x_n)$	$\|x\|_2=\sqrt{\sum_{i=1}^{n}x_i^2}$
dist_norm1()	两个向量之差的 1 范数，$\|\vec{a}-\vec{b}\|_1$	x vector $\vec{x}=(x_1,...,x_n)$ y vector $\vec{y}=(y_1,...,y_n)$	$\|x-y\|_1=\sum_{i=1}^{n}\|x_i-y_i\|$
dist_norm2()	两个向量之差的 2 范数，$\|\vec{a}-\vec{b}\|_2$	x vector $\vec{x}=(x_1,...,x_n)$ y vector $\vec{y}=(y_1,...,y_n)$	$\|x-y\|_2=\sqrt{\sum_{i=1}^{n}(x_i-y_i)^2}$
dist_pnorm()	两个向量之差的 p 范数，$\|\vec{a}-\vec{b}\|_p, p>0$	x vector $\vec{x}=(x_1,...,x_n)$ y vector $\vec{y}=(y_1,...,y_n)$ p Scalar p>0	$\|x-y\|_p=(\sum_{i=1}^{n}\|x_i-y_i\|^p)^{\frac{1}{p}}$
dist_inf_norm()	两个向量之差的无穷范数，$\|\vec{a}-\vec{b}\|_\infty$	x vector $\vec{x}=(x_1,...,x_n)$ y vector $\vec{y}=(y_1,...,y_n)$	$\|x-y\|_\infty=\max_{i=1}^{n}\|x_i-y_i\|$
squared_dist_norm2()	两个向量之差的 2 范数的平方，$\|\vec{a}-\vec{b}\|_2^2$	x vector $\vec{x}=(x_1,...,x_n)$ y vector $\vec{y}=(y_1,...,y_n)$	$\|x-y\|_2^2=\sum_{i=1}^{n}(x_i-y_i)^2$
cosine_similarity()	两个向量的余弦相似度（角距离），$\frac{\vec{a}\cdot\vec{b}}{\|\vec{a}\|_2\|\vec{b}\|_2}$	x vector $\vec{x}=(x_1,...,x_n)$ y vector $\vec{y}=(y_1,...,y_n)$	$\frac{(\vec{x},\vec{y})}{\|\vec{x}\|\cdot\|\vec{y}\|}$
dist_angle()	欧氏空间中两个向量之间的夹角，$\cos^{-1}(\frac{\vec{a}\cdot\vec{b}}{\|\vec{a}\|_2\|\vec{b}\|_2})$	x vector $\vec{x}=(x_1,...,x_n)$ y vector $\vec{y}=(y_1,...,y_n)$	$\arccos(\frac{(\vec{x},\vec{y})}{\|\vec{x}\|\cdot\|\vec{y}\|})$

（续表）

函数名称	描述	参数	返回值
dist_tanimoto()	两个向量间的谷本距离	x vector $\vec{x}=(x_1,...,x_n)$ y vector $\vec{y}=(y_1,...,y_n)$	$1-\dfrac{(\vec{x},\vec{y})}{\|\vec{x}\|^2 \cdot \|\vec{y}\|^2 - (\vec{x},\vec{y})}$
dist_jaccard()	两个字符向量集之间的杰卡德距离	x vector $\vec{x}=(x_1,...,x_n)$ y vector $\vec{y}=(y_1,...,y_n)$	$1-\dfrac{\|x \cap y\|}{\|x \cup y\|}$
get_row()	返回矩阵的行下标（2维数组）	Input 2-D array Index	二维数组的一行
get_col()	返回矩阵的列下表（2维数组）	Input 2-D array Index	二维数组的一列
avg()	计算向量的平均值	x Point x_i	$\dfrac{1}{n}\sum_{i=1}^{n} x_i$
normalized_avg()	计算向量的归一化平均值（欧氏空间中的单位向量）	x Point x_i	$\dfrac{\tilde{x}}{\|\tilde{x}\|}$
matrix_agg()	将向量合并进一个矩阵	x Point x_i	Matrix with columns $x_1,...,x_n$

（2）函数示例

- 范数与距离函数

创建包含两个向量列的数据库表，并添加数据。

```
drop table if exists two_vectors;
create table two_vectors(id integer, a float8[], b float8[]);
insert into two_vectors values
(1, '{3,4}', '{4,5}'),
(2, '{1,1,0,-4,5,3,4,106,14}', '{1,1,0,6,-3,1,2,92,2}');
```

范数函数：

```
dm=# select id, madlib.norm1(a), madlib.norm2(a) from two_vectors;
 id | norm1 |      norm2
----+-------+------------------
  1 |     7 |                5
  2 |   138 | 107.238052947636
(2 rows)
```

距离函数:

```
dm=# \x
Expanded display is on.
dm=# select id,
dm-#        madlib.dist_norm1(a, b),
dm-#        madlib.dist_norm2(a, b),
dm-#        madlib.dist_pnorm(a, b, 5) as norm5,
dm-#        madlib.dist_inf_norm(a, b),
dm-#        madlib.squared_dist_norm2(a, b) as sq_dist_norm2,
dm-#        madlib.cosine_similarity(a, b),
dm-#        madlib.dist_angle(a, b),
dm-#        madlib.dist_tanimoto(a, b),
dm-#        madlib.dist_jaccard(a::text[], b::text[])
dm-#   from two_vectors;
-[ RECORD 1 ]-----+--------------------
id                | 1
dist_norm1        | 2
dist_norm2        | 1.4142135623731
norm5             | 1.14869835499704
dist_inf_norm     | 1
sq_dist_norm2     | 2
cosine_similarity | 0.999512076087079
dist_angle        | 0.0312398334302684
dist_tanimoto     | 0.0588235294117647
dist_jaccard      | 0.666666666666667
-[ RECORD 2 ]-----+--------------------
id                | 2
dist_norm1        | 48
dist_norm2        | 22.6274169979695
norm5             | 15.585086360695
dist_inf_norm     | 14
sq_dist_norm2     | 512
cosine_similarity | 0.985403348449008
dist_angle        | 0.171068996592859
dist_tanimoto     | 0.0498733684005455
dist_jaccard      | 0.833333333333333
```

- 矩阵函数

创建包含矩阵列的数据库表。

```
drop table if exists matrix;
create table matrix(id integer, m float8[]);
insert into matrix values (1, '{{4,5},{3,5},{9,0}}');
```

调用矩阵函数：

```
dm=# select madlib.get_row(m, 1) as row_1,
dm-#        madlib.get_row(m, 2) as row_2,
dm-#        madlib.get_row(m, 3) as row_3,
dm-#        madlib.get_col(m, 1) as col_1,
dm-#        madlib.get_col(m, 2) as col_2
dm-#   from matrix;
 row_1 | row_2 | row_3 |  col_1  |  col_2
-------+-------+-------+---------+---------
 {4,5} | {3,5} | {9,0} | {4,3,9} | {5,5,0}
(1 row)
```

- 聚合函数

创建包含向量列的数据库表。

```
drop table if exists vector;
create table vector(id integer, v float8[]);
insert into vector values (1, '{4,3}'), (2, '{8,6}'), (3, '{12,9}');
```

调用聚合函数：

```
dm=# select madlib.avg(v), madlib.normalized_avg(v), madlib.matrix_agg(v)
dm-#   from vector;
  avg  | normalized_avg  |      matrix_agg
-------+-----------------+----------------------
 {8,6} | {0.8,0.6}       | {{4,3},{8,6},{12,9}}
(1 row)
```

3. 稀疏向量

（1）MADlib 的稀疏向量

MADlib 实现了一种稀疏向量数据类型，名为"svec"，能够为包含大量重复元素的向量提供压缩存储。浮点数组进行各种计算，有时会有很多的零或其他默认值，在科学计算、零售优化、文本处理等应用中，这是很常见的。每个浮点数在内存或磁盘中使用 8 字节存储，例如有如下 float8[]数据类型的数组：

```
'{0, 33,...40,000 zeros..., 12, 22 }'::float8[]
```

这个数组会占用 320KB 的内存或磁盘，而其中绝大部分存储的是 0 值。即使我们利用 null 位图，将 0 作为 null 存储，还是会得到一个 5KB 的 null 位图，内存使用效率还是不够高。何况在执行数组操作时，40000 个零列上的计算结果并不重要。为了解决这个向量存储问题，svec 类型使用行程长度编码（Run Length Encoding，RLE），即用一个数-值对数组表示稀疏向量。例如，上面的数组被存储为：

```
'{1,1,40000,1,1}:{0,33,0,12,22}'::madlib.svec
```

就是说 1 个 0、1 个 33、40000 个 0 等等，只使用 5 个整型和 5 个浮点数类型构成数组存储。除了节省空间，这种 RLE 表示也很容易实现向量操作，并使向量计算更快。SVEC 模块提供了相关的函数库。Madlib 1.10 版本仅支持 float8 稀疏向量类型。

（2）创建稀疏向量

步骤 01 直接使用常量表达式构建一个 SVEC，n1、n2、...、nk 指定值 v1、v2、...、vk 的个数。

```
select '{n1,n2,...,nk}:{v1,v2,...vk}'::madlib.svec;
```

步骤 02 将一个 float 数组转换成 SVEC。

```
select ('{v1,v2,...vk}'::float[])::madlib.svec;
```

步骤 03 使用聚合函数创建一个 SVEC。

```
select madlib.svec_agg(v1) from generate_series(1,10) v1;
```

步骤 04 利用 madlib.svec_cast_positions_float8arr() 函数创建 SVEC。

```
select madlib.svec_cast_positions_float8arr(
array[1,3,5], array[2,4,6], 10, 0.0);
```

生成的 SVEC 为：

```
svec_cast_positions_float8arr
-----------------------------
{1,1,1,1,1,5}:{2,0,4,0,6,0}
```

（3）稀疏向量示例

下面我们来看一个稀疏向量简单示例。

对 SVEC 类型可以应用<、>、*、**、/、=、+、SUM 等操作和运算，并且具有典型的向量操作的相关含义。例如，加法（+）操作是对两个向量中相同下标对应的元素进行相加。为了使用 SVEC 的操作符，需要将 madlib 模式添加到 search_path 中。

```
dm=# -- 将 madlib 模式添加到搜索路径中
dm=# set search_path="$user",public,madlib;
SET
dm=# -- 稀疏向量相加
dm=# select ('{0,1,5}'::float8[]::madlib.svec
dm(#         + '{4,3,2}'::float8[]::madlib.svec)::float8[];
 float8
---------
 {4,4,7}
(1 row)
```

如果最后不转换成 float8[]，结果是一个 SVEC 类型：

```
dm=# select ('{0,1,5}'::float8[]::madlib.svec
dm(#          + '{4,3,2}'::float8[]::madlib.svec);
 ?column?
--------------
 {2,1}:{4,7}
(1 row)
```

两个向量的点积（%*%）结果是 float8 类型，如(0*4 + 1*3 + 5*2) = 13：

```
dm=# select '{0,1,5}'::float8[]::madlib.svec
dm-#         %*% '{4,3,2}'::float8[]::madlib.svec;
 ?column?
----------
       13
(1 row)
```

有些聚合函数对 SVEC 也是可用的，如 SVEC_COUNT_NONZERO。

```
drop table if exists list;
create table list (a madlib.svec);
insert into list values
('{0,1,5}'::float8[]::madlib.svec), ('{10,0,3}'::float8[]::madlib.svec),
('{0,0,3}'::float8[]::madlib.svec),('{0,1,0}'::float8[]::madlib.svec);
```

SVEC_COUNT_NONZERO 函数统计 SVEC 中每一列非 0 元素的个数，返回计数的 SVEC。

```
dm=# select madlib.svec_count_nonzero(a)::float8[] from list;
 svec_count_nonzero
--------------------
 {1,2,3}
(1 row)
```

SVEC 数据类型中不应该使用 NULL，因为 NULL 会显式表示为 NVP（No Value Present）。

```
dm=# select '{1,2,3}:{4,null,5}'::madlib.svec;
      svec
-------------------
 {1,2,3}:{4,NVP,5}
(1 row)
```

含有 NULL 的 SVEC 相加，结果中显示 NVP。

```
dm=# select '{1,2,3}:{4,null,5}'::madlib.svec
dm-#         + '{2,2,2}:{8,9,10}'::madlib.svec;
        ?column?
---------------------------
 {1,2,1,2}:{12,NVP,14,15}
(1 row)
```

可以使用 SVEC_PROJ() 函数访问 SVEC 元素，该函数的参数为一个 SVEC 和一个元素下标。

```
dm=# select madlib.svec_proj('{1,2,3}:{4,5,6}'::madlib.svec, 1)
dm-#      + madlib.svec_proj('{4,5,6}:{1,2,3}'::madlib.svec, 15);
 ?column?
----------
        7
(1 row)
```

通过 SVEC_SUBVEC() 函数可以访问一个 SVEC 的子向量，该函数的参数为一个 SVEC，及其起止下标。

```
dm=# select madlib.svec_subvec('{2,4,6}:{1,3,5}'::madlib.svec, 2, 11);
   svec_subvec
-----------------
 {1,4,5}:{1,3,5}
(1 row)
```

SVEC 的元素/子向量可以通过 SVEC_CHANGE() 函数进行改变。该函数有三个参数：一个 m 维的 svec sv1，起始下标 j；一个 n 维的 svec sv2，其中 j + n - 1 <= m；返回类似 sv1 的 svec，但子向量 sv1[j:j+n-1] 被 sv2 所替换。

```
dm=# select madlib.svec_change('{1,2,3}:{4,5,6}'
dm(#       ::madlib.svec,3,'{2}:{3}'::madlib.svec);
     svec_change
---------------------
 {1,1,2,2}:{4,5,3,6}
(1 row)
```

还有处理 SVEC 的高阶函数。如 SVEC_LAPPLY 对应 R 语言中的 LAPPLY() 函数。

```
dm=# select madlib.svec_lapply('sqrt', '{1,2,3}:{4,5,6}'::madlib.svec);
                 svec_lapply
----------------------------------------------
 {1,2,3}:{2,2.23606797749979,2.44948974278318}
(1 row)
```

（4）扩展示例

下面的示例是对文档向量化为稀疏矩阵的说明，假设有一个由若干单词组成的文本数组：

```
drop table if exists features;
create table features (a text[]);
insert into features values
       ('{am,before,being,bothered,corpus,document,i,in,is,me,
         never,now,one,really,second,the,third,this,until}');
```

465

同时有一个文档集合，每个文档表示为一个单词数组：

```
drop table if exists documents;
create table documents(a int,b text[]);
insert into documents values
       (1,'{this,is,one,document,in,the,corpus}'),
       (2,'{i,am,the,second,document,in,the,corpus}'),
       (3,'{being,third,never,really,bothered,me,until,now}'),
       (4,'{the,document,before,me,is,the,third,document}');
```

现在有了字典和文档，我们要对每个文档中的出现单词的数量和比例应用向量运算，将文档进行分类。在开始处理前，需要找到每个文档中出现的字典中的单词。我们为每个文档创建一个稀疏特征向量（Sparse Feature Vector，SFV）。SFV 是一个 N 维向量，N 是字典单词的数量，SFV 中的每个元素是文档中对每个字典单词的计数。SVEC 有一个函数可以从文档创建 SFV：

```
dm=# select madlib.svec_sfv((select a from features limit 1),b)::float8[]
dm-#    from documents;
             svec_sfv
-----------------------------------------
 {0,0,0,0,1,1,0,1,1,0,0,0,1,0,0,1,0,1,0}
 {1,0,0,0,1,1,1,1,0,0,0,0,0,0,1,2,0,0,0}
 {0,0,1,1,0,0,0,0,0,1,1,1,0,1,0,0,1,0,1}
 {0,1,0,0,0,2,0,0,1,1,0,0,0,0,2,1,0,0}
(4 rows)
```

注意，madlib.svec_sfv()函数的输出是每个文档一个向量，元素值是相应字典顺序位置上单词在文档中出现的次数。通过对比特征向量和文档，更容易地理解这一点：

```
dm=# \x
Expanded display is on.
dm=# select madlib.svec_sfv((select a from features),b)::float8[], b
dm-#    from documents;
-[ RECORD 1 ]------------------------------------
svec_sfv | {0,0,0,0,1,1,0,1,1,0,0,0,1,0,0,1,0,1,0}
b        | {this,is,one,document,in,the,corpus}
-[ RECORD 2 ]------------------------------------
svec_sfv | {1,0,0,0,1,1,1,1,0,0,0,0,0,0,1,2,0,0,0}
b        | {i,am,the,second,document,in,the,corpus}
-[ RECORD 3 ]------------------------------------
svec_sfv | {0,0,1,1,0,0,0,0,0,1,1,1,0,1,0,0,1,0,1}
b        | {being,third,never,really,bothered,me,until,now}
-[ RECORD 4 ]------------------------------------
svec_sfv | {0,1,0,0,0,2,0,0,1,1,0,0,0,0,2,1,0,0}
b        | {the,document,before,me,is,the,third,document}
```

可以看到文档"i am the second document in the corpus",它的 SFV 为 {1,3*0,1,1,1,6*0,1,2,3*0}。单词"am"是字典中的第一个单词,并且在文档中只出现一次。单词"before"没有出现在文档中,所以它的值为 0,以此类推。函数 madlib.svec_sfv()能够将大量文档高速并行转换为对应的 SFV。

分类处理的其余部分都是向量运算。实际应用中很少使用实际计数值,而是将计数转为权重。最普通的权重叫做 tf/idf,对应术语是 Term Frequency / Inverse Document Frequency。对给定文档中给定单词的权重计算公式为:

```
{#Times in document} * log {#Documents / #Documents the term appears in}
```

例如,单词"document"在文档 A 中的权重为 1 * log (4/3),而在文档 D 中的权重为 2 * log (4/3)。在每个文档中都出现的单词的权重为 0,因为 log (4/4) = log(1) = 0。对于这部分处理,我们需要一个具有字典维数(19)的稀疏向量,元素值为:

```
log(#documents/#Documents each term appears in)
```

整个文档列表对应单一上述向量。#documents 是文档总数,本例中是 4,但对于每个字典单词都对应一个分母,其值为出现该单词的文档数。这个向量再乘以每个文档 SFV 中的计数,结果即为 tf/idf 权重。

```
drop table if exists corpus;
create table corpus
as (select a, madlib.svec_sfv((select a from features),b) sfv
    from documents);

drop table if exists weights;
create table weights
as (select a docnum, madlib.svec_mult(sfv, logidf) tf_idf
    from (select madlib.svec_log(madlib.svec_div(
count(sfv)::madlib.svec,madlib.svec_count_nonzero(sfv))) logidf
from corpus) foo, corpus order by docnum);
```

查询权重:

```
dm=# select * from weights;
 docnum | tf_idf
--------+----------------------------------------------------------------
------------------------------------------
      1 |
{4,1,1,1,2,3,1,2,1,1,1,1}:{0,0.693147180559945,0.287682072451781,0,0.693147180
559945,0,1.38629436111989,0,0.287
    682072451781,0,1.38629436111989,0}
      2 |
{1,3,1,1,1,1,6,1,1,3}:{1.38629436111989,0,0.693147180559945,0.287682072451781,
1.38629436111989,0.69314718055994
```

```
         5,0,1.38629436111989,0.575364144903562,0}
       3 |
 {2,2,5,1,2,1,1,2,1,1,1}:{0,1.38629436111989,0,0.693147180559945,1.386294361119
 89,0,1.38629436111989,0,0.6931471
         80559945,0,1.38629436111989}
       4 |
 {1,1,3,1,2,2,5,1,1,2}:{0,1.38629436111989,0,0.575364144903562,0,0.693147180559
 945,0,0.575364144903562,0.6931471
         80559945,0}
 (4 rows)
```

现在就可以使用文档向量的点积的 ACOS，获得一个文档与其他文档的"角距离"。下面计算第一个文档与其他文档的角距离：

```
dm=# select docnum, 180. *
dm-#        (acos(madlib.svec_dmin(1., madlib.svec_dot(tf_idf, testdoc)
dm(#        / (madlib.svec_l2norm(tf_idf)
dm(#        * madlib.svec_l2norm(testdoc))))/3.141592654) angular_distance
dm-#    from weights,
dm-#        (select tf_idf testdoc from weights where docnum = 1 limit 1) foo
dm-#    order by 1;
 docnum | angular_distance
--------+------------------
      1 |                0
      2 | 78.8235846096986
      3 | 89.9999999882484
      4 | 80.0232034288617
(4 rows)
```

可以看到文档 1 与自己的角距离为 0 度，而文档 1 与文档 3 的角距离为 90 度，因为它们之间没有任何相同的单词。

前面已经提到，SVEC 提供了从给定的位置数组和值数组声明一个稀疏向量的功能。下面再看一个例子。

```
dm=# select madlib.svec_cast_positions_float8arr(
dm(#        array[1,2,7,5,87],array[.1,.2,.7,.5,.87],90,0.0);
       svec_cast_positions_float8arr
-----------------------------------------------------------
 {1,1,2,1,1,1,79,1,3}:{0.1,0.2,0,0.5,0,0.7,0,0.87,0}
(1 row)
```

第一个整数数组表示第二个浮点数数组的位置，即结果数组的第 1、2、5、7、87 下标对应的值分别为 0.1、0.2、0.5、0.7、0.87。位置本身不需要有序，但要和值的顺序保持一致。第三个参数表示数组的最大维数。小于 1 最大维度将被忽略，此时数组的最大维度就是位置数组中的最大下标。最后的参数表示没有提供下标的位置上的值。

19.3.2 矩阵

矩阵是 MADlib 中数据的基本格式，通常是二维的。在 MADlib 中，数组的概念与向量类似，数组通常是一维的，是矩阵的一种特殊形式。

1. 矩阵表示

MADlib 支持稠密和稀疏两种矩阵表示形式，所有矩阵运算都以其中一种表示形式工作。

（1）稠密

矩阵被表示为一维数组的分布式集合，例如 3×10 的矩阵如下表：

```
row_id |          row_vec
-------+--------------------------
   1   | {9,6,5,8,5,6,6,3,10,8}
   2   | {8,2,2,6,6,10,2,1,9,9}
   3   | {3,9,9,9,8,6,3,9,5,6}
```

（2）稀疏

使用行列下标指示矩阵中每一个非零项，例如：

```
row_id | col_id | value
-------+--------+-------
   1   |   1    |   9
   1   |   5    |   6
   1   |   6    |   6
   2   |   1    |   8
   3   |   1    |   3
   3   |   2    |   9
```

2. 矩阵运算

（1）数组运算

MADlib 的数组运算模块提供了一组用 C 和 SQL 实现的基本数组操作，是需要快速数组操作的机器学习算法的支持模块。数组运算函数支持以下数据类型：

- SMALLINT
- INTEGER
- BIGINT
- REAL
- DOUBLE PRECISION（FLOAT8）
- NUMERIC（内部被转化为 FLOAT8，可能丢失精度）

另外，array_unnest_2d_to_1d() 函数还支持 TEXT 和 VARCHAR 数据类型。

① 数组运算函数

MADlib 提供了丰富的数组运算函数，函数列表及功能描述如表 19-2 所示。

表 19-2 MADlib 数组函数

函数	描述
sum()	向量元素求和，需要所有值非空，返回与输入相同的数据类型
array_add()	两个数组相加，需要所有值非空，返回与输入相同的数据类型
array_sub()	两个数组相减，需要所有值非空，返回与输入相同的数据类型
array_mult()	两个数组相乘，需要所有值非空，返回与输入相同的数据类型
array_div()	两个数组相除，需要所有值非空，返回与输入相同的数据类型
array_dot()	两个数组点积，需要所有值非空，返回与输入相同的数据类型
array_contains()	检查一个数组是否包含另一个数组。如果右边数组中的每个非零元素都等于左边数组中相同下标的元素，函数返回 TRUE
array_max()	返回数组中的最大值，忽略空值，返回与输入相同的数据类型
array_max_index()	返回数组中的最大值及其对应的下标，忽略空值，返回类型的格式为[max, index]，其元素类型与输入类型相同
array_min()	返回数组中的最小值，忽略空值，返回与输入相同的数据类型
array_min_index()	返回数组中的最小值及其对应的下标，忽略空值，返回类型的格式为[min, index]，其元素类型与输入类型相同
array_sum()	返回数组中值的和，忽略空值，返回与输入相同的数据类型
array_sum_big()	返回数组中值的和，忽略空值，返回 FLOAT8 类型。该函数的意思是当汇总值可能超出元素类型范围时，替换 array_sum()
array_abs_sum()	返回数组中绝对值的和，忽略空值，返回与输入相同的数据类型
array_abs()	返回由数组元素的绝对值组成的新数组，需要所有值非空
array_mean()	返回数组的均值，忽略空值
array_stddev()	返回数组的标准差，忽略空值
array_of_float()	该函数创建元素个数为参数值的 FLOAT8 数组，初始值为 0.0
array_of_bigint()	该函数创建元素个数为参数值的 BIGINT 数组，初始值为 0
array_fill()	该函数将数组每个元素设置为参数值
array_filter()	该函数只保留输入数组中符合指定标量运算符的元素。要求是一维数组，并且所有值非空。返回与输入相同的数据类型。默认时该函数移除所有 0 值
array_scalar_mult()	该函数将一个数组作为输入，元素与第二个参数指定的标量值相乘，返回结果数组。需要所有值非空，返回与输入相同的数据类型
array_scalar_add()	该函数将一个数组作为输入，元素与第二个参数指定的标量值相加，返回结果数组。需要所有值非空，返回与输入相同的数据类型
array_sqrt()	返回由数组元素的平方根组成的数组，需要所有值非空
array_pow()	该函数以数组和一个 float8 为输入，返回每个元素的乘幂（由第二个参数指定）组成的数组，需要所有值非空
array_square()	返回由数组元素的平方组成的数组，需要所有值非空
normalize()	该函数规范化一个数组，使它的元素平方和为 1。要求是一维数组，且所有值非空

② 数组运算示例

建立一个数据表，包含两个整型数组列，并添加数据。

```
drop table if exists array_tbl;
create table array_tbl (id integer, array1 integer[], array2 integer[]);
insert into array_tbl values
( 1, '{1,2,3,4,5,6,7,8,9}', '{9,8,7,6,5,4,3,2,1}' ),
( 2, '{1,1,0,1,1,2,3,99,8}','{0,0,0,-5,4,1,1,7,6}' );
```

查询 array1 列的最小值、最大值、均值和标准差。

```
dm=# select id, min1, max1, min_idx1, max_idx1,round(mean1::numeric,4) mean1,
dm-#        round(stddev1::numeric,4) stddev1
dm-#   from (select id,
dm(#            madlib.array_min(array1) min1,
dm(#            madlib.array_max(array1) max1,
dm(#            madlib.array_min_index(array1) min_idx1,
dm(#            madlib.array_max_index(array1) max_idx1,
dm(#            madlib.array_mean(array1) mean1,
dm(#            madlib.array_stddev(array1) stddev1
dm(#         from array_tbl) t1;
 id | min1 | max1 | min_idx1 | max_idx1 |  mean1  | stddev1
----+------+------+----------+----------+---------+---------
  1 |   1  |   9  |  {1,1}   |  {9,9}   |  5.0000 |  2.7386
  2 |   0  |  99  |  {0,3}   |  {99,8}  | 12.8889 | 32.3784
(2 rows)
```

执行数组加减。

```
dm=# select id,
dm-#        madlib.array_add(array1,array2),
dm-#        madlib.array_sub(array1,array2)
dm-#   from array_tbl;
 id |           array_add            |        array_sub
----+--------------------------------+-------------------------
  1 | {10,10,10,10,10,10,10,10,10}   | {-8,-6,-4,-2,0,2,4,6,8}
  2 | {1,1,0,-4,5,3,4,106,14}        | {1,1,0,6,-3,1,2,92,2}
(2 rows)
```

执行数组乘除。不包含 id=2 的行，因为有除数为 0，会报错 ERROR: division by zero is not allowed。

```
dm=# select id,
dm-#        madlib.array_mult(array1,array2),
dm-#        madlib.array_div(array1,array2)
dm-#   from array_tbl
```

```
dm-# where 0 != all(array2);
 id |          array_mult           |       array_div
----+-------------------------------+----------------------
  1 | {9,16,21,24,25,24,21,16,9}    | {0,0,0,0,1,1,2,4,9}
(1 row)
```

计算数组的点积，并根据点积定义验证结果。

```
dm=# select id,
dm-#        madlib.array_dot(array1, array2),
dm-#        madlib.array_sum(madlib.array_mult(array1,array2))
dm-#   from array_tbl;
 id | array_dot | array_sum
----+-----------+-----------
  1 |       165 |       165
  2 |       745 |       745
(2 rows)
```

数组元素乘标量值3。

```
dm=# select id, array1, madlib.array_scalar_mult(array1,3) from array_tbl;
 id |        array1         |      array_scalar_mult
----+-----------------------+----------------------------
  1 | {1,2,3,4,5,6,7,8,9}   | {3,6,9,12,15,18,21,24,27}
  2 | {1,1,0,1,1,2,3,99,8}  | {3,3,0,3,3,6,9,297,24}
(2 rows)
```

构造一个包含9个元素的数组，每个元素值设置为1.3。

```
dm=# select madlib.array_fill(madlib.array_of_float(9), 1.3::float);
              array_fill
---------------------------------------
 {1.3,1.3,1.3,1.3,1.3,1.3,1.3,1.3,1.3}
(1 row)
```

将二维数组列展开为一维数组集合。array_unnest_2d_to_1d 是 MADlib 1.11 版本的新增函数，用于将二维数组展开为一维数组。1.10 版本无次函数，但可以创建一个 UDF 实现。

```
create or replace function madlib.array_unnest_2d_to_1d(anyarray)
returns table(unnest_row_id int, unnest_result anyarray) as
$func$
select d1,array_agg(val)
  from (select $1[d1][d2] val,d1,d2
          from generate_series(array_lower($1,1), array_upper($1,1)) d1,
               generate_series(array_lower($1,2), array_upper($1,2)) d2
         order by d1,d2) t
```

```
group by d1
$func$ language sql immutable;
```

之后就可以调用函数展开二维数组：

```
dm=# select id, (madlib.array_unnest_2d_to_1d(val)).*
dm-#   from (select 1::int as id,
dm(#                array[[1.3,2.0,3.2],[10.3,20.0,32.2]]
dm(#                ::float8[][] as val
dm(#         union all
dm(#         select 2,
dm(#                array[[pi(),pi()/2],[2*pi(),pi()],[pi()/4,4*pi()]]
dm(#                ::float8[][]) t
dm-#  order by 1,2;
 id | unnest_row_id |            unnest_result
----+---------------+------------------------------------------
  1 |             1 | {1.3,2,3.2}
  1 |             2 | {10.3,20,32.2}
  2 |             1 | {3.14159265358979,1.5707963267949}
  2 |             2 | {6.28318530717959,3.14159265358979}
  2 |             3 | {0.785398163397448,12.5663706143592}
(5 rows)
```

如果调用函数时不用.*标记，函数将返回具有两个属性（行 ID 和对应的展开后一维数组）的复合记录类型。

（2）矩阵运算函数

矩阵运算函数支持的数据类型包括 SMALLINT、INTEGER、BIGINT、FLOAT8 和 NUMERIC（内部被转化为 FLOAT8，可能丢失精度）。

① 矩阵运算函数分类

矩阵运算函数可大致分成以下类型：

- 表示函数：

-- 转化为稀疏矩阵

matrix_sparsify(matrix_in, in_args, matrix_out, out_args)

-- 转化为稠密矩阵

matrix_densify(matrix_in, in_args, matrix_out, out_args)

-- 获取矩阵的维度

matrix_ndims(matrix_in, in_args)

- 算数函数：

-- 矩阵转置

matrix_trans(matrix_in, in_args, matrix_out, out_args)

-- 矩阵相加

matrix_add(matrix_a, a_args, matrix_b, b_args, matrix_out, out_args)

-- 矩阵相减

matrix_sub(matrix_a, a_args, matrix_b, b_args, matrix_out, out_args)

-- 矩阵乘法

matrix_mult(matrix_a, a_args, matrix_b, b_args, matrix_out, out_args)

-- 数组元素依次相乘

matrix_elem_mult(matrix_a, a_args, matrix_b, b_args, matrix_out, out_args)

-- 标量乘矩阵

matrix_scalar_mult(matrix_in, in_args, scalar, matrix_out, out_args)

-- 向量乘矩阵

matrix_vec_mult(matrix_in, in_args, vector)

- 提取函数：

-- 从行下标提取行

matrix_extract_row(matrix_in, in_args, index)

-- 从列下标提取列

matrix_extract_col(matrix_in, in_args, index)

-- 提取主对角线元素

matrix_extract_diag(matrix_in, in_args)

- 规约函数（跨指定维度的聚合）：

-- 获取维度最大值。如果 fetch_index = True，返回对应的下标。

matrix_max(matrix_in, in_args, dim, matrix_out, fetch_index)

-- 获取维度最小值。如果 fetch_index = True，返回对应的下标。

matrix_min(matrix_in, in_args, dim, matrix_out, fetch_index)

-- 获取维度的和

matrix_sum(matrix_in, in_args, dim)

-- 获取维度的均值

matrix_mean(matrix_in, in_args, dim)

-- 获取矩阵范数

matrix_norm(matrix_in, in_args, norm_type)

- 创建函数：

-- 创建一个指定矩阵，用1初始化为给定的行列维度。

matrix_ones(row_dim, col_dim, matrix_out, out_args)

-- 创建一个指定矩阵，用0初始化为给定的行列维度。

matrix_zeros(row_dim, col_dim, matrix_out, out_args)

-- 创建正方形恒等矩阵

matrix_identity(dim, matrix_out, out_args)

-- 用给定对角元素初始化矩阵

matrix_diag(diag_elements, matrix_out, out_args)

-- 用从分布中采样的值初始化矩阵。支持普通、均匀、伯努利分布。

matrix_random(distribution, row_dim, col_dim, in_args, matrix_out, out_args)

- 分解函数：

-- 矩阵求逆

matrix_inverse(matrix_in, in_args, matrix_out, out_args)

-- 广义逆矩阵

matrix_pinv(matrix_in, in_args, matrix_out, out_args)

-- 矩阵特征提取

matrix_eigen(matrix_in, in_args, matrix_out, out_args)

-- Cholesky 分解

matrix_cholesky(matrix_in, in_args, matrix_out_prefix, out_args)

-- QR 分解

matrix_qr(matrix_in, in_args, matrix_out_prefix, out_args)

-- LU 分解

matrix_lu(matrix_in, in_args, matrix_out_prefix, out_args)

-- 矩阵的核范数

matrix_nuclear_norm(matrix_in, in_args)

-- 矩阵的秩

matrix_rank(matrix_in, in_args)

分解函数仅基于内存操作实现。单一节点的矩阵数据被用于分解计算。这种操作只适合小型矩阵，因为计算不是分布到个多个节点的。

② 稠密矩阵运算示例

创建两个 4×4 的示例稠密矩阵表。

```
drop table if exists mat_a;
create table mat_a (row_id integer, row_vec integer[]);
insert into mat_a (row_id, row_vec) values
(1, '{9,6,5,8}'), (2, '{8,2,2,6}'), (3, '{3,9,9,9}'), (4, '{6,4,2,2}');

drop table if exists mat_b;
create table mat_b (row_id integer, vector integer[]);
insert into mat_b (row_id, vector) values
(1, '{9,10,2,4}'), (2, '{5,3,5,2}'), (3, '{0,1,2,3}'), (4, '{2,9,0,4}');
```

矩阵转置。matrix_trans 函数的第一个参数是源表名，第二个参数指定行、列字段名，第三个参数为输出表名。

```
dm=# drop table if exists mat_r;
DROP TABLE
dm=# select madlib.matrix_trans('"mat_b"', 'row=row_id, val=vector','mat_r');
 matrix_trans
---------------
 (mat_r)
(1 row)

dm=# select * from mat_r order by row_id;
 row_id |   vector
--------+-------------
      1 | {9,5,0,2}
      2 | {10,3,1,9}
      3 | {2,5,2,0}
      4 | {4,2,3,4}
(4 rows)
```

提取矩阵的主对角线。

```
dm=# select madlib.matrix_extract_diag('mat_b', 'row=row_id, val=vector');
 matrix_extract_diag
---------------------
 {9,3,2,4}
(1 row)
```

矩阵相加。

```
dm=# drop table if exists mat_r;
DROP TABLE
dm=# select madlib.matrix_add('mat_a', 'row=row_id, val=row_vec',
dm(#                          'mat_b', 'row=row_id, val=vector',
dm(#                          'mat_r', 'val=vector, fmt=dense');
 matrix_add
------------
 (mat_r)
(1 row)


dm=# select * from mat_r order by row_id;
 row_id |    vector
--------+---------------
      1 | {18,16,7,12}
      2 | {13,5,7,8}
      3 | {3,10,11,12}
      4 | {8,13,2,6}
```

```
(4 rows)
```

矩阵相乘。

```
dm=# drop table if exists mat_r;
DROP TABLE
dm=# select madlib.matrix_mult('mat_a', 'row=row_id, val=row_vec',
dm(#                           'mat_b', 'row=row_id, val=vector, trans=true',
dm(#                           'mat_r');
 matrix_mult
-------------
 (mat_r)
(1 row)

dm=# select * from mat_r order by row_id;
 row_id |     row_vec
--------+------------------
      1 | {183,104,40,104}
      2 | {120,68,24,58}
      3 | {171,105,54,123}
      4 | {106,56,14,56}
(4 rows)
```

创建对角矩阵。

```
dm=# drop table if exists mat_r;
DROP TABLE
dm=# select madlib.matrix_diag(array[9,6,3,10],
dm(#                           'mat_r', 'row=row_id, col=col_id, val=val');
 matrix_diag
-------------
 (mat_r)
(1 row)

dm=# select * from mat_r order by row_id::bigint;
 row_id | col_id | val
--------+--------+-----
      1 |      1 |   9
      2 |      2 |   6
      3 |      3 |   3
      4 |      4 |  10
(4 rows)
```

创建单位矩阵。

```
dm=# drop table if exists mat_r;
```

```
DROP TABLE
dm=# select madlib.matrix_identity(4, 'mat_r',
'row=row_id,col=col_id,val=val');
 matrix_identity
-----------------
 (mat_r)
(1 row)

dm=# select * from mat_r order by row_id;
 row_id | col_id | val
--------+--------+-----
      1 |      1 |   1
      2 |      2 |   1
      3 |      3 |   1
      4 |      4 |   1
(4 rows)
```

提取指定下标的行或列。

```
dm=# select madlib.matrix_extract_row('mat_a','row=row_id,val=row_vec',2) as row,
dm-#        madlib.matrix_extract_col('mat_a','row=row_id,val=row_vec',3) as col;
    row     |    col
------------+-----------
 {8,2,2,6}  | {5,2,9,2}
(1 row)
```

获取指定维度的最大最小值及其对应的下标。dim=2 表示计算每一行的最大最小值，返回一个列向量。

```
dm=# drop table if exists mat_max_r;
DROP TABLE
dm=# drop table if exists mat_min_r;
DROP TABLE
dm=# select madlib.matrix_max('mat_a', 'row=row_id, val=row_vec', 2,
dm(#                          'mat_max_r', true),
dm-#        madlib.matrix_min('mat_a', 'row=row_id, val=row_vec', 2,
dm(#                          'mat_min_r', true);
 matrix_max  | matrix_min
-------------+-------------
 (mat_max_r) | (mat_min_r)
(1 row)

dm=# select * from mat_max_r;
```

```
   index    |    max
------------+------------
 {1,1,2,1}  | {9,8,9,6}
(1 row)


dm=# select * from mat_min_r;
   index    |    min
------------+------------
 {3,2,1,3}  | {5,2,3,2}
(1 row)
```

用稀疏格式初始化矩阵。

```
dm=# drop table if exists mat_r;
DROP TABLE
dm=# select madlib.matrix_zeros(4,3,'mat_r','row=row_id,col=col_id,val=entry');
 matrix_zeros
---------------
 (mat_r)
(1 row)

dm=# select * from mat_r;
 row_id | col_id | entry
--------+--------+-------
      4 |      3 |     0
(1 row)
```

用稠密格式初始化矩阵。

```
dm=# drop table if exists mat_r;
DROP TABLE
dm=# select madlib.matrix_zeros(4, 3, 'mat_r', 'fmt=dense');
 matrix_zeros
---------------
 (mat_r)
(1 row)

dm=# select * from mat_r order by row;
 row |   val
-----+---------
   1 | {0,0,0}
   2 | {0,0,0}
   3 | {0,0,0}
   4 | {0,0,0}
```

```
(4 rows)
```

两个矩阵元素相乘。

```
dm=# drop table if exists mat_r;
_id;  DROP TABLE
dm=# select madlib.matrix_elem_mult('mat_a', 'row=row_id, val=row_vec',
dm(#                                'mat_b', 'row=row_id, val=vector',
dm(#                                'mat_r', 'val=vector');
 matrix_elem_mult
------------------
 (mat_r)
(1 row)

dm=# select * from mat_r order by row_id;
 row_id |    vector
--------+---------------
      1 | {81,60,10,32}
      2 | {40,6,10,12}
      3 | {0,9,18,27}
      4 | {12,36,0,8}
(4 rows)
```

按维度求和,本例中每行求和。

```
dm=# select madlib.matrix_sum('mat_a', 'row=row_id, val=row_vec', 2);
  matrix_sum
---------------
 {28,18,30,14}
(1 row)
```

获取维度均值。

```
dm=# select madlib.matrix_mean('mat_a', 'row=row_id, val=row_vec', 2);
   matrix_mean
------------------
 {7,4.5,7.5,3.5}
(1 row)
```

标量乘矩阵。

```
dm=# drop table if exists mat_r;
DROP TABLE
dm=# select madlib.matrix_scalar_mult('mat_a', 'row=row_id, val=row_vec',
dm(#                                  3, 'mat_r');
 matrix_scalar_mult
--------------------
```

```
 (mat_r)
(1 row)

dm=# select * from mat_r order by row_id;
 row_id |    row_vec
--------+----------------
      1 | {27,18,15,24}
      2 | {24,6,6,18}
      3 | {9,27,27,27}
      4 | {18,12,6,6}
(4 rows)
```

向量乘矩阵。

```
dm=# select madlib.matrix_vec_mult('mat_a', 'row=row_id, val=row_vec',
dm(#                               array[1,2,3,4]);
 matrix_vec_mult
-----------------
 {68,42,84,28}
(1 row)
```

获取矩阵的行列维度数。

```
dm=# select madlib.matrix_ndims('"mat_a"', 'row=row_id, val=row_vec');
 matrix_ndims
--------------
 {4,4}
(1 row)
```

③ 稀疏矩阵运算示例

稠密矩阵转为稀疏矩阵。

```
dm=# drop table if exists mat_b_sparse;
NOTICE:  table "mat_b_sparse" does not exist, skipping
DROP TABLE
dm=# select madlib.matrix_sparsify('mat_b', 'row=row_id, val=vector',
dm(#                               'mat_b_sparse', 'col=col_id, val=val');
 matrix_sparsify
-----------------
 (mat_b_sparse)
(1 row)


dm=# select * from mat_b_sparse order by row_id, col_id;
 row_id | col_id | val
--------+--------+-----
      1 |      1 |   9
```

```
    1  |    2  |  10
    1  |    3  |   2
    1  |    4  |   4
    2  |    1  |   5
    2  |    2  |   3
    2  |    3  |   5
    2  |    4  |   2
    3  |    2  |   1
    3  |    3  |   2
    3  |    4  |   3
    4  |    1  |   2
    4  |    2  |   9
    4  |    4  |   4
(14 rows)
```

再创建一个稀疏矩阵表。

```
drop table if exists mat_a_sparse;
create table mat_a_sparse
(rownum integer, col_num integer, entry integer);
insert into mat_a_sparse values
(1, 1, 9), (1, 2, 6), (2, 1, 8), (2, 3, 6),
(3, 1, 6), (3, 2, 3), (4, 1, 7), (4, 4, 8);
```

获取行列维度数。

```
dm=# select madlib.matrix_ndims('mat_a_sparse', 'row="rownum", val=entry');
 matrix_ndims
---------------
 {4,4}
(1 row)
```

添加两个稀疏矩阵，然后转换成稠密格式。

```
dm=# drop table if exists matrix_r_sparse;
NOTICE:  table "matrix_r_sparse" does not exist, skipping
DROP TABLE
dm=# drop table if exists matrix_r;
NOTICE:  table "matrix_r" does not exist, skipping
DROP TABLE
dm=# select madlib.matrix_add('mat_a_sparse', 'row=rownum, val=entry',
dm(#                          'mat_b_sparse', 'row=row_id, col=col_id, val=val',
dm(#                          'matrix_r_sparse', 'col=col_out');
   matrix_add
--------------------
 (matrix_r_sparse)
```

```
(1 row)

dm=# select madlib.matrix_densify('matrix_r_sparse',
dm(#                              'row=rownum, col=col_out, val=entry',
dm(#                              'matrix_r');
 matrix_densify
----------------
 (matrix_r)
(1 row)

dm=# select * from matrix_r order by rownum;
 rownum|    entry
-------+--------------
     1 | {18,16,2,4}
     2 | {13,3,11,2}
     3 | {6,4,2,3}
     4 | {9,9,0,12}
(4 rows)
```

矩阵相乘。

```
dm=# drop table if exists matrix_r;
m;DROP TABLE
dm=# select madlib.matrix_mult
dm-#        ('mat_a_sparse', 'row=rownum, col=col_num, val=entry',
dm(#         'mat_b_sparse', 'row=row_id, col=col_id, val=val, trans=true',
dm(#         'matrix_r');
 matrix_mult
--------------
 (matrix_r)
(1 row)

dm=# select * from matrix_r order by rownum;
 rownum|    entry
-------+----------------
     1 | {141,63,6,72}
     2 | {84,70,12,16}
     3 | {84,39,3,39}
     4 | {95,51,24,46}
(4 rows)
```

矩阵转置。

```
dm=# drop table if exists matrix_r_sparse;
DROP TABLE
```

```
dm=# select madlib.matrix_trans('mat_a_sparse', 'row=rownum, val=entry',
dm(#                            'matrix_r_sparse');
   matrix_trans
-------------------
 (matrix_r_sparse)
(1 row)

dm=# select rownum, col_num, entry from matrix_r_sparse order by col_num, rownum;
 rownum | col_num | entry
--------+---------+-------
      1 |       1 |     9
      2 |       1 |     6
      1 |       2 |     8
      3 |       2 |     6
      1 |       3 |     6
      2 |       3 |     3
      1 |       4 |     7
      4 |       4 |     8
(8 rows)
```

计算矩阵的 Euclidean 范数。

```
dm=# select madlib.matrix_norm('mat_a_sparse', 'row=rownum,
dm'#                            col=col_num, val=entry', '2');
  matrix_norm
---------------
 19.364916731
(1 row)
```

19.4 小结

在 HAWQ 中只使用 SQL 的查询语句就能进行简单的数据挖掘，这是通过整合 MADlib 实现的。MADlib 是一个开源机器学习库，它将算法封装成可在 SQL 中调用的函数，大大简化了数据挖掘的开发工作。本章说明了 MADlib 在 HAWQ 中的安装与卸载过程，介绍了 MADlib 中向量、数组、矩阵的概念，及其相关函数的使用方法，为后面章节的数据挖掘应用打下基础。

第 20 章

◀ 奇异值分解 ▶

矩阵分解的想法来自于矩阵补全，就是依据一个矩阵给定的部分数据，把缺失的值补全。一般假设原始矩阵是低秩的（矩阵中最大不相关向量的个数，称为矩阵的秩，可以简单理解为有秩序的程度），我们可以从给定的值来还原整个矩阵。由于直接求解矩阵从算法以及参数的复杂度来说效率很低，因此常用的方法是把原始矩阵分解为几个子矩阵相乘。MADlib 支持两种矩阵分解方法，分别是低秩矩阵分解和奇异值分解。本章介绍 MADlib 奇异值分解模型对应的函数，以及如何应用它实现典型的推荐算法。

20.1 奇异值分解简介

奇异值分解简称 SVD（singular value decomposition），可以理解为：将一个比较复杂的矩阵用更小更简单的三个子矩阵的相乘来表示，这三个小矩阵描述了大矩阵重要的特性。SVD 的用处有很多，比如：LSA（隐性语义分析）、推荐系统、数据降维、信号处理与统计等。

任何矩阵都可以使用 SVD 进行分解，对于一个 M×N（M>=N）的矩阵 M，存在以下的 SVD 分解：

$$M_{m \times n} = U_{m \times m} \sum\nolimits_{m \times n} (V_{n \times n})^T$$

∑是一个对角矩阵，其中的元素值就是奇异值，并且按照从大到小的顺序排列。很多情况下，前 10%甚至更少的奇异值的平方就占全部奇异值平方的 90%以上了，因此可以用前 k 个奇异值来近似描述矩阵：

$$M_{m \times n} \approx U_{m \times k} \sum\nolimits_{k \times k} (V_{n \times k})^T$$

k 的取值由下面的公式决定：

$$\frac{\sum_{i=1}^{k} \sigma_i^2}{\sum_{i=1}^{m} \sigma_i^2} \geq percentage$$

其中 percentage 称为"奇异值平方和占比的阈值",一般取 90%,k 是一个远小于 m 和 n 的值,这样也就达到了降维的目的。

20.2 MADlib 奇异值分解函数

MADlib 的 SVD 函数可以对稠密矩阵和稀疏矩阵进行奇异值因式分解,并且还提供了一个稀疏矩阵的本地高性能实现函数。

1. 稠密矩阵的 SVD 函数

(1)语法

```
svd( source_table,
    output_table_prefix,
    row_id,
    k,
    n_iterations,
    result_summary_table );
```

(2)参数

source_table:TEXT 类型,源表名(稠密矩阵)。表含有一个 row_id 列标识每一行,从数字 1 开始。其他列包含矩阵的数据。可以使用两种稠密格式的任何一个,例如下面示例的 2×2 矩阵。

```
格式一:
row_id    col1    col2
row1       1       1       0
row2       2       0       1


格式二:
row_id    row_vec
row1       1       {1, 0}
row2       2       {0, 1}
```

output_table_prefix:TEXT 类型,输出表的前缀。

row_id:TEXT 类型,代表行 ID 的列名。

k:INTEGER 类型,计算的奇异值个数。

n_iterations(可选):INTEGER 类型,运行的迭代次数,必须在[k, 列维度数]范围内,k 是奇异值个数。

result_summary_table(可选):TEXT 类型,存储结果摘要的表的名称。

2. 稀疏矩阵的 SVD 函数

表示为稀疏格式的矩阵使用此函数。为了高效计算,在奇异值分解操作之前,输入矩阵会被转换为稠密矩阵。

(1)语法

```
svd_sparse( source_table,
        output_table_prefix,
        row_id,
        col_id,
        value,
        row_dim,
        col_dim,
        k,
        n_iterations,
        result_summary_table );
```

(2)参数

source_table:TEXT 类型,源表名(稀疏矩阵)。稀疏矩阵使用行列下标指示矩阵的每个非零条目,非常适合含有很多零元素的矩阵。如下面所示的 4×7 矩阵,除去零值只有 6 行。矩阵的维度由行、列的最大值推导出来。需要注意最后一行,即使是 0 也要包含这一行,因为它标识了矩阵的维度,并暗示了第 4 行与第 7 列全是 0。

```
row_id | col_id | value
--------+--------+-------
     1 |      1 |     9
     1 |      5 |     6
     1 |      6 |     6
     2 |      1 |     8
     3 |      1 |     3
     3 |      2 |     9
     4 |      7 |     0
(6 rows)
```

output_table_prefix:TEXT 类型,输出表的前缀。
row_id:TEXT 类型,包含行下标的列名。
col_id:TEXT 类型,包含列下标的列名。
value:TEXT 类型,包含值的列名。
row_dim:INTEGER 类型,矩阵的行数。
col_dim:INTEGER 类型,矩阵的列数。
k:INTEGER 类型,计算的奇异值个数。

n_iterations（可选）：INTEGER 类型，运行的迭代次数，必须在[k，列维度数]范围内，k 是奇异值个数。

result_summary_table（可选）：TEXT 类型，存储结果摘要的表的名称。

3. 稀疏矩阵的本地实现 SVD 函数

此函数在计算 SVD 时使用本地稀疏表示（不跨节点），能够更高效地计算稀疏矩阵，适合高度稀疏的矩阵。

（1）语法

```
svd_sparse_native( source_table,
                   output_table_prefix,
                   row_id,
                   col_id,
                   value,
                   row_dim,
                   col_dim,
                   k,
                   n_iterations,
                   result_summary_table );
```

（2）参数

参数含义与 svd_sparse 函数相同。

4. 输出表

三个 SVD 函数的输出都是以下三个表：

左奇异值矩阵表：表名为<output_table_prefix>_u。

右奇异值矩阵表：表名为<output_table_prefix>_v。

奇异值矩阵表：表名为<output_table_prefix>_s。

左右奇异值向量表的格式为：

row_id：INTEGER 类型。每个特征值对应的 ID。

row_vec：FLOAT8[]类型。该 row_id 对应的特征向量元素，数组大小为 k。

由于只有对角线元素是非零的，奇异值表采用稀疏表格式，其中的 row_id 和 col_id 都是从 1 开始：

row_id：INTEGER 类型，第 i 个奇异值为 i。

col_id：INTEGER 类型，第 i 个奇异值为 i（与 row_id 相同）。

value：FLOAT8 类型，奇异值。

结果摘要表有以下列：

rows_used：INTEGER 类型，SVD 计算使用的行数。

exec_time：FLOAT8 类型，计算 SVD 使用的总时间。
iter：INTEGER 类型，迭代运行次数。
recon_error：FLOAT8 类型，质量得分（如近似精度）。计算公式为：

$$\sqrt{mean((X-USV^T)^2_{ij})}$$

relative_recon_error：FLOAT8 类型，相对质量分数。计算公式为：

$$\sqrt{mean(X^2_{ij})}$$

5. 联机帮助

可以执行下面的查询获得 SVD 函数的联机帮助。

```
select madlib.svd();
-- 用法
select madlib.svd('usage');
-- 示例
select madlib.svd('example');
```

20.3 奇异值分解实现推荐算法

1. 问题提出

假设要做一个音乐作品个性化推荐系统。业务收集到的原始用户行为数据是，每个用户为他所收听过的歌曲的打分。分数的定义为：单曲循环=5，分享=4，收藏=3，主动播放=2，听完=1，跳过=-2，拉黑=-5。在分析时能获得的实际评分矩阵 R，也就是输入矩阵大概是图 20-1 的样子。

	音乐1	音乐2	音乐3	音乐4	音乐5	音乐6	音乐7	音乐8	音乐9	音乐10	音乐11	音乐12	音乐13
用户1	5					-5			5	3		1	5
用户2			3						3				4
用户3			1		2	-5	4			-2	-2		-2
用户4		4	4	3			-2		-5			3	
用户5		5	-5		-5		3			4			
用户6			4			3		4					
用户7		-2				5				4		4	-2
用户8		-2				5		5		4			-2

图 20-1 用户评分矩阵

推荐系统的目标就是预测出空白对应位置的分值。推荐系统基于这样一个假设：用户对项目的打分越高，表明用户越喜欢。因此，预测出用户对未评分项目的评分后，根据分值大小排序，把分值高的项目推荐给用户。这是个非常稀疏的矩阵，因为大部分用户只听过全部音乐中很少一部分。下面就利用 MADlib 的奇异值分解模型分解图 20-1 所示的矩阵,并生成相应的推荐矩阵。

2. 建立输入表

(1) 建立索引表

从前面的解释可以看到,推荐矩阵的行列下标分别表示用户和音乐作品。然而在业务系统中,userid 和 musicid 很可能不是按从 1～N 的规则顺序生成的,因此需要建立矩阵下标值与业务表 ID 之间的映射关系,这里使用 HAWQ 的 BIGSERIAL 自增数据类型对应推荐矩阵的索引下标。

```sql
-- 用户索引表
drop table if exists tbl_idx_user;
create table tbl_idx_user (user_idx bigserial, userid varchar(10));
-- 音乐索引表
drop table if exists tbl_idx_music;
create table tbl_idx_music (music_idx bigserial, musicid varchar(10));
```

(2) 建立用户行为业务表

```sql
drop table if exists source_data;
create table source_data (
 userid varchar(10),      -- 用户 ID
 musicid varchar(10),     -- 作品 ID
 val float8               -- 分数
);
```

(3) 建立用户行为矩阵表

```sql
drop table if exists svd_data;
create table svd_data (
 row_id int,              -- 行 ID,从 1 开始,表示用户
 col_id int,              -- 列 ID,从 1 开始,表示作品
 val float8               -- 分数
);
```

3. 生成输入表数据

(1) 生成用户行为业务表数据

```sql
insert into source_data values
('u1', 'm1', 5), ('u1', 'm6', -5),
('u2', 'm4', 3),
('u3', 'm3', 1), ('u3', 'm5', 2), ('u3', 'm7', 4),
('u4', 'm2', 4), ('u4', 'm3', 4), ('u4', 'm4', 3), ('u4', 'm7', -2),
('u5', 'm2', 5), ('u5', 'm3', -5), ('u5', 'm5', -5),
('u5', 'm7', 4), ('u5', 'm8', 3),
```

```
('u6', 'm3', 4), ('u6', 'm6', 3),
('u7', 'm2', -2), ('u7', 'm6', 5),
('u8', 'm2', -2), ('u8', 'm6', 5), ('u8', 'm8', 5),
('u9', 'm3', 1), ('u9', 'm5', 2), ('u9', 'm7', 4);
```

（2）从业务表生成索引表数据

```
-- 用户表
insert into tbl_idx_user (userid)
select distinct userid from source_data order by userid;
-- 音乐表
insert into tbl_idx_music (musicid)
select distinct musicid from source_data order by musicid;
```

这里从业务数据生成有过打分行为的 9 个用户，以及被打过分的 8 个作品。注意查询中的排序子句，作用是便于业务 ID 与矩阵里的行列 ID 对应。

（3）从业务表和索引表生成矩阵表数据

```
insert into svd_data
select t1.user_idx, t2.music_idx, t3.val
  from tbl_idx_user t1, tbl_idx_music t2, source_data t3
 where t1.userid = t3.userid and t2.musicid = t3.musicid;
```

之所以要用用户业务表作为数据源，是因为矩阵中包含所有有过打分行为的用户和被打过分的作品，但不包括与没有任何打分行为相关的用户和作品。如果包含无行为记录的用户或作品，会在计算余弦相似度时出现除零错误或噪声数据。

4. 调用 svd_sparse_native 函数执行 SVD

```
drop table if exists svd_u, svd_v, svd_s, svd_summary cascade;
select madlib.svd_sparse_native
( 'svd_data',    -- 输入表
  'svd',         -- 输出表名前缀
  'row_id',      -- 行索引列名
  'col_id',      -- 列索引列名
  'val',         -- 矩阵元素值
  9,             -- 矩阵行数
  8,             -- 矩阵列数
  7,             -- 计算的奇异值个数，小于等于最小行列数
  NULL,          -- 使用默认的迭代次数
  'svd_summary'  -- 概要表名 );
```

说明：

- 选择 svd_sparse_native 函数的原因是测试数据比较稀疏，矩阵实际数据只占 1/3（25/72），该函数效率较高。

- 这里给出的行、列、奇异值个数分别为9、8、7。svd_sparse_native 函数要求行数大于等于列数，而奇异值个数小于等于列数，否则会报错。结果 U、V 矩阵的行数由实际的输入数据做决定，例如测试数据最大的行值为9，最大列值为8，则结果 U 矩阵的行数为9，V 矩阵的行数为8，而不论行、列参数的值是多少。U、V 矩阵的列数、S 矩阵的行列数均由奇异值个数参数所决定。

5. 查看 SVD 结果

```
dm=# select array_dims(row_vec) from svd_u;
 array_dims
------------
 [1:7]
 [1:7]
 [1:7]
 [1:7]
 [1:7]
 [1:7]
 [1:7]
 [1:7]
 [1:7]
(9 rows)

dm=# select * from svd_s order by row_id, col_id;
 row_id| col_id|      value
-------+-------+------------------
     1 |     1 | 10.6650887159422
     2 |     2 | 10.0400685494281
     3 |     3 | 7.26197376834848
     4 |     4 | 6.5227892843447
     5 |     5 | 5.11307075598297
     6 |     6 | 3.14838515537081
     7 |     7 | 2.67251694708376
     7 |     7 |
(8 rows)

dm=# select array_dims(row_vec) from svd_v;
 array_dims
------------
 [1:7]
 [1:7]
 [1:7]
 [1:7]
 [1:7]
```

```
     [1:7]
     [1:7]
     [1:7]
(8 rows)

dm=# select * from svd_summary;
 rows_used | exec_time (ms) | iter |   recon_error    | relative_recon_error
-----------+----------------+------+------------------+----------------------
         9 |        12277.5 |    8 |  0.116171249851  |    0.0523917951113
(1 row)
```

可以看到，结果 U、V 矩阵的维度分别是 9×7 和 8×7，奇异值是一个 7×7 的对角矩阵。

6. 对比不同奇异值个数的近似度

让我们按 k 的取值公式计算一下奇异值的比值，验证 k 设置为 6、8 时的近似程度。

```
-- k=8
drop table if exists svd8_u, svd8_v, svd8_s, svd8_summary cascade;
select madlib.svd_sparse_native
('svd_data', 'svd8', 'row_id', 'col_id', 'val', 9, 8, 8, NULL, 'svd8_summary');

-- k=6
drop table if exists svd6_u, svd6_v, svd6_s, svd6_summary cascade;
select madlib.svd_sparse_native
('svd_data', 'svd6', 'row_id', 'col_id', 'val', 9, 8, 6, NULL, 'svd6_summary');
```

对比近似度：

```
dm=# select * from svd6_summary;
 rows_used | exec_time (ms) | iter |   recon_error    |
relative_recon_error
-----------+----------------+------+------------------+----------------------
         9 |        5722.47 |    8 |  0.335700790666  |    0.151396899541
(1 row)

dm=# select * from svd_summary;
 rows_used | exec_time (ms) | iter |   recon_error    | relative_recon_error
-----------+----------------+------+------------------+----------------------
         9 |        12277.5 |    8 |  0.116171249851  |    0.0523917951113
(1 row)

dm=# select * from svd8_summary;
 rows_used | exec_time (ms) | iter |   recon_error    |
relative_recon_error
-----------+----------------+------+------------------+----------------------
```

```
---
         9 |     6956.24 |    8 | 1.55727734667e-15 | 7.02312799276e-16
(1 row)

dm=# select s1/s3, s2/s3
dm-#   from (select sum(value*value) s1 from svd6_s) t1,
dm-#        (select sum(value*value) s2 from svd_s) t2,
dm-#        (select sum(value*value) s3 from svd8_s) t3;
     ?column?      |      ?column?
-------------------+-------------------
 0.977078978809393 | 0.997255099805016
(1 row)
```

可以看到，随着 k 值的增加，误差越来越小。在本示例中，奇异值个数为 6、7 的近似度分别为 97.7%和 99.7%，当 k=8 时并没有降维，分解的矩阵相乘等于原矩阵。后面的计算都使用 k=7 的结果矩阵。

7. 基于用户的协同过滤算法 UserCF 生成推荐

所谓 UserCF 算法，简单说就是依据用户的相似程度形成推荐。

（1）定义计算余弦相似度函数

余弦相似度计算公式为：

$$\cos(\theta) = \frac{\sum_{i=1}^{n}(x_i \times y_i)}{\sqrt{\sum_{i=1}^{n}(x_i)^2} \times \sqrt{\sum_{i=1}^{n}(y_i)^2}}$$

madlib.cosine_similarity()函数返回两个向量的余弦相似度。

（2）定义基于用户的协同过滤函数

```
create or replace function fn_user_cf(user_idx int)
   returns table(r2 int, s float8, col_id int,
         val float8, musicid varchar(10)) as
$func$
   select r2, s, col_id, val, musicid
     from
    (select r2,s,col_id,val,
         row_number() over (partition by col_id order by col_id) rn
      from
     (select r2,s,col_id,val
       from
      (select r2,s
        from
       (select r2,s,row_number() over (order by s desc) rn
         from
```

```
            (select t1.row_id r1, t2.row_id r2, abs(madlib.cosine_similarity(v1, v2)) 
s     from
            (select row_id, row_vec v1 from svd_u where row_id = $1) t1,
            (select row_id, row_vec v2 from svd_u) t2
          where t1.row_id <> t2.row_id) t) t
          where rn <=5 and s < 1) t1, svd_data t2
          where t1.r2=t2.row_id and t2.val >=3) t
          where col_id not in (select col_id from svd_data where row_id = $1)) t1,
tbl_idx_music t2
          where t1.rn = 1 and t1.col_id = t2.music_idx
          order by t1.s desc, t1.val desc limit 5;
    $func$
    language sql;
```

说明：

最内层查询调用 madlib.cosine_similarity 函数返回指定用户与其他用户的余弦相似度。

```
select t1.row_id r1, t2.row_id r2, abs(madlib.cosine_similarity(v1, v2)) s
  from (select row_id, row_vec v1 from svd_u where row_id = $1) t1,
       (select row_id, row_vec v2 from svd_u) t2
 where t1.row_id <> t2.row_id
```

外面一层查询按相似度倒序取得排名。

```
select r2,s,row_number() over (order by s desc) rn from …
```

外面一层查询取得最相近的 5 个用户，同时排除相似度为 1 的用户，因为相似度为 1 说明两个用户的作品评分一模一样，而推荐的应该是用户没有打过分的作品。

```
select r2,s from … where rn <=5 and s < 1
```

外面一层查询取得相似用户打分在 3 及其以上的作品索引 ID。

```
select r2,s,col_id,val from … where t1.r2=t2.row_id and t2.val >=3
```

外面一层查询取得作品索引 ID 的排名，目的是去重，避免相同的作品推荐多次，并且过滤掉被推荐用户已经打过分的作品。

```
select r2,s,col_id,val,
       row_number() over (partition by col_id order by col_id) rn
 from … where col_id not in (select col_id from svd_data where row_id = $1)
```

最外层查询关联作品索引表取得作品业务主键，并按相似度和打分推荐前 5 个作品。

```
select r2, s, col_id, val, musicid …
 where t1.rn = 1 and t1.col_id = t2.music_idx
 order by t1.s desc, t1.val desc limit 5;
```

（3）定义接收用户业务 ID 的函数

```
create or replace function fn_user_recommendation(i_userid varchar(10))
   returns
 table (r2 int, s float8, col_id int, val float8, musicid varchar(10)) as    $func$
declare
   v_rec record;
   v_user_idx int:=0;
begin
   select user_idx into v_user_idx from tbl_idx_user where userid=i_userid;
   for v_rec in (select * from fn_user_cf(v_user_idx)) loop
      r2:=v_rec.r2;
      s:=v_rec.s;
      col_id:=v_rec.col_id;
      val :=v_rec.val;
      musicid:=v_rec.musicid;

      return next;
   end loop;

   return;
end;
$func$
language plpgsql;
```

通常输入的用户 ID 是业务系统的 ID，而不是索引下标，因此定义一个接收业务系统的 ID 函数，内部调用 fn_user_cf 函数生成推荐。

（4）测试推荐结果

```
dm=# select * from fn_user_recommendation('u1');
 r2 |          s          | col_id | val | musicid
----+---------------------+--------+-----+---------
  4 | 0.0264000585408379  |    3   |  4  | m3
  4 | 0.0264000585408379  |    2   |  4  | m2
  4 | 0.0264000585408379  |    4   |  3  | m4
  9 | 0.0083739912804568  |    7   |  4  | m7
(4 rows)

dm=# select * from fn_user_recommendation('u3');
 r2 |          s          | col_id | val | musicid
----+---------------------+--------+-----+---------
  7 | 0.0765016205578617  |    6   |  5  | m6
  2 | 0.0749416547815918  |    4   |  3  | m4
  4 | 0.0650665106555581  |    2   |  4  | m2
```

```
(3 rows)

dm=# select * from fn_user_recommendation('u9');
 r2 |          s           | col_id | val | musicid
----+----------------------+--------+-----+---------
  6 | 0.109930597010835    |      6 |   3 | m6
  2 | 0.0749416547815918   |      4 |   3 | m4
  4 | 0.0650665106555581   |      2 |   4 | m2
(3 rows)
```

可以看到，因为 u3 和 u9 的评分完全相同，相似度为 1，所以为他们生成的推荐也完全相同。

8. 基于作品的协同过滤算法 ItemCF 生成推荐

所谓 ItemCF 算法，简单说就是依据作品的相似程度形成推荐。

（1）定义基于物品的协同过滤函数

```
create or replace function fn_item_cf(user_idx int)
    returns table(r2 int, s float8, musicid varchar(10)) as
$func$
    select t1.r2, t1.s, t2.musicid
      from
    (select t1.r2,t1.s,row_number() over (partition by r2 order by s desc) rn
      from
    (select t1.*, row_number() over (partition by r1 order by s desc) rn
      from
    (select t1.row_id r1,t2.row_id r2,abs(madlib.cosine_similarity(v1, v2)) s
      from
    (select row_id, row_vec v1
       from svd_v
      where row_id in (select col_id from svd_data where row_id=$1)) t1,
    (select row_id, row_vec v2
       from svd_v
      where row_id not in (select col_id from svd_data where row_id=$1)) t2
      where t1.row_id <> t2.row_id) t1) t1
      where rn <=3) t1, tbl_idx_music t2
      where rn = 1 and t1.r2 = t2.music_idx
      order by s desc;
$func$
language sql;
```

说明：

最内层查询调用 madlib.cosine_similarity 函数返回指定用户打过分的作品与没打过分的作品的相似度。

```
select t1.row_id r1, t2.row_id r2, abs(madlib.cosine_similarity(v1, v2)) s
  from (select row_id, row_vec v1
```

```
       from svd_v
       where row_id in (select col_id from svd_data where row_id=$1)) t1,
   (select row_id, row_vec v2
       from svd_v
       where row_id not in (select col_id from svd_data where row_id=$1)) t2
 where t1.row_id <> t2.row_id
```

外面一层查询按相似度倒序取得排名。

```
select t1.*, row_number() over (partition by r1 order by s desc) rn …
```

外面一层查询取得与每个打分作品相似度排前三的作品,并以作品索引 ID 分区,按相似度倒序取得排名,目的是去重,避免相同的作品推荐多次。

```
select t1.r2,t1.s,row_number() over (partition by r2 order by s desc) rn
from … where rn <=3
```

最外层查询关联作品索引表取得作品业务主键并推荐。

```
select t1.r2, t1.s, t2.musicid
from ... where rn = 1 and t1.r2 = t2.music_idx order by s desc
```

(2)定义接收用户业务 ID 的函数

```
create or replace function fn_item_recommendation(i_userid varchar(10))
    returns table (r2 int, s float8, musicid varchar(10)) as
$func$
declare
    v_rec record;
    v_user_idx int:=0;
begin
    select user_idx into v_user_idx from tbl_idx_user where userid=i_userid;
    for v_rec in (select * from fn_item_cf(v_user_idx)) loop
        r2:=v_rec.r2;
        s:=v_rec.s;
        musicid:=v_rec.musicid;

        return next;
    end loop;

    return;
end;
$func$
language plpgsql;
```

通常输入的用户 ID 是业务系统的 ID,而不是索引下标,因此定义一个接收业务系统的 ID 函数,内部调用 fn_item_cf 函数生成推荐。

（3）测试推荐结果

```
dm=# select * from fn_item_recommendation('u1');
 r2 |        s         | musicid
----+------------------+---------
  5 | 0.25278871122462 | m5
  2 | 0.16791457694689 | m2
  3 | 0.120167300602806| m3
(3 rows)

dm=# select * from fn_item_recommendation('u3');
 r2 |        s         | musicid
----+------------------+---------
  2 | 0.444777562452136| m2
  1 | 0.25278871122462 | m1
  6 | 0.242084453944156| m6
(3 rows)

dm=# select * from fn_item_recommendation('u9');
 r2 |        s         | musicid
----+------------------+---------
  2 | 0.444777562452136| m2
  1 | 0.25278871122462 | m1
  6 | 0.242084453944156| m6
(3 rows)
```

可以看到，因为 u3 和 u9 的评分作品完全相同，相似度为 1，所以按作品相似度为他们生成的推荐也完全相同。

9. 为新用户寻找相似用户

假设一个新用户 u10 的评分向量为 '{0,4,5,3,0,0,-2,0}'，要利用已有的奇异值矩阵找出该用户的相似用户。

（1）添加业务数据。

```
insert into source_data
values ('u10','m2', 4), ('u10','m3', 5), ('u10','m4', 3), ('u10','m7', -2);

insert into tbl_idx_user (userid)
select distinct userid
  from source_data
 where userid not in (select userid from tbl_idx_user)
 order by userid;
```

（2）确认从评分向量计算 svd_u 向量的公式。

```
u10[1:8] x svd_v[8:7] x svd_s[7:7]^-1
```

（3）生成 u10 用户的向量表和数据。

```
drop table if exists mat_u10;
create table mat_u10(row_id int, row_vec float8[]);
insert into mat_u10 values (1, '{0,4,5,3,0,0,-2,0}');
```

（4）根据计算公式，先将前两个矩阵相乘。

```
drop table if exists mat_r_10;
select madlib.matrix_mult('mat_u10', 'row=row_id, val=row_vec',
                'svd_v', 'row=row_id, val=row_vec',
                'mat_r_10');
```

（5）根据公式，求奇异值矩阵的逆矩阵。

```
drop table if exists svd_s_10;
create table svd_s_10 as
select row_id, col_id,1/value val from svd_s where value is not null;
```

（6）根据公式，将（4）、（5）两步的结果矩阵相乘。

注意（4）的结果 mat_r_10 是一个稠密矩阵，（5）的结果 svd_s_10 是一个稀疏矩阵。

```
drop table if exists matrix_r_10;
select madlib.matrix_mult('mat_r_10', 'row=row_id, val=row_vec',
                'svd_s_10', 'row=row_id, col=col_id, val=val',
                'matrix_r_10');
```

（7）查询与 u10 相似的用户。

```
dm=# select t1.row_id r1, t2.row_id r2, abs(madlib.cosine_similarity(v1, v2))
sdm-#   from (select row_id, row_vec v1 from matrix_r_10 where row_id = 1) t1, dm-#
(select row_id, row_vec v2 from svd_u) t2
dm-#  order by s desc;
 r1 | r2 |           s
----+----+-----------------------
  1 |  4 | 0.989758250631095
  1 |  6 | 0.445518586781384
  1 |  7 | 0.253951334956948
  1 |  2 | 0.117185108937363
  1 |  9 | 0.0276611552976061
  1 |  3 | 0.0276611552976061
  1 |  5 | 0.00988637492741561
  1 |  8 | 0.00673214822878797
  1 |  1 | 0.00262000760517713
```

```
(9 rows)
```

可以看到，u10 与 u4 的相似度高达 99%，从原始的评分向量可以得到验证：

```
u4：'{0,4,4,3,0,0,-2,0}'
u10：'{0,4,5,3,0,0,-2,0}'
```

（8）将结果向量插入 svd_u 矩阵。

```
insert into svd_u
select user_idx, row_vec from matrix_r_10, tbl_idx_user where userid = 'u10';
```

20.4 小结

奇异值分解将原始矩阵逼近为三个矩阵的乘积，是一种降维处理方法，常被应用于推荐系统。MADlib 提供了三个奇异值分解模型函数：稠密矩阵的 SVD 函数、稀疏矩阵的 SVD 函数、稀疏矩阵的本地实现 SVD 函数。我们使用第三个函数，演示了一个典型推荐系统的实现过程，对比了不同 k 值对分解准确度的影响，并展现了 userCF 和 itemCF 两种协同过滤算法。

第 21 章

主成分分析

数据挖掘中经常会遇到多个变量的问题,而且在多数情况下,多个变量之间常常存在一定的相关性。例如,网站的"浏览量"和"访客数"往往具有较强的相关关系,而电商应用中的"下单数"和"成交数"也具有较强的相关关系。这里的相关关系可以直观理解为当浏览量较高(或较低)时,应该很大程度上认为访客数也较高(或较低)。在这个简单的例子中只有两个变量,当变量个数较多且变量之间存在复杂关系时,会显著增加分析问题的复杂性。主成分分析方法可以将多个变量综合为少数几个代表性变量,使这些变量既能够代表原始变量的绝大多数信息又互不相关,这种方法有助于对问题的分析和建模。

MADlib 提供了两组主成分分析函数:训练函数与投影函数。训练函数以原始数据为输入,输出主成分。投影函数将原始数据投影到主成分上,实现线性无关降维,输出降维后的数据矩阵。本章介绍 MADlib 主成分分析模型对应的函数,并以一个示例说明如何利用这些函数解决数据的去相关性和降维问题。

21.1 主成分分析简介

1. PCA 的基本思想

主成分分析采取一种数学降维的方法,其所要做的就是设法将原来众多具有一定相关性的变量,重新组合为一组新的相互无关的综合变量来代替原来的变量。通常,数学上的处理方法就是将原来的变量做线性组合,作为新的综合变量,转换后的变量叫主成分。变换的定义方法是用 F1(选取的第一个线性组合,即第一个综合指标)的方差来表达,即 $Var(F1)$ 越大,表示 F1 包含的信息越多。因此在所有的线性组合中选取的 F1 应该是方差最大的,故称 F1 为第一主成分。如果第一主成分不足以代表原来 P 个指标的信息,再考虑选取 F2 即选第二个线性组合,为了有效地反映原始信息,F1 已有的信息不需要再出现在 F2 中,用数学语言表达就是要求 $Cov(F1, F2)=0$,称 F2 为第二主成分,以此类推,可以构造出第三、第四、……、第 P 个主成分。Cov 表示统计学中的协方差。

2. PCA 的计算步骤

这里关于 PCA 方法的理论推导不再赘述，我们将重点放在如何应用 PCA 解决实际问题上。下面先简单介绍一下 PCA 的典型步骤。

（1）对原始数据进行标准化处理。

假设样本观测数据矩阵为：

$$X = \begin{bmatrix} x_{11} & x_{12} & \cdots & x_{1p} \\ x_{21} & x_{22} & \cdots & x_{2p} \\ \vdots & \vdots & \vdots & \vdots \\ x_{n1} & x_{n2} & \cdots & x_{np} \end{bmatrix}$$

那么可以按照如下方法对原始数据进行标准化处理：

$$x_{ij}^* = \frac{x_{ij} - \overline{x}_j}{\sqrt{\text{var}(x_j)}}, i = 1,2,\dots,n; j = 1,2,\dots,p$$

其中，$\overline{x}_j = \frac{1}{n}\sum_{i=1}^n x_{ij}$，$\text{var}(x_j) = \frac{1}{n-1}\sum_{i=1}^n (x_{ij} - \overline{x}_j)^2$ $(j = 1,2,\dots,p)$

（2）计算样本相关系数矩阵。

为方便，假定原始数据标准化后仍用 X 表示，则经标准化处理后数据的相关系数为：

$$R = \begin{bmatrix} r_{11} & r_{12} & \cdots & r_{1p} \\ r_{21} & r_{22} & \cdots & r_{2p} \\ \vdots & \vdots & \vdots & \vdots \\ r_{p1} & r_{p2} & \cdots & r_{pp} \end{bmatrix}$$

其中，$r_{ij} \frac{\text{cov}(x_i, x_j)}{\sqrt{\text{var}(x_1)}\sqrt{\text{var}(x_2)}} = \frac{\sum_{k=1}^{k=n}(x_{ki} - \overline{x}_i)(x_{kj} - \overline{x}_j)}{\sqrt{\sum_{k=1}^{k=n}(x_{ki} - \overline{x}_i)^2}\sqrt{\sum_{k=1}^{k=n}(x_{kj} - \overline{x}_j)^2}}, n > 1$。

（3）计算相关矩阵 R 的特征值（λ_1，λ_2，…，λ_p）和相应的特征向量：

$$a_i = (a_{i1}, a_{i2}, \dots, a_{ip}), i = 1,2,\dots,p$$

（4）选择重要的主成分，并写出主成分表达式。

主成分分析可以得到 p 个主成分，但是，由于各个主成分的方差是递减的，包含的信息量也是递减的，所以实际分析时，一般不是选取 p 个主成分，而是根据各个主成分累计贡献率的大小选取前 k 个主成分，这里的贡献率是指某个主成分的方差占全部方差的比重，实际也就是某个特征值占全部特征值合计的比重，即：

$$\text{贡献率} = \frac{\lambda_i}{\sum_{i=1}^p \lambda_i}$$

贡献率越大，说明该主成分所包含的原始变量的信息越强。主成分个数 k 的选取，主要根据主成分的累计贡献率来决定，即一般要求累计贡献率达到 85%以上，这样才能保证综合变量能包括原始变量的绝大多数信息。

另外，在实际应用中，选择了重要的主成分后，还要注意主成分实际含义的解释。主成分分析中一个很关键的问题是如何给主成分赋予新的意义，给出合理的解释。一般而言，这个解释是根据主成分表达式的系数结合定性分析来进行的。主成分是原来变量的线性组合，在这个线性组合中各变量的系数有大有小，有正有负，有的大小相当，因而不能简单地认为这个主成分是某个原变量的属性的作用。线性组合中各变量系数的绝对值大者表明该主成分主要综合了绝对值大的变量，有几个变量系数大小相当时，应认为这一主成分是这几个变量的总和。这几个变量综合在一起应赋予怎样的实际意义，就要结合具体的实际问题和专业，给出恰当的解释，进而才能达到深刻分析的目的。

（5）计算主成分得分。

根据标准化的原始数据，按照各个样品，分别代入主成分表达式，就可以得到各主成分下的各个样品的新数据，即为主成分得分。具体形式如下：

$$\begin{bmatrix} F_{11} & F_{12} & \cdots & F_{1k} \\ F_{21} & F_{22} & \cdots & F_{2k} \\ \vdots & \vdots & \vdots & \vdots \\ F_{n1} & F_{n2} & \cdots & F_{nk} \end{bmatrix}$$

其中，$F_{ij} = a_{j1}x_{i1} + a_{j2}x_{i2} + \cdots + a_{jp}x_{ip}$（$i = 1,2,\ldots,n; j = 1,2,\ldots,k$）。

（6）依据主成分得分数据，进一步对问题进行后续的分析和建模。

后续分析和建模常见的形式有主成分回归、变量子集合的选择、综合评价等。

21.2 MADlib 的 PCA 相关函数

1. 训练函数

MADlib 中 PCA 的实现是使用一种分布式的 SVD（奇异值分解）找出主成分，而不是直接计算方差矩阵的特征向量。设 x 为与原始数据矩阵，为 x 的列平均值向量。PCA 首先将原始矩阵标准化为矩阵：

$$\hat{X} = X - \vec{e}\hat{x}^T$$

其中 $\vec{e}$ 是所有的行向量。然后 Madlib PCA 函数对矩阵进行 SVD 分解：

$$\hat{X} = U \sum V^T$$

其中 $\sum$ 是对角矩阵，特征值为 $\sum/(\sqrt{N-1})$ 的条目，主成分是 V 的行。最后使用 Bessel's correction 用 N-1 代替 N 计算协方差。

这种实现的重要前提是假设用户只使用具有非零特征值的主成分，因为 SVD 计算用的是 Lanczos 算法，它并不保证含有零特征值的奇异向量的正确性。

（1）语法

稠密矩阵和稀疏矩阵的训练函数有所不同。稠密矩阵训练函数为：

```
pca_train (source_table,
           out_table,
           row_id,
           components_param,
           grouping_cols,
           lanczos_iter,
           use_correlation,
           result_summary_table)
```

稀疏矩阵训练函数为：

```
pca_sparse_train (source_table,
                  out_table,
                  row_id,
                  col_id,              -- 只针对稀疏矩阵
                  val_id,              -- 只针对稀疏矩阵
                  row_dim,             -- 只针对稀疏矩阵
                  col_dim,             -- 只针对稀疏矩阵
                  components_param,
                  grouping_cols,
                  lanczos_iter,
                  use_correlation,
                  result_summary_table )
```

（2）参数

source_table：TEXT 类型，PCA 训练数据的输入表名。输入的数据矩阵应该具有 N 行 M 列，N 为数据点的数量，M 为每个数据点的特征数。稠密输入表可以使用两种标准的 MADlib 稠密矩阵格式：

```
{TABLE|VIEW} source_table (
    row_id INTEGER,
    row_vec FLOAT8[],
    …)
```

或者：

```
{TABLE|VIEW} source_table (
    row_id INTEGER,
    col1 FLOAT8,
    col2 FLOAT8,
    …)
```

注意 row_id 作为入参是输入矩阵的行标识，必须是从 1 开始且连续的整数。PCA 的稀疏

矩阵输入表的格式如下，其中 row_id 和 col_id 列指示矩阵下标，是正整数，val_id 列定义非 0 的矩阵元素值。

```
{TABLE|VIEW} source_table (
    ...
    row_id INTEGER,
    col_id INTEGER,
    val_id FLOAT8,
    ...)
```

out_table：TEXT 类型，输出表的名称。有两种可能的输出表：主输出表和均值输出表。主输出表（out_table）包含特征值最高的 k 个主成分的特征向量，k 值直接由用户参数指定，或者根据方差的比例计算得出。主输出表包含以下四列：

- row_id: 特征值倒序排名。
- principal_components: 包含主成分元素的向量（特征向量）。
- std_dev: 每个主成分的标准差。
- proportion: 每个主成分标准差所占的比例。

均值输出表（out_table_mean）包含列的均值，只有一列：

- column_mean: 包含输入矩阵的列的均值。

row_id：TEXT 类型，输入表中表示行 ID 的列名。该列应该为整型，值域为 1~N，对于稠密矩阵格式，该列应该包含从 1~N 的连续整数。

col_id：TEXT 类型，稀疏矩阵中表示列 ID 的列名。列应为整型，值域为 1 ~ M。该参数只用于稀疏矩阵。

val_id：TEXT 类型，稀疏矩阵中表示非零元素值的列名。该参数只用于稀疏矩阵。

row_dim：INTEGER 类型，矩阵的实际行数，指的是当矩阵转换为稠密矩阵时所具有的行数。该参数只用于稀疏矩阵。

col_dim：INTEGER 类型，矩阵的实际列数，指的是当矩阵转换为稠密矩阵时所具有的列数。该参数只用于稀疏矩阵。row_dim 和 col_dim 实际上可以从稀疏矩阵推断出，当前是为了向后兼容而存在，将来会被移除。这两个值大于矩阵的实际值时会补零。

components_param：INTEGER 或 FLOAT 类型，该参数控制如何从输入数据确定主成分的数量。如果为 INTEGER 类型，代表需要计算的主成分的个数。如果为 FLOAT 类型，算法将返回足够的主成分向量，使得累积特征值大于此参数（标准差比例）。components_param 的值域为正整数或(0.0,1.0]。这里要注意整型和浮点数的区别，如果 components_param 指定为 1，则返回一个主成分，而指定为 1.0 时，返回所有的主成分，因为此时方差比例为 100%。还要注意一点，主成分数量是全局的。在分组时（由 grouping_cols 参数指定）可能选择标准差比例更好，因为这可以使不同分组具有不同的主成分数量。

grouping_cols（可选）：TEXT 类型，默认值为 NULL。指定逗号分隔的列名，使用此参数的所有列分组，对每个分组独立计算 PCA。稠密矩阵的各分组大小可能不同，而稀疏矩阵的每个分组大小都一样，因为稀疏矩阵的 row_dim 和 col_dim 是跨所有组的全局参数。

lanczos_iter（可选）：INTEGER 类型，默认值为 k+40 与最小矩阵维度的较小者，k 是主成分数量。此参数定义计算 SVD 时的 Lanczos 迭代次数，迭代次数越大精度越高，同时计算越花时间。迭代次数不能小于 k 值，也不能大于最小矩阵维度。如果此参数设置为 0，则使用默认值。如果 lanczos_iter 和 components_param 参数同时设置，在确定主成分数量时，优先考虑 lanczos_iter。

use_correlation（可选）：BOOLEAN 类型，默认值为 FALSE。指定在计算主成分时，是否使用相关矩阵代替协方差矩阵。当前该参数仅用于向后兼容，因此必须设置为 false。

result_summary_table（可选）：TEXT 类型，默认值为 NULL。指定概要表的名称，NULL 时不生成概要表。概要表具有下面的列。

- rows_used：INTEGER 类型，输入数据点的个数（行数）。
- exec_time (ms)：FLOAT8 类型，函数执行的毫秒数。
- iter：INTEGER 类型，SVD 计算时的迭代次数。
- recon_error：FLOAT8 类型，SVD 近似值的绝对误差。
- relative_recon_error：FLOAT8 类型，SVD 近似值的相对误差。
- use_correlation：BOOLEAN 类型，表示使用的是相关矩阵。

（3）联机帮助

可以执行下面的查询获得 PCA 培训函数的联机帮助。

```
select madlib.pca_train('usage');
select madlib.pca_sparse_train('usage');
```

2. 投影函数

给定包含主成分 P 的输入数据矩阵 X，对应的降维后的低维度矩阵为，其计算公式为：

$$\hat{X} = X - \vec{e}\hat{x}^T$$
$$X' = \hat{X}P$$

其中 $\hat{x}$ 是的列平均值，$\vec{e}$ 是所有的行向量。这步计算的结果近似于原始数据，保留了绝大部分的原始信息。

残余表用于估计降维后的矩阵与原始输入数据的近似程度，其计算公式为：

$$R = \hat{X} - X'P^T$$

如果残余矩阵的元素接近于零，则表示降维后的信息丢失很少，基本相当于原始数据。残差范数表示为：

$$r = \|R\|_F$$

其中 $\|\cdot\|_F$ 是 Frobenius 范数。相对残差范数的计算公式为：

$$r' = \frac{\|R\|_F}{\|X\|_F}$$

（1）语法

稠密矩阵和稀疏矩阵的投影函数有所不同。稠密矩阵投影函数为：

```
madlib.pca_project (source_table,
                    pc_table,
                    out_table,
                    row_id,
                    residual_table,
                    result_summary_table)
```

稀疏矩阵的投影函数为：

```
madlib.pca_sparse_project (source_table,
                    pc_table,
                    out_table,
                    row_id,
                    col_id,              -- 只针对稀疏矩阵
                    val_id,              -- 只针对稀疏矩阵
                    row_dim,             -- 只针对稀疏矩阵
                    col_dim,             -- 只针对稀疏矩阵
                    residual_table,
                    result_summary_table)
```

（2）参数

source_table：TEXT 类型，等同于 PCA 训练函数，指定源表名称。输入数据矩阵应该有 N 行 M 列，N 为数据点个数，M 为每个数据点的特征数。与 PCA 训练函数类似，pca_project 函数的输入表格式，应该为 MADlib 两种标准稠密矩阵格式之一，而 pca_sparse_project 函数的输入表应该为 MADlib 的标准稀疏矩阵格式。

pc_table：TEXT 类型，主成分表名，通常使用 PCA 训练函数的主输出表。

out_table：TEXT 类型，输入数据降维后的输出表名称。out_table 是一个投影到主成分上的稠密矩阵，具有以下几列：

- row_id：输出矩阵的行 ID。
- col_id：同 PCA 训练函数。
- val_id：同 PCA 训练函数。
- row_dim：同 PCA 训练函数。
- col_dim：同 PCA 训练函数。

residual_table（可选）：TEXT 类型，默认值为 NULL，残余表的名称。residual_table 表现为一个稠密矩阵，具有以下两列：

- row_id：输出矩阵的行 ID。
- row_vec：包含残余矩阵行元素的向量。

result_summary_table（可选）：TEXT 类型，可选值为 NULL，结果概述表的名称。result_summary_table 中含有 PCA 投影函数的性能信息，具有以下三列：

- exec_time：函数执行所用的时间（毫秒）。
- residual_norm：绝对误差。
- relative_residual_norm：相对误差。

（3）联机帮助

可以执行下面的查询获得 PCA 投影函数的联机帮助。

```
select madlib.pca_project('usage');
select madlib.pca_sparse_project('usage');
```

21.3 PCA 应用示例

我们用一个企业综合实力排序的例子来说明 MADlib PCA 的用法。为了系统地分析某 IT 类企业的经济效益，选择了 8 个不同的利润指标，对 15 家企业进行了调研，并得到如表 21-1 所示的数据。现在需要根据这些数据对 15 家企业进行综合实例排序。

表 21-1 企业综合实例评价表

企业编号	净利润率（%）	固定资产利润率（%）	总产值利润率（%）	销售收入利润率（%）	产品成本利润率（%）	物耗利润率（%）	人均利润（千元/人）	流动资产利润率（%）
1	40.4	24.7	7.2	6.1	8.3	8.7	2.442	20
2	25	12.7	11.2	11	12.9	20.2	3.542	9.1
3	13.2	3.3	3.9	4.3	4.4	5.5	0.578	3.6
4	22.3	6.7	5.6	3.7	6	7.4	0.176	7.3
5	34.3	11.8	7.1	7.1	8	8.9	1.726	27.5
6	35.6	12.5	16.4	16.7	22.8	29.3	3.017	26.6
7	22	7.8	9.9	10.2	12.6	17.6	0.847	10.6
8	48.4	13.4	10.9	9.9	10.9	13.9	1.772	17.8
9	40.6	19.1	19.8	19	29.7	39.6	2.449	35.8
10	24.8	8	9.8	8.9	11.9	16.2	0.789	13.7
11	12.5	9.7	4.2	4.2	4.6	6.5	0.874	3.9
12	1.8	0.6	0.7	0.7	0.8	1.1	0.056	1
13	32.3	13.9	9.4	8.3	9.8	13.3	2.126	17.1
14	38.5	9.1	11.3	9.5	12.2	16.4	1.327	11.6
15	26.2	10.1	5.6	15.6	7.7	30.1	0.126	25.9

由于本问题中涉及 8 个指标，这些指标间的关联关系并不明确，而且各指标数值的数量级也有差异，为此这里首先借助 PCA 方法对指标体系进行降维处理，然后根据 PCA 打分结果实现对企业的综合实力排序。

1. 创建原始稠密矩阵表并添加数据

```
drop table if exists mat;
create table mat (id integer, row_vec double precision[]);
insert into mat values
(1,  '{40.4, 24.7,  7.2,  6.1,  8.3,  8.7, 2.442,   20}'),
(2,  '{  25, 12.7, 11.2,   11, 12.9, 20.2, 3.542,  9.1}'),
(3,  '{13.2,  3.3,  3.9,  4.3,  4.4,  5.5, 0.578,  3.6}'),
(4,  '{22.3,  6.7,  5.6,  3.7,    6,  7.4, 0.176,  7.3}'),
(5,  '{34.3, 11.8,  7.1,  7.1,    8,  8.9, 1.726, 27.5}'),
(6,  '{35.6, 12.5, 16.4, 16.7, 22.8, 29.3, 3.017, 26.6}'),
(7,  '{  22,  7.8,  9.9, 10.2, 12.6, 17.6, 0.847, 10.6}'),
(8,  '{48.4, 13.4, 10.9,  9.9, 10.9, 13.9, 1.772, 17.8}'),
(9,  '{40.6, 19.1, 19.8,   19, 29.7, 39.6, 2.449, 35.8}'),
(10, '{24.8,    8,  9.8,  8.9, 11.9, 16.2, 0.789, 13.7}'),
(11, '{12.5,  9.7,  4.2,  4.2,  4.6,  6.5, 0.874,  3.9}'),
(12, '{ 1.8,  0.6,  0.7,  0.7,  0.8,  1.1, 0.056,    1}'),
(13, '{32.3, 13.9,  9.4,  8.3,  9.8, 13.3, 2.126, 17.1}'),
(14, '{38.5,  9.1, 11.3,  9.5, 12.2, 16.4, 1.327, 11.6}'),
(15, '{26.2, 10.1,  5.6, 15.6,  7.7, 30.1, 0.126, 25.9}');
```

2. 调用 PCA 训练函数生成特征向量矩阵

```
drop table if exists result_table, result_table_mean;
select madlib.pca_train('mat',              -- 原始表
                        'result_table',     -- 输出表
                        'id',               -- 源表 ID 列
                        3                   -- 主成分个数 );
```

3. 查看输出表

```
dm=# select row_id id, std, prop,
dm-#        lpad(p1,6,' ')||','||lpad(p2,6,' ')||','||
dm-#        lpad(p3,6,' ')||','||lpad(p4,6,' ')||','||
dm-#        lpad(p5,6,' ')||','||lpad(p6,6,' ')||','||
dm-#        lpad(p7,6,' ')||','||lpad(p8,6,' ') principal_components
dm-#   from (select row_id,
dm(#                round(p[1]::numeric,3) p1,
dm(#                round(p[2]::numeric,3) p2,
dm(#                round(p[3]::numeric,3) p3,
dm(#                round(p[4]::numeric,3) p4,
dm(#                round(p[5]::numeric,3) p5,
```

```
dm(#                    round(p[6]::numeric,3) p6,
dm(#                    round(p[7]::numeric,3) p7,
dm(#                    round(p[8]::numeric,3) p8,
dm(#                    round(std_dev::numeric,3) std,
dm(#                    round(proportion::numeric,3) prop
dm(#             from (select row_id, principal_components p, std_dev, proportion
dm(#                   from result_table) t) t
dm-#  order by row_id;
 id |  std   | prop  |                 principal_components
----+--------+-------+------------------------------------------------------------
  1 | 19.487 | 0.744 | -0.550,-0.221,-0.222,-0.234,-0.324,-0.460,-0.035,-0.477
  2 |  9.067 | 0.161 | -0.679,-0.258, 0.092, 0.224, 0.255, 0.590,-0.019, 0.009
  3 |  5.063 | 0.050 |  0.293,-0.117, 0.358, 0.036, 0.371, 0.070, 0.056,-0.791
(3 rows)
```

可以看到，主成分数量为 3 时，累积标准差比例为 95.5，即反映了 95.5%的原始信息，而维度已经从 8 个降低为 3 个。

4. 调用 PCA 投影函数生成降维后的数据表

```
drop table if exists residual_table, result_summary_table, out_table;
select madlib.pca_project( 'mat',
                           'result_table',
                           'out_table',
                           'id',
                           'residual_table',
                           'result_summary_table');
```

5. 查看投影函数输出表

```
dm=# select * from out_table order by row_id;
 row_id |                          row_vec
--------+------------------------------------------------------------
      1 | {7.08021173666004,-17.6113408380368,3.62504928877282}
      2 | {-0.377499074086429,5.25083911315586,-6.06667264391957}
      3 | {-24.3659516926199,2.69294552046529,-0.854680487274518}
      4 | {-15.235298685282,-2.7756923532236,-1.48789032627869}
      5 | {4.64264829479035,-9.80192214158058,9.97441166441563}
      6 | {23.6146598612176,7.91277187194796,-1.70125446716029}
      7 | {-4.25445515316499,6.71053107113929,-3.63489574437095}
      8 | {12.8547303317577,-15.2151276724561,-4.53202062778529}
      9 | {40.4531114732088,11.566606363421,0.33351408976578}
     10 | {-2.39187210257759,3.48063922820141,-1.53633678788746}
     11 | {-22.6173674430242,2.15970955881415,0.0711392924992467}
     12 | {-37.2273102800874,6.50778045364591,3.06216108712084}
     13 | {2.45676837959725,-5.55018275237518,0.715863146049782}
```

```
    14 | {5.05828673790116,-5.6726215744102,-7.79762716115411}
    15 | {10.3093376158273,10.3450641508798,9.82923967771456}
(15 rows)

dm=# select * from result_summary_table;
   exec_time    | residual_norm  | relative_residual_norm
----------------+----------------+------------------------
 7834.64002609  | 17.8804330669  |      0.0999460602087
(1 row)

dm=# select row_id,
dm-#        lpad(round(row_vec[1]::numeric,3),6,' ')||',  '||
dm-#        lpad(round(row_vec[2]::numeric,3),6,' ')||',  '||
dm-#    lpad(round(row_vec[3]::numeric,3),6,' ')||',  '||
dm-#        lpad(round(row_vec[4]::numeric,3),6,' ')||',  '||
dm-#        lpad(round(row_vec[5]::numeric,3),6,' ')||',  '||
dm-#        lpad(round(row_vec[6]::numeric,3),6,' ')||',  '||
dm-#        lpad(round(row_vec[7]::numeric,3),6,' ')||',  '||
dm-#        lpad(round(row_vec[8]::numeric,3),6,' ')  row_vec
dm-#   from residual_table order by row_id;
 row_id|                        row_vec
-------+----------------------------------------------------------------
     1 | -2.253,  7.276, -0.316, -0.494,  1.005,  0.433,  0.599, -1.523
     2 | -0.863,  3.954, -0.237,  0.678, -1.408,  1.207,  1.860, -1.401
     3 |  0.308, -1.423, -0.116,  0.357,  0.447, -0.586, -0.021,  0.445
     4 |  0.490, -1.375, -0.164, -1.179,  0.250,  0.294, -0.884,  0.337
     5 |  0.153, -3.813,  1.675, -0.442,  1.857, -2.405,  0.477,  2.049
     6 | -0.359, -1.362,  0.957,  0.324,  1.662, -1.994,  0.792,  1.175
     7 | -0.028,  0.002,  0.058,  0.547,  0.078, -0.300, -0.534,  0.012
     8 |  1.811, -3.726, -1.035,  1.121, -1.902,  0.993, -0.685, -0.048
     9 | -1.533,  2.229,  1.013, -2.063,  2.932, -1.451, -0.181,  0.700
    10 |  0.169, -1.288,  0.593, -0.389,  0.377, -0.506, -0.602,  0.592
    11 | -1.443,  4.345,  0.176,  0.001,  0.560, -0.011,  0.256, -0.817
    12 | -0.284, -0.759,  0.585, -0.944,  1.492, -1.045,  0.202,  0.850
    13 | -0.471,  0.949,  0.756, -0.019, -0.154, -0.154,  0.516, -0.023
    14 |  1.721, -3.462, -0.954,  0.290, -1.723,  1.225, -0.855, -0.029
    15 |  2.584, -1.547, -2.992,  2.213, -5.472,  4.300, -0.938, -2.318
(15 rows)
```

out_table 为降维后，投影到主成分的数据表。residual_table 中的数据表示与每个原始数据项对应的误差，越接近零说明误差越小。result_summary_table 表中包含函数执行概要信息。

6. 按主成分总得分降序排列得到综合实力排序

```
dm=# select row_id id, row_vec, round(madlib.array_sum(row_vec)::numeric,4) r
```

```
dm-#    from out_table
dm-#    order by r desc;
 id |                       row_vec                                      |     r
----+--------------------------------------------------------------------+-----------
  9 | {40.4531114732088,11.566606363421,-0.333514089765778}              |  51.6862
  6 | {23.61465986122176,7.91277187194796,1.70125446716029}              |  33.22869
  2 | {-0.377499074086431,5.25083911315585,6.06667264391957}             |  10.9400
 15 | {10.3093376158273,10.3450641508798,-9.82923967771456}              |  10.8252
 14 | {5.05828673790117,-5.6726215744102,7.79762716115411}               |   7.1833
  7 | {-4.254455153165,6.71053107113929,3.63489574437095}                |   6.0910
 10 | {-2.3918721025776,3.48063922820141,1.53633678788746}               |   2.6251
  8 | {12.8547303317577,-15.2151276724561,4.53202062778528}              |   2.1716
 13 | {2.45676837959725,-5.55018275237518,-0.715863146049783}            |  -3.8093
  1 | {7.08021173666005,-17.6113408380368,-3.62504928877282}             | -14.1562
  5 | {4.64264829479035,-9.80192214158058,-9.97441166441563}             | -15.1337
  4 | {-15.235298685282,-2.7756923532236,1.48789032627869}               | -16.5231
 11 | {-22.6173674430242,2.15970955881415,-0.0711392924992431}           | -20.5288
  3 | {-24.3659516926199,2.69294552046528,0.85468048727452}              | -20.8183
 12 | {-37.2273102800874,6.50778045364591,-3.06216108712083}             | -33.7817
(15 rows)
```

从该结果可知，第 9 家企业的综合实力最强，第 12 家企业的综合实力最弱。row_vec 中的三列为各个主成分的得分。以上应用示例比较简单，真实场景中，PCA 方法还要根据实际问题和需求灵活使用。

21.4 小结

主成分分析简称 PCA，是一种常用的多变量分析方法，它有两个主要作用，一是降维；二是去掉变量之间的相关性。主成分分析作为基础的数学分析方法，其实际应用十分广泛，比如在人口统计学、数学建模、数理分析等领域中均有应用。MADlib 提供了两组 PCA 函数，训练函数与投影函数。训练函数以原始数据（行是数据点，列是特征）作为输入，输出主成分特征向量，投影函数将训练函数输出的主成分作为入参，输出逼近原始信息的降维后的矩阵表。我们用 MADlib 的 PCA 函数实现了一个企业综合实力排序的简单需求。

第 22 章
◀ 关联规则方法 ▶

数据仓库或数据挖掘从业者一定对"啤酒与尿布"的故事不会陌生。这就是一个使用关联规则的经典案例。根据对超市顾客购买行为的数据挖掘发现,男顾客经常一起购买啤酒和尿布,于是经理决定将啤酒与尿布放置在一起,让顾客很容易在货架上看到,从而使销售额大幅度增长。关联规则挖掘在多个领域得到了广泛应用,包括互联网数据分析、生物工程、电信和保险业的错误校验等。本章将介绍关联规则方法、Apriori 算法和 MADlib 的 Apriori 相关函数。之后我们用一个示例说明如何使用 MADlib 的 Apriori 函数发现关联规则。

22.1 关联规则简介

关联规则挖掘的目标是发现数据项集之间的关联关系,是数据挖掘中一个重要的课题。关联规则最初是针对购物篮分析(Market Basket Analysis)问题提出的。假设超市经理想更多地了解顾客的购物习惯,特别是想知道,哪些商品顾客可能会在一次购物时同时购买?为回答该问题,可以对商店的顾客购买记录进行购物篮分析。该过程通过发现顾客放入"购物篮"中的不同商品之间的关联,分析顾客的购物习惯。这种关联的发现可以帮助零售商了解哪些商品频繁地被顾客同时购买,从而帮助他们开发更好的营销策略。

为了对顾客的购物篮进行分析,1993 年,Agrawal 等首先提出关联规则的概念,同时给出了相应的挖掘算法 AIS,但是性能较差。1994 年,又提出了著名的 Apriori 算法,至今仍然作为关联规则挖掘的经典算法被广泛讨论。

Apriori 数据挖掘算法使用事务数据。每个事务事件都具有唯一标识,事务由一组项目(或项集)组成。购买行为被认为是一个布尔值(买或不买),这种实现不考虑每个项目的购买数量。MADlib 的关联规则函数假设数据存储在事务 ID 与项目两列中。具有多个项的事务将扩展为多行,每行一项目,如:

```
 trans_id| product
--------+---------
      1 | 1
      1 | 2
      1 | 3
      1 | 4
      2 | 3
```

```
        2 | 4
        2 | 5
        3 | 1
        3 | 4
        3 | 6
...
```

关联规则挖掘涉及以下一些基本概念。

1. 项目与项集

数据库中不可分割的最小单位信息，称为项目，用符号 i 表示。项目的集合称为项集。设集合 I={i1,i2,...ik} 是项集，I 中项目的个数为 k，则集合 I 称为 k-项集。例如，集合{啤酒,尿布,牛奶}是一个 3-项集。

2. 事务

设 I={i1,i2,...ik} 是由数据库中所有项目构成的集合，一次处理所含项目的集合用 T 表示，T={t1,t2,...tn}。每个 ti 包含的项集都是 I 的子集。例如，如果顾客在商场里同一次购买多种商品，这些购物信息在数据库中有一个唯一的标识，用以表示这些商品是同一顾客同一次购买的。称该用户的本次购物活动对应一个数据库事务。

3. 关联规则

关联规则是形如 X=>Y 的蕴含式，意思是"如果 X 则 Y"，其中 X、Y 都是 I 的子集，且 X 与 Y 交集为空。X、Y 分别称为规则的前提和结果，或者规则的左、右。关联规则反映 X 中的项目出现时，Y 中的项目也跟着出现的规律。

4. 项集的频数（Count）

对于任何给定的项集 X，包含 X 的事务数，称为 X 的频数。

5. 支持度（Support）

包含项集 X 的事务数与总事务数之比，称为 X 的支持度，记为：

$$S(X) = \frac{TotalX}{Total transactions}$$

6. 关联规则的支持度

关联规则的支持度是事务集中同时包含 X 和 Y 的事务数与所有事务数之比，其实也就是两个项集{X Y}出现在事务库中的频率，记为：

$$S(X => Y) = \frac{Total(X \cup Y)}{Total transactions}$$

7. 关联规则的置信度（Confidence）

关联规则的置信度是事务集中包含 X 和 Y 的事务数与所有包含 X 的事务数之比，也就是当项集 X 出现时，项集 Y 同时出现的概率，记为：

$$\text{conf}(X => Y) = \text{supp}(X \cup Y)/\text{supp}(X)$$

8. 关联规则的提升度（Lift）

提升度表示含有 X 的条件下，同时含有 Y 的概率，与不含 X 的条件下却含有 Y 的概率之比。这个值越大，越表明 X 和 Y 有较强的关联度。关联规则的提升度定义为：

$$\text{lift}(X => Y) = \frac{\text{supp}(X \cup Y)}{\text{supp}(X) \times \text{supp}(Y)}$$

9. 关联规则的确信度（Conviction）

确信度表示 X 出现而 Y 不出现的概率，也就是规则预测错误的概率。确信度也用来衡量 X 和 Y 的独立性，这个值越大，X、Y 越关联。关联规则的确信度定义为：

$$\text{conv}(X => Y) = \frac{1 - \text{supp}(Y)}{1 - \text{conf}(X => Y)}$$

10. 最小支持度与最小置信度

通常用户为了达到一定的要求，需要指定规则必须满足的支持度和置信度阈值，当 support(X=>Y)、confidence(X=>Y)分别大于等于各自的阈值时，认为 X=>Y 是有趣的，此两个值称为最小支持度阈值（min_sup）和最小置信度阈值（min_conf）。其中，min_sup 描述了关联规则的最低重要程度，min_conf 规定了关联规则必须满足的最低可靠性。

11. 频繁项集

设 U={u1,u2,...,un}为项目的集合，且 U I, U≠∅，对于给定的最小支持度 min_sup，如果项集 U 的支持度 support(U)>=min_sup，则称 U 为频繁项集，否则 U 为非频繁项集。

12. 强关联规则

support(X=>Y)>=min_sup 且 confidence(X=>Y)>=min_conf,称关联规则 X=>Y 为强关联规则，否则称 X=>Y 为弱关联规则。

下面用一个简单的例子来帮助理解这些定义。假设表 22-1 是顾客购买记录的数据库 D，包含 6 个事务。

表 22-1 购买事务记录

TID	网球拍	网球	运动鞋	羽毛球
1	1	1	1	0
2	1	1	0	0
3	1	0	0	0
4	1	0	1	0
5	0	1	1	1
6	1	1	0	0

项集 I={网球拍,网球,运动鞋,羽毛球}。考虑关联规则：网球拍=>网球，事务 1、2、3、4、6 包含网球拍，事务 1、2、6 同时包含网球拍和网球，支持度 support=3/6，置信度 confidence=3/5，提升度 lift=(3/6)/((5/6)*(4/6))=9/10，确信度 conviction=(1-4/6)/(1-3/5)=5/6。若给定最小支持度 α=0.5，最小置信度 β=0.5，关联规则网球拍=>网球是有趣的，认为购买网球拍和购买网球之间存在强关联规则。但是，由于提升度 Lift 小于 1，就是说是否购买网球，与有没有购买网球拍关联性很小。当提升度 Lift(X=>Y)>1 时，则规则"X=>Y"是有效的强关联规则。否则，规则"X=>Y"是无效的强关联规则。特别地，如果 Lift(X=>Y)=1，则表示 X 与 Y 相互独立。因此规则网球拍=>网球是无效的强关联规则。

22.2 Apriori 算法

22.2.1 Apriori 算法基本思想

关联规则挖掘分为两步：（1）找出所有频繁项集；（2）由频繁项集产生强关联规则。其总体性能由第一步决定。在搜索频繁项集时，最简单、最基本的算法就是 Apriori 算法。算法的名字基于这样一个事实：使用频繁项集的先验知识。Apriori 使用一种被称作逐层搜索的迭代方法，k-项集用于搜索(k+1)-项集。首先，通过扫描数据库，积累每个项目的计数，并收集满足最小支持度的项目，找出频繁 1-项集的集合。该集合记作 L1。然后，L1 用于找频繁 2-项集的集合 L2，L2 用于找 L3，如此下去，直到不能再找到频繁 k-项集。找每个 Lk 需要一次数据库全扫描。

Apriori 核心算法思想中有两个关键步骤：连接和剪枝。

1. 连接

为找出 Lk（频繁 k-项集），通过 Lk-1 与自身连接，产生候选 k-项集，该候选项集记作 Ck；其中 Lk-1 的元素是可连接的。

2. 剪枝

Ck 是 Lk 的超集，即它的成员可以是也可以不是频繁的，但所有的频繁项集都包含在 Ck 中。扫描数据库，确定 Ck 中每一个候选项的计数，从而确定 Lk（计数值不小于最小支持度计数的所有候选是频繁的，从而属于 Lk）。然而，Ck 可能很大，这样所涉及的计算量就很大。为了压缩 Ck，使用 Apriori 性质：任一频繁项集的所有非空子集也必须是频繁的，反之，如果某个候选的非空子集不是频繁的，那么该候选肯定不是频繁的，从而可以将其从 Ck 中删除。例如，如果{A,B,C}是一个 3 项的频繁项集，则其子集{A,B}、{B,C}、{A,C}也一定是 2 项的频繁项集。剪枝事先对候选集进行过滤，以减少访问外存的次数，而这种子集测试本身可以使用所有频繁项集的散列树快速完成。

22.2.2 Apriori 算法步骤

假设给定最小支持度和最小置信度，Apriori 算法的主要步骤如下：

（1）扫描全部数据，产生候选 1-项集的集合 C1；
（2）根据最小支持度，由候选 1-项集集合 C1 产生频繁 1-项集的集合 L1；
（3）对 k>1，重复执行步骤（4）、（5）、（6）；
（4）由 Lk 执行连接和剪枝操作，产生候选(k+1)-项集的集合 Ck+1；
（5）根据最小支持度，由候选(k+1)-项集的集合 Ck+1，产生频繁(k+1)-项集的集合 Lk+1；
（6）若 L≠∅，则 k=k+1，跳往步骤（4）；否则跳往步骤（7）；
（7）设产生的频繁项集为 A，A 的所有真子集为 B，根据最小置信度，产生 B=>（A-B）的强关联规则，结束。

22.3 MADlib 的 Apriori 算法函数

MADlib 的 assoc_rules 函数实现 Apriori 算法，用于生成所有满足给定最小支持度和最小置信度的关联规则。

1. 语法

```
assoc_rules (support,
             confidence,
             tid_col,
             item_col,
             input_table,
             output_schema,
             verbose,
             max_itemset_size );
```

2. 参数

support：最小支持度。
confidence：最小置信度。
tid_col：事务 ID 列名。
item_col：项目对应的列名。
input_table：包含输入数据的表名。输入表的结构为：

```
{TABLE|VIEW} input_table (
    trans_id INTEGER,
    product TEXT )
```

该算法将产品名称映射到从 1 开始的连续的整型事务 ID 上。如果输入数据本身已经结构化为这种形式，则事务 ID 保持不变。

output_schema：存储最终结果的模式名称，调用函数前，模式必须已创建。如果此参数为 NULL，则输出到当前模式。结果存储在输出模式中的 assoc_rules 表中，具有以下列：

```
Column      |  Type
------------+-------------------
ruleid      | integer
pre         | text[]
post        | text[]
count       | integer
support     | double precision
confidence  | double precision
lift        | double precision
conviction  | double precision
```

在 HAWQ 中，assoc_rules 表通过 ruleid 列哈希分布存储。pre 和 post 列分别是相应关联规则的左右项集。count、support、confidence、lift 和 conviction 列分别对应关联规则的频数、支持度、置信度、提升度和确信度。

verbose：BOOLEAN 类型，默认为 false，指示是否详细打印算法过程中每次迭代的结果。

max_itemset_size：INTEGER 类型，该参数值必须大于等于 2，指定用于产生关联规则的频繁项集的大小，默认值是产生全部项集。当项集太大时，可用此参数限制数据集的大小，以减少运行时长。

22.4 Apriori 应用示例

本节我们就用"啤酒与尿布"的经典示例，人为模拟一些购买记录作为输入数据，然后调用 madlib.assoc_rules 函数生成关联规则。我们将对比控制台的打印信息，一步步说明该函数获取关联规则的计算过程，并对最终结果进行分析。

1. 创建输入数据集

```
drop table if exists test_data;
create table test_data (
    trans_id int,
    product text
);
insert into test_data values
(1, 'beer'), (1, 'diapers'), (1, 'chips'),
(2, 'beer'), (2, 'diapers'),
(3, 'beer'), (3, 'diapers'),
(4, 'beer'), (4, 'chips'),
```

```
(5, 'beer'),
(6, 'beer'), (6, 'diapers'), (6, 'chips'),
(7, 'beer'), (7, 'diapers');
```

2. 调用 madlib.assoc_rules 函数生成关联规则

设置最小支持度 min(support) = 0.25，最小置信度 min(confidence) = 0.5。输出模式设置为 NULL，输出到当前模式。verbose 设置为 TRUE，这样就能观察和验证函数执行的过程。

```
select * from madlib.assoc_rules
( .25,            -- 最小支持度
  .5,             -- 最小置信度
  'trans_id',     -- 事务 ID 列名
  'product',      -- 产品列名
  'test_data',    -- 输入数据表
  null,           -- 在当前模式创建输出表
  true            -- 打印详细信息
);
```

控制台打印的输出信息如下：

```
INFO:  finished checking parameters
CONTEXT:  PL/Python function "assoc_rules"
INFO:  finished removing duplicates
CONTEXT:  PL/Python function "assoc_rules"
INFO:  finished encoding items
CONTEXT:  PL/Python function "assoc_rules"
INFO:  finished encoding input table: 4.55814695358
CONTEXT:  PL/Python function "assoc_rules"
INFO:  Beginning iteration #1
CONTEXT:  PL/Python function "assoc_rules"
INFO:  3 Frequent itemsets found in this iteration
CONTEXT:  PL/Python function "assoc_rules"
INFO:  Completed iteration # 1. Time: 0.756361961365
CONTEXT:  PL/Python function "assoc_rules"
INFO:  Beginning iteration # 2
CONTEXT:  PL/Python function "assoc_rules"
INFO:  time of preparing data: 0.784854888916
CONTEXT:  PL/Python function "assoc_rules"
INFO:  3 Frequent itemsets found in this iteration
CONTEXT:  PL/Python function "assoc_rules"
INFO:  Completed iteration # 2. Time: 1.89147591591
CONTEXT:  PL/Python function "assoc_rules"
INFO:  Beginning iteration # 3
CONTEXT:  PL/Python function "assoc_rules"
INFO:  time of preparing data: 0.577459096909
CONTEXT:  PL/Python function "assoc_rules"
```

```
INFO:  1 Frequent itemsets found in this iteration
CONTEXT:  PL/Python function "assoc_rules"
INFO:  Completed iteration # 3. Time: 1.32646298409
CONTEXT:  PL/Python function "assoc_rules"
INFO:  Beginning iteration # 4
CONTEXT:  PL/Python function "assoc_rules"
INFO:  time of preparing data: 0.31945681572
CONTEXT:  PL/Python function "assoc_rules"
INFO:  0 Frequent itemsets found in this iteration
CONTEXT:  PL/Python function "assoc_rules"
INFO:  Completed iteration # 4. Time: 0.751129865646
CONTEXT:  PL/Python function "assoc_rules"
INFO:  begin to generate the final rules
CONTEXT:  PL/Python function "assoc_rules"
INFO:  7 Total association rules found. Time: 1.1944539547
CONTEXT:  PL/Python function "assoc_rules"
 output_schema | output_table | total_rules |   total_time
---------------+--------------+-------------+-----------------
 public        | assoc_rules  |           7 | 00:00:10.566625
(1 row)
```

madlib.assoc_rule 函数产生频繁项集及关联规则的过程如下：

（1）验证参数、去除重复数据、对输入数据编码（生成从 1 开始的连续的事务 ID，本例原始数据已经满足要求，因此不需要编码）。

（2）首次迭代，生成所有支持度大于等于 0.25 的 1 阶项集作为初始项集，如表 22-2 所示。因为三个 1 阶项集的支持度都大于 0.25，所以初始项集包含所有三个项集。

表 22-2　1 阶项集

项集	频数	支持度
beer	7	7/7
diapers	5	5/7
chips	3	3/7

（3）最大阶数为 3，因此开始第二次迭代。迭代结果的频繁项集如表 22-3 所示。因为三个 2 阶项集的支持度都大于 0.25，所以本次迭代结果的频繁项集包含所有三个项集。

表 22-3　2 阶项集

项集	频数	支持度
beer, diapers	5	5/7
beer, chips	3	3/7
diapers，chips	2	2/7

(4）第三次迭代，得到的频繁项集如表 22-4 所示。因为唯一一个 3 阶项集的支持度大于 0.25，所以本次迭代结果的频繁项集包含一个项集。

表 22-4　3 阶项集

项集	频数	支持度
beer, diapers, chips	2	2/7

(5）第四次迭代，因为阶数为 3，所以本次迭代的结果为空集。
(6）产生关联规则，置信度大于等于 0.5 的关联规则如表 22-5 中粗体行所示。

表 22-5　关联规则

前提	结果	支持度	置信度
beer	diapers	5/7=0.71428571	(5/7)/(7/7)= 0.71428571
diapers	beer	5/7=0.71428571	(5/7)/(5/7)=1
beer	chips	3/7=0.42857143	(3/7)/(7/7)= 0.42857143
chips	beer	3/7=0.42857142	(3/7)/(3/7)=1
diapers	chips	2/7=0.28571429	(2/7)/(5/7)=0.4
chips	diapers	2/7=0.28571429	(2/7)/(3/7)=0.66666667
beer	diapers, chips	2/7=0.28571429	(2/7)/(7/7)= 0.28571429
beer, diapers	chips	2/7=0.28571429	(2/7)/(5/7)=0.4
beer, chips	diapers	2/7=0.28571429	(2/7)/(3/7)= 0.66666667
diapers	beer, chips	2/7=0.28571429	(2/7)/(5/7)=0.4
diapers, chips	beer	2/7=0.28571429	(2/7)/(2/7)=1
chips	beer, diapers	2/7=0.28571429	(2/7)/(3/7)= 0.66666667

关联规则存储在 assoc_rules 表中，查询结果与表 22-5 所反映的相同：

```
dm=# select ruleid id,
dm-#        pre,
dm-#        post,
dm-#        count c,
dm-#        round(support::numeric,4) sup,
dm-#        round(confidence::numeric,4) conf,
dm-#        round(lift::numeric,4) lift,
dm-#        round(conviction::numeric,4) conv
dm-#   from assoc_rules
dm-#  order by support desc, confidence desc;
 id |     pre      |      post       | c |  sup   |  conf  |  lift  |  conv
----+--------------+-----------------+---+--------+--------+--------+-------
  4 | {diapers}    | {beer}          | 5 | 0.7143 | 1.0000 | 1.0000 | 0.0000
```

```
 2 | {beer}         | {diapers}      | 5 | 0.7143 | 0.7143 | 1.0000 | 1.0000
 3 | {chips}        | {beer}         | 3 | 0.4286 | 1.0000 | 1.0000 | 0.0000
 5 | {diapers,chips}| {beer}         | 2 | 0.2857 | 1.0000 | 1.0000 | 0.0000
 1 | {chips}        | {diapers}      | 2 | 0.2857 | 0.6667 | 0.9333 | 0.8571
 7 | {chips}        | {diapers,beer} | 2 | 0.2857 | 0.6667 | 0.9333 | 0.8571
 6 | {beer,chips}   | {diapers}      | 2 | 0.2857 | 0.6667 | 0.9333 | 0.8571
(7 rows)
```

3. **限制生成关联规则的项集大小为 2，再次执行 madlib.assoc_rules 函数**

注意，madlib.assoc_rules 函数总会新创建 assoc_rules 表。如果要保留多个关联规则表，需要在运行函数前备份已存在的 assoc_rules 表。

```
select * from madlib.assoc_rules
( .25,              -- 最小支持度
  .5,               -- 最小置信度
  'trans_id',       -- 事务 ID 列名
  'product',        -- 产品列名
  'test_data',      -- 输入数据表
  null,             -- 在当前模式创建输出表
  true,             -- 打印详细信息
  2                 -- 最大项集数
);
```

这次生成的关联规则如下：

```
dm=# select ruleid id,
dm-#        pre,
dm-#        post,
dm-#        count c,
dm-#        round(support::numeric,4) sup,
dm-#        round(confidence::numeric,4) conf,
dm-#        round(lift::numeric,4) lift,
dm-#        round(conviction::numeric,4) conv
dm-#   from assoc_rules
dm-#  order by support desc, confidence desc;
 id |   pre     |   post    | c |  sup   |  conf  |  lift  |  conv
----+-----------+-----------+---+--------+--------+--------+--------
  4 | {diapers} | {beer}    | 5 | 0.7143 | 1.0000 | 1.0000 | 0.0000
  2 | {beer}    | {diapers} | 5 | 0.7143 | 0.7143 | 1.0000 | 1.0000
  3 | {chips}   | {beer}    | 3 | 0.4286 | 1.0000 | 1.0000 | 0.0000
  1 | {chips}   | {diapers} | 2 | 0.2857 | 0.6667 | 0.9333 | 0.8571
(4 rows)
```

4. **按需过滤关联规则**

例如，如果想得到规则左端项集只有一个项目，并且规则右端项集为"beer"：

```
dm=# select ruleid, pre, post, count c, round(support::numeric,4) sup,
dm-#        confidence, lift, conviction
dm-#   from assoc_rules
dm-#  where array_upper(pre,1) = 1 and post = array['beer'];
 ruleid |   pre     |  post  | c |  sup   | confidence | lift | conviction
--------+-----------+--------+---+--------+------------+------+------------
      3 | {chips}   | {beer} | 3 | 0.4286 |          1 |    1 |          0
      4 | {diapers} | {beer} | 5 | 0.7143 |          1 |    1 |          0
(2 rows)
```

5. 分析关联规则

madlib.assoc_rules 函数是根据用户设置的支持度与置信度阈值生成关联规则的。我们以支持度和置信度阈值限制得到了 7 个关联规则，但这些规则的有效性如何呢？ 前面提到，满足最小支持度和最小置信度的规则，叫做"强关联规则"。然而，强关联规则里，也分有效的强关联规则和无效的强关联规则。

从提升度来看，提升度大于 1，则规则是有效的强关联规则，否则是无效的强关联规则。如果提升度=1，说明前提与结果彼此独立，没有任何关联，如果<1，说明前提与结果是相斥的。以规则 5 为例，提升度为 0.93，说明购买了 diapers 的顾客就不会购买 chips。因此用提升度作为度量，结果中的 7 个规则都是无效的。

22.5 小结

关联规则挖掘的目标在于发现数据项集之间的关联关系，该方法常被用于购物篮分析，著名的"啤酒与尿布"案例就是关联规则的典型应用。Apriori 是关联规则挖掘的经典算法，特点是比较简单，易于理解和实现。MADlib 提供的 assoc_rules 函数实现 Apriori 算法，用于生成所有满足给定最小支持度和最小置信度阈值的关联规则。使用该函数生成强关联规则后，还需要分析提升度判断其有效性。

第 23 章
◀ 聚类方法 ▶

"物以类聚，人以群分"，其核心思想就是聚类。所谓聚类，就是将相似的事物聚集在一起，而将不相似的事物划分到不同的类别的过程，是数据分析中十分重要的一种手段。比如古典生物学中，人们通过物种的形貌特征将其分门别类，可以说就是一种朴素的人工聚类。如此，我们就可以将世界上纷繁复杂的信息，简化为少数方便人们理解的类别，因此聚类可以说是人类认知这个世界的最基本方式之一。通过聚类，人们能意识到密集和稀疏的区域，发现全局的分布模式，以及数据属性之间有趣的相互关系。

聚类起源于分类学，在古老的分类学中，人们主要依靠经验和专业知识来实现分类，很少利用数学工具定量分类。随着科学技术的发展，对分类的要求越来越高，以至有时仅凭经验难以确切地分类，于是人们逐渐把数学工具引用到了分类学中，形成了数值分析学，之后又将多元分析技术引入进数值分类学，从而形成了聚类。在实践中，聚类往往为分类服务，即先通过聚类来判断事物的合适类别，然后在利用分类技术对新的样本进行分类。

聚类算法大都是几种最基本的方法，如 k-means、层次聚类、SOM 等，以及它们的许多改进变种。MADlib 提供了一种 k-means 算法的实现。本章主要介绍 MADlib 的 k-means 算法相关函数和应用案例。

23.1 聚类方法简介

1. 聚类的概念

将物理或抽象对象的集合分成由类似的对象组成的多个类或簇（Cluster）的过程被称为聚类（Clustering）。由聚类所生成的簇是一组数据对象的集合，这些对象与同一个簇中的对象相似度较高，与其他簇中的对象相似度较低。相似度是根据描述对象的属性值来度量的，距离是经常采用的度量方式。分析事物聚类的过程称为聚类分析或群分析，是研究样品或指标分类问题的一种统计分析方法。

在数据分析的术语中，聚类和分类是两种技术。分类是指已经知道了事物的类别，需要从样品中学习分类规则，对新的、无标记的对象赋予类别，是一种有监督学习。而聚类则没有事先预定的类别,而是依据人为给定的规则进行训练,类别在聚类过程中自动生成,从而得到分类,是一种无监督学习。作为一个数据挖掘的功能,聚类可当作独立的工具来获得数据分布情况,观察每个簇的特点,集中对特定簇做进一步的分析。此外,聚类分析还可以作为其他算法的预处理步骤,简少计算量,提高分析效率。

2. 类的度量方法

虽然类的形式各有不同,但总的来说,一般用距离作为类的度量方法。设 x、y 是两个向量 $\vec{x}=(x_1,\dots,x_n)$ 和 $\vec{y}=(y_1,\dots,y_n)$,聚类分析中常用的距离有以下几种。

(1) 曼哈顿距离

x、y 的曼哈顿距离定义为:

$$\|x-y\|_1 = \sum_{i=1}^{n} |x_i - y_i|$$

(2) 欧氏距离

x、y 的欧氏距离定义为:

$$\|x-y\|_2 = \sqrt{\sum_{i=1}^{n}(x_i-y_i)^2}$$

(3) 欧氏平方距离

x、y 的欧氏平方距离定义为:

$$\|x-y\|_2^2 = \sum_{i=1}^{n}(x_i-y_i)^2$$

(4) 角距离

x、y 的角距离定义为:

$$\arccos\left(\frac{(\vec{x},\vec{y})}{\|\vec{x}\|\cdot\|\vec{y}\|}\right)$$

分母是 x、y 两个向量的 2 范数乘积。

(5) 谷本距离

x、y 的谷本距离定义为:

$$1 - \frac{(\vec{x},\vec{y})}{\|\vec{x}\|^2 \cdot \|\vec{y}\|^2 - (\vec{x},\vec{y})}$$

23.2 k-means 方法

在数据挖掘中,k-means 算法是一种广泛使用的聚类分析算法,也是 MADlib 1.10.0 官方文档中唯一提及的聚类算法。

23.2.1 基本思想

k-means 聚类划分方法的基本思想是：将一个给定的有 N 个数据记录的集合，划分到 K 个分组中，每一个分组就代表一个簇，K<N。而且这 K 个分组满足下列条件：

- 每一个分组至少包含一个数据记录。
- 每一个数据记录属于且仅属于一个分组。

算法首先给出一个初始的分组，以后通过反复迭代的方法改变分组，使得每一次改进之后的分组方案都较前一次好，而所谓好的标准就是：同一分组中对象的距离越近越好（已经收敛，反复迭代至组内数据几乎无差异），而不同分组中对象的距离越远越好。

23.2.2 原理与步骤

k-means 算法的工作原理是：首先随机从数据集中选取 K 个点，每个点初始地代表每个簇的中心，然后计算剩余各个样本到中心点的距离，将它赋给最近的簇，接着重新计算每一簇的平均值作为新的中心点，整个过程不断重复，如果相邻两次调整没有明显变化，说明数据聚类形成的簇已经收敛。本算法的一个特点是在每次迭代中都要考察每个样本的分类是否正确。若不正确，就要调整，在全部样本调整完后，再修改中心点，进入下一次迭代。这个过程将不断重复直到满足某个终止条件，终止条件可以是以下任何一个：

- 没有对象被重新分配给不同的聚类。
- 聚类中心不再发生变化。
- 误差平方和局部最小。

k-means 算法是很典型的基于距离的聚类算法，采用距离作为相似性的评价指标，即认为两个对象的距离越近，其相似度就越大。该算法认为簇是由距离靠近的对象组成，因此把得到紧凑且独立的簇作为最终目标。

k-means 算法的输入是聚类个数 k，以及 n 个数据对象，输出是满足误差最小标准的 k 个聚簇。其处理流程为：

（1）从 n 个数据对象中任意选择 k 个对象作为初始中心。
（2）计算每个对象与这些中心对象的距离，并根据最小距离对相应的对象进行划分。
（3）重新计算每个有变化聚类的均值作为新的中心。
（4）循环步骤（2）、（3）直到每个聚类不再发生变化为止。终止条件一般为最小化对象到其聚类中心的距离的平方和：

$$\min \sum_{i=1}^{K} \sum_{x \in C_i} dist(c_i, x)^2$$

23.2.3 k-means 算法

k-means 算法接受输入量 k，然后将 n 个数据对象划分为 k 个簇以便使得所获得的簇满足：

同一簇中的对象相似度较高，而不同簇中的对象相似度较低。簇相似度是利用各簇中对象的均值所获得的中心对象来进行计算的。为了便于理解 k-means 算法，可以参考图 23-1 所示的二维向量的例子。

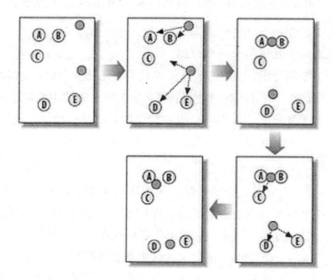

图 23-1   k-means 聚类算法

从图中我们可以看到 A、B、C、D、E 五个点。而灰色的点是初始中心点，也就是用来找簇的点。有两个中心点，所以 K=2。

k-means 的算法如下：

（1）随机在图中取 K（这里 K=2）个初始中心点。

（2）对图中的所有点求到这 K 个中心点的距离，假如点 Pi 离种子点 Si 最近，那么 Pi 属于 Si 聚类。图 23-1 中，我们可以看到 A、B 属于上面的中心点，C、D、E 属于下面中部的中心点。

（3）移动中心点到属于它的簇的中心，作为新的中心点，见图 23-1 上的第三步。

（4）重复第（2）和第（3）步，直到中心点没有移动，可以看到图 23-1 中的第四步上面的中心点聚合了 A、B、C，下面的中心点聚合了 D、E。

二维坐标中两点之间距离公式如下：

$$|AB| = \sqrt{(x_1 - x_2)^2 + (y_1 - y_2)^2}$$

公式中(x1，y1)，(x2，y2)分别为 A、B 两个点的坐标。求聚类中心点的算法可以简单使用各个点的 X/Y 坐标的平均值。

k-means 主要有两个重大缺陷，并且都和初始值有关：

- K 是事先给定的，这个 K 值的选定是非常难以估计的。很多时候，事先并不知道给定的数据集应该分成多少个类别才最合适（ISODATA 算法通过类的自动合并和分裂，得到较为合理的类型数目 K）。

- k-means 算法以初始随机中心点为基础，这个随机中心点非常重要，不同的随机中心点会有得到完全不同的结果。k-means++算法就是用来解决这个问题，它可以有效地选择初始点。

k-means++算法步骤：

（1）先从输入数据对象中随机挑一个作为中心点。

（2）对于每个数据对象 x，计算其和最近的一个中心点的距离 D(x)并保存在一个数组里，然后把这些距离加起来得到 Sum(D(x))。

（3）再取一个随机值，用取权重的方式来计算下一个中心点。这个算法的实现是，先取一个能落在 Sum(D(x))中的随机值 Random，然后用 Random -= D(x)，直到其<=0，此时的 x 就是下一个中心点。

（4）重复第（2）和第（3）步直到所有的 K 个中心点都被选出来。

（5）进行 k-means 算法。

23.3 MADlib 的 k-means 相关函数

形式上，我们希望最小化以下目标函数：

$$(c_1, \ldots, c_k) \to \sum_{i=1}^{n} \min_{1 \leq j \leq k} \text{dist}(x_i, c_j)$$

其中 $x_1, \ldots, x_n$ 是 n 个数据对象，$c_1, \ldots, c_k$ 是 k 个中心点，常见的情况下，距离使用欧氏平方距离。这个问题在计算上很困难（NP-hard 问题），但由于局部启发式搜索算法在实践中表现得相当好，如今被普遍采用，其中之一就是前面讨论的 k-means 算法。MADlib 提供了三组 k-means 算法相关函数，分别是训练函数、簇分配函数和轮廓系数函数。

1. 训练函数

（1）语法

MADlib 提供了以下四个 k-means 算法训练函数。使用随机中心点方法，语法如下：

```
kmeans_random (rel_source,
         expr_point,
         k,
         fn_dist,
         agg_centroid,
         max_num_iterations,
         min_frac_reassigned
)
```

使用 kmeans++中心点方法，语法如下：

```
kmeanspp( rel_source,
         expr_point,
         k,
         fn_dist,
         agg_centroid,
         max_num_iterations,
         min_frac_reassigned,
         seeding_sample_ratio
       )
```

由 rel_initial_centroids 参数提供一个包含初始中心点的表名，语法如下：

```
kmeans( rel_source,
        expr_point,
        rel_initial_centroids,
        expr_centroid,
        fn_dist,
        agg_centroid,
        max_num_iterations,
        min_frac_reassigned
      )
```

由 initial_centroids 参数提供的数组表达式，指定一个初始中心点集合，语法如下：

```
kmeans( rel_source,
        expr_point,
        initial_centroids,
        fn_dist,
        agg_centroid,
        max_num_iterations,
        min_frac_reassigned
      )
```

（2）参数

rel_source：TEXT 类型，含有输入数据对象的表名。数据对象和预定义中心点（如果使用的话）应该使用一个数组类型的列存储，如 FLOAT[]或 INTEGER[]。调用任何以上四种函数进行数据分析时，都会跳过具有 non-finite 值的数据对象，non-finite 值包括 NULL、NaN、infinity 等。

expr_point：TEXT 类型，包含数据对象的列名。

k：INTEGER 类型，指定要计算的中心点的个数。

fn_dist（可选）：TEXT 类型，默认值为 squared_dist_norm2，指定计算数据对象与中心点距离的函数名称。可以使用以下距离函数，括号内为均值计算方法：

- dist_norm1: 1 范数/曼哈顿距离（元素中位数）。
- dist_norm2: 2 范式/欧氏距离（元素平均数）。

- squared_dist_norm2：欧氏平方距离（元素平均数）。
- dist_angle：角距离（归一化数据的元素平均数）。
- dist_tanimoto：谷本距离（归一化数据的元素平均数）。
- 具有 DOUBLE PRECISION[] x, DOUBLE PRECISION[] y → DOUBLE PRECISION 参数形式的用户自定义函数。

agg_centroid（可选）：TEXT 类型，默认值为 avg。确定中心点使用的聚合函数名，可以使用以下聚合函数：

- avg：平均值（默认）。
- normalized_avg：归一化平均值。

max_num_iterations（可选）：INTEGER 类型，默认值为 20，指定执行的最大迭代次数。

min_frac_reassigned（可选）：DOUBLE PRECISION 类型，默认值为 0.001。相邻两次迭代所有中心点相差小于该值时计算完成。

seeding_sample_ratio（可选）：DOUBLE PRECISION，默认值为 1.0。kmeans++将扫描数据 k 次，对大数据集会很慢。此参数指定用于确定初始中心点所使用的原始数据集样本比例。当此参数大于 0 时（最大值为 1.0），初始中心点在数据均匀分布的随机样本上。注意，k-means 算法最终会在全部数据集上执行。此参数只是为确定初始中心点建立一个子样本，并且只对 kmeans++有效。

rel_initial_centroids：TEXT 类型，包含初始中心点的表名。

expr_centroid：TEXT 类型，rel_initial_centroids 指定的表中包含中心点的列名。

initial_centroids：TEXT 类型，包含初始中心点的 DOUBLE PRECISION 数组表达式的字符串。

（3）输出格式

k-means 模型的输出具有以下列的复合数据类型：

- centroids：DOUBLE PRECISION[][]类型，最终的中心点。
- cluster_variance：DOUBLE PRECISION[]类型，每个簇的方差。
- objective_fn：DOUBLE PRECISION 类型，方差合计。
- frac_reassigned：DOUBLE PRECISION 类型，在最后一次迭代的误差。
- num_iterations：INTEGER 类型，迭代执行的次数。

2. 簇分配函数

（1）语法

得到中心点后，可以调用以下函数为每个数据对象进行簇分配：

```
closest_column( m, x )
```

（2）参数

m：DOUBLE PRECISION[][]类型，训练函数返回的中心点。

x：DOUBLE PRECISION[]类型，输入数据。

（3）输出格式

column_id：INTEGER 类型，簇 ID，从 0 开始。

distance：DOUBLE PRECISION 类型，数据对象与簇中心点的距离。

3. 轮廓系数函数

轮廓系数（Silhouette Coefficient），是聚类效果好坏的一种评价方法。作为 k-means 模型的一部分，MADlib 提供了一个轮廓系数方法的简化版本函数，该函数结果值处于-1～1 之间，值越大，表示聚类效果越好。注意，对于大数据集，该函数的计算代价很高。

（1）语法

```
simple_silhouette( rel_source,
                   expr_point,
                   centroids,
                   fn_dist
                 )
```

（2）参数

rel_source：TEXT 类型，含有输入数据对象的表名。

expr_point：TEXT 类型，数据对象列名。

centroids：TEXT 类型。中心点表达式。

fn_dist（可选）：TEXT 类型，计算数据点到中心点距离的函数名，默认值为'dist_norm2'。

23.4　k-means 应用示例

1. 问题提出

RFM 模型是在做用户价值细分时常用的方法，主要涵盖的指标有最近一次消费时间 R（Recency）、消费频率（Frequency），消费金额（Monetary）。我们用 R、F、M 三个指标作为数据对象属性，应用 MADlib 的 k-means 模型相关函数对用户进行聚类分析，并得出具有实用性和解释性的结论。

2. 建立测试数据表并装载原始数据

```
-- 创建原始数据表
drop table if exists t_source;
create table t_source
(cust_id int,
 amount decimal(10 , 2),
```

```
    quantity int,
    dt date);

-- 添加100条数据
insert into t_source (cust_id,amount,quantity,dt) values
(567,1100.51,2,'2017-07-20'),(568,2003.47,2,'2017-07-20'),
(569,297.91,2,'2017-07-14'),(570,300.02,2,'2017-07-12'),
...
(663,954.77,2,'2017-06-27'),(664,6006.78,3,'2017-06-22'),
(665,25755.7,2,'2017-06-06'),(666,60201.48,2,'2017-07-11');
```

3. 数据预处理

（1）将最近一次访问日期处理成最近一次访问日期到当前日期的间隔天数，代表该用户是否最近有购买记录（即目前是否活跃）。

（2）因为 k-means 受异常值影响很大，并且金额变异比较大，所以去除该维度的异常值。

（3）使用 PCA 方法消除维度之间的相关性。

（4）0-1 归一化处理。

```
-- 去掉异常值
drop table if exists t_source_change;
create table t_source_change
(row_id serial,
 cust_id int,
 amount decimal(10 , 2 ),
 quantity int,
 dt int);

insert into t_source_change (cust_id,amount,quantity,dt)
select cust_id,
       amount,
       quantity,
       current_date-dt dt
  from t_source
 where amount < (select percentile_cont (0.99) within group (order by amount)
                   from t_source);

select * from t_source_change order by cust_id;
```

查询结果为：

```
...
    94 |   660 | 11594.24 |   10 |  2
    95 |   661 | 12039.49 |    2 | 30
    96 |   662 |  1494.97 |    2 | 39
```

```
        97 |     663 |   954.77 |      2 | 25
        98 |     664 |  6006.78 |      3 | 30
        99 |     665 | 25755.70 |      2 | 46
(99 rows)
```

可以看到，因为 cust_id=666 用户的金额不在 99%的范围内，所以 t_source_change 表中去掉了该条记录。在此去除异常并非这个用户异常，而是为了改善聚类结果。最后需要给这些"异常用户"做业务解释。

```
-- PCA 去掉相关性
drop table if exists mat;
create table mat (id integer,
                  row_vec double precision[] );

insert into mat
select row_id,
       string_to_array(amount||','||quantity||','||dt,',')::double precision[] row_vec
   from t_source_change;

-- PCA 培训
drop table if exists result_table, result_table_mean;
select madlib.pca_train('mat',              -- source table
                        'result_table',     -- output table
                        'id',               -- row id of source table
                        3                   -- number of principal components
                        );

-- PCA 投影
drop table if exists residual_table, result_summary_table, out_table;
select madlib.pca_project( 'mat',
                           'result_table',
                           'out_table',
                           'id',
                           'residual_table',
                           'result_summary_table'
                           );
-- 0-1 归一化
drop table if exists t_source_change_nor;
create table t_source_change_nor
as
select row_id,
string_to_array(amount_nor||','||quantity_nor||','||dt_nor,',')::double precision[] row_vec
    from
    (
    select row_id,
           (row_vec[1] - min_amount)/(max_amount - min_amount) amount_nor,
           (row_vec[2] - min_quantity)/(max_quantity - min_quantity) quantity_nor,
```

```
            (max_dt - row_vec[3])/(max_dt - min_dt) dt_nor
      from out_table,
           (select max(row_vec[1]) max_amount,
                   min(row_vec[1]) min_amount,
                   max(row_vec[2]) max_quantity,
                   min(row_vec[2]) min_quantity,
                   max(row_vec[3]) max_dt,
                   min(row_vec[3]) min_dt
              from out_table) t) t;

select * from t_source_change_nor order by row_id;
```

查询结果为:

```
...
    94 | {0.558470357737996,0.954872666162949,0.296935710714377}
    95 | {0.54122257689463,0.482977156688704,0.81244230552888}
    96 | {0.949697477408967,0.385844448834949,0.65901807391295}
    97 | {0.970623648952883,0.62014760223173,0.704941708880569}
    98 | {0.774918367989914,0.513405499602443,0.666993533505089}
    99 | {0.00988267286683593,0.150872332720288,0.908966781310526}
(99 rows)
```

4. k-means 聚类

（1）调用 kmeanspp 函数执行聚类

```
drop table if exists km_result;
create table km_result as
select * from madlib.kmeanspp
( 't_source_change_nor',         -- 源数据表名
  'row_vec',                     -- 包含数据点的列名
  3,                             -- 中心点个数
  'madlib.squared_dist_norm2',   -- 距离函数
  'madlib.avg',                  -- 聚合函数
  20,                            -- 迭代次数
  0.001                          -- 停止迭代条件 );

\x on;
select centroids[1][1]||', '||centroids[1][2]||', '||centroids[1][3] cent1,
       centroids[2][1]||', '||centroids[2][2]||', '||centroids[2][3] cent2,
centroids[3][1]||', '||centroids[3][2]||', '||centroids[3][3] cent3,
       cluster_variance,
       objective_fn,
       frac_reassigned,
       num_iterations
  from km_result;
```

查询结果如下:

```
-[ RECORD 1 ]----+--------------------------------------------------
cent1            | 0.872433445942, 0.0724942318135, 0.318094096598
cent2            | 0.890144445443, 0.546835465582, 0.333554735766
```

```
     cent3            | 0.238390106949, 0.449997152636, 0.267439867941
     cluster_variance | {1.33448519773,2.05461238207,1.83212942768}
     objective_fn     | 5.22122700748
     frac_reassigned  | 0
     num_iterations   | 8
```

（2）调用 simple_silhouette 函数评价聚类质量

```
select * from madlib.simple_silhouette
( 't_source_change_nor',
  'row_vec',
  (select centroids
     from madlib.kmeanspp('t_source_change_nor',
                 'row_vec',
                 3,
                 'madlib.squared_dist_norm2',
                 'madlib.avg',
                 20,
                 0.001)),
  'madlib.dist_norm2' );
```

结果如下：

```
-[ RECORD 1 ]-----+------------------
simple_silhouette | 0.640471849127657
```

（3）调用 closest_column 函数执行簇分配

```
\x off;

select cluster_id,
       round(count(cust_id)/99.0,4) pct,
       round(avg(amount),4) avg_amount,
       round(avg(quantity),4) avg_quantity,
       round(avg(dt),2) avg_dt
  from
(
select t2.*,
    (madlib.closest_column(centroids, row_vec)).column_id as cluster_id
  from t_source_change_nor as t1, km_result, t_source_change t2
 where t1.row_id = t2.row_id) t
 group by cluster_id;
```

查询结果为：

```
 cluster_id | pct    | avg_amount | avg_quantity | avg_dt
------------+--------+------------+--------------+--------
          2 | 0.1919 | 5439.9795  |       2.0526 |  48.79
          1 | 0.4848 | 3447.5631  |       2.4375 |  29.56
          0 | 0.3232 | 5586.0203  |       4.0313 |   5.56
(3 rows)
```

5. 解释聚类结果

表 23-1 对聚类结果分成的三类用户进行了说明。

表 23-1　聚类形成的三类用户

类别	占比	描述
第一类：高价值用户	32.3%	购买频率高（平均 4 次）；消费金额较高（平均 5586 元）；最近一周有过购买行为，这部分用户需要大力发展
第二类：中价值用户	48.5%	购买频率中等（平均 2.4 次）；消费金额不高（平均 3447）；最近一个月有个别购买行为，这部分用户可以适当诱导购买
第三类：高价值挽留用户	19.2	购买频率一般（平均 2 次）；消费金额较高（平均 5439 元）；较长时间没有购买行为，这部分客户需要尽量挽留

23.5　小结

聚类方法是根据给定的规则进行训练，自动生成类别的数据挖掘方法，属于无监督学习范畴。聚类已经被应用在模式识别、数据分析、图像处理、市场研究等多个领域。虽然类的形式各不相同，但一般都用距离作为类的度量方法。聚类算法有很多种，其中 k-means 是应用最广泛、适应性最强的聚类算法，也是 MADlib 唯一支持的聚类算法。MADlib 提供了 4 个 k-means 训练函数、一个簇分配函数、一个轮廓系数函数。我们利用 MADlib 提供的这些函数，实现了一个按照 RFM 模型对用户进行细分的示例需求。

第 24 章

回归方法

当人们对研究对象的内在特性和各因素间的关系有比较充分的认识时,一般用机理分析方法建立数学模型。如果由于客观事物内部规律的复杂性及人们认识程度的限制,无法分析实际对象内在的因果关系,建立合乎机理规律的数学模型,那么通常的办法是搜集大量数据,基于对数据的统计分析建立模型。数据挖掘正是这种处理数据的技术,本章将讨论数据挖掘中用途非常广泛的一类方法——回归方法。

MADlib 中定义了丰富的回归模型,其中包括聚类方差、Cox 比例风险回归、弹性网络正则化、广义线性模型、线性回归、逻辑回归、边际效应、多项式回归、有序回归、稳健方差等,它们都属于有监督学习。我们将以逻辑回归为例演示 MADlib 的实现,并以示例说明 MADlib 逻辑回归函数的使用方法。

24.1 回归方法简介

回归指研究一组随机变量$(y_1, y_2, \cdots, y_i)$和另一组$(x_1, x_2, \cdots, x_k)$变量之间关系的统计分析方法,又称多重回归分析。通常前者叫做因变量,后者叫做自变量。事物之间的关系可以抽象为变量之间的关系。变量之间的关系可以分为两类:一类叫确定关系,也叫函数关系,其特征是一个变量随着其他变量的确定而确定。另一类关系叫相关关系,变量之间的关系很难用一种精确的方法表示出来。例如,通常人的年龄越大血压越高,但人的年龄和血压之间没有确定的数量关系,人的年龄和血压之间的关系就是相关关系。回归方法是处理变量之间相关关系的一种数学方法。其解决问题的大致方法、步骤如下:

(1)收集一组包含因变量和自变量的数据。
(2)选定因变量和自变量之间的模型,即一个数学定量关系式,利用数据按照一定准则(如最小二乘法)计算模型中的系数。
(3)利用统计分析方法对不同的模型进行比较,找出效果最好的模型。
(4)判断得到的模型是否适合于这组数据。
(5)利用模型对因变量做出预测或解释。

回归在数据挖掘中是最为基础的方法，也是应用领域和应用场景最多的方法，只要是量化型问题，一般都会先尝试用回归方法来研究或分析。

24.2 Logistic 回归

回归分析中，因变量 y 可能有两种情形：（1）y 是一个定量的变量，这时就用通常的回归函数对 y 进行回归；（2）y 是一个定性的变量，比如 y=0 或 1，这时就不能用通常的回归函数进行回归，而是使用所谓的 Logistic 回归。Logistic 方法主要应用在研究某些现象发生的概率 P，比如股票涨跌、公司成败的概率。Logistic 回归模型的基本形式为：

$$p(Y=1|x_1,x_2,...,x_k) = \frac{\exp(\beta_0+\beta_1 x_1+\cdots+\beta_k x_k)}{1+\exp(\beta_0+\beta_1 x_1+\cdots+\beta_k x_k)}$$

其中，$\beta_0,\beta_1,...,\beta_k$ 类似于多元线性回归模型中的回归系数。该式表示当自变量为 $x_1,x_2,...,x_k$ 时，因变量 P 为 1 的概率。对该式进行对数变换，可得：

$$\ln\frac{p}{1-p} = \beta_0+\beta_1 x_1+...+\beta_k x_k$$

至此，我们会发现，只要对因变量 P 按照 ln(p/(1-p)) 的形式进行对数变换，就可以将 Logistic 回归问题转化为线性回归问题，此时就可以按照多元线性回归的方法会得到回归参数。但对于定性实践，P 的取值只有 0 和 1（二分类），这就导致 ln(p/(1-p)) 形式失去意义。为此，在实际应用 Logistic 模型的过程中，常常不是直接对 P 进行回归，而是先定义一种单调连续的概率 π，令：

$$\pi = p(Y=1|x_1,x_2,...,x_k), 0<\pi<1$$

有了这样的定义，Logistic 模型就可变形为：

$$\ln\frac{\pi}{1-\pi} = \beta_0+\beta_1 x_1+\cdots+\beta_k x_k, 0<\pi<1$$

虽然形式相同，但此时的 π 为连续函数。然后只需要对原始数据进行合理的映射处理，就可以用线性回归方法得到回归系数。最后再由 π 和 P 的映射关系进行反映射而得到 P 的值。

24.3 MADlib 的 Logistic 回归相关函数

MADlib 中的二分类 Logistic 回归模型，对双值因变量和一个或多个预测变量之间的关系建模。因变量可以是布尔值，或者是可以用布尔表达式表示的分类变量。在该模型中，训练函数作为预测变量的函数，描述一次训练可能结果的概率。

1. 训练函数

（1）语法

Logistic 回归训练函数形式如下：

```
logregr_train (source_table,
        out_table,
        dependent_varname,
        independent_varname,
        grouping_cols,
        max_iter,
        optimizer,
        tolerance,
        verbose )
```

（2）参数

source_table：TEXT 类型，包含训练数据的表名。

out_table：TEXT 类型，包含输出模型的表名。由 Logistic 回归训练函数生成的输出表如表 24-1 所示。

表 24-1 logregr_train 函数输出表列

列名	数据类型	描述
<...>	TEXT	分组列，取决于 grouping_col 输入，可能是多个列
Coef	FLOAT8[]	回归系数向量
log_likelihood	FLOAT8	对数似然比 l(c)
std_err	FLOAT8[]	系数的标准方差向量
z_stats	FLOAT8[]	系数的 z-统计量向量
p_values	FLOAT8[]	系数的 P 值向量
odds_ratios	FLOAT8[]	比值比 exp(ci)
condition_no	FLOAT8	X*X 矩阵的条件数。高条件数说明结果中的一些数值不稳定，产生的模型不可靠。这通常是由于底层设计矩阵中有相当多的共线性造成的，在这种情况下可能更适合使用其他回归技术
num_iterations	INTEGER	实际迭代次数。如果提供了 tolerance 参数，并且算法在所有迭代完成之前收敛，此列的值将会与 max_iter 参数的值不同
num_rows_processed	INTEGER	实际处理的行数，等于源表中的行数减去跳过的行数
num_missing_rows_skipped	INTEGER	训练时跳过的行数。如果自变量名是 NULL 或者包含 NULL 值，则该行被跳过

训练函数在产生输出表的同时，还会创建一个名为<out_table>_summary 的概要表，如表 24-2 所示。

表 24-2 logregr_train 函数输出概要表列

列名	数据类型	描述
source_table	TEXT	源数据表名称
out_table	TEXT	输出表名
dependent_varname	TEXT	因变量名
independent_varname	TEXT	自变量名
optimizer_params	TEXT	包含所有优化参数的字符串,形式是" optimizer=..., max_iter=..., tolerance=..."
num_all_groups	INTEGER	用 Logistic 模型拟合了多少组数据
num_failed_groups	INTEGER	有多少组拟合过程失败
num_rows_processed	INTEGER	用于计算的总行数
num_missing_rows_skipped	INTEGER	跳过的总行数

dependent_varname：TEXT 类型，训练数据中因变量列的名称（BOOLEAN 兼容类型），或者一个布尔表达式。

independent_varname：TEXT 类型，评估使用的自变量的表达式列表，一般显式地由包括一个常数 1 项的自变量列表提供。

grouping_cols（可选）：TEXT 类型，默认值为 NULL。和 SQL 中的 "GROUP BY" 类似，是一个将输入数据集分成离散组的表达式，每个组运行一个回归。此值为 NULL 时，将不使用分组，并产生一个单一的结果模型。

max_iter（可选）：INTEGER 类型，默认值为 20，指定允许的最大迭代次数。

optimizer（可选）：TEXT 类型，默认值为 'irls'，指定所使用的优化器的名称：

- 'newton' 或 'irls'：加权迭代最小二乘。
- 'cg'：共轭梯度法。
- 'igd'：梯度下降法。

tolerance（可选）：FLOAT8 类型，默认值为 0.0001，连续的迭代次数的对数似然值之间的差异。零不能作为收敛准则，因此当连续两次的迭代差异小于此值时停止执行。

verbose（可选）：默认值为 FALSE，指定是否提供训练的详细输出结果。

2. 预测函数

（1）语法

MADlib 提供了两个预测函数，预测因变量的布尔值，或预测因变量是"真"的概率值。两个函数语法相同。预测因变量的布尔值的函数：

```
logregr_predict(coefficients, ind_var)
```

预测因变量是"真"的概率值的函数：

```
logregr_predict_prob(coefficients, ind_var)
```

（2）参数

coefficients：DOUBLE PRECISION[]类型，来自 logregr_train()的模型系数。

ind_var：自变量构成的 DOUBLE 数组，其长度应该与调用 logregr_train()函数时，由 independent_varname 参数所赋值的数组相同。

24.4 Logistic 回归示例

1. 问题提出

企业到金融商业机构贷款，金融商业机构需要对企业进行评估。设评估结果为 0 或 1 两种形式，0 表示企业两年后破产，将拒绝贷款，而 1 表示企业两年后具备还款能力，可以贷款。在表 24-3 中，已知 20 家企业（编号 1~20）的三项评价指标值和评估结果，试建立模型对其他 5 家企业（编号 21~25）进行评估。

表 24-3　企业还款能力评价表

企业编号	X1	X2	X3	Y
1	-62.8	-89.5	1.7	0
2	3.3	-3.5	1.1	0
3	-120.8	-103.2	2.5	0
4	-18.1	-28.8	1.1	0
5	-3.8	-50.6	0.9	0
6	-61.2	-56.2	1.7	0
7	-20.3	-17.4	1.0	0
8	-194.5	-25.8	0.5	0
9	20.8	-4.3	1.0	0
10	-106.1	-22.9	1.5	0
11	43.0	16.4	1.3	1
12	47.0	16.0	1.9	1
13	-3.3	4.0	2.7	1
14	35.0	20.8	1.9	1
15	46.7	12.6	0.9	1
16	20.8	12.5	2.4	1
17	33.0	23.6	1.5	1
18	26.1	10.4	2.1	1
19	68.6	13.8	1.6	1
20	37.3	33.4	3.5	1
21	-49.2	-17.2	0.3	?
22	-19.2	-36.7	0.8	?
23	40.6	5.8	1.8	?
24	34.6	26.4	1.8	?
25	19.9	26.7	2.3	?

对于该问题,很明显可以用 Logistic 模型来求解,已知的三项评价指标为自变量,能否贷款的评价结果是因变量。我们可以调用 **madlib.logregr_train** 函数,用已知的 20 条数据进行训练,然后调用 **madlib.logregr_predict** 函数对其他 5 条数据执行预测,还可以用 madlib.logregr_predict_prob 函数得到预测值为"真"的概率。

2. 建立测试数据表并装载原始数据

通常训练数据与被预测数据是不同的数据集合,因此这里分别建立两个表。

```
drop table if exists source_data;
create table source_data
(id integer not null, x1 float8, x2 float8, x3 float8, y int);

copy source_data from stdin with delimiter '|';
  1 |      -62.8 |     -89.5 |    1.7 |   0
  2 |        3.3 |      -3.5 |    1.1 |   0
  3 |     -120.8 |    -103.2 |    2.5 |   0
  4 |      -18.1 |     -28.8 |    1.1 |   0
  5 |       -3.8 |     -50.6 |    0.9 |   0
  6 |      -61.2 |     -56.2 |    1.7 |   0
  7 |      -20.3 |     -17.4 |    1   |   0
  8 |     -194.5 |     -25.8 |    0.5 |   0
  9 |       20.8 |      -4.3 |    1   |   0
 10 |     -106.1 |     -22.9 |    1.5 |   0
 11 |         43 |      16.4 |    1.3 |   1
 12 |         47 |        16 |    1.9 |   1
 13 |       -3.3 |         4 |    2.7 |   1
 14 |         35 |      20.8 |    1.9 |   1
 15 |       46.7 |      12.6 |    0.9 |   1
 16 |       20.8 |      12.5 |    2.4 |   1
 17 |         33 |      23.6 |    1.5 |   1
 18 |       26.1 |      10.4 |    2.1 |   1
 19 |       68.6 |      13.8 |    1.6 |   1
 20 |       37.3 |      33.4 |    3.5 |   1
\.

drop table if exists source_data_predict;
create table source_data_predict
(id integer not null, x1 float8, x2 float8, x3 float8, y int);

copy source_data_predict from stdin with delimiter '|' NULL AS '';
 21 |      -49.2 |     -17.2 |    0.3 |
 22 |      -19.2 |     -36.7 |    0.8 |
 23 |       40.6 |       5.8 |    1.8 |
```

```
 24 |       34.6 |   26.4 |  1.8 |
 25 |       19.9 |   26.7 |  2.3 |
\.
```

3. 训练回归模型

```
drop table if exists loan_logregr, loan_logregr_summary;
select madlib.logregr_train ('source_data',
                             'loan_logregr',
                             'y',
                             'array[1, x1, x2, x3]',
                             null,
                             20,
                             'irls' );
```

注意本例中我们从列名动态创建自变量数组。如果自变量的数目很大，以至于超过了 PostgreSQL 对于每个表中最多列数的限制时（一个表中的列不能超过 1600 个，这是个硬限制），应该建立自变量数组，并存储于一个单一列中。

4. 查看回归结果

```
dm=# \x off
Expanded display is off.
dm=# select round(unnest(coef)::numeric,4) as coefficient,
dm-#        round(unnest(std_err)::numeric,4) as standard_error,
dm-#        round(unnest(z_stats)::numeric,4) as z_stat,
dm-#        round(unnest(p_values)::numeric,4) as pvalue,
dm-#        round(unnest(odds_ratios)::numeric,4) as odds_ratio
dm-#   from loan_logregr;
 coefficient | standard_error | z_stat  | pvalue | odds_ratio
-------------+----------------+---------+--------+------------
    -20.3054 |       1101.1738 | -0.0184 | 0.9853 |     0.0000
      0.1347 |         24.8599 |  0.0054 | 0.9957 |     1.1442
      1.2877 |         49.3232 |  0.0261 | 0.9792 |     3.6243
     10.7682 |        581.7361 |  0.0185 | 0.9852 | 47486.3813
(4 rows)
```

5. 使用 Logistic 回归预测因变量

```
dm=# \x off
Expanded display is off.
dm=# select p.id, madlib.logregr_predict(coef, array[1, x1, x2, x3])
dm-#   from source_data_predict p, loan_logregr m
dm-#   order by p.id;
 id | logregr_predict
----+-----------------
 21 | f
```

```
 22 | f
 23 | t
 24 | t
 25 | t
(5 rows)
```

预测的结果是 21、22 两家企业应拒绝贷款,其他三家企业可以贷款。

6. 预测因变量为"真"的概率

```
dm=# \x off
Expanded display is off.
dm=# select p.id, madlib.logregr_predict_prob(coef, array[1, x1, x2, x3])
dm-#   from source_data_predict p, loan_logregr m
dm-# order by p.id;
 id | logregr_predict_prob
----+----------------------
 21 | 1.22296014464276e-20
 22 | 1.88777536644339e-27
 23 |     0.999993946936041
 24 |                     1
 25 |                     1
(5 rows)
```

21、22 为"真"的概率几乎为 0,其他三个为"真"的概率为 1。

7. 在训练数据上执行预测函数

```
dm=# \x off
Expanded display is off.
dm=# select p.id, madlib.logregr_predict(coef, array[1, x1, x2, x3]), p.y
dm-# from source_data p, loan_logregr m
dm-# order by p.id;
 id | logregr_predict | y
----+-----------------+---
  1 | f               | 0
  2 | f               | 0
  3 | f               | 0
  4 | f               | 0
  5 | f               | 0
  6 | f               | 0
  7 | f               | 0
  8 | f               | 0
  9 | f               | 0
 10 | f               | 0
 11 | t               | 1
```

```
 12 | t        | 1
 13 | t        | 1
 14 | t        | 1
 15 | t        | 1
 16 | t        | 1
 17 | t        | 1
 18 | t        | 1
 19 | t        | 1
 20 | t        | 1
(20 rows)
```

可以看到，Logistic 模型预测的结果与训练数据完全一致。

实际应用中，以下因素对 Logistic 回归分析预测模型的可靠性有较大影响：

- 样本量问题

Logistic 回归分析中，到底样本量多大才算够，这一直是个令许多人困惑的问题。尽管有人从理论角度提出了 Logistic 回归分析中的样本含量估计，但从使用角度来看多数并不现实。直到现在，这一问题尚无广为接受的答案。一般认为，如果样本量小于 100，Logistic 回归的最大似然估计可能有一定的风险，如果大于 500 则显得比较充足。当然，样本大小还依赖于变量个数、数据结构等条件。每一个自变量至少要 10 例结局保证估计的可靠性。注意：这里是结局例数，而不是整个样本例数。

- 混杂因素的影响

混杂因素一般可以通过三个方面确定：一是该因素对结局有影响；二是该因素在分析因素中的分布不均衡；三是从专业角度来判断，该因素是分析因素与结局中间的一个环节。也就是说，分析因素引起该因素，通过该因素再引起结局。

- 交互作用的影响

交互作用有时也叫效应修饰，是指在该因素的不同水平（不同取值），分析因素与结局的关联大小有所不同。在某一水平上（如取值为 0）可能分析因素对结局的效应大，而在另一个水平上（如取值为 1）可能效应小。

24.5 小结

回归是重要的数据挖掘方法之一。MADlib 提供了多种回归模型，本章主要讨论了其中的逻辑回归方法，并用一个企业贷款评估的例子，演示了如何使用 MADlib 的逻辑回归函数进行预测。逻辑回归方法主要应用在预测某些现象发生的概率。

第 25 章

◀ 分类方法 ▶

分类是一种重要的数据挖掘技术。分类的目的是根据数据集的特点构造一个分类函数或分类模型（也常常称作分类器），该模型能把未知类别的样本映射到给定的类别中。分类方法是解决分类问题的方法，是数据挖掘、机器学习和模式识别中一个重要的研究领域。分类算法通过对已知类别训练集的分析，从中发现分类规则，以此预测新数据的类别。分类算法的应用非常广泛，包括风险评估、客户分类、文本检索等。本章介绍分类的基本概念、MADlib 的决策树分类模型及应用示例。

25.1 分类方法简介

1. 分类的概念

数据挖掘中分类的目的是学会一个分类函数或分类模型，该模型能把数据库中的数据项映射到给定类别中的某一个。分类可描述如下：输入数据，或称训练集（Training Set），是由一条条数据库记录（Record）组成的。每一条记录包含若干个属性（Attribute），组成一个特征向量。训练集的每条记录还有一个特定的类标签（Class Label）与之对应。该类标签是系统的输入，通常是以往的一些经验数据。一个具体样本的形式可为样本向量：$(v1,v2,...,vn;c)$，在这里 vi 表示字段值，c 表示类别。分类的目的是：分析输入数据，通过在训练集中的数据表现出来的特征，为每一个类找到一种准确的描述或模型。由此生成的类描述用来对未来的测试数据进行分类。尽管这些测试数据的类标签是未知的，我们仍可以由此预测这些新数据所属的类。注意是预测，而不是肯定，因为分类的准确率不能达到百分之百。我们也可以由此对数据中的每一个类有更好的理解。也就是说：我们获得了对这个类的知识。

所以分类（Classification）也可以定义为：对现有的数据进行学习，得到一个目标函数或规则，把每个属性集 x 映射到一个预先定义的类标号 y。目标函数或规则也叫分类模型（Classification Model），它有两个主要作用：一是描述性建模，即作为解释性工具，用于区分不同类的对象；二是预测性建模，即用于预测未知记录的类标号。

2. 分类的原理

分类方法是一种根据输入数据建立分类模型的系统方法，这些方法都是使用一种学习算法

（Learning Algorithm）确定分类模型，使该模型能够很好地拟合输入数据中类标号和属性集之间的联系。学习算法得到的模型不仅要很好地拟合输入数据，还要能够正确地预测未知样本的类标号。因此，训练算法的主要目标就是建立具有很好泛化能力的模型，即建立能够准确预测未知样本类标号的模型。图 25-1 展示了解决分类问题的一般方法。首先，需要一个训练集，它由类标号已知的记录组成。使用训练集建立分类模型，该模型随后将运用于检验集（Test Set），检验集由类标号未知的记录组成。

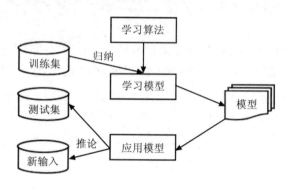

图 25-1　分类原理示意图

通常分类学习所获得的模型可以表示为分类规则形式、决策树形式或数学公式形式。例如，给定一个顾客信用信息数据库，通过学习所获得的分类规则可用于识别顾客是否具有良好的信用等级或一般的信用等级。分类规则也可用于对今后未知所属类别的数据进行识别判断，同时还可以帮助了解数据库中的内容。

构造模型的过程一般分为训练和测试两个阶段。在构造模型之前，要求将数据集随机地分为训练数据集和测试数据集。在训练阶段，使用训练数据集，通过分析由属性描述的数据库元组来构造模型，假定每个元组属于一个预定义的类，由一个称作类标号的属性来确定。训练数据集中的单个元组也称作训练样本。由于提供了每个训练样本的类标号，该阶段也被称为有指导的学习。在测试阶段，使用测试数据集来评估模型的分类准确率，如果认为模型的准确率可以接受，就可以用该模型对其他数据元组进行分类。一般来说，测试阶段的代价远远低于训练阶段。

为了提高分类的准确性、有效性和可伸缩性，在进行分类之前，通常要对数据进行预处理，包括：

（1）数据清理。其目的是消除或减少数据噪声，处理空缺值。

（2）相关性分析。由于数据集中的许多属性可能与分类任务不相关，若包含这些属性可能会减慢或误导学习过程。相关性分析的目的就是删除这些不相关或冗余的属性。

（3）数据变换。数据可以概化到较高层概念。比如，连续值属性"收入"的数值可以概化为离散值：低、中、高。又比如，标称值属性"市"可以概化到高层概念"省"。此外，数据也可以规范化，规范化将给定属性的值按比例缩放，落入较小的区间，比如[0,1]等。

25.2 决策树

常用的分类方法有 K-邻近、贝叶斯、神经网络、逻辑分类、判别分析、支持向量机、决策树七种。下面主要介绍决策树方法的基本概念和原理。

25.2.1 决策树的基本概念

决策树（Decision Tree）又叫分类树（Classification Tree），是使用最为广泛的归纳推理算法之一，处理类别型或连续型变量的分类预测问题，可以用图形或 if-then 的规则表示模型，可读性较高。决策树模型通过不断地划分数据，使因变量的差别最大，最终目的是将数据分类到不同的组织或不同的分枝，在因变量的值上建立最强的归类。

决策树是一种监督式的学习方法，产生一种类似流程图的树结构（可以是二叉树或非二叉树）。其每个非叶节点表示一个特征属性上的测试，每个分支代表这个特征属性在某个值域上的输出，而每个叶节点存放一个类别。使用决策树进行决策的过程就是从根节点开始，测试待分类项中相应的特征属性，并按照其值选择输出分支，直到到达叶子节点，将叶子节点存放的类别作为决策结果。

25.2.2 决策树的构建步骤

决策树构建的主要步骤有三个：第一是选择适当的算法训练样本构建决策树，第二是适当的修剪决策树，第三则是从决策树中萃取知识规则。

1. 决策树的分割

决策树是通过递归分割（Recursive Partitioning）建立而成，递归分割是一种把数据分割成不同大小的部分的迭代过程。构建决策树的归纳算法如下：

（1）将训练样本的原始数据放入决策树的树根。

（2）将原始数据分成两组，一部分为训练数据；另一部分为测试组资料。

（3）使用训练样本来建立决策树，在每一个内部节点以信息论作为选择哪一个属性继续做分隔的依据。

（4）使用测试数据来进行决策树修剪，修剪到决策树的每个分类都只有一个节点，以提升预测能力与速度。也就是经过节点分割后，判断这些内部节点是否为树叶节点，如果不是，则以新内部节点为分枝的树根来建立新的次分枝。

（5）不断递归第（1）至第（4）步，一直到所有内部节点都是树叶节点为止。当决策树完成分类后，可将每个分枝的树叶节点萃取为知识规则。

如果有以下情况发生，决策树将停止分割：

（1）训练数据的每一笔数据都已经归类到同一类别。

（2）训练数据已经没有办法再找到新的属性来进行节点分割。

（3）训练数据已经没有任何尚未处理的数据。

2. 决策树的剪枝

在实际构造决策树时，大都要进行剪枝，这是为了处理由于数据中的噪声和离群点导致的过分拟合问题。剪枝有两种方法：先剪枝——在构造过程中，当某个节点满足剪枝条件，则直接停止此分支的构造；后剪枝——先构造完成完整的决策树，再通过某些条件遍历树进行剪枝。也可以交叉使用先剪枝和后剪枝形成组合式，后剪枝所需的计算比先剪枝多，但通常会产生较为可靠的决策树。

3. 决策树算法

决策树算法基本上是一种贪心算法，采取由上至下的逐次搜索方式，渐次产生决策树模型结构。划分数据集的最大原则是：使无序的数据变得有序。如果一个训练数据中有 20 个特征，那么选取哪个作为划分依据？这时必须采用量化的方法来判断，常用的量化划分方法是"信息论度量信息分类"。基于信息论的决策树算法有 ID3、C4.5 和 CART 等算法，其中 C4.5 和 CART 两种算法从 ID3 算法中衍生而来。

C4.5 和 CART 支持数据特征为连续分布时的处理，主要通过使用二元切分来处理连续型变量，即求一个特定的值——分裂值：特征值大于分裂值就走左子树，否则就走右子树。这个分裂值的选取原则是使得划分后的子树中的"混乱程度"降低，具体到 C4.5 和 CART 算法则有不同的定义方式。

ID3 算法由 Ross Quinlan 发明，建立在"奥卡姆剃刀"的基础上：越是小型的决策树越优于大的决策树（be simple 简单理论）。ID3 算法中根据信息论的信息增益评估和选择特征，每次选择信息增益最大的特征做判断属性。ID3 算法可用于划分标称型数据集，没有剪枝的过程，为了去除过度数据匹配的问题，可通过裁剪合并相邻的无法产生大量信息增益的叶子节点（例如设置信息增益阀值）。使用信息增益有一个缺点，那就是它偏向于具有大量值的属性，就是说在训练集中，某个属性所取的不同值的个数越多，那么越有可能拿它来作为分裂属性，而这样做有时候是没有意义的，最典型的就是自增 ID 序列。另外 ID3 不能处理连续分布的数据特征，于是就有了 C4.5 算法。

C4.5 是 ID3 的一个改进算法，它继承了 ID3 算法的优点。C4.5 算法用信息增益率来选择属性，克服了用信息增益选择属性时偏向选择取值多的属性的不足，在树构造过程中进行剪枝；能够完成对连续属性的离散化处理；也能对不完整数据进行处理。C4.5 算法产生的分类规则易于理解、准确率较高，但效率低，因树构造过程中，需要对数据集进行多次的顺序扫描和排序。也是因为必须多次数据集扫描，C4.5 只适合于能够驻留于内存的数据集。

CART 算法的全称是 Classification And Regression Tree，采用的是 Gini 指数（选 Gini 指数最小的特征 s）作为分裂标准，同时它也包含后剪枝操作。ID3 算法和 C4.5 算法虽然在对训练样本集的学习中可以尽可能多地挖掘信息，但其生成的决策树分支较大。为了简化决策树的规模，提高生成决策树的效率，就出现了根据 Gini 系数来选择测试属性的决策树算法 CART。MADlib 中的决策树训练函数使用的就是 CART 算法。

25.3 MADlib 的决策树相关函数

MADlib 中有三个决策树函数，分别为训练函数、预测函数和显示函数。训练函数接收输入的训练数据进行学习，生成决策树模型。预测函数用训练函数生成的决策树模型预测数据的所属分类。显示函数用来显示决策树模型。

1. 训练函数

（1）语法

```
tree_train
( training_table_name,
  output_table_name,
  id_col_name,
  dependent_variable,
  list_of_features,
  list_of_features_to_exclude,
  split_criterion,
  grouping_cols,
  weights,
  max_depth,
  min_split,
  min_bucket,
  num_splits,
  pruning_params,
  surrogate_params,
  verbosity )
```

（2）参数

training_table_name：TEXT 类型，训练数据输入表名。

output_table_name：TEXT 类型，包含决策树模型的输出表名，如果表已经存在则报错。训练函数生成的模型表如表 25-1 所示。

表 25-1 tree_train 函数输出模型表列

列名	数据类型	描述
<...>	TEXT	当提供了 grouping_cols 入参时，该列存储分组列，依赖于 grouping_cols 入参的值，可能有多列，类型与训练表相同
tree	BYTEA8	二进制格式存储的决策树模型
cat_levels_in_text	TEXT[]	分类变量的层次
cat_n_levels	INTEGER[]	每个分类变量的层号
tree_depth	INTEGER	剪枝前的决策树最大深度（根的深度为 0）
pruning_cp	FLOAT8[]	用于剪枝决策树的复杂性成本参数。如果使用交叉验证，该值应与 pruning_params 入参的值不同

生成模型表的同时还会生成一个名为<model_table>_summary 的概要表，如表 25-2 所示。

表 25-2 tree_train 函数输出概要表列

列名	数据类型	描述
method	TEXT	值为 tree_train
is_classification	BOOLEAN	用于分类时为 TRUE，用于回归时为 FALSE
source_table	TEXT	源表名
model_table	TEXT	模型表名
id_col_name	TEXT	ID 列名
dependent_varname	TEXT	因变量
independent_varname	TEXT	自变量
cat_features	TEXT	逗号分隔字符串，分类特征名称列表
con_features	TEXT	逗号分隔字符串，连续特征名称列表
grouping_col	TEXT	分组列名
num_all_groups	INTEGER	训练决策树时的总分组数
num_failed_groups	INTEGER	训练决策树时失败的分组数
total_rows_processed	BIGINT	所有分组处理的总行数
total_rows_skipped	BIGINT	所有分组中因为缺少值或失败而跳过的总行数
dependent_var_levels	TEXT	对于分类，因变量的不同取值
dependent_var_type	TEXT	因变量类型
input_cp	FLOAT8[]	交叉验证前，用于剪枝决策树的复杂度参数。与 pruning_params 入参输入的值相同
independent_var_types	TEXT	逗号分隔字符串，自变量类型

　　id_col_name：TEXT 类型，训练数据中，含有 ID 信息的列名。这是一个强制参数，用于预测和交叉验证。每行的 ID 值应该是唯一的。

　　dependent_variable：TEXT 类型，包含用于训练的输出列名。分类的输出列是 boolean、integer 或 text 类型，回归的输出列是 double precision 类型。决策树的因变量可以为多个，训练函数的时间和空间复杂度，会随着因变量数量的增加呈线性增长。

　　list_of_features：TEXT 类型，逗号分隔字符串，用于预测的特征列名，也可以用 "*" 表示所有列都用于预测（除下一个参数中的列名外）。特征列的类型可以是 boolean、integer、text 或 double precision。

　　list_of_features_to_exclude：TEXT 类型，逗号分隔字符串，不用于预测的列名。如果自变量是一个表达式（包括列的类型转换），那么这个列表中应该包括用于自变量表达式的所有列名，否则那些列将被包含在特征中。

　　split_criterion：TEXT 类型，默认值为 gini，用于分类，而 mse 用于回归。不纯度函数计算用于分裂的特征值。分类树支持的标准有 gini 、 entropy 或 misclassification，回归树的分裂标准总是使用 mse 。

grouping_cols（可选）：TEXT 类型，默认值为 NULL，逗号分隔字符串，分组的列名。将为每个分组产生一棵决策树。

weights（可选）：TEXT 类型，权重列名。

max_depth（可选）：INTEGER 类型，默认值为 10。最终决策树的最大深度，根的深度为 0。

min_split（可选）：INTEGER 类型，默认值为 20。一个试图被分裂的节点中，必须存在的元组的最小数量。此参数的最佳值取决于数据集的元组数目。

min_bucket（可选）：INTEGER 类型，默认值为 min_split/3。任何叶节点对应的最小元组数量。如果 min_split 和 min_bucket 只指定一个，那么 min_split 设置成 min_bucket*3，或者 min_bucket 设置成 min_split/3。

num_splits（可选）：INTEGER 类型，默认值为 100。为计算分割边界，需要将连续特征值分成离散型分位点。此全局参数用于计算连续特征的分割点，值越大预测越准，处理时间也越长。

pruning_params（可选）：TEXT 类型，逗号分隔的键-值对，用于决策树剪枝，当前接受的值为：

- cp: 默认值为 0。cp 全称为 complexity parameter，指某个点的复杂度，对每一步拆分，模型的拟合优度必须提高的程度。试图分裂一个节点时，分裂增加的精确度必须提高 cp，才进行分裂，否则剪枝该节点。该参数值用于在运行检查验证前，创建一棵初始树。

- n_folds: 默认值为 0。用于计算 cp 最佳值的交叉验证褶皱数。为执行交叉验证，n_folds 的值应该大于 2。执行交叉验证时，会产生一个名为<model_table>_cv 的输出表，其中包含估计的 cp 值和交叉验证错误。输出表中返回的决策树对应具有最少交叉错误的 cp（如果多个 cp 值有相同的错误数，取最大的 cp）。

surrogate_params：TEXT 类型，逗号分隔的键值对，控制替代分裂点的行为。替代变量是与主预测变量相关的另一种预测变量，当主预测变量的值为 NULL 时使用替代变量。此参数当前接受的值为：

- max_surrogates 默认值为 0，每个节点的替代变量数。

verbosity（可选）：BOOLEAN 类型，是否提供训练结果的详细输出，默认值为 FALSE。

（3）提示

MADlib 决策树训练函数的很多参数设计与流行的 R 语言函数 rpart 相似。它们的一个重要区别是，对特征和分类变量，MADlib 使用整型作为变量值类型，而 rpart 认为它们是连续的。

不使用替代变量时（max_surrogates=0），用于训练的特征值为 NULL 的行，在训练和预测时都被忽略。

不使用交叉验证时（n_folds=0），决策树依赖输入的 cost-complextity（cp）进行剪枝。使用交叉验证时，所有节点 cp 都要大于参数 cp。在进行交叉验证时，训练函数使用 cp 入参建立一个初始树，并探索所有可能的子树（直到单节点树），计算每个节点的 cp 进行剪枝，得到优化的子树。优化的子树及其相应的 cp 被放在输出表中，分别对应输出表的 tree 和 pruning_cp 列。

影响内存使用的参数主要是树的深度、特征的数量和每个特征的不同值的数量。如果遇到 VMEM 限制，要考虑降低这些参数的值。

2. 预测函数

（1）语法

```
tree_predict(tree_model,
        new_data_table,
        output_table,
        type)
```

（2）参数

tree_model：TEXT 类型，包含决策树模型的表名，应该是决策树训练函数的输出表。

new_data_table：TEXT 类型，包含被预测数据的表名。该表应该和训练表具有相同的特征，也应该包含用于标识每行的 id_col_name。

output_table：TEXT 类型，预测结果的输出表名，如果表已经存在则报错。表中包含标识每个预测的 id_col_name 列，以及每个因变量的预测列。

- 如果 type = 'response'，表有单一预测列。此列的类型依赖于训练时使用的因变量的类型。
- 如果 type = 'prob'，每个因变量对应多列，每列表示因变量的一个可能值。列标识为'estimated_prob_dep_value'，其中 dep_value 表示对应的因变量值。

type（可选）：TEXT 类型，默认值为'response'。对于回归树，输出总是因变量的预测值。对于分类树，变量类型可以是'response'或'prob'。

3. 显示函数

显示函数输出一个决策树的格式化表示。输出可以是 dot 格式，或者是一个简单的文本格式。dot 格式可以使用 GraphViz 等程序进行可视化。

（1）语法

```
tree_display(tree_model, dot_format, verbosity)
```

还有一个显示函数输出为每个内部节点选择的替代分裂点：

```
tree_surr_display(tree_model)
```

(2)参数

tree_model：TEXT 类型，含有决策树模型的表名。

dot_format：BOOLEAN 类型，默认值为 TRUE，使用 dot 格式，否则输出文本格式。

verbosity：BOOLEAN 类型，默认值为 FALSE。如果设置为 TRUE，dot 格式输出中包含附加信息，如不纯度、样本大小、每个因变量的权重行数、被剪枝的分类或预测等。输出总是返回文本形式，对于 dot 格式，可以重定向输出到客户端文件，然后使用可视化程序处理。

25.4 决策树示例

1. 问题描述

本示例取自维基百科中的"决策树"条目，问题描述如下：

小王是一家著名高尔夫俱乐部的经理。但是他被雇员数量问题搞得心情十分不好。某些天好像所有人都来玩高尔夫，以至于所有员工都忙得团团转还是应付不过来，而有些天不知道什么原因却一个人也不来，俱乐部为雇员数量浪费了不少资金。小王的目的是通过下周天气预报寻找什么时候人们会打高尔夫，以适时调整雇员数量。因此首先他必须了解人们决定是否打球的原因。在两周时间内可以得到以下记录：天气状况有晴、云和雨；华氏温度表示的气温；相对湿度百分比；是否有风。当然还有顾客是不是在这些日子光顾俱乐部。最终得到了表 25-3 所示的 14 行 5 列的数据。

表 25-3 两周天气与是否打高尔夫球数据

自变量				因变量
天气情况	华氏温度	相对湿度	是否有风	是否打球
sunny	85	85	FALSE	Don't Play
sunny	80	90	TRUE	Don't Play
overcast	83	78	FALSE	Play
rain	70	96	FALSE	Play
rain	68	80	FALSE	Play
rain	65	70	TRUE	Don't Play
overcast	64	65	TRUE	Play
sunny	72	95	FALSE	Don't Play
sunny	69	70	FALSE	Play
rain	75	80	FALSE	Play
sunny	75	70	TRUE	Play
overcast	72	90	TRUE	Play
overcast	81	75	FALSE	Play
rain	71	80	TRUE	Don't Play

我们将利用 MADlib 的决策树相关函数来解决此问题。

2. 准备输入数据

创建 dt_golf 表，将 14 条数据插入 dt_golf 表中。

```
drop table if exists dt_golf;
create table dt_golf
( id integer not null,
    "outlook" text,
    temperature double precision,
    humidity double precision,
    windy text,
    class text );

copy dt_golf (id,"outlook",temperature,humidity,windy,class) from stdin with delimiter '|';
1|sunny|85|85|'false'|'Don''t Play'
2|sunny|80|90|'true'|'Don''t Play'
3|overcast|83|78|'false'|'Play'
4|rain|70|96|'false'|'Play'
5|rain|68|80|'false'|'Play'
6|rain|65|70|'true'|'Don''t Play'
7|overcast|64|65|'true'|'Play'
8|sunny|72|95|'false'|'Don''t Play'
9|sunny|69|70|'false'|'Play'
10|rain|75|80|'false'|'Play'
11|sunny|75|70|'true'|'Play'
12|overcast|72|90|'true'|'Play'
13|overcast|81|75|'false'|'Play'
14|rain|71|80|'true'|'Don''t Play'
\.
```

3. 运行决策树训练函数

```
drop table if exists train_output, train_output_summary, train_output_cv;
select madlib.tree_train(
 'dt_golf',           -- 训练数据表
 'train_output',      -- 输出模型表
 'id',                -- id 列
 'class',             -- 因变量是分类
 '"outlook", temperature, humidity, windy',
 -- 四个属性是特征，注意加了双引号的 outlook，要区分大小写
 null::text,          -- 没有排除列
 'gini',              -- 分裂标准使用 gini
 null::text,          -- 无分组
 null::text,          -- 无权重
 5,                   -- 因为只有 4 个特征，设置最大深度为 5，防止过拟合
 3,                   -- 最小分裂元组数
 1,                   -- 最小桶数，min_split/3
 10,                  -- 每个连续性变量的离散型分位点数量
 'cp=0,n_folds=6'     -- 初始 cp=0，6 折验证
 );
```

4. 查看输出

(1) 查询概要表

```
dm=# \x
Expanded display is on.
dm=# select * from train_output_summary;
-[ RECORD 1 ]----------+----------------------------------------------
method                 | tree_train
is_classification      | t
source_table           | dt_golf
model_table            | train_output
id_col_name            | id
dependent_varname      | class
independent_varnames   | "outlook", windy, temperature, humidity
cat_features           | "outlook",windy
con_features           | temperature,humidity
grouping_cols          |
num_all_groups         | 1
num_failed_groups      | 0
total_rows_processed   | 14
total_rows_skipped     | 0
dependent_var_levels   | "'Don't Play' ","'Play' "
dependent_var_type     | text
input_cp               | 0
independent_var_types  | text, text, double precision, double precision
```

说明：

- is_classification 为 t，指的是函数用于预测而不是回归。
- "outlook",windy 是分类特征，temperature,humidity 是连续特征。
- 因变量为文本类型，有 'Don't Play' 和 'Play' 两种取值。
- 通过 pruning_params 参数设置剪枝使用的初始 cp 为 0，交叉检验折皱数为 6。

(2) 查询决策树表

```
dm=# \x
Expanded display is off.
dm=# select madlib.tree_display('train_output', false);
                    tree_display
-----------------------------------------------------------------------
 - Each node represented by 'id' inside ().
Each internal nodes has the split condition at the end, while each      leaf
node has a * at the end.                                       - For each internal
node (i), its child nodes are indented by 1 level              with ids (2i+1) for
True node and (2i+2) for False node.                           - Number of (weighted) rows
for each response variable inside [].'                         The response label order is
given as ['"\'Don\'t Play\'"', '"\'Play\'"'].                  For each leaf, the prediction
is given after the '-->'

 ----------------------------------------
```

```
(0)[5 9]   "outlook" in {overcast}                                       (1)[0 4] 
 *  --> "'Play'"                                                         (2)[5 5] 
temperature <= 75                                                        (5)[3 5] 
temperature <= 65                                                (11)[1 0]   * --> 
"'Don't Play'"                                             (12)[2 5]   temperature <= 
70                                                         (25)[0 3]   * --> "'Play'"
(26)[2 2]   temperature <= 72                                            (53)[2 0] 
 * --> "'Don't Play'"                                            (54)[0 2]   * --> 
"'Play'"                                       (6)[2 0]   * --> "'Don't Play'"
 -------------------------------------
(1 row)
```

说明：

- 输出为简单文本格式。
- 节点号以 0 开始，代表根节点。如果树没有剪枝，后面的节点号应该是连续的 1、2、……、n。而树被剪枝后，节点号是不连续的，正如上面查询结果中所显示的。带*号的节点是叶子节点，其他是内部测试节点。
- 顺序值[x y]表示其测试节点上["Don't play" "Play"]所占的行数。例如，在根节点 0，有 5 行 "Don't play"，9 行 "Play"。如果以特征 "outlook"=overcast 作为测试条件，结果有 0 行 "Don't play"，4 行 "Play"，节点 1 为叶子结点。剩下 5 行 "Don't play" 与 5 行 "Play" 在节点 2 上测试 temperature<=75。以此类推，从上到下解读输出结果。
- 虽然我们的输入数据中提供了四个特征自变量，但训练函数输出的结果决策树中只体现出天气状况和气温，相对湿度与是否有风没有作为测试条件。

（3）以 dot 格式显示决策树

```
\t                 -- 不显示表头
\o test.dot        -- 将查询结果输出到文件
select madlib.tree_display('train_output', true, true);   -- 输出决策树详细信息
\o
\t
```

生成 dot 文件后，使用第三方图形软件 GraphViz 画出决策树，执行以下 shell 命令：

```
# 安装 GraphViz
yum -y install graphviz
# 将 dot 文件转成 jpg 文件输出
dot -Tjpg test.dot -o test.jpg
```

生成的决策树如图 25-2 所示。

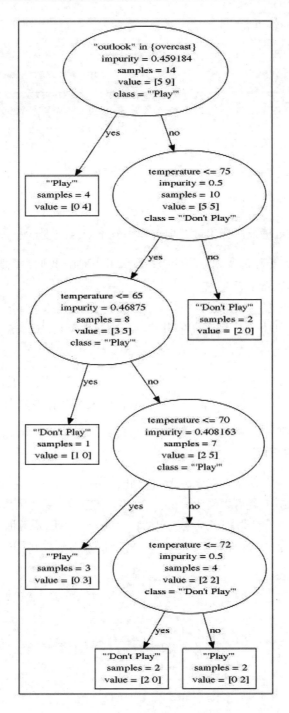

图 25-2　图形化显示 dot 格式决策树

图中显示的决策树与文本的输出一致，矩形为叶子节点，椭圆形为内部测试节点。除了文本输出的信息外，图中还多了一个 impurity，代表不纯度，是指将来自集合中的某种结果随机应用在集合中，某一数据项的预期误差率。不纯度越小，集合的有序程度越高，分类的效果越好。叶子节点的不纯度为 0。

（4）查询交叉验证结果表

```
dm=# select * from train_output_cv;
 cp  |     cv_error_avg       |            cv_error_stddev
-----+------------------------+-------------------------------------------
   0 | 0.548611111111111111   | 0.182775782126589549951534508979822238903
 0.2 | 0.583333333333333333   | 0.229734145868170362800232510882382939349
(2 rows)
```

可以看到，随着 cp 值的增大，剪枝增多，精度降低。

5. 分析决策树

分类树算法可以通过特征，找出最好地解释因变量 class（是否打高尔夫球）的方法。

（1）变量 outlook 的范畴被划分为两组：多云和其他。我们得出第一个结论：如果天气是多云，人们总是选择玩高尔夫。

（2）在不是多云天气时，以气温划分，当气温高于华氏 75 度或低于华氏 65 度时，没人玩高尔夫；65～70 度之间人们选择玩高尔夫，70～72 度之间没人玩高尔夫；72～75 度会玩高尔夫。

这就通过分类树给出了一个解决方案。小王在晴天、雨天并且气温过高或过低时解雇了大部分员工，因为这种天气不会有人玩高尔夫。而其他的天气会有很多人玩高尔夫，因此可以雇用一些临时员工来工作。

6. 用决策树模型进行预测

从交叉验证结果看，即便是初始 cp=0，得到的标准差仍然较大，这和我们的样本数据过少有关。从本示例的分析中就可以看到，65～70 度、72～75 度会玩高尔夫，而 70～72 度之间没人玩，这个预测显然有悖常理。现在就用此模型预测一下原始数据，再和实际情况对比一下。这里只是演示一下如何用模型进行预测，实践中训练数据集与预测数据集相同意义不大。

```
drop table if exists prediction_results;
select madlib.tree_predict
('train_output',          -- 决策树模型
'dt_golf',                -- 被预测的数据表
'prediction_results',     -- 预测结果表
'response');              -- 预测结果
select t1.*,t2.class
  from prediction_results t1, dt_golf t2
 where t1.id=t2.id order by id;
```

查询结果如下：

```
 id | estimated_class |    class
----+-----------------+--------------
  1 | 'Don't Play'    | 'Don't Play'
```

```
  2 | 'Don't Play'    | 'Don't Play'
  3 | 'Play'          | 'Play'
  4 | 'Play'          | 'Play'
  5 | 'Play'          | 'Play'
  6 | 'Don't Play'    | 'Don't Play'
  7 | 'Play'          | 'Play'
  8 | 'Don't Play'    | 'Don't Play'
  9 | 'Play'          | 'Play'
 10 | 'Play'          | 'Play'
 11 | 'Play'          | 'Play'
 12 | 'Play'          | 'Play'
 13 | 'Play'          | 'Play'
 14 | 'Don't Play'    | 'Don't Play'
(14 rows)
```

25.5 小结

分类是数据挖掘的重要方法之一，到目前为止，已有多种基于各种思想和理论基础的分类算法，实际应用也已趋于成熟。本章简要介绍了决策树方法的相关概念和原理，并用一个实际的例子，详细说明了 MADlib 的决策树相关函数的用法。

第 26 章

◀ 图算法 ▶

图算法指利用特制的线条算图求得答案的一种简便算法。无向图、有向图和网络能运用很多常用的图算法，其中主要包括各种遍历算法（这些遍历类似于树的遍历），寻找最短路径的算法，寻找网络中最低代价路径的算法。这些算法常被用以回答一些与图相关的问题，诸如图是否是连通的，图中两个顶点间的最短路径是什么等。在数据挖掘领域中，图算法可应用到多种场合，以解决特定问题，如管道优化、路由选择、快递服务、网站通信等。

本章介绍图的基本概念和表示方法，并简要说明一些常用的图算法。MADlib 文档中只列出了一种图算法模型，即单源最短路径，因此我们将详细描述该算法及其相关函数。同样也会用一个简单示例，说明 MADlib 单源最短路径函数的用法。

26.1 图算法简介

1. 图的概念

在计算中,常将运算方程或实验结果绘制成由若干有标尺的线条所组成的图,称为"算图"。计算时根据已知条件，从有关线段上一点开始，连接相关线段上的点，连线与表示所求量线段的交点即为答案。图算法是对树的拓展，树是自上而下的数据结构，除根节点外，其他每个结点都有一个父结点，从上向下排列。而图没有了父子节点的概念，图中的结点都是平等关系，结果更加复杂。

定义图 G=(V,E)，其中 V 代表顶点 Vertex，E 代表边 Edge，一条边就是一个定点对(U,V)，其中(U,V)∈V。图分有向图和无向图。在无向图中，如果(U,V)（表示 U 到 V 的路径）联通，那么(U,V)也联通，例如"1"到"2"联通，"2"到"1"也联通。但是在有向图中"1"到"2"联通，但是"2"到"1"是不联通的。图 26-1 与图 26-2 分别表示一个无向图和一个有向图。

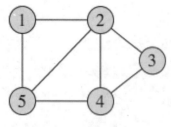

图 26-1　无向图

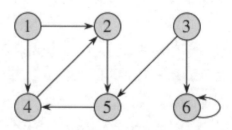

图 26-2　有向图

在图的概念中，除了顶点和边的概念外，还经常涉及权值，表示一个顶点到另一个顶点的"代价"，如果顶点不联通，可以认为权值无限大。如果不涉及权值，那么可以认为联通的顶点权值都为 1。

2. 图的表示

数据结构中经常用邻接表和邻接矩阵表示图。

（1）邻接表

图 26-3 即为图 26-2 所示有向图的邻接表，表中的一个节点对应图中的一个顶点，节点后面的链表是与这个节点联通的节点。

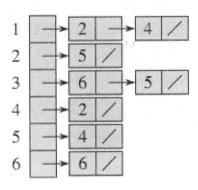

图 26-3　邻接表

邻接表常用于表示稀疏图，即结点的边数 |E| 远小于 |V|^2 。对于有向图，邻接表存储所占空间为 |V|+|E|，对于无向图为 |V|+2|E|，因为每条边在邻接表中出现两次。邻接表在存储上占优势，但是在判断两个结点(U,V)是否联通时，要首先在邻接表中找到 u，然后再遍历 u 后面的链表。

（2）邻接矩阵

图 26-4 是图 26-1 所示无向图的邻接矩阵表示。邻接矩阵是一个|V|×|V|的矩阵 GMatr，如果(U,V)联通，那么 GMatr[u][v]=1。如果图是加权的话，GMatr[u][v]=权值。

	1	2	3	4	5	6
1	0	1	0	1	0	0
2	0	0	0	0	1	0
3	0	0	0	0	1	1
4	0	1	0	0	0	0
5	0	0	0	1	0	0
6	0	0	0	0	0	1

图 26-4　邻接矩阵

可以看出，邻接矩阵表示方法所占空间为 $O(V^2)$，但是在判断两个结点是否联通时，只需 $O(1)$。当图比较小时更多采用邻接矩阵，因为它更明了。如果图没有加权，可以用一个二进制位来表示两个图是否联通。

3. 常用图算法

（1）图的遍历

图的遍历是指从图中的任一顶点出发，对图中的所有顶点访问一次且只访问一次。遍历操作是图的一种基本操作，图的许多操作都建立在遍历的基础之上。在遍历图时，为保证图中各顶点在遍历过程中被访问且仅一次，需要为每个顶点设计一个访问标记，设置一个数组，用于标识图中哪个顶点被访问过。数组元素的初始值全部为 0，表示顶点均未被访问过。某个顶点被访问后，将相应访问标志数组中的值设为 1，以表示该顶点已经被访问。通常图的遍历有两种：深度优先遍历搜索和广度优先遍历搜索。

深度优先遍历是尽可能"深"的遍历图。假设从结点 v0 开始遍历，遍历与 v0 联通的且未被遍历过的结点 v1，再遍历与 v1 联通的且未被遍历过的结点 v2……。如果遍历到 vn 后无结点可以遍历，那么退回到 v(n-1)再去找结点遍历，以此类推，直到图中所有节点都被遍历过。可以看出图的深度优先遍历可以借助堆栈实现。

a. 把结点 v 放入堆栈，标记 v。
b. 若堆栈为空则结束，否则取出栈顶节点 u。
c. 找出与 u 联通的且未被标记的结 w1,w2,……，并入栈，转到第二步。

图的广度优先遍历有点像树的层次遍历，是一个分层搜索的过程。假设从 v0 结点开始遍历，首先遍历与 v0 节点联通的点 w1,w2,……，再遍历与 w1 联通的点 u1,u2,……，与 w2 联通的点 q1……。在遍历过程中，要注意不要重复遍历一个节点，往往在遍历过一个结点后就对这个节点做标记。广度优先遍历常常借助队列实现，步骤如下：

a. 把结点 v 放入队列，标记 v。
b. 若队列为空则结束，否则取出队列头节点 u。
c. 找出与 u 联通的节点 w1,w2,……，若未被遍历则遍历，然后标记、入队，转到上一步。

（2）最小生成树

对于有 n 个顶点的无向连通图，至少有 n-1 条边，而生成树恰好有 n-1 条边，所以生成树是图的极小连通子图。如果无向连通图是一个网，那么它的所有生成树中必有一棵边的权值总和最小的生成树，称这颗生成树为最小生成树。

最小生成树是通过贪心算法来构建，通过局部最优来达到整体最优。设 G(V, E)是一个无向联通图，其权值函数为 w。A 是最小生成树的子集，初始为空。通过循环迭代，每次往 A 中加入一条边，且确保加入边后，A 仍是最小生成树的子集，那么加入的这条边就叫做"安全边（safe edge）"。直到把所有的节点都加入到 A 中，循环结束。

最小生成树可以用 Kruskal 算法或 Prim 算法求出。在 Kruskal 算法中，A 是一个森林，将权值进行排序，选取权值最小的边，若选取的边不形成回路，则为安全边，把它添加到正在生长的森林中。在 Prim 算法中，A 中的边形成单树，每次循环向 A 中添加一个顶点（权值最小的边连接的顶点）。在算法实现中用到一个最小优先级队列，不在树中的顶点都放在基于权值 key 的最小优先级队列 Q 中，对于顶点 v 来说，key[v]的值是与树 A 中某一顶点连接的某一条边的最小权值，如果不连接，那么 key[v]= ∞。

（3）最短路径

此问题求从一个源点到其他各点的最短路径。求解单源最短路径的算法主要有 Dijkstra 算法和 Bellman-Ford 算法，其中 Dijkstra 算法用来解决所有边的权为非负的单源最短路径问题，而 Bellman-Ford 算法可以适用于更一般的问题，图中边的权值可以为负。MADlib 的单源最短路径函数就是使用 Bellman-Ford 算法实现的。如果要得到每一对顶点之间的最短路径，可使用 Floyd 算法来求解。

26.2 单源最短路径

（1）问题描述

给定一个带权有向图 G=(V,E)，其中每条边的权值是一个非负实数。另外，还给定 V 中的一个顶点，称为源。现在我们要计算从源到所有其他各顶点的最短路径长度。这里的长度是指路上各边权值之和。这个问题通常称为单源最短路径问题。

（2）Dijkstra 算法

Dijkstra 算法是一种典型最短路径算法，用于计算一个节点到其他所有节点的最短路径。不过，它针对的是非负权值边。其主要特点是以起始点为中心向外层层扩展，直到扩展到终点为止。Dijkstra 算法能得出最短路径的最优解，但由于它遍历计算的节点很多，所以效率较低。

Dijkstra 算法的输入包含了一个有权重的有向图 G，以及 G 中的一个来源顶点 s。我们以 V 表示 G 中所有顶点的集合，以 E 表示 G 中所有边的集合。表示从顶点 u 到 v 有路径相连，而边的权重则由权重函数 w:E→[0, ∞] 定义。因此，w(u,v)就是从顶点 u 到顶点 v 的非负成本值（cost），边的成本可以想像成两个顶点之间的距离。任两点间路径的成本值，就是该路径上所有边的成本值总和。

已知有 V 中有顶点 s 及 t，Dijkstra 算法可以找到 s 到 t 的最低成本路径（最短路径）。这个算法也可以在一个图中，找到从一个顶点 s 到任何其他顶点的最短路径。

（3）Bellman-Ford 算法

Dijkstra 算法无法判断含有负权边图的最短路径。如果遇到负权值，在没有负权回路（回路的权值和为负，即便有负权的边）存在时，可以采用 Bellman-Ford 算法正确求出最短路径。Bellman-Ford 算法能在更普遍的情况下（存在负权边）解决单源点最短路径问题。对于给定的带权（有向或无向）图，其源点为 s，加权函数 w 是边集 E 的映射。对图 G 运行 Bellman-Ford 算法的结果是一个布尔值，表明图中是否存在着一个从源点 s 可达的负权回路。若不存在这样的回路,算法将给出从源点 s 到图 G 的任意顶点 v 的最短路径 d[v]。Bellman-Ford 算法寻找单源最短路径的时间复杂度为 O(V*E)。

Bellman-Ford 算法描述:

① 初始化:将除源点外的所有顶点的最短距离估计值,d[v]→+∞,d[s]→0;

② 迭代求解:反复对边集 E 中的每条边进行松弛操作,使得顶点集 V 中的每个顶点 v 的最短距离估计值逐步逼近其最短距离(运行｜v｜-1 次);

③ 检验负权回路:判断边集 E 中的每一条边的两个端点是否收敛。如果存在未收敛的顶点,则算法返回 false,表明问题无解;否则算法返回 true,并且从源点可达的顶点 v 的最短距离保存在 d[v]中。

26.3 MADlib 的单源最短路径相关函数

1. 单源最短路径函数

(1) 语法

```
graph_sssp( vertex_table,
            vertex_id,
            edge_table,
            edge_args,
            source_vertex,
            out_table )
```

(2) 参数

vertex_table:TEXT 类型,包含图中顶点数据的表名。

vertex_id:TEXT 类型,默认值为 id,vertex_table 表中包含顶点的列名。顶点列必须是 INTEGER 类型,并且数据不能重复,但不要求连续。

edge_table:TEXT 类型,包含边数据的表名。边表必须包含源顶点、目标顶点和边长三列。边表中允许出现回路,并且构成回路的权重可以不同。

edge_args:TEXT 类型,是一个逗号分隔字符串,包含多个"name=value"形式的参数,支持的参数如下:

- src: INTEGER 类型,边表中包含源顶点的列名,默认值为 'src'。
- dest: INTEGER 类型,边表中包含目标顶点的列名,默认值为 'dest'。
- weight: FLOAT8 类型,边表中包含边长的列名,默认值为 'weight'。

source_vertex:INTEGER 类型,算法的起始顶点。此顶点必须在 vertex_table 表的 vertex_id 列中存在。

out_table:TEXT 类型,存储单源最短路径的表名,表中的每一行对应一个 vertex_table

表中的顶点，具有以下列：

- vertex_id：目标顶点 ID，使用 vertex_id 入参的值作为列名。
- weight：从源顶点到目标顶点最短路径边长合计，使用 weight 入参的值作为列名。
- parent：在最短路径上，本顶点的上一节点，列名为 'parent'。

2. 路径检索函数

路径检索函数返回从源顶点到指定目标顶点的最短路径。

（1）语法

```
graph_sssp( sssp_table,
            dest_vertex )
```

（2）参数

sssp_table：TEXT 类型，单源最短路径函数的输出表名。
dest_vertex：INTEGER 类型，指定的目标顶点。

26.4 单源最短路径示例

单源最短路径问题是图算法的经典问题，在现实中有很多应用，比如在地图中找出两个点之间的最短距离、最小运费等。社交网络中出现的"六度人脉"功能，可以查看到一个用户和一个陌生人之间可以通过哪几个人认识，也就是所谓的六度关系。这个问题也可抽象为一个单源最短路径问题。将用户作为顶点，用户之间的好友关系作为边，"六度关系"就是两个用户之间的最短路径。在这个特殊场景下，所有边的权重都可认为是 1。当然，如果用户量巨大，用户好友关系将变得非常复杂，单纯的最短路径算法可能存在性能问题，需要进行改进与优化。

1. 建立表示图的顶点表和边表

```
drop table if exists vertex, edge;
create table vertex( id integer );
create table edge( src integer, dest integer, weight float8 );

insert into vertex values
(0), (1), (2), (3), (4), (5), (6), (7);

insert into edge values
(0, 1, 1.0), (0, 2, 1.0), (0, 4, 10.0), (1, 2, 2.0),
(1, 3, 10.0), (2, 3, 1.0), (2, 5, 1.0), (2, 6, 3.0),
(3, 0, 1.0), (4, 0, -2.0), (5, 6, 1.0), (6, 7, 1.0);
```

2. 计算从 0 顶点到各顶点的最短路径

```
drop table if exists out;
select madlib.graph_sssp
( 'vertex',        -- 顶点表
  null,            -- 顶点列名，这里使用默认值 'id'
  'edge',          -- 边表
  null,            -- 边参数，这里全部使用默认列名
  0,               -- 计算最短路径的起始顶点
  'out');          -- 输出表名
select * from out order by id;
```

查询结果如下：

```
id | weight | parent
----+--------+--------
 0 |   0    |   0
 1 |   1    |   0
 2 |   1    |   0
 3 |   2    |   2
 4 |  10    |   0
 5 |   2    |   2
 6 |   3    |   5
 7 |   4    |   6
(8 rows)
```

3. 获得从 0~6 的最短路径

```
dm=# select madlib.graph_sssp_get_path('out',6) as spath;
   spath
------------
 {0,2,5,6}
(1 row)
```

4. 使用非默认列名

```
drop table if exists vertex_alt, edge_alt;
create table vertex_alt as select id as v_id from vertex;
create table edge_alt as select src as e_src, dest, weight as e_weight from edge;
```

5. 计算从 1 顶点到各顶点的最短路径

```
drop table if exists out_alt;
select madlib.graph_sssp
( 'vertex_alt',                      -- 顶点表
  'v_id',                            -- 顶点列名
  'edge_alt',                        -- 边表
  'src=e_src, weight=e_weight',      -- 边参数，指定顶点和边长的列名
```

```
    1,                                -- 计算最短路径的起始顶点
    'out_alt');                       -- 输出表名
select * from out_alt order by v_id;
```

结果：

```
 v_id | e_weight| parent
------+---------+--------
    0 |      4  |    3
    1 |      0  |    1
    2 |      2  |    1
    3 |      3  |    2
    4 |     14  |    0
    5 |      3  |    2
    6 |      4  |    5
    7 |      5  |    6
(8 rows)
```

26.5 小结

图算法是一类特殊的数据挖掘方法，常被用于解决确定图连通性、寻找最短路径等相关问题。实际应用中，图算法广泛用于社交网络分析（如 Community Detection）、互联网（如 PageRank）、计算生物学（如研究分子活动路径）、电子工程（如集成电路设计）、科学计算（如图划分）、安全领域（如安全事件分析）等很多方面。图算法主要包括图遍历、图匹配、最小生成树、最短路径等几大类，每一类中有多种算法。MADlib 仅提供了一种图算法模型，即单源最短路径模型，它是使用 Bellman-Ford 算法实现的。

第 27 章 模型验证

验证是评估数据挖掘模型对实际数据执行情况的过程。在将挖掘模型部署到生产环境之前，必须通过了解模型的质量和特征来对其进行验证，评估模型的准确性、可靠性和可用性。可以使用多种方法评估数据挖掘模型的质量和特征：

- 使用统计信息有效性的各种度量值来确定数据或模型中是否存在问题。
- 将数据划分为定型集和测试集，以测试预测的准确性。
- 请求商业专家查看数据挖掘模型的结果，以确定发现的模式在目标商业方案中是否有意义。

所有这些方法在数据挖掘方法中都非常有用，创建、测试和优化模型来解决特定问题时，可以反复使用这些方法。没有一个全面的规则可以说明什么时候模型已足够好，或者什么时候具有足够的数据。本章介绍最常用的交叉验证方法，以及 MADlib 中交叉验证函数的用法。

27.1 交叉验证简介

数据挖掘技术在应用之前使用的"训练+检验"模式，通常被称作"交叉验证"，如图 27-1 所示。实际上在第 25 章的决策树函数中，我们已经接触过交叉验证，当 n_folds 参数大于 0 时，决策树函数在构造模型过程中就会进行交叉验证。

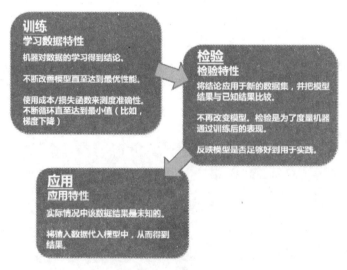

图 27-1 交叉验证过程

1. 预测模型的稳定性

我们通过一个例子来理解模型的稳定性问题,如图 27-2 所示。

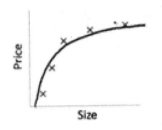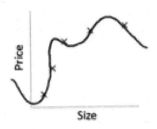

图 27-2　尺寸与价格模型图

此处我们试图找到尺寸(size)和价格(price)的关系。三个模型各自做了如下工作:

- 第一个模型使用了线性等式。对于训练用的数据点,此模型有很大误差。这是"拟合不足(Under fitting)"的一个例子。此模型不足以发掘数据背后的趋势。
- 第二个模型发现了价格和尺寸的正确关系,此模型误差低,概括程度高。
- 第三个模型对于训练数据几乎是零误差。这是因为此关系模型把每个数据点的偏差(包括噪声)都纳入了考虑范围,也就是说,这个模型太过敏感,甚至会捕捉到只在当前数据训练集出现的一些随机模式。这是"过度拟合(Over fitting)"的一个例子。

在应用中,常见的做法是对多个模型进行迭代,从中选择表现更好的一个。然而,最终的数据是否会有所改善依然未知,因为我们不确定这个模型是更好的发掘出潜在关系,还是过度拟合了。为解答这个难题,需要使用交叉验证(cross validation)技术,它能帮我们得到更有概括性的数据模型。实际上,数据挖掘关注的是通过训练集训练后的模型对测试样本的学习效果,我们称之为泛化能力。左右两图的泛化能力就表现不好。具体到数据挖掘中,对偏差和方差的权衡是数据挖掘理论着重解决的问题。

2. 交叉验证步骤

交叉验证意味着需要保留一个样本数据集,不用来训练模型。在最终完成模型前,用这个数据集验证模型。交叉验证包含以下步骤:

(1)保留一个样本数据集,即测试集。
(2)用剩余部分(训练集)训练模型。
(3)用保留的数据集(测试集)验证模型。

这样做有助于了解模型的有效性。如果当前模型在此测试数据集也表现良好,说明模型的泛化能力较好,可以用来预测未知数据。

3. 交叉验证的常用方法

交叉验证有很多方法,下面介绍其中三种。

(1)"验证集"法

保留 50% 的数据集用作验证，剩下 50% 训练模型。之后用验证集测试模型表现。这个方法的主要缺陷是，由于只使用了 50% 数据训练模型，原数据中一些重要的信息可能被忽略，也就是说，会有较大偏误。

（2）留一法交叉验证（LOOCV）

这种方法只保留一个数据点用作验证，用剩余的数据集训练模型。然后对每个数据点重复这个过程。该方法有利有弊：

- 由于使用了所有数据点，所以偏差较低。
- 验证过程重复了 n 次（n 为数据点个数），导致执行时间很长。
- 由于只使用一个数据点验证，该方法导致模型有效性的差异更大。得到的估计结果深受此点的影响。如果这是个离群点，会引起较大偏差。

（3）K 折交叉验证（K-fold cross validation）

从以上两个验证方法中，我们知道：

- 应该使用较大比例的数据集来训练模型，否则会导致失败，最终得到偏误很大的模型。
- 验证用的数据点，其比例应该恰到好处。如果太少，会影响验证模型有效性时，得到的结果波动较大。
- 训练和验证过程应该重复多次（迭代）。训练集和验证集不能一成不变，这样有助于验证模型的有效性。

是否有一种方法可以兼顾这三个方面？答案是肯定的！这种方法就是"K 折交叉验证"。

4. K 折交叉验证简要步骤

K 折交叉验证方法的简要步骤如下：

（1）把整个数据集随机分成 K "层"。

（2）对于每一份数据来说：一要以该份作为测试集，其余作为训练集，也就是说用其中 K-1 层训练模型，然后用第 K 层验证。二要在训练集上得到模型。三要在测试集上得到生成误差。

（3）重复这个过程，直到每"层"数据都做过验证集。这样对每一份数据都有一个预测结果，记录从每个预测结果获得的误差。

（4）记录下的 K 个误差的平均值，被称为交叉验证误差（cross-validation error）。可以被用做衡量模型表现的标准。

（5）取误差最小的那个模型。

此算法的缺点是计算量较大，当 K=10 时，K 层交叉验证示意图如图 27-3 所示。

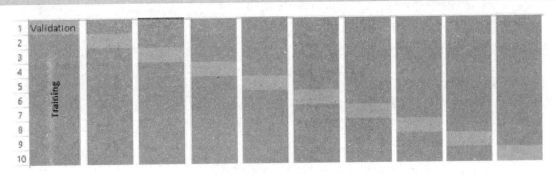

图 27-3　10 折交叉验证

一个常见的问题是：如何确定合适的 K 值？K 值越小，偏误越大，所以不推荐。另一方面，K 值太大，所得结果会变化多端。K 值小，则会变得像"验证集法"，K 值大，则会变得像"留一法"（LOOCV），因此通常建议的经验值是 K=10。

5. 衡量模型的偏误/变化程度

K 层交叉检验之后，我们得到 K 个不同的模型误差估算值（e1, e2,…, ek）。理想情况是，这些误差值相加的结果值为 0。计算模型的偏误时，我们把所有这些误差值相加再取平均值，平均值越低，模型越好。模型表现变化程度的计算与之类似。取所有误差值的标准差，标准差越小说明模型随训练数据的变化越小。

应该试图在偏误和变化程度间找到一种平衡。降低变化程度、控制偏误可以达到这个目的，这样会得到更好的数据模型。进行这个取舍，通常会得出复杂程度较低的预测模型。

27.2　MADlib 的交叉验证相关函数

决策树例子中的交叉验证，是内嵌在决策树训练函数中的。MADlib 还提供了独立的交叉验证函数，可对大部分 MADlib 的预测模型进行交叉验证。

交叉验证可以估计一个预测模型在实际中的执行精度，还可用于设置预测目标。MADlib 提供的交叉验证函数非常灵活，不但可以选择已经支持的交叉验证算法，用户还可以编写自己的验证算法。从交叉验证函数输入需要验证的训练、预测和误差估计函数规范。这些规范包括三部分：函数名称、传递给函数的参数数组、参数对应的数据类型数组。

训练函数使用给定的自变量和因变量数据集产生模型，模型存储于输出表中。预测函数使用训练函数生成的模型，并接收不同于训练数据的自变量数据集，产生基于模型的对因变量的预测，并将预测结果存储在输出表中。预测函数的输入中应该包含一个表示唯一 ID 的列名，便于预测结果与验证值作比较。注意，有些 MADlib 的预测函数不将预测结果存储在输出表中，这种函数不适用于 MADlib 的交叉验证函数。误差度量函数比较数据集中已知的因变量和预测结果，用特定的算法计算误差度量，并将结果存入一个表中。其他输入包括输出表名、K 折交叉验证的 K 值等。

1. 语法

```
cross_validation_general( modelling_func,
                  modelling_params,
                  modelling_params_type,
                  param_explored,
                  explore_values,
                  predict_func,
                  predict_params,
                  predict_params_type,
                  metric_func,
                  metric_params,
                  metric_params_type,
                  data_tbl,
                  data_id,
                  id_is_random,
                  validation_result,
                  data_cols,
                  fold_num )
```

2. 参数

modelling_func：VARCHAR 类型，模型训练函数名称。

modelling_params：VARCHAR[]类型，训练函数参数数组。

modelling_params_type：VARCHAR[]类型，训练函数参数对应的数据类型名称数组。

param_explored：VARCHAR 类型，被寻找最佳值的参数名称，必须是 modelling_params 数组中的元素。

explore_values：VARCHAR 类型，候选的参数值。如果为 NULL，只运行一轮交叉验证。

predict_func：VARCHAR 类型，预测函数名称。

predict_params：VARCHAR[]类型，提供给预测函数的参数数组。

predict_params_type：VARCHAR[]类型，预测函数参数对应的数据类型名称数组。

metric_func：VARCHAR 类型，误差度量函数名称。

metric_params：VARCHAR[]类型，提供给误差度量函数的参数数组。

metric_params_type：VARCHAR[]类型，误差度量函数参数对应的数据类型名称数组。

data_tbl：VARCHAR 类型，包含原始输入数据表名，表中数据将被分成训练集和测试集。

data_id：VARCHAR 类型，表示每一行唯一 ID 的列名，可以为空。理想情况下，数据集中的每行数据都包含一个唯一 ID，这样便于将数据集分成训练部分与验证部分。id_is_random 参数值告诉交叉验证函数 ID 值是否是随机赋值。如果原始数据不是随机赋的 ID 值，验证函数为每行生成一个随机 ID。

id_is_random：BOOLEAN 类型，为 TRUE 时表示提供的 ID 是随机分配的。

validation_result：VARCHAR 类型，存储交叉验证函数输出结果的表名，具有以下列：

- param_explored：被寻找最佳值的参数名称。与 cross_validation_general() 函数的 param_explored 入参相同。
- average error：误差度量函数计算出的平均误差。
- standard deviation of error：标准差。

data_cols：逗号分隔的用于计算的数据列名。为 NULL 时，函数自动计算数据表中的所有列。只有当 data_id 参数为 NULL 时才会用到此参数，否则忽略。如果数据集没有唯一 ID，交叉验证函数为每行生成一个随机 ID，并将带有随机 ID 的数据集复制到一个临时表。设置此参数为自变量和因变量列表，通过只复制计算需要的数据，最小化复制工作量。计算完成后临时表被自动删除。

fold_num：INTEGER 类型，K 值，默认值为 10，指定验证轮数，每轮验证使用 1/fold_num 数据做验证。

训练、预测和误差度量函数的参数数组中可以包含以下特殊关键字：

- %data%：代表训练/验证数据。
- %model%：代表训练函数的输出，即预测函数的输入。
- %id%：代表唯一 ID 列（用户提供的或函数生成的）。
- %prediction%：代表预测函数的输出，即误差度量函数的输入。
- %error%：代表误差度量函数的输出。

27.3 交叉验证示例

我们将调用交叉验证函数，量化弹性网络正则化回归模型的准确性，并找出最佳的正则化参数。关于弹性网络正则化的说明参见 https://en.wikipedia.org/wiki/Elastic_net_regularization。

1. 准备输入数据

```
drop table if exists houses;
-- 房屋价格表
create table houses (
    id serial not null,    -- 自增序列
    tax integer,           -- 税金
    bedroom real,          -- 卧室数
    bath real,             -- 卫生间数
    price integer,         -- 价格
    size integer,          -- 使用面积
    lot integer            -- 占地面积
);

insert into houses(tax, bedroom, bath, price, size, lot) values
( 590, 2,   1,  50000,  770, 22100),
```

```
(1050, 3,   2,  85000, 1410, 12000),
(  20, 3,   1,  22500, 1060,  3500),
( 870, 2,   2,  90000, 1300, 17500),
(1320, 3,   2, 133000, 1500, 30000),
(1350, 2,   1,  90500,  820, 25700),
(2790, 3, 2.5, 260000, 2130, 25000),
( 680, 2,   1, 142500, 1170, 22000),
(1840, 3,   2, 160000, 1500, 19000),
(3680, 4,   2, 240000, 2790, 20000),
(1660, 3,   1,  87000, 1030, 17500),
(1620, 3,   2, 118600, 1250, 20000),
(3100, 3,   2, 140000, 1760, 38000),
(2070, 2,   3, 148000, 1550, 14000),
( 650, 3, 1.5,  65000, 1450, 12000);
```

2. 创建函数执行交叉验证

```
create or replace function check_cv()
returns void as $$
begin
    execute 'drop table if exists valid_rst_houses';
    perform madlib.cross_validation_general(
    -- 训练函数
    'madlib.elastic_net_train',
    -- 训练函数参数
    '{%data%, %model%, (price>100000), "array[tax, bath, size, lot]",
    binomial, 1, lambda, true, null, fista,
"{eta = 2, max_stepsize = 2, use_active_set = t}",
null, 2000, 1e-6}'::varchar[],
    -- 训练函数参数数据类型
    '{varchar, varchar, varchar, varchar, varchar, double precision,
    double precision, boolean, varchar, varchar, varchar, varchar, integer,
    double precision}'::varchar[],
    -- 被考察参数
    'lambda',
    -- 被考察参数值
    '{0.04, 0.08, 0.12, 0.16, 0.20, 0.24, 0.28, 0.32, 0.36}'::varchar[],
    -- 预测函数
    'madlib.elastic_net_predict',
    -- 预测函数参数
    '{%model%, %data%, %id%, %prediction%}'::varchar[],
    -- 预测函数参数数据类型
    '{text, text, text, text}'::varchar[],
    -- 误差度量函数
```

```
        'madlib.misclassification_avg',
        -- 误差度量函数参数
        '{%prediction%, %data%, %id%, (price>100000), %error%}'::varchar[],
        -- 误差度量函数参数数据类型
        '{varchar, varchar, varchar, varchar, varchar}'::varchar[],
        -- 数据表
        'houses',
        -- ID列
        'id',
        -- id是否随机
        false,
        -- 验证结果表
        'valid_rst_houses',
        -- 数据列
        '{tax,bath,size,lot, price}'::varchar[],
        -- 折数
        3
        );
end;
$$ language plpgsql volatile;
```

3. 执行函数并查询结果

```
select check_cv();
select * from valid_rst_houses order by lambda;
```

结果如下:

```
lambda |      error_rate_avg       |          error_rate_stddev
-------+---------------------------+------------------------------------------
  0.04 | 0.2666666666666666667     | 0.1154700538379251529018297561003914911294
  0.08 | 0.3333333333333333333     | 0.1154700538379251529018297561003914911294
  0.12 | 0.3333333333333333333     | 0.1154700538379251529018297561003914911294
  0.16 | 0.5333333333333333333     | 0.2309401076758503058036595122007829822590
   0.2 | 0.6000000000000000000     | 0.2000000000000000000000000000000000000000
  0.24 | 0.6000000000000000000     | 0.2000000000000000000000000000000000000000
  0.28 | 0.6666666666666666667     | 0.2309401076758503058036595122007829822590
  0.32 | 0.6666666666666666667     | 0.2309401076758503058036595122007829822590
  0.36 | 0.7333333333333333333     | 0.1154700538379251529018297561003914911294
(9 rows)
```

上面的查询结果表示，随着正则化参数不断加大，平均误差也会增加，而且当正则化参数较小时标准差也较小。因此得出结论，用 0.04 作为正则化参数，将得到较好的预测模型。

27.4 小结

验证对于由训练数据集生成的数据挖掘预测模型的准确性非常重要。在模型正式投入使用前必须经过验证过程。交叉验证是常用一类的模型验证评估方法，其中"K 折交叉验证"法重复多次执行训练和验证过程，每次训练集和验证集发生变化，有助于验证模型的有效性。MADlib 提供的 K 折交叉验证函数，可用于大部分 MADlib 的预测模型。